IUTAM-IAHR

FLOW-INDUCED STRUCTURAL VIBRATIONS

SYMPOSIUM KARLSRUHE (GERMANY)
AUGUST 14-16, 1972

EDITOR

EDUARD NAUDASCHER

WITH 360 FIGURES

SPRINGER-VERLAG BERLIN HEIDELBERG NEW YORK 1974

Professor Dr.-Ing. Eduard Naudascher
Institut für Hydromechanik
Universität Karlsruhe
Karlsruhe, Germany

ISBN 978-3-642-48281-6 ISBN 978-3-642-48279-3 (eBook)
DOI 10.1007/978-3-642-48279-3

Softcover reprint of the hardcover 1st edition 1974

Library of Congress Catalog Card Number: 73-18746

Preface

In recent years there has been an unmistakable tendency for research in the fields of hydraulics and fluid mechanics to involve ever more sophisticated study of increasingly detailed aspects of problems, while in the process becoming further and further detached from the realities faced by practising engineers. At the same time, too many hydraulic engineers have continued to conduct model studies with the goal of obtaining ad-hoc solutions for specific problems, while directing little if any of their attention toward elucidation of the underlying mechanics and toward generalization of their results for future application. It was against this background that the International Union of Theoretical and Applied Mechanics (IUTAM) and the International Association for Hydraulic Research (IAHR) were approached and asked to cosponsor the Symposium on Flow-Induced Structural Vibrations, which would have as its objective the promotion of interaction between the active researchers and the practitioners, and of communication between the different disciplines concerned. To this end, invitations were extended to individuals who together represented the full range of the investigation-application spectrum, and contributions were solicited from scientists and engineers who, in general, approach vibration problems along different avenues. In soliciting contributions to the Symposium it was emphasized that priority would be given to papers which shed light on the constituent phenomena involved in vibration problems and thereby enhance our understanding of the basic vibration-inducing mechanisms.

Included herein are the 49 papers and short contributions selected from among the more than 100 manuscripts submitted to the Scientific Committee, together with the 5 General Lectures and 5 Workshop Reports. The material is presented more or less in the order given in the Symposium program, and in general with only minor editorial changes. Preprints of all papers accepted for the conference were prepared and distributed well in advance of the meeting to the 210 participants, from 27 countries, who preregistered for the Symposium. The many prepared discussions which were submitted as a consequence added significantly to the stimulating atmosphere of the technical sessions. A separate volume containing all contributions to the discussions at the technical sessions is available from the Institute of Hydromechanics, University of Karlsruhe, Karlsruhe, Germany.

For many participants, one of the highlights of the Symposium were the workshops organized around the general theme "Research Priorities in Flow-Induced Vibrations". In the workshop reports, which were presented during a Panel Discussion, one will find recommendations extending far beyond research topics to a variety of non-technical questions, including improved communication between researchers and practitioners, long-range cooperation programs between different disciplines, and the necessary structuring and generalization of the rapidly growing and diversifying body of knowledge relevant to flow-induced vibrations. The general consensus that future conferences similar to the one reported herein would be a major step toward achieving these difficult goals was most gratifying to the Symposium organizers.

As chairman of the Scientific Committee, I wish to express my sincere gratitude to all promoters and supporters of the Symposium: to the members of the Scientific Committee for their help in the program planning, for selecting the papers, and for their cooperation in the Panel Discussion; to the Presidents of the sponsoring associations,

Professors H. Görtler (IUTAM) and T. Hayashi (IAHR), and to the President of the University of Karlsruhe, Professor H. Draheim, for their addresses during the opening session; to all Session Chairmen and Co-Chairmen for presiding over the meetings and leading the discussions; and, of course, to the many active participants who supplied a wealth of new ideas and information. Financial support was received not only from IUTAM and IAHR, but also very generously from the German Science Foundation and the Ministry of Cultural Affairs of Baden-Württemberg. Grateful acknowledgement is also given for the financial and other support received from the City of Karlsruhe, the Karlsruhe Alumni Association, the University of Karlsruhe, the German Academic-Exchange Service, the Klein, Schanzlin & Becker Foundation, the Landesgewerbeamt Karlsruhe, and the U.S. Army Karlsruhe Community Leader.

Last, but by no means least, I am happy to have the opportunity here to express special thanks to all collaborators and to my associates at the Institute of Hydromechanics, University of Karlsruhe, who worked with me during the more than two years of preparation that went into making the Symposium and its supporting program a success. I wish to single out for special mention two of the many individuals involved: Mr. Andreas Richter, who like no one else devoted time and effort to this endeavor; and my wife, Brigitte, who carried the responsibility for the ladies program.

Karlsruhe, February 1974 E. Naudascher

Scientific Committee

G. Coupry, France

A.G. Davenport, Canada

L.I. Sedov, U.S.S.R.

G.H. Toebes, U.S.A.

E. Naudascher, Germany (Chairman)

Local Organizing Committee

R. Armstrong, J.R. Chaplin, F.-D. Heidt,
A. List, E. Plate, A. Richter,
O.H. Schiele, C. Zimmermann,
and E. Naudascher (Chairman)

List of Participants

Australia: H.R. Graze

Austria: M. Pflügl

Belgium: J. Berlamont, M. de Somer, M. Geradin, G.J. Sander

Bulgaria: V.A. Dzhupanov, K. Kusow

Canada:
W.D. Baines, K.R. Cooper, I.G. Currie, A.G. Davenport, G.I. Fekete, A.D. Hogg, W.W. Martin, V.J. Modi, M. Novak, M.P. Païdoussis, G.V. Parkinson, C.F.M. Twigge-Molecey, D.S. Weaver

ČSSR: V. Feranec

Denmark: S.E. Larsen, R. Sigbjörnsson, K.E. Widell

France:
M. Aureille, M. Baylac, G. Binder, M. Bolvin, F. Boulot, R. Brepson, M. Chenal, M. Chevalier, G. Coupry, R.M. Curtet, R. Dat, A. Daubert, M. Debaene, J. Driviere, J.P. Gregoire, M. Guesdon, P. Léon, M. Levin, H. Loiseau, B.V. Martins, J. Michel, M. Milan, J.C. Pennanech, M. Sagner, P. Schmitt, P. Sonneville, E. Szechenyi, P. Villard

Germany:
R. Armstrong, H. Bardowicks, E. Berger, P. Bublitz, F. Buesching, P. Čolak-Antić, E. Dietrich, H. Dorer, H. Draheim, F. Durst, O. Eichler, K. Eimer, H. Fiedler, H. Försching, H. Fröhner, H. Görtler, R.U. Hartermann, F.-D. Heidt, G. Hirsch, E. Huthloff, K. Kraemer, R. Landl, O. Mahrenholtz, H.U. Meier, H.E. Minor, E. Mosonyi, E. Naudascher, A. Naumann, M. Padmanabhan, H.W. Partenscky, E. Plate, H. Quadflieg, J. Renner, A. Richter, G. Rouvé, Ch. Rückauer, R.H. Rueff, H. Rusche-

weyh, O.H. Schiele, H. Schlichting, B. Schmidt, M. Seidel, H. Stadie, B. Treiber, F.W. Treplin, J. Ure, H. Vollmers, H. Wagner, J.R. Weske, B. Westrich, R. Wille, G. Wonik, C. Zimmermann, G. Zimmermann

Hungary: Z.G. Hankó, O. Haszpra

Israel: N.M. Kuluva, A. Lavee, A. Rozena

Italy: G. Diana, G. Di Silvio, F. Giordana

Japan: F. Hara, T. Hayashi, M. Hino, N. Shiraishi

Mexico: S. Tirado Ledesma

The Netherlands:
A. de Bruijn, P.A. Kolkman, H. Tijdeman, H. van Koten, J.W.G. van Nunen

Norway: J. Langseth, L. Vinnogg

Portugal: M.I.F.M. Baptista, C.A.M. Ramos

Rhodesia: M.E. Greenway

Rumania: P. Stoenescu

South Africa: R. Rösel

Sweden:
G.B. Benkö, K.N. Handa, J.E.V. Nilsson, B.B. Oddving, O.E. Söderberg

Switzerland:
B. Chaix, Y.N. Chen, H. Grein, R. Grütter, G. Gyarmathy, H. Thomann

U.K.:
P.W. Bearman, H.H. Bruun, W.B. Bryce, J.R. Calvert, J.R. Chaplin, D.H. Cooper, T. Cousins, M.E. Davies, N. Ellis, E.R. France, F.R. Fricke, G.D. Galletly, J.H. Gerrard, J.M.R. Graham, J.D. Hardwick, P. Holmes, J.C.R. Hunt, D.J. Johns, C.H. Jones, P.F. Kilty, R. King, W.A. Mair, D.J. Maull, M.W. Parkin, D.D.G. Petty, M.J. Prosser, A. Ratcliffe, R.A. Sawyer, C. Scruton, T.L. Shaw, A.F. Taylor, M.H. Warner, J.S. Wharmby, R.H. Wilkinson, C.J. Wood, L.R. Wootton, G.A.J. Young, M.M. Zdravkovich

U.S.A.:
H.N. Abramson, K.S. Budlong, S.S. Chen, R.F. Dominguez, J.H. Douma, H.M. Fitzpatrick, G.J. Franz, O.M. Griffin, H. Halle, W.J. Heilker, G. Herrmann, R.M. Kanazawa, M.V. Morkovin, J.H. Nath, A. Roshko, D.W. Sallet, R.H. Scanlan, C.C. Shih, Ch.E. Smith, A.O.St. Hilaire, G.H. Toebes, P. Tong, R.L. Wiegel, K.H. Yang

U.S.S.R.:
A.S. Abelev, G.A. Chepajkin, L.L. Dolnikov, L.A. Goncharov, V.M. Ljatkher, P.E. Lyssenko, V.M. Semenkov, I.S. Sheinin, M.F. Skladnev

Yugoslavia:
J. Grcič, J. Muškatirovic, M. Ridjanovic

List of Contributors

Abelev, A.S.
All-Union Scientific Research Institute of Hydrotechnics, Gzhatskaya 21, Leningrad K-220, U.S.S.R.

Abramson, H.N.
Department of Mechanical Sciences, Southwest Research Institute, 8500 Culebra Road, San Antonio, TX 78284, U.S.A.

Angrilli, F.
Institute of Hydraulics, University of Padua, Via Loredan 20, 35100 Padova, Italy

Baines, W.D.
Department of Mechanical Engineering, University of Toronto, Toronto 5, Canada

Bearman, P.W.
Department of Aeronautics, Imperial College of Science and Technology, Prince Consort Road, London SW7 2BY, U.K.

Binder, G.
Laboratoire de Mécanique des Fluides, Université de Grenoble, B.P. 53, Centre de Tri, 38041 Grenoble-Cédex, France

Boulot, F.
Electricité de France, 6, Quai Watier, 78 Chatou, France

Brepson, R.
Technique des Fluides, B.P. 75, Centre de Tri, 38041 Grenoble-Cédex, France

Bublitz, P.
Deutsche Forschungs- und Versuchsanstalt für Luft- und Raumfahrt, AVA Göttingen, 34 Göttingen, Bunsenstr. 10, Germany

Budlong, K.S.
Department of Civil Engineering, Princeton University, Princeton, NJ 08540, U.S.A.

Chaplin, J.R.
Rendel, Palmer & Tritton, London, 61 Southwark Street, London SE1 1SA, U.K.

Chepajkin, G.A.
Scientific Research Centre, Institute "Hydroproject", P.O. Box 393, Moscow D-362, U.S.S.R.

Chen, S.S.
Argonne National Laboratory, 9700 South Cass Avenue, Argonne, IL 60439, U.S.A.

Chen, Y.N.
Sulzer Brothers Ltd., 8401 Winterthur, Switzerland

Cooper, K.R.
National Research Council of Canada, National Aeronautical Establishment, Ottawa K1A OR6, Canada

Cooper, D.H.
British Transport Docks Board Research Station, Hayes Road, Southall, Middlesex, U.K.

Coupry, G.
Office National d'Études et Recherches Aérospatiales (O.N.E.R.A.), 29, Avenue de la Division Leclerc, 92 Châtillon, France

Currie, I.G.
Department of Mechanical Engineering, University of Toronto, Toronto 5, Canada

Curtet, R.M.
Laboratoires de Mécanique des Fluides, Université de Grenoble B.P. 53, Centre de Tri, 38041 Grenoble-Cédex, France

Dat, R.
Office National d'Études et de Recherches Aérospatiales (O.N.E.R.A.), 29, Avenue de la Division Leclerc, 92 Châtillon, France

Daubert, A.
Electricité de France, 6, Quai Watier, 78 Chatou, France

Davenport, A.G.
Boundary Layer Wind Tunnel Laboratory, Faculty of Engineerin Science, The University of Western Ontario, 1151 Richmond Street, London, Ontario N6A 3K7, Canada

Davies, P.O.A.L.
Institute of Sound and Vibration Research, University of Southampton, Southampton, U.K.

Di Silvio, G.
Institute of Hydraulics, University of Padua, Via Loredan 20, 35100 Padova, Italy

Dolnikov, L.L.
All-Union Scientific Research Institute of Hydrotechnics, Gzhatskaya 21, Leningrad K-220, U.S.S.R.

Dominguez, R.F.
Department of Civil Engineering, Texas A and M University, College Station, TX 77801, U.S.A.

Douma, J.H.
Department of the Army, Office of the Chief of Engineers, Washington, DC 20314, U.S.A.

Durgin, F.H.
Department of Aeronautics and Astronautics, Massachusetts Institute of Technology, Cambridge, MA 02139, U.S.A.

Dzhupanov, V.A.
Institute of Mathematics and Mechanics, Bulgarian Academy of Sciences, 36 Street, Block VIII, Sofia 13, Bulgaria

Favre-Marinet, M.
Laboratoire de Mécanique des Fluides, Université de Grenoble, B.P. 53, Centre de Tri, 38041 Grenoble-Cédex, France

Fricke, F.R.
Department of Building Science, Sheffield University, Sheffield, Yorkshire, S10 2TN, U.K.

Gerrard, J.H.
Department of the Mechanics of Fluids, University of Manchester, Manchester, M13 9PL, U.K.

Goncharov, L.A.
Scientific Research Centre, Institute "Hydroproject", P.O. Box 393, Moscow D-362, U.S.S.R.

Griffin, O.M.
Naval Research Laboratory, Ocean Technology Division, Code 8441, Washington, DC 20375, U.S.A.

Hara, F.
Science University of Tokyo, 1-3 Kagurazaka, Shinjuku, Tokyo, Japan

Hartlen, R.T.
Ontario Hydro, Toronto, Canada

Hogan, J.M.
Department of Mechanics and Mechanical and Aerospace Engineering, Illinois Institute of Technology, Chicago, IL 60616, U.S.A.

Hunt, J.C.R.
Department of Applied Mathematics and Theoretical Physics, University of Cambridge, Silver Street, Cambridge, CB3 9EW, U.K.

Johns, D.J.
Department of Transport Technology, University of Technology, Loughborough, Leics., LE11 3TU, U.K.

King, R.
BHRA Fluid Engineering, Cranfield, Bedford, MK43 0AJ, U.K.

Kolkman, P.A.
Delft Hydraulics Laboratoy, P.O. Box 177, Delft, The Netherlands

Kouwen, N.
Department of Civil Engineering, University of Waterloo, Waterloo, Ontario, Canada

Léon, P.
Technique des Fluides, B.P. 75, Centre de Tri, 38041 Grenoble-Cédex, France

Lyssenko, P.E.
Scientific Research Centre, Institute "Hydroproject", P.O. Box 393, Moscow D-362, U.S.S.R.

Ljatkher, V.M.
Committee for the U.S.S.R., Participation in International Power Conferences, 11 Gorky Street, Moscow K-9, U.S.S.R.

Loiseau, H.
Office National d'Études et de Recherches Aérospatiales (O.N.E.R.A.) 29, Avenue de la Division Leclerc, 92 Châtillon, France

Mansour, W.M.
Department of Mechanical Engineering, University of Waterloo, Waterloo, Ontario, Canada

Martins, B.V.
Laboratoires de Mécanique des Fluides, Université de Grenoble, B.P. 53, Centre de Tri, 38041 Grenoble-Cédex, France

Martin, W.W.
Department of Mechanical Engineering, University of Toronto, Toronto 5, Canada

Maull, D.J.
University Engineering Department, Trumpington Street, Cambridge CB2 1PZ, U.K.

Modi, V.J.
Department of Mechanical Engineering, University of British Columbia, Vancouver 8, Canada

Morkovin, M.V.
Department of Mechanics and Mechanical and Aerospace Engineering, Illinois Institute of Technology, Chicago, IL 60616, U.S.A.

Nath, J.H.
School of Oceanography, Oregon State University, Corvallis, OR 97331, U.S.A.

Naumann, A.
Aerodynamisches Institut, Technische Universität Aachen, 51 Aachen, Templergraben 55, Germany

Novak, M.
Faculty of Engineering Science, University of Western Ontario, London 72, Ontario, Canada

Païdoussis, M.P.
Department of Mechanical Engineering, McGill University, P.O. Box, 6070, Station A, Montreal H3C 3G1, Québec, Canada

Parkin, M.W.
Reactor Development Laboratories, U.K.A.E.A., Windscale Works, Seascale, Cumberland, U.K.

Parkinson, G.V.
Department of Mechanical Engineering, University of British Columbia, Vancouver 8, Canada

Plate, E.
Institut für Wasserbau III, Universität Karlsruhe, 75 Karlsruhe 1, Kaiserstraße 12, Germany

Prosser, M.J.
BHRA Fluid Engineering, Cranfield, Bedford, MK43 0AJ, U.K.

Pugnet, L.
Electricité de France, 6, Quai Watier, 78 Chatou, France

Quadflieg, H.
Aerodynamisches Institut, Technische Universität Aachen, 51 Aachen, Templergraben 55, Germany

Sallet, D.W.
Department of Mechanical Engineering, University of Maryland, College Park, MD 20472, U.S.A.

Sawyer, R.A.
Department of Mechanical Engineering, University of Salford, Salford, Lancs., M5 4WT, U.K.

Scanlan, R.H.
School of Engineering and Applied Science, Princeton University, Princeton, NJ 08540, U.S.A.

Semenkov, V.M.
Scientific Research Centre, Institute "Hydroproject", P.O. Box 393, Moscow D-362, U.S.S.R.

Shaw, T.L.
Department of Civil Engineering, University of Bristol, Queen's Building, University Walk, Bristol, BS8 1TR, U.K.

Sharma, C.B.
Department of Transport Technology, University of Technology, Loughborough, Leics., LE11 3TU, U.K.

Sheinin, I.S.
All-Union Research Institute of Hydrotechnics, Gzhatskaja 21, Leningrad K-220, U.S.S.R.

Shibata, H.
Institute of Industrial Science, University of Tokyo, 7-22-1 Roppongi, Tokyo, Japan

Shigeta, T.
Institute of Industrial Science, University of Tokyo, 7-22-1 Roppongi, Tokyo, Japan

Shiraishi, N.
Department of Civil Engineering, Kyoto University, Kyoto, Japan

Skladnev, M.F.
All-Union Research Institute of Hydrotechnics, Gzhatskaja 21, Leningrad K-220, U.S.S.R.

Slater, J.E.
Department of Mechanical Engineering, University of British Columbia, Vancouver 8, Canada

Smith, Ch.E.
Department of Mechanical and Metallurgical Engineering, Oregon State University, Corvallis, OR 97331, U.S.A.

Szechenyi, E.
Office National d'Études et de Recherches Aérospatiales (O.N.E.R.A.), 29, Avenue de la Division Leclerc, 92 Châtillon, France

Toebes, G.H.
School of Civil Engineering, Purdue University, Lafayette, IN 47907, U.S.A.

Tong, P.
Department of Aeronautics and Astronautics, Massachusetts Institute of Technology, Cambridge, MA 02139, U.S.A.

Treiber, B.
Institut für Wasserbau III, Universität Karlsruhe, 75 Karlsruhe 1, Kaiserstraße 12, Germany

Twigge-Molecey, C.F.M.
Department of Mechanical Engineering, University of Toronto, Toronto 5, Canada

van Nunen, J.W.G.
National Aerospace Laboratory NLR, Sloterweg 145, Amsterdam (17), The Netherlands

Unny, T.E.
Department of Civil Engineering, University of Waterloo, Waterloo, Ontario, Canada

Warner, M.H.
British Transport Docks Board Research Station, Hayes Road, Southall, Middlesex, U.K.

Weaver, D.S.
Department of Mechanical Engineering, McMaster University, Hamilton, Ontario, L8S 4L7, Canada

Wiegel, R.L.
Department of Civil Engineering, University of California, 412 O'Brien Hall, University of California, Berkeley, CA 94720, U.S.A.

Wilkinson, R.H.
Department of Civil Engineering, Bristol University, Queen's Building, University Walk, Bristol BS8 1TR, U.K.

Wille, R.
Institut für Thermo- und Fluiddynamik, Technische Universität Berlin, 1 Berlin 12, Straße des 17. Juni 135, Germany

Wootton, L.R.
Atkins Research and Development, Woodcote Grove, Ashley Road, Epsom, Surrey, U.K.

Young, R.A.
University Engineering Department, Trumpington Street, Cambridge CB2 1PZ, U.K.

Zanardo, A.
Institute of Hydraulics, University of Padua, Via Loredan 20, 35100 Padova, Italy

Zdravkovich, M.M.
Department of Mechanical Engineering, University of Salford, Salford M5 4WT, U.K.

Contents*

TECHNICAL SESSION A

"Generation of Oscillatory Flows"
Monday, August 14, 1972, Morning

Chairman: H. Schlichting, Germany
Co-Chairman: B. Westrich, Germany

---

* A separate volume containing all contributions to the discussions at the technical sessions is available from the Institute of Hydromechanics, University of Karlsruhe, Karlsruhe, Germany.

TECHNICAL SESSION B

"Mathematical Models of Flow-Induced Vibrations"

Monday, August 14, 1972, Afternoon

Chairman: A.G. Davenport, Canada

Co-Chairman: O.H. Schiele, Germany

TECHNICAL SESSION C

"Flow-Induced Vibrations of Hydraulic Structures"

Tuesday, August 15, 1972, Morning

Chairman: H.W. Partenscky, Germany

Co-Chairman: B. Treiber, Germany

TECHNICAL SESSION D

"Flow-Induced Vibrations of Beams and Bridge-Decks"

Tuesday, August 15, 1972, Morning

Chairman: W.D. Baines, Canada

Co-Chairman: W.W. Martin, Canada

TECHNICAL SESSION E

"Flow-Induced Vibrations of Bluff Bodies"

Tuesday, August 15, 1972, Afternoon

Chairman: A. Roshko, U.S.A.

Co-Chairman: J. Renner, Germany

PANEL DISCUSSION

"Research Priorities in Flow-Induced Vibrations"

Wednesday, August 16, 1972

Chairman: E. Naudascher, Germany

TECHNICAL SESSION F

"Flow-Induced Vibrations of Marine Structures"

Wednesday, August 16, 1972, Morning

Chairman: M. Hino, Japan

Co-Chairman: C. Zimmermann, Germany

TECHNICAL SESSION G

"Flow-Induced Vibrations of Shells and Pipes"

Wednesday, August 16, 1972, Morning

Chairman: A. Daubert, France

Co-Chairman: F.-D. Heidt, Germany

TECHNICAL SESSION H

"Flow-Induced Vibrations Relating to Buildings"

Wednesday, August 16, 1972, Afternoon

Chairman: C. Scruton, U.K.

Co-Chairman: E. Plate, Germany

TECHNICAL SESSION A

"Generation of Oscillatory Flows"

Chairman: H. Schlichting, Germany

Co-Chairman: B. Westrich, Germany

GENERAL LECTURE

Generation of Oscillatory Flows

By

Rudolf Wille

Berlin, Germany

Summary

Kronauer quoted by Bearman [2] introduced the concept that a vortex-street wake built-up from alternately shed vortices behind an obstacle has to do with drag rather than stability of the vortex-street. Computer based studies of the von Kármán vortex-street have the tendency of over-emphasizing the role of "induction". An argument is brought forward that it is the base pressure and the irrotational flow past the body which brings about the alternating vortex shedding, thus producing an oscillating wake.

Introduction

The subject of this paper is embedded in the wide theme of flow induced vibrations. The general problem is that of the feedback between a vibrating body or structure and the oscillating flow field. In order to master problems of aeroelasticity or hydrodynamically excited time dependent forces on solid matter one has to consider the obstacle plus the flow. The analysis of this complexity is complicated, and it is true to state that only a few groping steps have been made in the right direction. Hence, one has to fall back on Bacon of Verulam's motto "dissecare naturam" which under the present heading means the attempt to describe the origin of an oscillatory flow without looking at the feedback to the solid matter to which it is linked in one way or the other.

The limits of the subject I want to treat may yet be drawn in another way: An oscillating flow of any origin can act upon a body which stands in its path or an obstacle within a smooth flow may itself be the generator of an oscillatory flow, see Fig. 1. The first case has been treated exhaustively by E. Naudascher [15], and it is the second case I chose to take as a starting point; again stressing deliberately the omission of a feedback. In other words, my contribution is devoted to problems of wakes having a distinct oscillatory character.

Vortex-Street Wake as Example of an Oscillatory Flow

It should be remembered that von Kármán [8, 9] in his original treatises on the stability of an arrangement of line vortices thought of the resulting flow field induced by line vortices as the reason for drag, i.e., a force acting upon a body in the direction of flow. He derived a general formula for the resistance using the characteristic of a wake produced by point vortices in staggered relationship. Figure 2 is a reproduction of von Kármán's well known paper.

The drag F_w is a function of the circulation Γ, of the velocity U of the onflow, the velocity u of the vortex system and of the spacing ratio of the point vortices, h/l:

$$F_w = \rho\Gamma(U - 2u)\,\frac{h}{l} + \rho\,\frac{\Gamma^2}{2\pi l}\,.$$

This formula is the result of the application of the law of momentum to a control surface moving with the velocity of the vortex-street whose momentum balance has been calculated. The concept is that the point-vortices produce by induction an influx into the control surface from the rear end which is a surplus over the influx through the front face. The resulting change of momentum within the control surface is equal to the force acting upon the solid body in the direction of the main flow. The conservation of mass within the control surface is verified by an outflow through the lateral boundaries, not adding to the momentum in flow direction.

The consequent step in von Kármán's theory of hydrodynamic drag was to introduce the numerical data for the vortex arrangement found by linear stability theory. He found to the first order a stable array of staggered point vortices with the velocity of the system: $u = \Gamma / l\sqrt{8}$ and the spacing ratio: $h/l = 0.281$. Side by side with mathematical analysis, von Kármán proved by towing tank experiments that circular cylinders and flat plates in two-dimensional flow have vortex-street wakes with a staggered arrangement of vortices.

But in water and in air, the spacing ratio would not be found to be a constant along the street. Over a wide range of the Reynolds number, $40 < Re < 10^3$, the spacing ratio begins near the cylinder to be $h/l = 0.2$ going up to about 0.5 at the end of the life time of a vortex triple. Moreover, experimentalists during the 50 to 60 years after von Kármán detected that the appearance of a clearly visible and reproducible "street" of vortices is limited to the range of the Reynolds number of $40 < Re < 190$, this Reynolds number being based on the undisturbed velocity of onflow and a measure of the obstacle across the flow, i.e., the diameter in the case of a circular cylinder. Hence, for a time of five decades the idea of linking the generation of an oscillating vortex-street wake with a drag-force was not pursued - mainly because at the "interesting" ranges of Reynolds number the wake became turbulent and concentrated vortices forming a "street" did not exist.

On the other hand, all experiments proved that periodic vortex shedding exists to Reynolds number up to 10^7. A survey of these and related phenomena has been given recently by Berger and Wille [3]. Periodic or oscillating vortex shedding from a solid body is accompanied by a periodic change of the circulation encircling the body which leads to an alternating lift force crosswise to the direction of the onflow. The vector sum of this and of the drag gives a resultant alternating force F_{res} as shown in Fig. 3. This leads to the swinging motion of slender chimney-stacks and other tall upstanding cylindrical bodies like the Baseline Orbiters of the 1980's. The cross-wise oscillation of the historical submarine "Schnorchel" is another typical example. The various cases of aerodynamic excitation, like vortex-shedding and galloping excitation, have been treated comprehensively by Scruton [20].

Coming back to the problem of drag and vortex generation, it is possible to record the date when fifty years after von Kármán a new approach was made. At the IUTAM-Symposium on Concentrated Vortex Motion in Fluids, Ann Arbor, Michigan 1964, Kronauer gave a paper which, although it did not appear in print, through the consequent activities of Bearman [2] became a sort of a "missing link" between base pressure, resp. drag and vortex formation. In brief, Kronauer postulated that it is not the induction stability of vortex rows which builds up the street, but that it is the vortex velocity and the spacing ratio which adjust themselves in such a way that a minimum drag is obtained for the body.

In the meantime, we speak of the Kronauer criterion, writing:

$$\left(\frac{\partial C_D}{\partial h/l}\right)_{u/U = \mathrm{const}} \rightarrow 0 .$$

In words: The change of the drag coefficient C_D with spacing ratio h/l for a constant relation of vortex speed u and velocity U of the onflow towards the obstacle tends to go to zero. Indeed, using a drag formula like that of von Kármán stated above, the significant property of which is, that it contains characteristic measures of the vortex-street and of the body, one is able to plot the drag coefficient versus the spacing ratio for various parameters u/U. Kronauer did this and provided us with a diagram like in Fig. 4, taken from Bearman [2]. From this we have a coordination of u/U and h/l for the minima of the drag. Schiele [19] found a similar result from evaluating the energy of the vortex-street.

In an experiment it is easy to measure the longitudinal spacing l of vortices of one row and the shedding frequency f. From the obvious relation

$$U - u = f \cdot l$$

the quotient u/U can be computed and under the auspices of the Kronauer criterion, using the minimum-line in Fig. 4, h/l and consequently h, the lateral spacing of the vortex rows can be found.

Next, Bearman [2] formed a new wake frequency number

$$f_B = \frac{f \cdot h}{U_s}$$

where f is the vortex shedding frequency, h the lateral spacing of the vortex rows found like stated above and U_s the "separation velocity" found from the measured base pressure coefficient c_p by writing: $U_s = \sqrt{1 - c_p}\ U_\infty$.

If f_B is plotted versus the base pressure coefficient, a constant $f_B = 0.18$ is found for a wide variety of geometrical shapes including those having splitter plates or base bleeding. Thus, the Kronauer criterion has gained credit although some more corroboration is needed. In the context of this paper, though, Bearmans argument served to substantiate the connection between base pressure or drag with vortex shedding. Earlier papers of Roshko, e.g., Refs. [17] and [18], prepared these arguments.

Oscillatory Flow and Concentration of Vorticity

The classical approach to oscillatory periodic or non-periodic flow phenomena is that of stability theory. In flows containing vorticity in restricted areas, e.g. along a sheet of vorticity in a separated shear flow or "free" boundary layer, it is the law of induction which determines whether a vortex-street having suffered by a random occurrence a "small" disturbance - meaning a displacement of a limited section of vorticity-filled flow - is growing, decaying, or remaining neutral.

Induction instability invited the treatment of "inviscid models", and potential flow theory suggested two-dimensional or rotational symmetrical cases. Such cases are: the free-jet boundary layers and the separated boundary layers of cylindrical blunt bodies.

There is quite a number of papers in which a boundary layer having continuous vorticity has been approximated by one single row of discreet point vortices. After having introduced a disturbance, e.g. in the form of a sinusoidal wave, the effect of the induction of all vortices on all vortices is being calculated. Rosenhead [16] made the start and one generation later Birkhoff and Fisher [4] and Hama and Burke [7] computed the same problem, but introducing a special disturbance and more finaly spaced vortices by the help of a machine.

Michalke [10] improved this model, by introducing a layer with finite thickness in which the continuous vorticity distribution is approximated by three parallel rows of point vortices spaced apart. On the other hand he approximated the velocity distribution by a so called "broken-line" profile. With a disturbance derived from stability theory the rolling-up of this layer by the action of mutual induction could be formed.

A single layer of vorticity and its rolling-up process can also be treated under a different aspect. Michalke [11] used for the free boundary layer a so called "hyperbolic-tangent" profile of the velocity and its continuous distribution of vorticity. Using linearized stability theory the spatially changing pattern of five streaklines could be calculated. Later Michalke and Freymuth [12] proved that experiments done with smoke filaments and the results of inviscid linearized stability theory are in better agreement as had been expected. Figure 5, reproduced from Michalke and Freymuth, shows these results.

The upper part represents the rolling-up streaklines calculated for the hyperbolic-tangent velocity profile at maximum amplification of the disturbance. The dots are indicating fluid particles which initially at $x = 0$ had been one on top of the other along the y-axis. The distribution of vorticity is indicated by the data for Ω assigned to the streaklines. The lower photograph shows a band of streaklines, visualized by introducing smoke into the boundary layer of a nozzle from which a free jet escapes. Although the two cases under comparison have not exactly the same conditions - the mathematical model holds for two-dimensional inviscid flow and the experiment was done in air with a nozzle of circular cross section - the similarity of the patterns is striking. For the experiment it should be mentioned that the nozzle had an exit of 75 mm diameter whereas the boundary-layer thickness was less than 1 mm; an interaction across the jet core, therefore, could not be expected.

Summarizing, one finds that the mathematical treatment of inviscid flow using either the concept of induction among point vortices or the concept of linearized stability calculation for a layer having continuous distribution of vorticity gives a description of what can be said to be vortex formation. "Vortex formation" is the process of the concentration of vorticity within "clouds". The vorticity had been produced in the thin layer separating from a body or a nozzle. Adding to this, the concept of Michalke leads to the insight into the structure of a "cloud": The formation process includes the wrapping-in of irrotational flow. Vorticity is attached to matter and hence the widened and and recurrant streaklines according to the calculation of Michalke are indicating areas in which the content of vorticity is growing. On the other hand the calculation as well as the smoke photograph indicate the narrowing of the neighboring streaklines between two consecutive patches of vorticity. These thin strands of vorticity are in nature quickly consumed by viscosity. This explains that in photographs of the development further downstream only the "clouds" or "big vortices" are detectable. A periodic or quasi periodic oscillating flow can be produced from the instability of one single shear layer.

For the reason of computing the mutual interaction of two free shear layers the simulation of a continuous layer of vorticity by finely spaced line vortices has also been successfully used.

Abernathy and Kronauer [1] proved that the interaction induces contractions of vorticity in "clouds" and leads to a wake pattern comparable to the von Kármán vortex-street. Recently R.R. Clements [5] used another approximation based on an array of line vortices and computed the vortex-wake development behind a long body with a sharply cut-off base. He finds concentration of vorticity in clouds and a periodic sequance of these clouds. The overall picture appears different from that found by Abernathy and Kronauer as Clements considers the flow along the body whereas Abernathy and Kronauer started from existing vortex sheets.

In the calculations the alternating formation of clouds of vorticity stem from the fact that either a deliberately or randomly introduced disturbance appears at one time in one of the two shear layers. The ensueing concentration of vorticity then triggers by induction a disturbance in the other shear layer, and a concentration process follows there. The repeated sequence of this mutual interaction leads to the von Kármán vortex-street. This mathematical demonstration of a vortex-street has no need of an obstacle or a wall in the flow where vorticity is being generated in a boundary layer.

If we look at results of experiments done with the flow past an obstacle in the windtunnel or with a body towed in a water tank we find striking changes of the oscillatory wake flow with increasing Reynolds number. One cannot do better than referring to Morkovin [13] from which the drawings in Fig.6 have been adapted. It is difficult to understand that it is one and the same mechanism which leads to big vortices in both cases. Pattern d appears in the range of the Reynolds number of $40 < Re < 200$ to 300 whereas wakes of the type e are being produced up to $Re \simeq 10^5$. At such high Reynolds numbers the calculated inviscid vortex-streets and those found in experiment should be very similar. On the other hand, it is hardly conceivable that real fluid induction acts effectively across a flow region the width of which is one to two orders of magnitude larger than the layer of vorticity. Gerrard [6] overcome this difficulty by the concept that, in the first stage, a concentration of vorticity takes place at one shear layer. Quotation: "The growing vortex continues to be fed by circulation from the shear layer until the vortex becomes strong enough to draw the other shear layer across the wake. The approach of oppositely signed vorticity in sufficient concentration cuts off further supply of circulation to the vortex which then ceases to increase in strength. We may speak of the vortex as being shed from the body at this stage". Figure 7, adapted from Gerrard [6], explains the quotation. Region A is the growing vortex which has become strong enough to draw region B across the wake. The arrowhead b points to the locus where vorticity stemming from region B weakens the vorticity going into region A and thus cuts the feeding line of vortex A which then moves away to the right. In the next scene region B takes the role of the growing vortex and the play is repeated.

This shuttle-mechanism explains very well the alternating vortex shedding and the strong oscillation of the wake. The arrows a, b and c show the paths along which fluid from outside the shear layers is drawn into the wake region. This is irrotational fluid, but its wrapping-in into the growing vortex is not of the same kind as having been described by means of Fig. 5 which shows the rolling-up of one single shear layer. In the case of the double row, the irrotational fluid from the outside carries the deformed sheet of vorticity across the wake, and this transfer of circulation of opposite sign goes into the rolling-up vortex.

One more question is whether the rolling-up of a growing vortex starts in the same way as rolling-up of a single shear-layer, described before. It is suggested that this is not so, and that it is the base pressure which comes into play.

The argument is as follows (the sketches in Fig. 8 have the sole purpose of explaining the step of the argument; the sketches do not describe a development in time): The two free boundary layers separating from the wall of an obstacle in two-dimensional flow are dragging fluid between them downstream. Consequently the outer flow field begins to narrow down the region between the two shear layers, and the streamlines of the outer flow are curving inwards. In order to balance the pressure across the curved streamlines, a centrifugal force field has to be built up in the interior of the body's wake. At high Reynolds numbers, this is done by an inward-spiralling motion of the outer flow. It is this reaction to the curvature of the outer irrotational flow field which carries vorticity adhering to the fluid particles across the wake. Then the periodic process, described by Gerrard, has been triggered. Gerrard's expression "the vortex ... to draw the other shearlayer ... " suggests either friction or induction. The new argument is that it is none of these but that it is the potential flow which reacts to the base pressure by spiralling inwards. By this motion of fluid from outside the wake the two layers of oppositely signed vorticity are being brought to close neighborhood and are being stretched to such an extent that vorticity is dissipated by viscosity: "... and so down to viscosity". In a real fluid we cannot speak or think in terms of annulling vorticity by an approach of vorticity of opposite sign. This is only possible in the sense of the net-vorticity of a "cloud" made by the computer. Hence, the growing vortex is being forced across the wake and the breakdown of the line along which vorticity is being fed is the consequence of the action of viscosity. The overall significance of the role of the pressure fields in wake flows has been pointed out by Morkovin [14].

Also, the alternating process of vortex growth is the only possible way to shed vorticity and at the same time to get a back-flow of irrotational fluid to the base. A simultaneous growth of two vorticity-containing spirals has no possible mechanism to weaken or stop the influx of vorticity, as there is no possibility of thinning out vorticity. This is being illustrated in Fig. 9.

Films taken by a high-speed camera lead to this argument. Figure 10 represents two scenes taken randomly from a film taken by F. Etzold at the Hermann-Föttinger-Institut. The Reynolds number, based on cylinder diameter was $Re \simeq 10^5$ and the film speed was about 1000 fps. The flow is coming from the left. The thin vertical white line is a moving thread of wool carrying continuously a smoking liquid to the wake region. Scene "a" shows that irrotational flow from below is bending upwards and is entrained into the area immediately behind the vortex cloud under formation behind the cylinder the rear contour of which is clearly visible. In scene "b" the irrotational flow comes from above and is being entrained downwards. The technique used for visualisation can be described in the following way: The moving acid-impregnated thread produces a thin smoke screen which is lit by a strong light slanting upon it from the front. At areas of spiralling flow, i.e., where the vortex clouds are being formed, the downstream mean motion is being slowed down relative to the main flow; hence, the film "sees" these areas more clearly than the smoke-screen.

Conclusion

A mechanism is explained by which at high Reynolds numbers alternating vortex shedding can be controlled by base pressure and entrainment of potential flow.

References

1. Abernathy, F.H., Kronauer, R.E.: The Formation of Vortex-Streets. J. Fluid Mech. 13, 1-20 (1962).

2. Bearman, P.W.: On Vortex-Street Wakes. J. Fluid Mech. 28, 625-641 (1967).

3. Berger, E., Wille, R.: Periodic Flow Phenomena. Annual Review of Fluid Mechanics (1971).

4. Birkhoff, G.D., Fischer, J.: Do Vortex-Streets Roll up? Rendi. Circ. Math. Palermo, Ser. 2, 8 (1951).

5. Clements, R.R.: An Inviscid Model of Two-Dimensional Vortex Shedding. Submitted for publication to J. Fluid Mech. (1972).

6. Gerrard, J.H.: The Mechanics of the Formation Region of Vortices behind Bluff Bodies. J. Fluid Mech. 25 pt. I, 401-413 (1966).

7. Hama, F.R., Burke, E.R.: On the Rolling up of a Vortex-Street. Univ. of Maryland Techn. Note, BN - 220 (1960).

8. v. Kármán, Th.: Über den Mechanismus des Widerstandes, den ein bewegter Körper in einer Flüssigkeit erfährt. Göttinger Nachrichten, math.-phys. Kl. 547 (1911).

9. -, Th., Rubach, H.: Über den Mechanismus des Flüssigkeits- und Luftwiderstandes. Phys. Z. 13, No. 2 (1912).

10. Michalke, A.: Zur Instabilität und nichtlinearen Entwicklung einer gestörten Scherschicht. Ing.-Arch. 33, 264 (1964).

11. -, A.: On Spatially Growing Disturbances in an Inviscid Shear Layer. J. Fluid Mech. 23, 521 (1965).

12. -, A., Freymuth, P.: The Instability and the Formation of Vortices in a Free Boundary Layer. AGARD Conference Proceedings, No. 4, Separated Flows, Part 2, 575-595 (1966).

13. Morkovin, M.V.: Flow Around Circular Cylinders. A Caleidoscope of Challenging Fluid Phenomena. ASNE Symp. on Fully Separated Flows, Philadelphia, 102-118 (1964).

14. -, M.V.: The Interaction of Vorticity and Pressure Fields. Proc. Symp. on Flow-Induced Vibrations in Heat Exchangers. ASME Dec. 1970.

15. Naudascher, E.: From Flow Instability to Flow- Induced Excitations. Journal of the Hydraulic Divisions. Proc. of the Am. Soc. of Civil Engineers (1967).

16. Rosenhead, L.: The Formation of Vortices from a Surface of Discontinuity. Proc. Roy. Soc. A 134 (1931).

17. Roshko, A.: On the Development of Turbulent Wakes from Vortex-Streets. NACA Rep. 1191, 25 (1954).

18. -, A.: Experiments on the Flow Past a Circular Cylinder at very High Reynolds Numbers. J. Fluid Mech. 10, 345-356 (1961).

19. Schiele, O.: Ein Beitrag zur Theorie instationär und periodisch arbeitender Propulsionsorgane. 23. Einzelmitteilung des Instituts für Strömungslehre und Strömungsmaschinen an der Technischen Hochschule Fridericiana, Karlsruhe 1961.

20. Scruton, C., Rogers, E.W.E.: Steady and Unsteady Wind Loading of Buildings and Structures. Proc. Roy. Soc. A 269, 353 (1971).

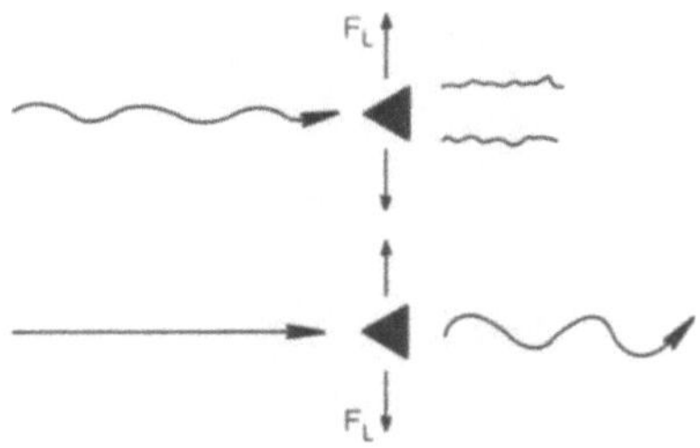

Fig. 1. Relations of an oscillating flow to a vibrating body.

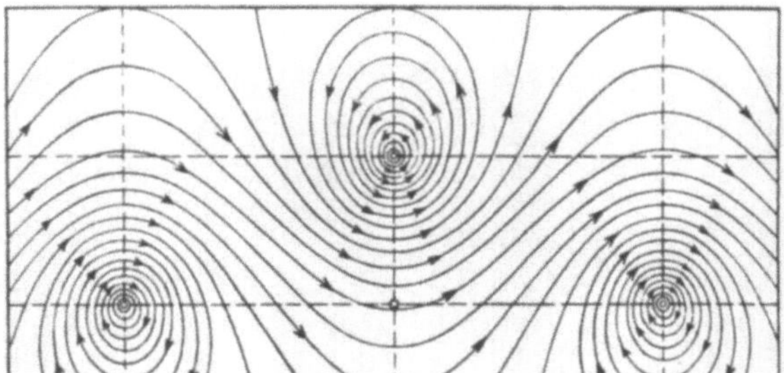

Fig. 2. Streamlines induced by a von Kármán point-vortex street.

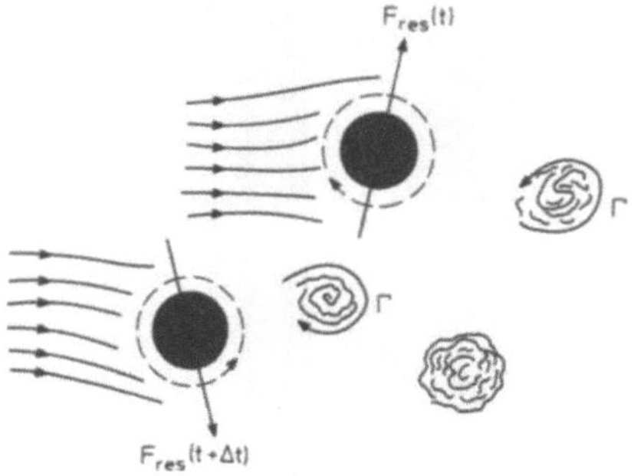

Fig. 3. Resultant force on a cylinder as consequence of vortex shedding.

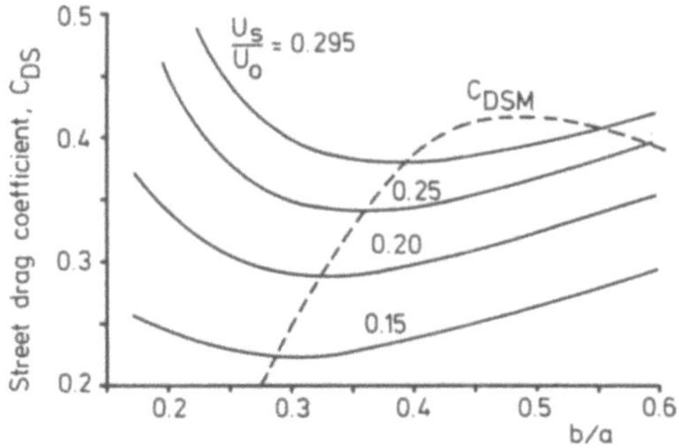

Fig. 4. Demonstration of the Kronauer criterion. Drag coefficient C_D of a cylinder plotted versus the spacing ratio h/l of a vortex street wake following von Kármán's drag formula.

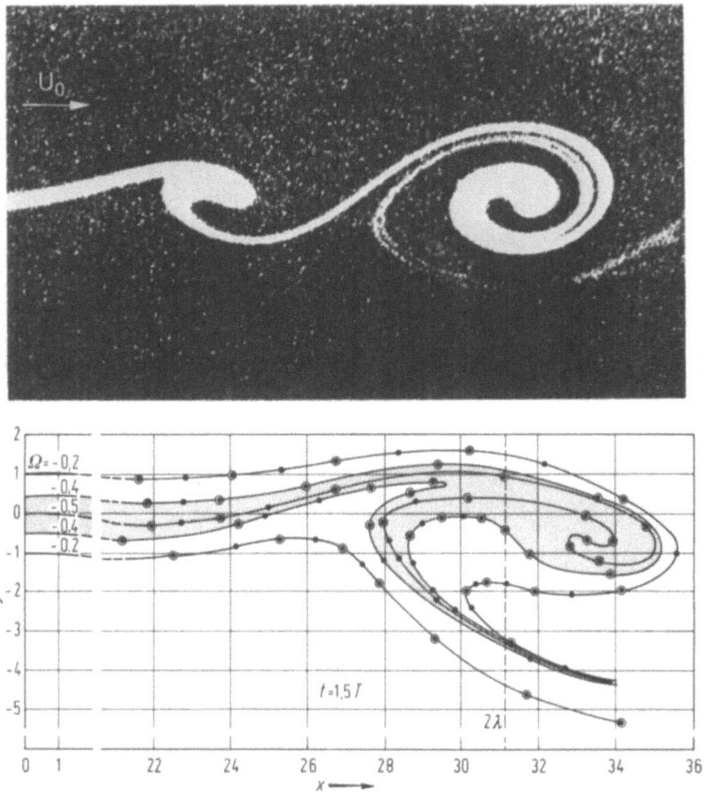

Fig. 5. The rolling-up of streaklines calculated for the hyperbolic tangent profile at maximum amplification [11], and the rolling-up of a free boundary layer into vortices visualized by smoke [12].

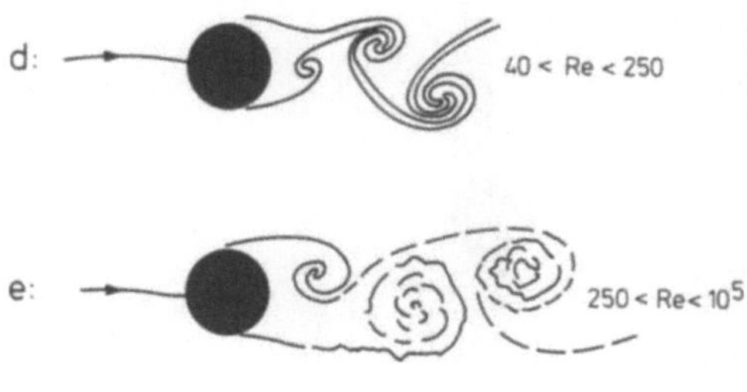

Fig.6. Sketches of vortex street pattern associated to certain ranges of the Reynolds Number (adapted from Morkovin 1964, Ref. [13]).

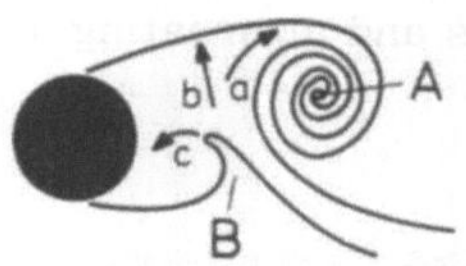

Fig.7. Sketch of flow behind a circular cylinder (adapted from Ref. [6]).

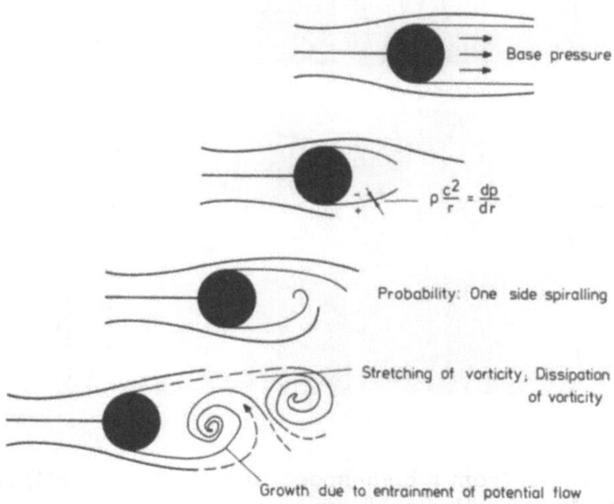

Fig.8. Potential streamlines and base pressure in close neighborhood to an obstacle.

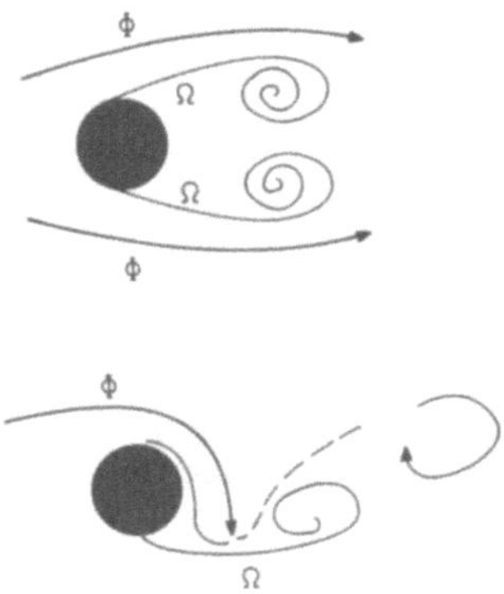

Fig. 9. Simultaneous and alternating vortex formation.

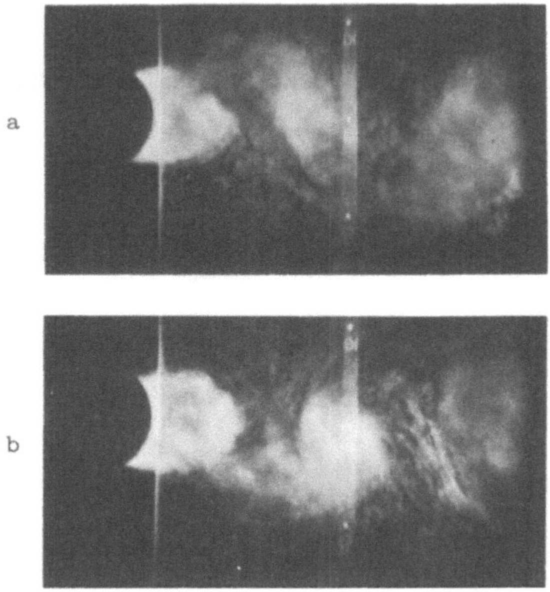

Fig. 10. Two scenes from a high-speed film at about 1000 fps, Re $\sim 10^5$. Diameter of cylinder 16 cm.

CONTRIBUTIONS

Instability of a Vertical Water-Curtain Closing an Air-Chamber

By

P. A. Kolkman

Delft, The Netherlands

Summary

A study is presented on the water-curtain/air-chamber system. The aim of the study was to get an insight into the feedback mechanism, and to detect whether factors are involved on which scale effects in physical models have to be expected. The theoretical part of the study consists of the behavior of the curtain when a periodic air pressure variation is assumed, and conditions are derived for the cushion rigidity and damping of the air chamber for which the feedback mechanism can occur. Experiments in a small model show results which can be partly explained by the theory, but questions on the air compressibility still remain. Tests with an elastic membrane in the wall indicate that a physical model on which the air compressibility is too high gives results on the unsafe side for the prediction of instabilities.

Introduction

Interest in the problem of instabilities in a water curtain has tended to be limited because early studies showed that splitting the water curtain is a simple remedy to stop the vibrations and to ensure a safe design of the overflow type of gate. But in some cases the splitting of the curtain is not technically attractive (ice problems etc.) and aeration tubes are considered, the capacity of which is determined by model studies. The more recent interest (Schwarz [1, 2] and Partenscky [3, 4]) has been more directed towards the fundamentals of the feedback mechanism which can be of importance for other vibration problems also.

The work of Schwarz showed that an analytical approach is in principle possible and leads to good results in the determination of critical frequencies. Recently, Partenscky [4] has shown also how the elasticity of the gate can be introduced. However, still missing at this moment is the explanation of the overall mechanism of the instability and the explanation why this instability of only the water curtain (rigid gate) can occur even when the air rigidity in test-installations is higher than acceptable in the computations [1]. As the compressibility of air is incorrect in small-scale models it is important to get information about this discrepancy.

To start with a disillusion: our recent work has not produced a complete solution either. Its interest is mainly connected with the fact that the theoretical and experimental set-ups are relatively simple to overlook because there is only a vertical water sheet enclosing a rectangular air-cushion (Fig.1).

Theoretical Study

When it is assumed that outside of the chamber (Fig.1) the pressure is constant (P_o) and inside it is varying ($P_o + p$), then the horizontal acceleration of the falling water particle depends on the mass ($\rho\delta$) and the pressure variation (p). The inside pressure is supposed to be equal over the height of the curtain. Because the effect of p on the fall velocity has been neglected, the system is linear.

When the water curtain and the air cushion form a feedback vibration, the air pressure will have the characteristic $p = \hat{p} \cdot e^{zt}$, in which z is complex. When the real part of z is positive, the amplitude increases and the system is unstable and when negative the system is damped. With this approach, however, the equations became too complicated to find the sign of the real part of z.

Thus, the following method was used. When a stationary vibration is assumed, the air pressure variation $\hat{p} \cos \omega t$ determines the behavior of the curtain (Eq. (4a) and Fig.3), and the variation of the volume ΔV can be computed. The volume variation can be written as

$\Delta V = A(\hat{p}, \omega\tau_H, v/g\tau_H) \sin\omega t - B(\hat{p}, \omega\tau_H, v/g\tau_H) \cos\omega t$ (Eq. (6), (Appendix I). Then a formulation is possible of the condition to be fulfilled for ΔV causing $\hat{p} \cos\omega t$ (Appendix II).

This leads to a "necessary rigidity" and a "necessary damping" which can be either positive or negative. When the "necessary damping" is positive for the stationary vibration ($A/B < 0$), and the real damping of the air is smaller, the real system will have a surplus of energy and will amplify.

The computations were made for variations of the height (H) and frequency (ω), and it appears (Appendix I and II) that the spring rigidity of the air, expressed in L with the assumption of $p \cdot V =$ = constant (isotherm) can be found from:

$$\Psi = P_o \cdot V_o^2 / L \cdot g^2 \cdot \rho_w \cdot \delta_o = \Psi(\omega\tau_H, v/g\tau_H),$$

in which τ_H is the time of fall ($H = V_o\tau_H + \frac{1}{2}g\tau_H^2$). The value of Ψ is to be found in Fig.2.

When a real situation is given, Ψ and $V/g\tau_H$ are known, and a number of values will be found for $\omega\tau_H$, of which some have a positive "necessary damping" (Fig.2). At these values of $\omega\tau_H$ vibrations can occur.

Theoretically the amplitude in the case of instability of a linear system grows indefinitely. Only the speed of increase is coupled to the "necessary damping" (log. incr. = $-\pi A/B$, Eq. (11)). The limit of the amplitude is only due to non-linear effects.

The use of this linear theory is limited to infinitely small amplitudes, which is to be seen from an energy balance. The incoming water has a certain energy (kinetic + potential). The energy of the curtain increases due to the horizontal (vibration) velocities and also, for negative A/B values, energy is "pumped" into the air. This means that the fall energy has to decrease to obtain a correct energy balance.

Experimental Study

With a small model it was checked how far the computations are valuable. The total height over which the curtain could fall (H) was up to 0.5 m, the width 0.28 m, and the length L was variable up to 0.3 m (Fig.4). H could be varied by regulation of the downstream waterlevel.

On the back wall, a rubber membrane was braced weakly and could be taken out of work by a valve inside of the air-chamber. The membrane rigidity was not calibrated, but the relations between the amplitudes of pressures and of the membrane movement during vibration indicated that 1 mm of water column gave 40 to 60 cm^3 volume variation. This means roughly that the weakest aircushion during the test (H = = 0.5 m, L = 0.3 m) became ten times weaker with the membrane.

The fall velocity V_o was determined accurately by photographing oildrops, injected intermittently at the same frequency as the frequency of a stroboscope lamp.

The test results are presented in Fig.5; the measured frequencies differ by a maximum 5% from the average lines which are presented in Fig.5b.

The main conclusions which are drawn from a comparison between the measured and calculated values are:

1. F r e q u e n c i e s . The frequencies of vibration as presented in Fig.5b are, for the first harmonic, reasonably described by $\omega\tau_H \sim 8.2$; for the values of H, where vibrations could only occur with help of the membrane, this goes down to 6.8. For the second harmonic $\omega\tau_H$ is about 14.5 without and down to 12.5 with membrane. The measurement of the third harmonic was only incidentally obtained, with a value of $\omega\tau_H = 22.3$. In the region where vibration could occur without membrane, practically no influence was seen when the membrane was used.

The $\omega\tau_H$ values do not really correspond with the maxima of the (-A/B) values as was expected from the computations (Fig.2), nor do the

Ψ values of Fig. 2 agree with the Ψ values of the model (mentioned in Fig.5a). The $\omega\tau_H$ values correspond better with the frequencies where the pressure variation $\hat{p}\cos\omega\tau$ has a maximum energy transfer from the curtain into the air; this is when (-A) reaches the maximum value. To demonstrate this, the A curves are also plotted in Fig.2. The $\omega\tau_H$ values correspond with the $2\pi\left(n+\frac{1}{4}\right)$ criterium of Schwarz: this would result in $\omega\tau_H = 7.9$, 14, and 20.4 for the first three harmonics.

The measurement of the third harmonic with $\omega\tau_H = 22.3$ comes just where A becomes positive. As this value was only taken at one measurement, no premature conclusions should be drawn.

2. The Membrane. As is seen in Fig.5, vibrations occurred at much greater ranges of H with the use of the membrane. Also, the vibration started much easier. This indicates that a model on a reduced scale is on the unsafe side for the study of vibrations, because then the air rigidity is relatively too high. In some cases, the membrane had to be used to stimulate the vibrations; then, the vibrations remained also without it. In Fig.2, those cases are presented as vibrations without membrane.

3. The Rigidity of the Air Cushion. The results, as discussed under Point 1, are not at all related to the theoretical Ψ value of Fig. 2. The rigidity of the aircushion in the model was much too large compared with the theoretical value, and discrepancies are more than 100-fold. Different possibilities, apart from the curtain itself, can be considered:

(a) The downstream water-level under the airchamber is another elastic element. With the static deflection due to a pressure variation, the total rigidity can come in the right order of magnitude. However, wave studies indicate that this only occurs at very low frequencies. When they are high, the water acts as mass or as spring, depending on the frequency. Moreover, the quasi-rigidity of the water becomes much higher at the frequencies being dealt with here. Tests were carried out with the downstream water-level replaced by a sloping bottom. As the vibrations remained, it is concluded that the water cushion is definitely not the determining parameter for the rigidity of the system.

(b) The downstream water-level is supposed to be a damper, being in series with the air "spring". This is analyzed in Appendix II; The resulting rigidity can be (A^2/B^2+1) higher than the Ψ on Fig.2, where the damper was supposed to be in series with the spring. This results from Eq.(10). For the range of measurements, the maximum of A/B is 1.2, so the theoretical Ψ value becomes 1 to 2.5 times the Ψ value of Fig.2.

4. Pressure Variations in the Air Chamber. Simultaneous pressure measurements were made on one side wall. Depending on L and H, the pick-ups could be placed on 2-12 locations. All pressure variations had the same amplitude without phase shift. Only at the corner where the curtain falls into the water, p was a little greater. This shows that for this model the concept of a simple air cushion was correct. When the dimensions of the air cushion are greater, resonance in the air can play a role; in this case, a small-scale model will also have scale effects.

5. Magnitudes of Pressures. Pressures have only been measured for the steady-state vibration. Roughly it can be said that for tests I and III $\hat{p}$ is proportional to H^{-1}, and for the first harmonic the average value for H = 0.5 m varies from 0.65 mm of water column to 0.8 mm for test I and from 1.0 to 1.4 mm for test III (so roughly increasing proportional to δ_o). Lowest amplitudes are found for small L, the greatest for L = 0.3 m and with membrane. The second harmonic gives 1.3 to 1.6 times greater $\hat{p}$ values. At smaller H values, near where the vibration ends, $\hat{p}$ decreases to zero.

Test II has low $\hat{p}$ values compared with test III (same δ_o value but lower V_o), and decreases rapidly from 0.8 mm for H = 0.5 m w.c. to 0.2 mm for H = 0.43 m.

For tests I and III computations show that 2 to 4 % of the (kinetic + potential) energy of the curtain is transferred to the air. For test II, this is 0.3 to 1 %.

6. Shape of the Slit. Shapes with round and sharp corners (stable and unstable separation) have been tested, without producing any effect.

7. Aeration of the Air Chamber. When the curtain is split or not attached to the side walls, there is no vibration. For thin curtains (δ = 0.6 to 1.2 mm), aeration through holes in the side wall was possible up to 10 % of the wall area. Without membrane this was somewhat less. For test III only 2 % was acceptable, for test II nothing could be opened.

8. Test with an L of 1.5 m but the Back Wall Removed. Vibrations appeared to occur after some initial excitation; no further measurements were made. Theoretically, this means that Ψ can be negative.

9. Influence of the Ψ Values of the Physical Model. As is seen from Fig.5a, there is a tendency that a decrease of air rigidity (by greater L or membrane) gives rise to lower harmonics. This agrees with theory (Fig.2). On the other hand, test II with a low Ψ value only vibrated at the second harmonic. As it is concluded that the Ψ is not the relevant parameter, the effect of V_o and δ_o is discussed separately.

10. Influence of the Initial Velocity V_o. Greater V_o eases the vibration; this agrees with Fig.2 because A (or - A/B) increases at higher $V/g\tau_H$ values.

11. Influence of δ_o. No vibrations were obtained for δ_o greater than 2.5 to 3 mm. The limit of vibration depends on L and the membrane.

The elimination of the interference of the downstream water-level and the apparent discrepancy in the Ψ value indicates that the linear theory is too rough an approximation. Another indication for the importance of non-linear effects is the initial stimulation which is sometimes necessary to obtain the steady-state vibration.

Non-linear effects, of which the most important are related to the influence of p on the fall velocity, can affect, for instance, the apparent Ψ value. But the magnitude of this influence will then be related to the amplitude of vibration. This makes it more understandable that, for instance, different $\hat{p}$ values have been measured after the cushion rigidity has been varied. As the amplitude of the vibration is related to $\hat{p}/\delta_0$ the remark under point 5 that $\hat{p}$ is proportional to δ_0 also indicates that the influence of p on the fall velocity is the determining parameter.

Conclusion

The linear approach for the calculation of the water curtain is valid to predict the correct possible frequencies of vibrations, that is, those frequencies where maximum energy transfer from the curtain into the air occurs. Discrepancy between theory and experiment is found on the point of the air rigidity, but qualitative agreement is found on the fact that greater rigidity gives rise to higher harmonics and that higher initial velocity eases the vibration. No information is found for conditions where lower harmonics transform into higher, or where vibration stops. Indications are given that probable non-linear effects by periodic changes in the fall velocity have to be taken into account.

Experiments with a membrane in the wall show that a model with a too stiff air cushion is on the unsafe side for the prediction of vibrations.

Acknowledgement

The author is very grateful to W.C. Bisschoff van Heemskerk, Professor at the Delft Technical University, and to N. Schoemakers. The former stimulated the set-up of this study and allowed Mr. Schoemakers to do the experiments and a part of the elaborate computations for his engineering-degree work.

References

1. Schwarz, H.I.: Nappe Oscillation. J. Hydraul. Div. ASCE 90, 129-143 (1964).

2. Schwarz, H.I.: Edgetones and Nappe Oscillation. J. Acoust. Soc. of Amer. (1966/3).
3. Partenscky, H.W.; Sar Khloeung I.: Oscillations de lames déversantes non aérées. Seminar S 6, 12th Congress Int. Ass. of Hydr. Res., Ft. Collins, 1967
4. Partenscky, H.W.; Swain A.: Theoretical Study of Flap-Gate Oscillation. Paper B 26 of 14th Congress Int. Ass. of Hydr. Res., Paris, 1971.

Appendix I. Calculation of the Momentary Curtain Shape due to a Harmonic Pressure Variation

The particle which is at time t on place Y left the top at time $t - \tau_Y$ (with $Y = v_o\tau_y + \frac{1}{2}g\tau_y^2$). While $v = v_o + g\tau$, the thickness of the curtain during falling, becomes

$$\delta = \frac{\delta_o v_o}{v_o + g\tau} . \tag{1}$$

(The notations are followed according to Fig.1)

Horizontal acceleration due to pressure difference $p = \hat{p}\cos\omega t$:

$$\frac{d^2x}{d\tau^2} = \frac{p}{\rho_w \cdot \delta} = \frac{(v+g\tau)}{\rho_w v_o \delta_o} \cdot \hat{p}\cos\omega(t - \tau_y + \tau) . \tag{2}$$

The horizontal velocity at the moment τ after leaving the top:

$$\frac{dx}{d\tau} = \int_0^\tau \frac{d^2x}{d\tau^2}\, d\tau . \tag{3}$$

The horizontal displacement X of the particle being at Y at the moment t (being τ_y after leaving the top):

$$X(y) = \int_0^{\tau_y} \frac{dx}{d\tau}\, d\tau . \tag{4}$$

In complete form, Eq.(4) becomes:

$$X(y) = X(\tau_y) = \frac{\hat{p}}{\rho_w \delta_o \omega^2} \Big[(\omega\tau_y \sin\omega\tau_y + \cos\omega\tau_y - 1)\cos\omega t +$$

$$+ (\sin\omega\tau_y - \omega\tau_y \cos\omega\tau_y)\sin\omega t\Big] +$$

$$+ \frac{\hat{p}g}{v_o \rho_w \delta_o \omega^3} \Big[(2\sin\omega\tau_y - \omega\tau_y \cos\omega\tau_y - \omega\tau_y)\cos\omega t +$$

$$+ (2 - \omega\tau_y \sin\omega\tau_y - 2\cos\omega\tau_y)\sin\omega t\Big]. \qquad (4a)$$

One example is represented in Fig.3.

The variation of volume per unit of width becomes:

$$\Delta V = \int_o^H X dy = \int_o^{\tau_H} X(v + g\tau_y) d\tau_y \qquad (5)$$

or:

$$\Delta V = \frac{\hat{p}g^2}{\delta_o \rho_w v_o \omega^5} \Big[\Big\{\omega^2\tau_H^2 + 4 + (\omega^2\tau_H^2 - 4)\cos\omega\tau_H - 4\omega\tau_H \sin\omega\tau_H\Big\}\sin\omega t -$$

$$- \Big\{\frac{1}{3}\omega^3\tau_H^3 + 4\omega\tau_H \cos\omega\tau_H + (\omega^2\tau_H^2 - 4)\sin\omega\tau_H\Big\}\cos\omega t\Big] +$$

$$+ \frac{\hat{p}v_o(v_o + g\tau_H)}{\delta_o \rho_w v_o \omega^3} \Big[\Big\{2 - 2\cos\omega\tau_H - \omega\tau_H \sin\omega\tau_H\Big\}\sin\omega t -$$

$$- \Big\{\omega\tau_H + \omega\tau_H \cos\omega\tau_H - 2\sin\omega\tau_H\Big\}\cos\omega t\Big] = A\sin\omega t - B\cos\omega t. \qquad (6)$$

In the evaluation $\alpha = \frac{v_o}{g\tau_H}$ is used, and

$$\frac{\hat{p}g^2}{\delta_o \rho_w v_o \omega^5} \cdot (\alpha^2 + \alpha)(\omega\tau_H)^2 \text{ instead of } \frac{\hat{p}v_o(v_o + g\tau_H)}{\delta_o \rho_w v_o \omega^3}.$$

Appendix II. Cushion Rigidity and Damping, Related to Harmonic Oscillation

It is supposed that the oscillation is steady, and that this is reached by the property of the air-cushion.

The air-cushion is compressed by a volume reduction $(-\Delta V)$ and this causes an extra pressure of

$$\Delta p = c(-\Delta V) + R\, d(-\Delta V)/dt \tag{7}$$

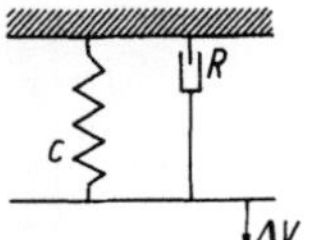

a

It is assumed that the damping is in the air, so rigidity c and damping R work parallel. In reality, d and R will be positive; in the following computation these can be either positive or negative.

When Δp equals the value necessary to produce the volume variation ΔV (Appendix I), equilibrium is reached and steady oscillation is obtained. Eq. (6) of Appendix I is used.

$$\Delta p = \hat{p}\cos \omega t \text{ produces } \Delta V = A \sin \omega t - B \cos \omega t. \tag{8}$$

Equation (7) results in:

$$\Delta p = \hat{p}\cos \omega t = -cA \sin \omega t + cB \cos \omega t - R\omega A \cos \omega t - R\omega B \sin \omega t. \tag{9}$$

This results in:

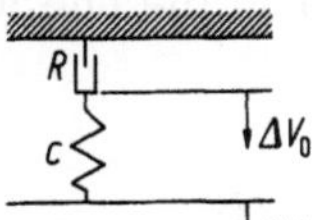

b

$$c = \frac{\hat{p}\cdot B}{A^2 + B^2} \quad \text{and} \tag{10}$$

$$\frac{\omega R}{c} = -\frac{A}{B}. \tag{11}$$

For comparison: with a free-damped motion of a mechanical oscillator $(m\ddot{y} + R\dot{y} + cy = 0)$ the logarithmic decrement equals $\pi\omega R/c = -\pi A/B$.

For the water-curtain/air-cushion the value $\pi\omega R/c = -\pi A/B$ will, when the cushion is undamped, represent the logarithmic increment of the self-excited oscillation.

The damping R can also be in series with the rigidity c (when the damping is in the downstream water, or when the air-cushion has a leakage with a linear resistance).

$$\Delta p = R \frac{d(-\Delta V_o)}{dt} \qquad (12)$$

and

$$\Delta p = c\ (-\Delta V + \Delta V_o)\ . \qquad (13)$$

Integration of (12) and application of (8) results in:

$$V_o = \frac{-\hat{p}}{R} \sin \omega t \qquad (14)$$

Equations (13) and (14) finally result in:

$$c = \frac{\hat{p}}{B} = \frac{\hat{p}B}{A^2 + B^2}\left(\frac{A^2}{B^2} + 1\right), \qquad (15)$$

$$\frac{\omega R}{c} = -\frac{B}{A}\ . \qquad (16)$$

With the A/B values of Fig.2, it is seen that for the region where $A/B < 0$, the A^2/B^2 becomes maximum $(1.2)^2$; this means the cushion rigidity can at a maximum be 2.5 times greater than the one found with Fig.2.

The value of the logarithmic increment remains $-\pi A/B$.

Explanation of Fig.2

Calculations are made with the assumption of a damped air-cushion (rigidity c parallel to damping R). Equation (11) is represented in the lower graph of Fig.2. When $A/B < 0$ the system will be unstable and self-excitation can occur.

The upper graph of Fig. 2 has been determined for two conditions:

(A) Condition of isothermal compression of the air-cushion. Pressure x volume = constant (Volume per unit width), so $dP \cdot Vol + dV \cdot P_o = 0$

$$c_I = \frac{-dP}{dV} = \frac{P_o}{Vol} = \frac{P_o}{LH} = \frac{P_o}{L(v_o\tau_H + \frac{1}{2}g\tau_H^2)} = \frac{P_o}{Lg\tau_H^2(\alpha+\frac{1}{2})} \; . \qquad (17)$$

(B) Equation (10) is $c_2 = \hat{p}\,B/(A^2 + B^2)$
From Eq. (6), Appendix I, it follows:

$$A = \frac{\hat{p}g^2}{\delta_o\rho_w v_o\omega^5} \cdot f_A(\omega\tau_H, \alpha) \text{ and } B = \frac{\hat{p}g^2}{\delta_o\rho_w v_o\omega^5} \cdot f_B(\omega\tau_H\,\alpha).$$

c_2 can be written as $c_2 = \dfrac{\delta_o\rho_w v_o\omega^5}{g^2} f_{(AB)}(\omega\tau_H, \alpha)$.

The condition for a steady oscillation will arise when $c_I = c_2$.

$$c_I = c_2 = \frac{P_o}{Lg\tau_H^2(\alpha+\frac{1}{2})} = \frac{\delta_o\rho_w v_o\omega^5}{g^2} f_{AB}(\omega\tau_H, \alpha), \text{ and it follows:}$$

$$\frac{P_o \cdot g \cdot}{L\cdot\tau_H^2\cdot\delta_o\cdot\rho_w\cdot v_o\cdot\omega^5} \equiv \frac{P_o v_o^2}{L\cdot\delta_o\rho_w\cdot g^2} \cdot \frac{1}{\alpha^3\cdot(\omega\tau_H)^5} = (\alpha+\tfrac{1}{2}) f_{ab}(\omega\tau_H, \alpha),$$

and consequently:

$$\frac{P_o v_o^2}{L\cdot\delta_o\rho_w g^2} = \Psi\,(\omega\cdot\tau_H, \alpha). \qquad (18)$$

Ψ is presented in Fig. 2.

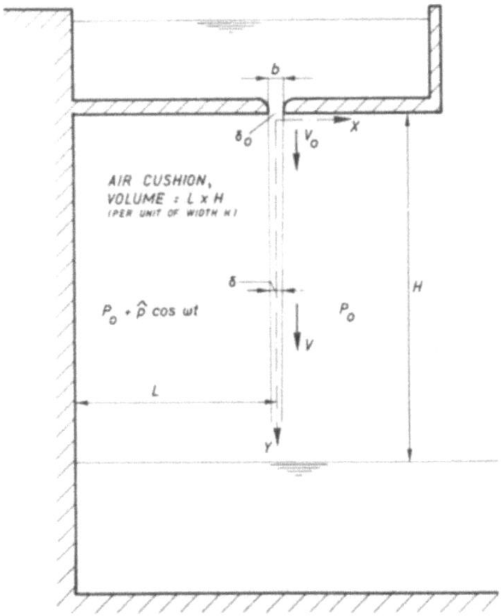

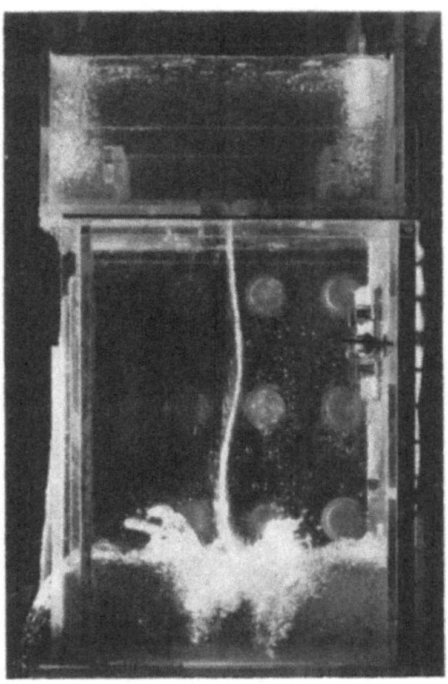

Fig.1. Symbols.

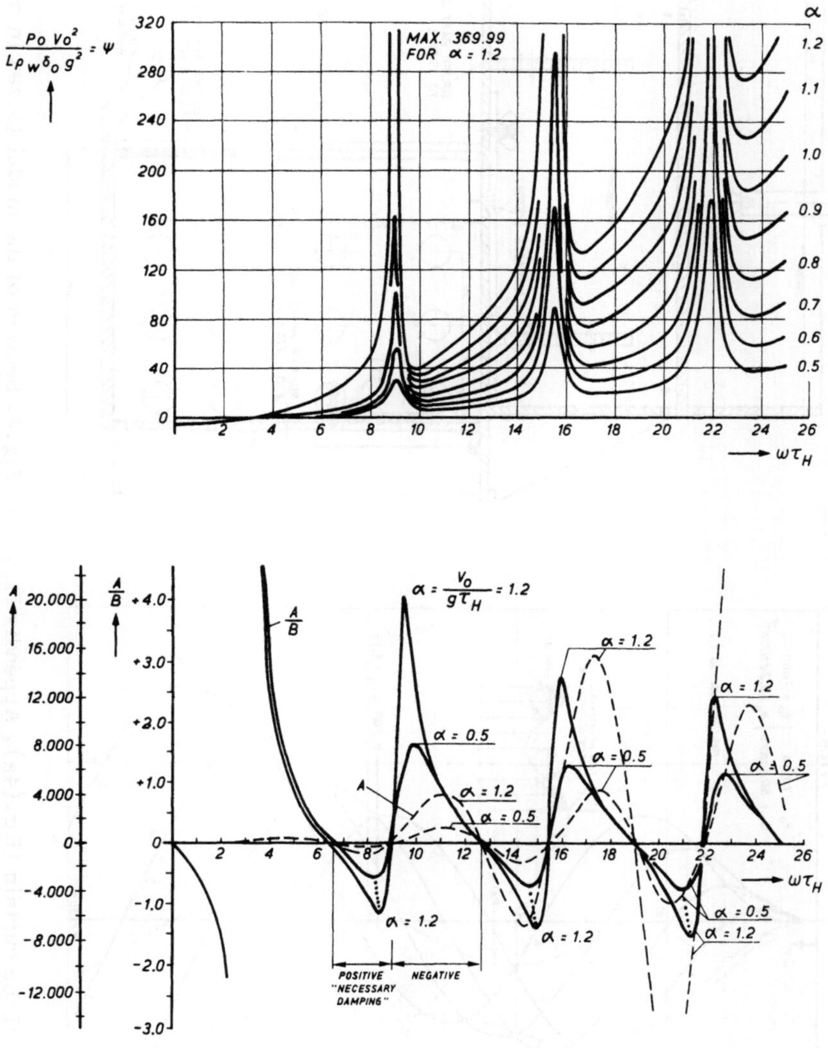

Fig.2. Theoretical relation of Ψ and A/B with $\omega\tau_H$ (from Appendix I and II).

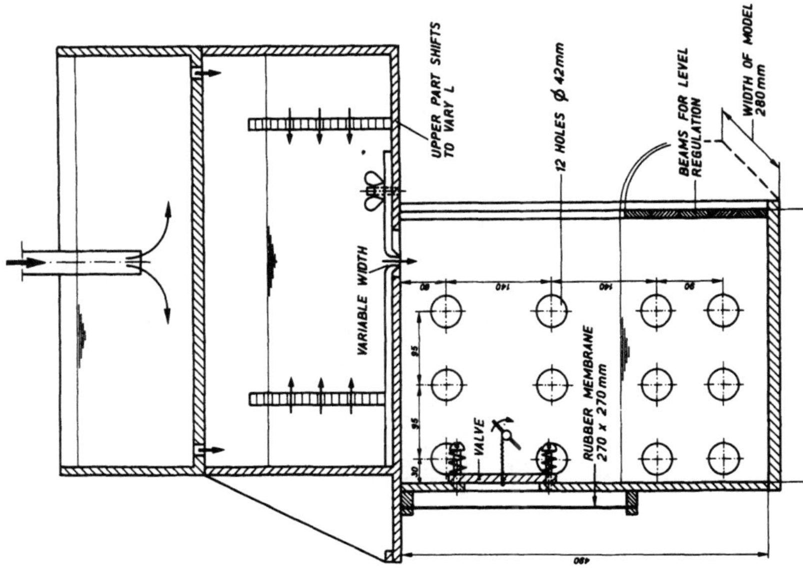

Fig.4. Section of the model (sizes in mm).

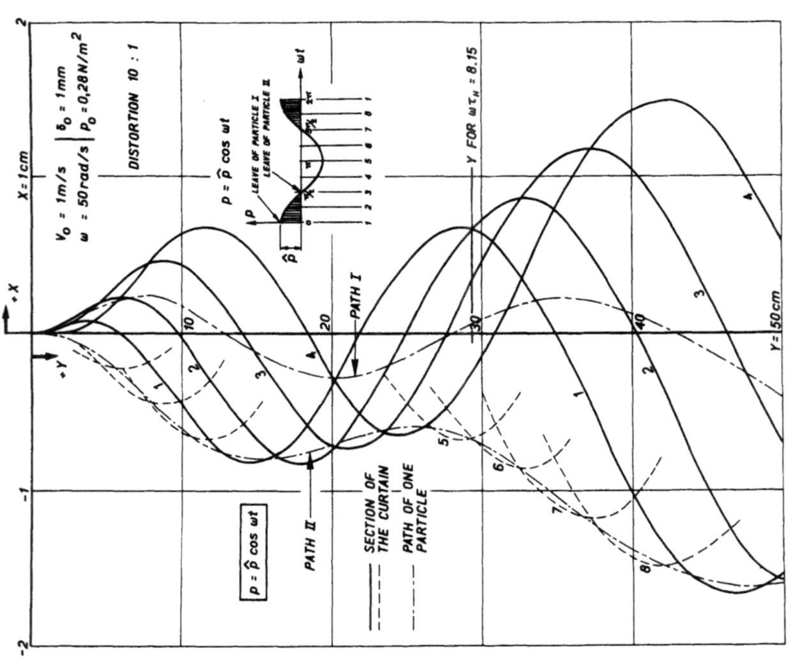

Fig.3. Vibration of the curtain (Eq.(4a), Appendix I).

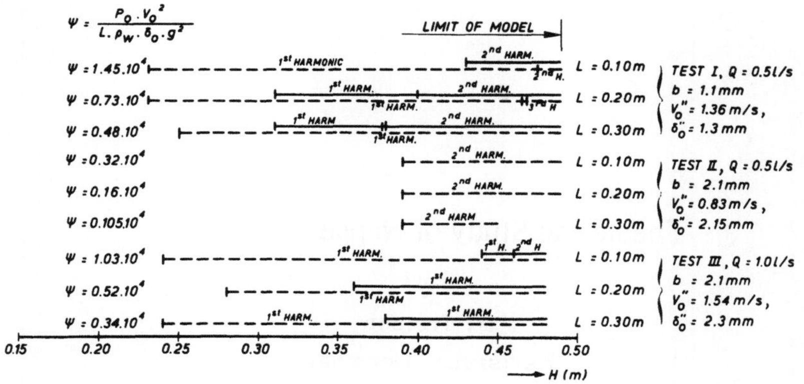

Fig. 5a. Ranges of H where vibrations occur.

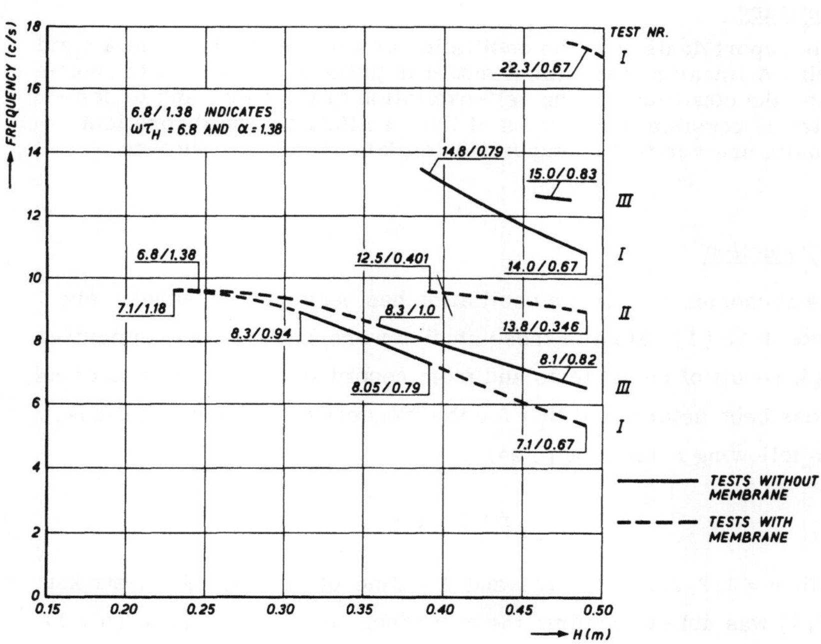

Fig. 5b. Measured frequencies.

Theoretical Study of Nappe Oscillation

By

B. Treiber

Karlsruhe, Germany

Summary

This report deals with the oscillation of a nappe falling from a rigid weir. A linear mathematical model is presented and used to determine the condition for the self-excitation of the nappe and to compute the possible frequencies of this oscillation. The theoretical results are verified by experimental data of several authors.

Introduction

The phenomenon of nappe oscillation has been studied extensively since 1930 [1]. Most of these studies were experimental in nature. As a result of model tests and from recorded observations on dams, it has been determined that for the frequency f of the oscillations, the following relation applies:

$$f \cdot T = n + \delta$$

with $n = 1, 2, 3 \ldots$, $\delta \approx 1/4$ and T = time of fall [5,7,8]. Petrikat [2,3] was able to confirm the occurrence of the first mode ($n = 1$) by a mathematical model. But because he treated the nappe as a solid body with one degree of freedom, he could not calculate the higher modes with his theory. To the author, no other theory is known with which the higher mode frequencies observed in experiments can be predicted. In this paper, a linear mathematical model is presented and used to compute all possible frequencies of nappe oscillation and to define a condition for the self-excitation of the nappe.

The Problem

A nappe falling from a rigid weir under the force of gravity alone has a parabolic profile. Under certain conditions the nappe and the air enclosed behind it can be excited to oscillate periodically. The falling water particles are forced by periodic pressure variations in the enclosed air volume to oscillate across the parabolic path. It is this oscillating motion of the water particles which in turn produces the pressure variations in the air by changing the volume.

Periodic oscillations are only possible if the pressure variations $\Delta p_1(t)$ exciting the nappe and the pressure variations $\Delta p_2(t)$ caused by the oscillating nappe are equal in frequency and if they are in phase (Fig.2). According to experimental results it is reasonable to assume that, to the first approximation, the pressure variations are harmonic:

$$\Delta p_1(t) = A \cdot \sin \omega t, \tag{1}$$

where A = amplitude of the pressure fluctuations and ω = cycle-frequency. The volume change $\Delta V(t)$ due to $\Delta p_1(t)$ can be calculated. Since this change of the volume of air is small and nearly isothermic, $\Delta p_2(t)$ can be computed from the equation

$$\Delta p_2(t) = -\frac{p_0}{V_0} \Delta V(t), \tag{2}$$

where p_0 = atmospheric pressure, and V_0 = air volume behind the parabolic nappe under atmospheric pressure.

The condition

$$\Delta p_1(t) = \Delta p_2(t) \tag{3}$$

is used to determine the possible frequencies of the oscillations, and to define regions where periodic oscillations can be excited.

The Equation of Motion for a Falling Water Particle

The parabolic path of a water particle falling from the weir is defined by Eqs. (4) and (5). The velocity of the particles leaving the weir is u_0 and θ is the angle between the flow direction at the origin and the horizontal (Fig.3a). For $t = 0$, $y = x = 0$.

$$\bar{y} = u_0 t \cdot \sin\theta + \frac{1}{2} g t^2, \qquad (4)$$

$$\bar{x} = u_0 t \cdot \cos\theta . \qquad (5)$$

With Eq. (4) and the condition $y(T) = L$, the time of fall from the weir to the tail-water is given by

$$T = \sqrt{\left(\frac{u_0 \cdot \sin\theta}{g}\right)^2 + \frac{2L}{g}} - \frac{u_0 \sin\theta}{g} . \qquad (6)$$

Due to pressure variations $\Delta p_1(t)$, the fluid particles oscillate across the parabolic path. This additional motion can be described in terms of coordinates x, y moving along the parabola with the mean flow velocity.

At each point, the pressure variations $\Delta p_1(t)$ act normal to the surface of the nappe accelerating the fluid particles in this direction (Fig.3b). Thus

$$\ddot{n}(t) \cdot dm = \Delta p_1(t) \cdot dF, \qquad (7)$$

and since $dF = 1 \cdot ds$, and $dm = \rho \cdot ds \cdot 1 \cdot d(t)$,

$$\ddot{n}(t) = \frac{\Delta p_1(t)}{\rho \cdot d(t)} . \qquad (8)$$

The equation of continuity demands that

$$q = u(t) \cdot d(t), \qquad (9)$$

where $u(t)$ is the velocity of the falling particle as a function of t and $d(t)$ is the thickness of the nappe as a function of t. With Eq.(9) one obtains

$$\ddot{n}(t) = \frac{\Delta p_1(t) \cdot u(t)}{\rho \cdot q} . \qquad (10)$$

The acceleration $\ddot{n}$ can be separated into components in the x,y directions (Fig.3c)

$$\ddot{y}(t) = \ddot{n}(t) \cdot \cos\alpha, \qquad (11)$$

$$\ddot{x}(t) = \ddot{n}(t) \cdot \sin\alpha, \qquad (12)$$

Fig.3d shows that

$$\sin\alpha = \frac{u_0 \cdot \sin\theta + g \cdot t + F_y}{u(t)} \qquad (13)$$

$$\cos\alpha = \frac{u_0 \cdot \cos\theta + F_x}{u(t)} , \qquad (14)$$

where F_x and F_y take into account the deviations from the parabolic path due to the oscillation. Therefore, one obtains from Eqs.(10) to (14),

$$\ddot{y}(t) = \frac{u_0 \cdot \cos\theta + F_x}{\rho \cdot q} \Delta p_1(t), \qquad (15)$$

$$\ddot{x}(t) = \frac{u_0 \cdot \sin\theta + gt + F_y}{\rho \cdot q} \Delta p_1(t). \qquad (16)$$

If the oscillations of the water particles across the parabolic path are small, the direction of the pressure $\Delta p_1(t)$ can be assumed normal to the surface of the undisturbed parabolic path. Hence, the influence of F_x and F_y in Eqs.(15) and (16) can be neglected. Thus

$$\ddot{y}(t) = \frac{u_0 \cos\theta}{\rho \cdot q} \Delta p_1(t), \qquad (17)$$

$$\ddot{x}(t) = \frac{u_0 \sin\theta + gt}{\rho \cdot q} \Delta p_1(t) . \tag{18}$$

The Oscillating Motion of a Falling Water Particle

As a fluid particle first leaves the weir, the following initial conditions apply:

$$t = 0 \begin{cases} x = \dot{x} = 0 \\ y = \dot{y} = 0 \end{cases} . \tag{19}$$

The oscillating motion of a particle can only be determined if the pressure $\Delta p_1(t)$ at $t = 0$ is known. Therefore, one introduces in Eq. (1) the zero-phase angle β.

$$\Delta p_1(t) = A \cdot \sin(\omega t + \beta) . \tag{20}$$

With Eq. (20) one obtains instead of Eqs. (17) and (18),

$$\ddot{y}(t) = \frac{u_0 \cdot \cos\theta}{\rho \cdot q} \cdot A \cdot \sin(\omega t + \beta) , \tag{21}$$

$$\ddot{x}(t) = \frac{u_0 \cdot \sin\theta + gt}{\rho \cdot q} \cdot A \cdot \sin(\omega t + \beta) . \tag{22}$$

Integrating these equations with respect to time and using the initial conditions, one obtains:

$$y(t) = \frac{A}{\rho \cdot q} \left[\frac{u_0 \cos\theta}{\omega^2} \left(-\sin(\omega t + \beta) + \sin\beta + \omega t \cdot \cos\beta \right) \right] , \tag{23}$$

$$x(t) = \frac{A}{\rho \cdot q} \left[-\frac{1}{\omega^2} (u_0 \cdot \sin\theta + gt) \cdot \sin(\omega t + \beta) - \frac{2g}{\omega^3} \cos(\omega t + \beta) \right. \tag{24}$$

$$\left. + \frac{1}{\omega^2} (u_0 \sin\theta - gt) \sin\beta + \left(\frac{2g}{\omega^3} + \frac{u_0 \cdot t \cdot \sin\theta}{\omega} \right) \cdot \cos\beta \right] .$$

The Oscillating Motion of the Nappe

The instantaneous profile of the nappe at time τ can be determined by Eqs.(23) and (24), in which β must now be considered as variable.

$$\beta = \omega(\tau - t). \tag{25}$$

This is necessary, because at any time τ the nappe consists of fluid particles which have left the weir at different moments and, therefore, with different zerophase angles β.

With Eq.(25), one obtains from Eqs.(23) and (24),

$$y(t,\tau) = \frac{A}{\rho \cdot q}\left[\frac{u_0 \cos\theta}{\omega^2}\left(-\sin\omega\tau + \sin\omega(\tau - t) + \omega t \cdot \cos\omega(\tau - t)\right)\right], \tag{26}$$

$$\begin{aligned} x(t,\tau) = \frac{A}{\rho \cdot q}\Bigg[& -\frac{1}{\omega^2}(u_0 \sin\theta + gt)\sin\omega\tau - \frac{2g}{\omega^3}\cos\omega\tau \\ & + \frac{1}{\omega^2}(u_0 \cdot \sin\theta - gt)\sin\omega(\tau - t) \\ & + \left(\frac{2g}{\omega^3} + \frac{u_0 \cdot t \cdot \sin\theta}{\omega}\right)\cos\omega(\tau - t)\Bigg]. \end{aligned} \tag{27}$$

Volume Change and Pressure Variations Produced by Nappe Oscillation

The volume change $\Delta V(t)$ can be computed by integration of x and y over the total length of the nappe.

$$\Delta V(\tau) = \int x(\tau, t)\, d\bar{y} + \int y(\tau, t)\, d\bar{x}. \tag{28}$$

Differentiation of Eqs.(4) and (5) results in

$$d\bar{y} = (u_0 \cdot \sin\theta + gt)\,dt, \quad d\bar{x} = u_0 \cdot \cos\theta \cdot dt. \tag{29}$$

The interval of integration is now $0 \leqslant t \leqslant T$.

Thus

$$\Delta V(\tau) = \int_0^T x(\tau,t)\cdot(u_0 \sin\theta + gt)\,dt + \int_0^T y(\tau,t)\cdot u_0\cdot \cos\theta\,dt. \quad (30)$$

Substitution of $x(\tau,t)$ and $y(\tau,t)$ from Eqs. (26) and (27) and integration yields

$$\Delta V(\tau) = \frac{-A}{\rho\cdot q}\cdot\frac{g^2\cdot T^2}{\omega^3}\,[K_1 \sin\omega\tau + K_2 \cos\omega\tau] \quad (31)$$

with

$$K_1 = \sin\omega T\left(1-\left(\frac{2}{\omega T}\right)^2 - 2b\right) + \cos\omega T\left(\frac{4}{\omega T} + b\omega T\right) + \left(\frac{1}{3}+b\right)\omega T, \quad (32)$$

$$K_2 = \sin\omega T\left(-b\omega T - \frac{4}{\omega T}\right) + \cos\omega T\left(1-\left(\frac{2}{\omega T}\right)^2 - 2b\right) \quad (33)$$

$$+ 1 + 2b + \left(\frac{2}{\omega T}\right)^2,$$

and

$$b = \left(\frac{u_0}{g\cdot T}\right)^2 + \frac{u_0}{g\cdot T}\cdot\sin\theta. \quad (34)$$

The pressure variation $\Delta p_2(\tau)$ can be computed from Eq. (2) and Eq. (31):

$$\Delta p_2(\tau) = \frac{p_0\cdot A\cdot g^2\cdot T^2}{\rho\cdot q\cdot V_0\cdot\omega^3}\,[K_1\cdot \sin\omega\tau + K_2\cdot\cos\omega\tau]. \quad (35)$$

Determination of Frequencies

The condition for periodic nappe oscillations is

$$\Delta p_1(\tau) = \Delta p_2(\tau). \quad (3)$$

Because $\Delta p_1(\tau)$ contains no cosine term, this Eq. (3) can only be satisfied if the last term in Eq. (35) is zero. This is only possible for all values of τ if

$$K_2 = 0.$$

This means,

$$K_2 = 0 = \sin\omega T\left(b\omega T + \frac{4}{\omega T}\right) + \cos\omega T\left(2b + \left(\frac{2}{\omega T}\right)^2 - 1\right) - 1 - 2b - \left(\frac{2}{\omega T}\right)^2 . \quad (36)$$

Solving this transcendent equation for a constant value of b, one obtains an infinite number of solutions for ωT. The continuous and broken lines in Fig.4 show the solutions of Eq. (36) for variable values of b in the range $f \cdot T = \frac{\omega T}{2\pi} < 6$. Besides, several experimental data are included. For $b \to \infty$, the force of gravity has no influence and so the thickness of the nappe remains constant. For that limiting case, one obtains two sets of solutions for Eq. (36).

$$b \to \infty, \quad \left.\begin{matrix} f \cdot T \to n \\ f \cdot T \to n + \frac{1}{2} \end{matrix}\right\} n = 1,2,3....$$

For $b \to 0$ one obtains only one set of solutions of $f \cdot T$:

$$b \to 0, \; f \cdot T \to n + \frac{1}{2} \quad \text{with } n = 1,2,3....$$

The Condition for Self-Excited Oscillations

Because damping effects are inevitable in oscillating systems, self-excited oscillations are only possible if the pressure variations in the air volume are amplified by the nappe-air system. This means, that the amplitude of $\Delta p_2(\tau)$ must be greater than the amplitude of $\Delta p_1(\tau)$, or with Eqs. (1) and (35)

$$\frac{p_0 \cdot A \cdot g^2 \cdot T^2 \cdot K_1}{\rho \cdot q \cdot V_0 \cdot \omega^3} > A \quad (37)$$

or

$$\frac{\rho \cdot q \cdot V_0}{p_0 \cdot g^2 \cdot T^5} < \frac{K_1}{(\omega T)^3} \, . \tag{38}$$

One can show by calculation that this condition is satisfied for all oscillations observed in experiments as shown by Fig.4.

For a defined air-nappe system, the expression on the left of Eq.(38) is constant, whereas the expression on the right depends on the frequency and decreases if the frequency increases. So, Eq.(38) yields an upper limit for the frequencies determined by Eq.(36).

Discussion of Results

The continuous lines in Fig.4 are verified by the experiments, whereas frequencies belonging to the broken lines were not measured. This can be explained by considering the phase plane of the pressure variations $\Delta p(t)$ for a defined air-nappe system (b = const). A periodic pressure variation is represented in the phase plane by a closed trajectory. Because the pressure variations are harmonic, the phase trajectory of a pressure variation with a defined frequency is an ellipse. Fig.5 shows qualitatively the phase trajectories for several frequencies determined by Eq.(36). According to Eq.(38), the ellipses become smaller if the frequency increases. The continuous ellipses in Fig.5 belong to frequencies lying on the continuous lines in Fig.4. The broken ellipses in Fig.5 belong to frequencies lying on the broken lines in Fig.4. The dotted line in Fig.5 indicates the limit between excitation and damping determined by Eq.(38). Inside this line, an oscillation of the nappe is always damped. If a periodic oscillation, presented in the phase plane by an ellipse, is disturbed, the pressure oscillation leaves the ellipse. But because periodic oscillations are only possible along the ellipses, after some time the oscillation has to come back to the ellipse from which it started, or it has to join another ellipse. Thus these closed trajectories are called "limit cycles" of the oscillation. What really happens is dependent on the kind and intensity of the disturbance. If the disturbance is small, the oscillation returns to the ellipse from which it started. The change from one to another ellipse or frequency occurs if the disturbance is strong. This change was also observed in experiments. Between two ellipses or frequencies which are stable limit cycles for small disturbances, there must exist a limiting case corresponding to an unstable limit cycle. This suggests the conclusion that the broken lines in Fig.4 and the broken ellipses in Fig.5 represent these unstable limit cycles. The corresponding frequencies were not measured in experiments, because in physical systems small disturbances are inevitable, and hence an unstable process cannot be realized.

The presented linear mathematical model allows the determination of the limit between damping and self-excitation. But it does not permit to determine the amplitudes of the oscillations. Without damping in the linear model, the amplitudes would increase indefinitely. Naturally, the linear model is only valid for small amplitudes. In reality, the amplitudes do not increase indefinitely but are limited by several non-linear effects. Due to these non-linear effects the pressure variations also contain higher harmonics. Hence, the ellipses in Fig.5 are only first approximations for the real phase trajectories.

References

1. Müller, O.: Das bei Überfall schwingende Wehr als selbsterregtes, gekoppeltes System unter Berücksichtigung gewisser Analogien zum Röhrensender und zur Zungenpfeife. Mitteilungen der Preuß. Versuchsanstalt für Wasserbau und Schiffbau, Berlin, Heft 33 (1937).

2. Petrikat, K.: Die schwingungsanfachenden Kräfte im Wehrbau. M.A.N. Forschungsheft (1953).

3. Petrikat, K.: Vibration Tests on Weirs and Bottom Gates. Water Power (February and March 1958).

4. Schwartz, H.I.: A Study of the Trajectory of a Two-Dimensional Nappe of Projected Liquid. The Civil Engineer in South Africa (January 1963).

5. Schwartz, H.I.: A Contribution to the Study of Nappe Oscillation. The Civil Engineer in South Africa (September 1963).

6. Schwartz, H.I.: Projected Nappes Subject to Harmonic Pressures. Proceedings, Inst. Civ. Engrs. (London) 28 (1964).

7. Schwartz, H.I.: Nappe Oscillation. Journal of the Hydraulic Division, Proceedings ASCE HY6 (November 1964).

8. Naudascher, E.: Discussion of Nappe Oscillation by H.I. Schwartz. Journal of the Hydraulic Division, ASCE HY3 (May 1965).

9. Partenscky, H.-W., Khloeung, I.S.: Etude des vibrations des lames déversantes, rapport soumis au conseil national de recherches, Ottawa/Canada, Août 1967.

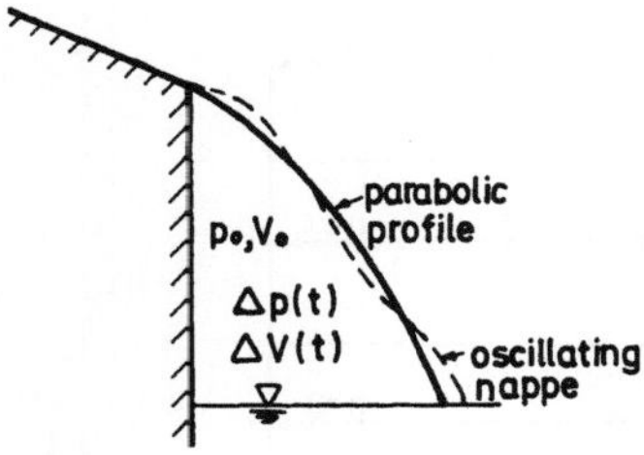

Fig.1

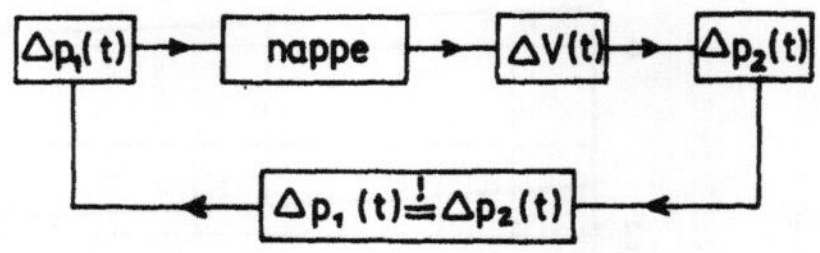

Fig.2

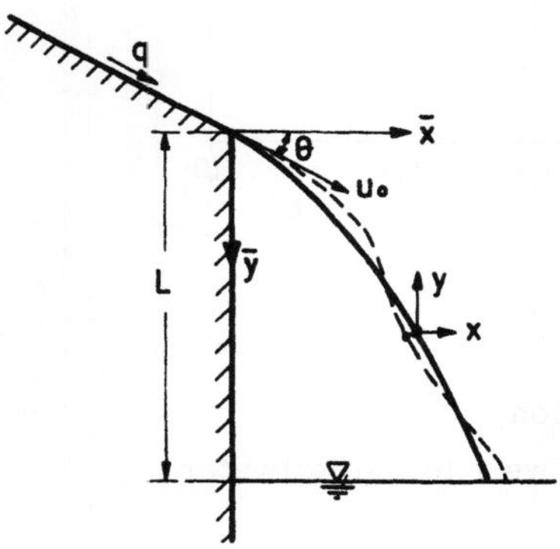

Fig.3a

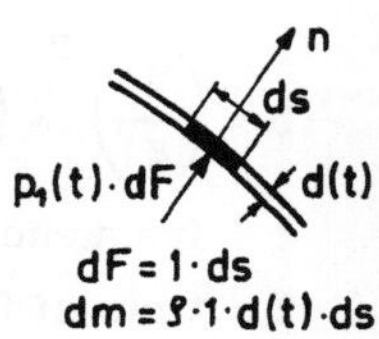

Fig.3b

Fig.3c

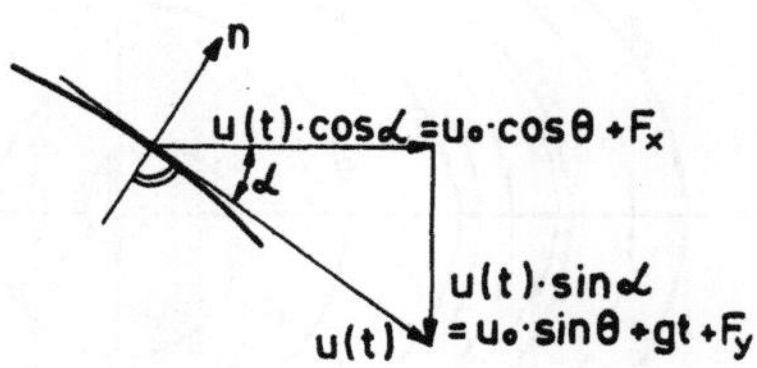

Fig.3d

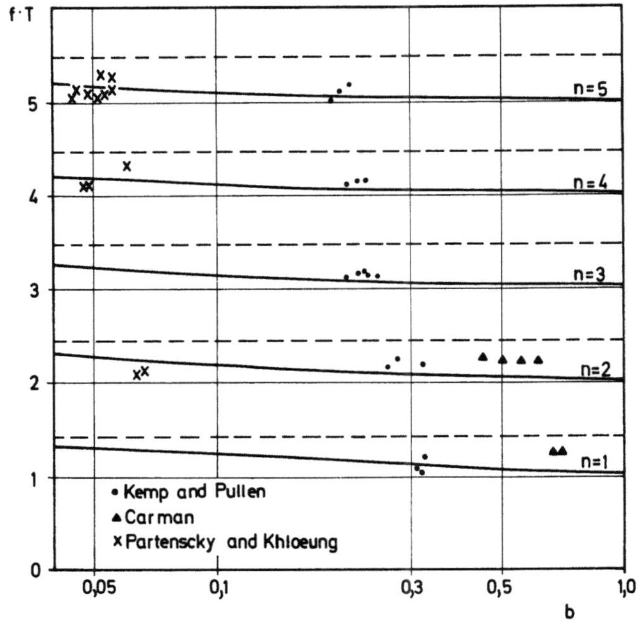

Fig. 4. $b = \left(\frac{u_0}{gT}\right)^2 + \left(\frac{u_0}{gT}\right)\sin\theta$,

f = frequency of oscillation,

T = time of fall from the weir to the tailwater,

$f \cdot T = \frac{\omega T}{2\pi}$.

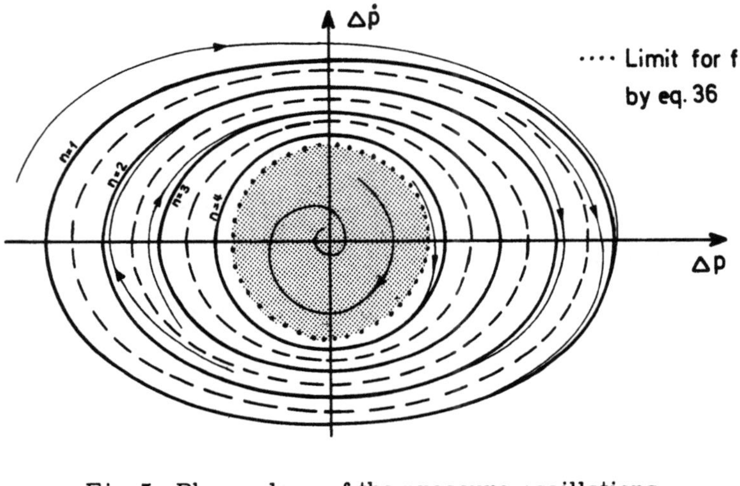

Fig. 5. Phase plane of the pressure oscillations.

On the Response of Separated Pockets to Modulations of the Free Stream

By

J.M. Hogan, M.V. Morkovin

Chicago, Ill., USA

Many practical problems arise where the free stream oscillates about a mean value: from turbomachinery to structures located in wakes of larger structures. Yet our theoretical understanding of such unsteady flows with boundary layers is essentially confined to the type of problems pioneered by Lighthill [1] and others (see the review in Schlichting [2]). For turbulent boundary layers we have only experimental guidance, e.g. Karlsson's [3]. The search of published literature discloses even less illumination of the phenomena in modulated flows where separated flow pockets are present.

During a discussion of this state of affairs, A. Roshko suggested to us that we tackle the simplest and the cleanest case - that of a backward facing step on a flat plate. Here we share with you the up-to-date insights gained from the study of this geometry which is still in progress. We shall illustrate them in the case of laminar boundary layers to which we had to retreat before we glimpsed the three modular ingredients of the complex flow field. These same ingredients - (1) pressure diffraction around the step, (2) puff-like release of lower velocity fluid which travels wave-like downstream at roughly half the free-stream speed, and (3) the instability and new turbulence production in the region of the traveling puffs - are becoming discernible for turbulent boundary layers.

We may yet encounter surprises - the generality of these coupled flow modules remains to be demonstrated. After all, the number of governing parameters grows with complexity of the geometry and with the character of the unsteadiness - h: step height; δ: boundary layer thickness at the step location; l_0: distance from the leading edge of the flat plate to the step; x: distance from step, positive in streamwise direction; f: dominant frequency of the modulation; f_i: other frequencies present in the stream; U_∞: mean free stream speed; ΔU: amplitude of dominant frequency, etc. Clearly as the dimensionless amplitude $N_A = \Delta U/U_\infty$ becomes very small we approach the acoustic case and the "puffing" reduces to generation of minute unsteady vorticity in the separating shear layer[1]. For turbulent layers these "minipuffs" are of no consequence while in the laminar case they may breed amplifying Rayleigh-Tollmien-Schlichting waves, a phenomenon distinct from module (3) above. Besides a finite amplitude N_A, the puffing phenomenon probably requires sufficient time for its development. Preliminary indications suggest that when the dimensionless (reduced) frequency $F = fh/U_\infty$ grows above some threshold on the order of 0.05, the puff initiation involves less and less of the temporarily near-stagnant flow of the breathing, separated pocket and the successive puffs degenerate into "corrugations" of the mixing layer which straddles the mean dividing streamline. Again, for very low Reynolds numbers Re_h based on the step height, the puffs apparently may diffuse without the instability module (3) and the downstream flow should then approach the aforementioned Lighthill type of unsteady boundary layer, despite the step. Anyway, once we know what to look for, we can delineate better the regions of dominance of the three modular flow ingredients in the space of our primary dimensionless variables N_A, F, Re_h, and Re_δ.

Having stated the main observations and circumscribed their probable applicability we must refer the reader to Hogan's thesis [5] for details of the experimental techniques and to Section 4 of [6] for the problems in generating periodic changes in the free stream. Fig. 1 portrays the main tools: simultaneous smoke visualization

---

[1] The module of the pressure diffraction, (1), persists in this case. According to Morkovin and Paranjape [4] the sharpness of the corner thus enhances the amplitude of the seeded vorticity fluctuations.

of two successive puffs[1] in the top picture (available only at lower speeds) and the corresponding "instantaneous" velocity profile at the right. This profile is traced on a memory oscilloscope by luminous dots triggered at the same phase of the periodic driving cycle, as the hot-wire probe (visible on top) traverses the boundary layer. Here the horizontal x location (19.3h downstream of the step) is such that the local horizontal velocity fluctuation $u_P(x,y,t)$ imposed by the puff is in phase with the $u(x,y,t)$ induced by the free-stream pressure gradient and its diffraction by the step. This phase coincidence is evident on the hot-wire trace at lower left which was taken for $y = 0.10$ in. = 2.54 mm. Because of the phase coincidence the instantaneous profile betrays no unsteady or peculiar features (except below $y = .01$ in. = 0.254 mm, where the hot-wire indications must be disregarded anyway because of wall interference). We note for comparison that at the low Reynolds number of Fig. 1, the steady flow reattachment should occur upstream of $x = 12$ h according to Goldstein et al [7].

The same phase coincidence occurs for $x = 4.0$ in. = 104 mm = 80h in Fig. 2, where the horizontal velocity fluctuations are compared at selected x,y positions for $N_A = 0.09$, $f = 9$ Hz, $U_\infty = 6.9$ ft/sec = = 2.1 m/sec, $h = 0.05$ in. = 1.27 mm. (The amplitudes can be compared by ratioing the peak-to-peak distances; the absolute vertical distances were chosen for display purposes and do not indicate the total u values.) At other x locations, the puff signal u_P is out of phase with the driven-diffracted signal. The u_P field clearly travels to the right at a speed which is essentially independent of y. The measurements of this propagation speed (made more difficult by the gradual x-development of u_P) scatter nonpreferentially between the extremes of $0.35\ U_\infty$ and $0.57\ U_\infty$ (Fig. 14 of [5]) for F between 0.005 and 0.038. As stated earlier, the puffs tend to lose their definition for higher F values, the distance between them becoming less than the mean separated-pocket length. At the first loca-

---

[1] The smoke is photographed at grazing incidence to the unpolished flat plate which nevertheless acts as an excellent mirror. Note the dark line at $y = 0$, which marks the surface of the plate.

cation of Fig. 2, x = 1.0 in. = 25.4 mm, i.e. full twenty step heights downstreams, the u_P disturbance remains confined below the level of the step height and then proceeds to spread over a thickness of 2h even though the combination of the Reynolds number and N_A is below the aforementioned instability threshold in this case and u_P decays steadily in amplitude.

Similar downstream propagation of the puffs or wounds at the fixed height h is observed for larger N_A and Re_h values as in Fig. 3 (where N_A was increased to 0.14 and U_∞ to 10.6 ft/sec = 3.23 m/sec). The signals at y = h evolve and give indications of instability near x = 8 in. = 203 mm. A hashy turbulent patch (spanwise belt) is evident by the time the puff reaches x = 9.0 in. = 229 mm. Thereafter the turbulent belt spreads steadily into the neighboring laminar regions. Measurements of $\overline{u^2}$ and $\overline{uv}$ in the mixing layer generated by turbulent boundary layers (with steady U_∞), flowing over downstream-facing steps, made by Tani et al [8], suggest that new instability associated with the inflected mean profiles can take place. It is plausible that the puffs, present for an unsteady boundary layer which is turbulent upstream of the step, may undergo a similar downstream instability for a sufficiently large N_A. Preliminary, limited evidence suggests that turbulent puffs propagate at a speed in excess of 0.6 U_∞.

Returning to laminar cases we focus on the instantaneous profiles on the right of Fig. 4, for which h and f remain the same (0.05 in. = = 1.27 mm and 9.5 Hz) but N_A and U_∞ are slightly higher, at 0.16 and 11.2 ft/sec = 3.42 m/sec, respectively. At x = 5.5 in. = 140 mm = = 110 h the u_P defect is centered around y = 0.13 in. = 3.5 mm = 2.6 h and the wound appears quite spectacular in both the upper pictures. The lower picture shows the development of the puff signature as the hot-wire probe travels downstream at this fixed height. It well illustrates how difficult it is to separate the fields induced by the free-stream pressure gradients and by the puff field even in the laminar case. Since the fundamental frequency of both these fields is the same, a rational decomposition, if possible, apparently must rely on the instantaneous profiles and some elements of theory.

Attempts at instantaneous pressure measurements near and away from the step are under way. Hot-wire measurements u p s t r e a m of the step indicate non-Lighthillian velocity profiles for $|x|$ as much as 10 h to 15 h. For an estimate of the step influence we might assume that the unsteady component diffracts in an inviscid manner as it essentially would for superposed acoustic waves of small amplitude N_A. Then, utilizing ideas of asymptotic matching à la Amiet and Sears [9], we would conclude that Milne-Thomson's steady results for the inviscid step problem (Chapter 10 of [10]) should give us an indication of the extent of the pressure perturbation for the unsteady pressure diffraction. Such estimates suggest that the inviscid velocity is within 5 % of its undiffracted value for $|x|$ of approximately 8 h. Of course, we should correct for the presence of viscosity and of the separated pocket, somehow... Even with such correction, the diffraction field of the unsteady flow modulation is likely to be of consequence. The cited upstream influence of the step and the substantial effects portrayed in Figs. 2-4 further support this conjecture when we reveal that for these cases the step is buried deep in the boundary layer, δ_0/h being approximately seven! It is indeed desirable to get an experimental indication of the local unsteady pressure loads.

We conclude with Fig. 5 as a stimulus for the reader's intuition for unsteady streak lines. Here the step height h is once more 0.05 in.= = 1.27 mm but U_∞ = 8 ft/sec = 2.44 m/sec, rather high for smoke visualization. Of special interest are the lower three pictures showing the development of the streak field with 90° phase differences. The instantaneous profiles for the phase at minimum free-stream velocity indicate that the flow within the region $0 < y < h$, $o < x < 15h$ is essentially stagnant. For steady flow the reattachment should occur near x/h of 12 according to [7]!

We believe that the evidence for the modular flow ingredients: (1) diffraction of pressure around the step even when the step is buried deep in the boundary layer, (2) localized puff-like flow field $u_P(x,y,t)$ traveling downstream at roughly $0.5\,U_\infty$, and (3) localized instability and new turbulence production in the region of the propagating, distorting puffs, is sufficiently strong to utilize these concepts for diagnostics of other periodically modulated flows with separated pockets.

This research was supported in part under NSF Grant GK-1903 and in part under USAF OSR-Themis Contract F44620-69-0022.

References

1. Lighthill, M.J.: The Response of Laminar Skin Friction and Heat Transfer to Fluctuations in the Stream Velocity. Proc. Roy. Soc. A, 224, 1 (1954).

2. Schlichting, H.: Non-Steady Boundary Layers. Chapter XV of Boundary Layer Theory, McGraw-Hill 1968.

3. Karlsson, S.K.F.: An Unsteady Turbulent Boundary Layer. Ph. D. Thesis, Dept. of Aeronautics, the John Hopkins Univ. (1958); abbreviated in J. Fluid Mech. 5, 622 (1959).

4. Morkovin, M.V., Paranjape, S.V.: On Acoustic Excitation of Shear Layers. Z. Flugwissen. 19, 328 (1971).

5. Hogan, J.M.: Flow Over a Downstream Facing Step in Response to Periodic Changes in the Free-Stream Speed. Master's Thesis, Dept. of Mechanics and Mechanical and Aerospace Engineering, Illinois Institute of Technology, May 1972.

6. Morkovin, M.V., Loehrke, R.I., Fejer, A.A.: On the Response of Laminar Boundary Layers to Periodic Changes in Free-Stream Speed, Proc. 1972 IUTAM Sympos. on Unsteady Boundary Layers, Eichelbrenner, E., Editor, Laval University Press 1972.

7. Goldstein, R.J., Eriksen, V.L., Olson, R.M., Eckert, E.R.G.: Laminar Separation, Reattachment, and Transition of the Flow Over a Downstream-Facing Step. Journal of Basic Engineering, Trans. ASME, Series D, 92, 732 (1970).

8. Tani, I., Iuchi, M., Komoda, H.: Experimental Investigation of Flow Separation Associated with a Step or a Groove. Aero. Res. Inst., Univ. Tokyo, Rept. No. 364, 1961.

9. Amiet, R., Sears, W.R.: The Aerodynamic Noise of Small-Perturbation Subsonic Flows. J. Fluid Mech. 44, 227 (1970).

10. Milne-Thomson, L.M.: Theoretical Hydrodynamics, McMillan 1968.

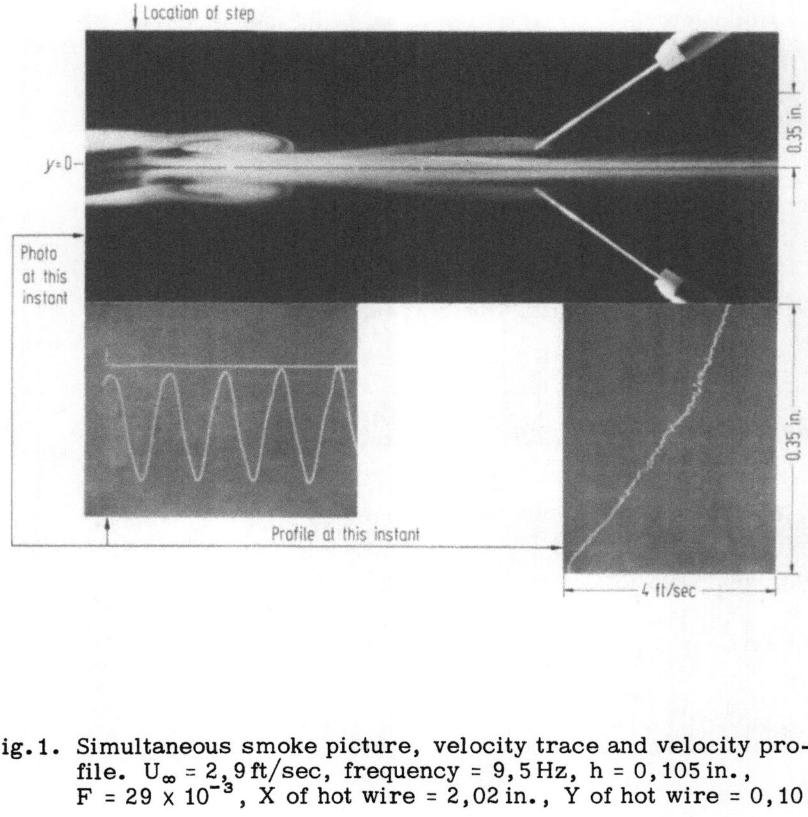

Fig. 1. Simultaneous smoke picture, velocity trace and velocity profile. U_∞ = 2,9 ft/sec, frequency = 9,5 Hz, h = 0,105 in., F = 29 x 10^{-3}, X of hot wire = 2,02 in., Y of hot wire = 0,10 in.

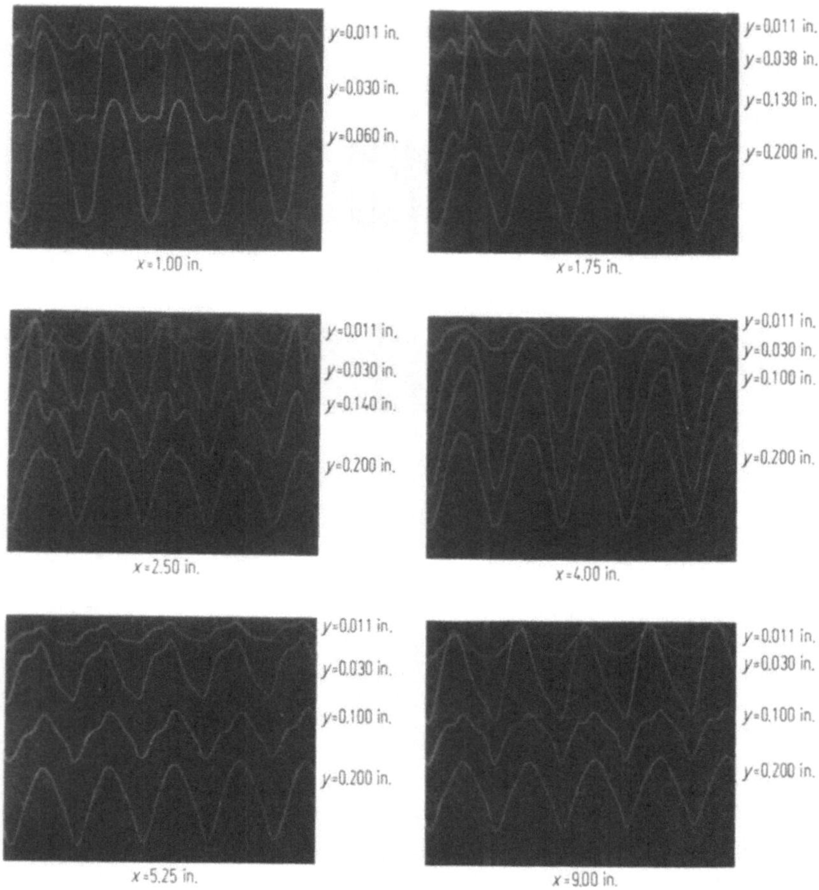

Fig. 2. Oscilloscope traces of velocity behind a 0,050 in. step for various values of X and Y. $U_\infty = 6,9$ ft/sec, frequency = 9,5 Hz, $N_A = 0,09$.

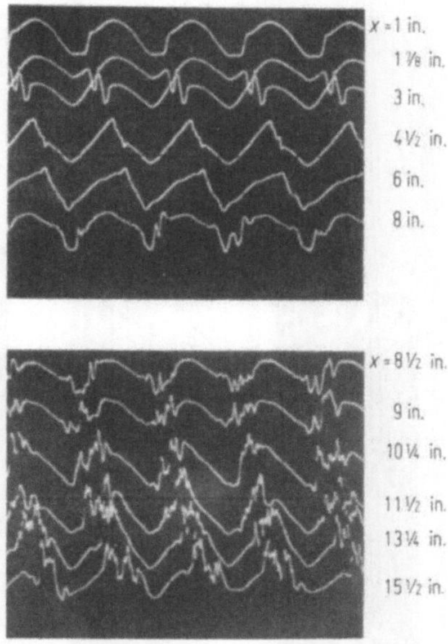

Fig. 3. Oscilloscope traces of velocity behind a 0,050 in. step for Y=0,050 in. and various values of X, $U_\infty = 10,6$ ft/sec, frequency = 9,5 Hz, $N_A = 0,14$. All traces start at same freestream phase angle.

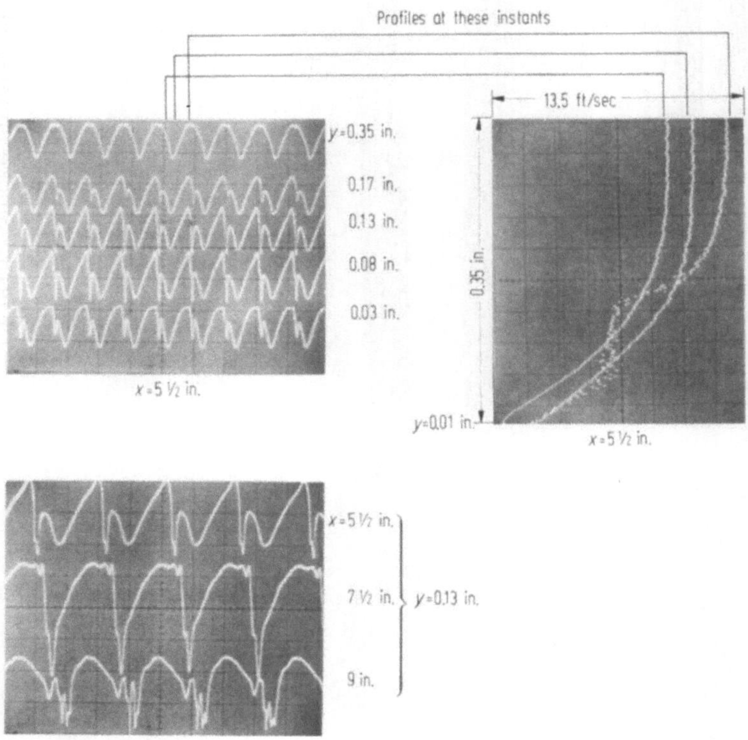

Fig. 4. Oscilloscope traces and instantaneous profiles behind 0,050 in. step, $U_\infty = 11,2$ ft/sec, frequency = 9,5 Hz, $N_A = 0,16$.

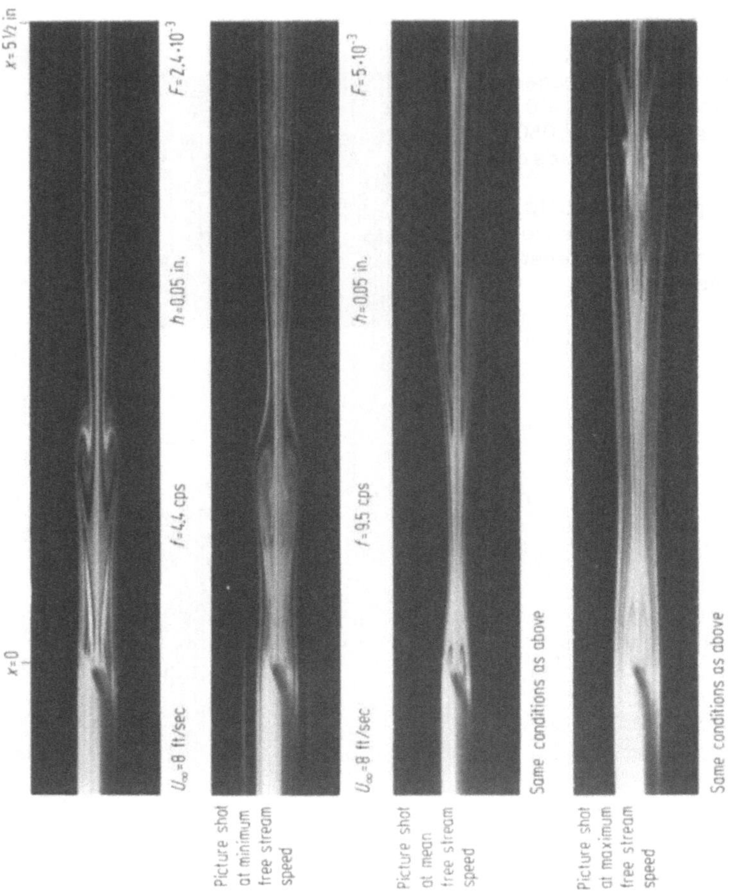

Fig. 5. Smoke pictures of flow behind step.

Flapping Jets

By

G. Binder, M. Favre-Marinet

Grenoble, France

Summary

A device to generate large flapping jets is described. The characteristics reported show some striking differences between flapping and steady jets.

Generation

The design of the flapping device makes use of the Coanda effect. The main elements of the nozzle (Fig.1) are two hollow slotted cylinders. For proper values of d_0/R, the air jet is in a bistable situation, i.e., it cannot stay in the middle and must attach to one cylinder. If a sufficiently strong air pulse is sent into this cylinder, the jet is blown away and flips over to the other one from where it may in turn be blown off in similar fashion. Alternate blowing causes then the jet to flap back and forth. Such a blowing is generated by a specially designed rotating cylindrical double-valve on which one port is blocked while the other one is open. This valve - the only moving mechanical part of the device - is driven by a variable speed motor which imposes the flapping frequency f.

The instantaneous velocity at the nozzle exit is approximatly:

$$u_0 = \overline{u_0} \cos\varphi, \quad v_0 = \overline{u_0} \sin\varphi, \qquad \text{where: } \varphi = \varphi_0 \sin(2\pi f t),$$

$\overline{u_0}$ being the velocity modulus. The flapping angle φ_0 depends upon $\overline{u_0}$ or p_0 (generating pressure of main flow), d_0/R, θ (orientation of blowing slots) f, p_1 (generating pressure of blowing), parameters which are all adjustable on the set-up.

Measurements were made with a linearized constant temperature hot-wire anemometer and a digital integrating voltmeter which can determine mean values and the quadratic mean of fluctuations.

Flapping was achieved for a minimum blowing pressure p_1 = 16 cm watercolumn while p_0 varied from about 8 to 23 cm watercolumn.

Flow Kinematics in the Initial Region $x/d_0 < 6$

In the region near the nozzle, the flow is dominated by the periodic flapping imposed on the jet whereas further downstream it is dominated by the turbulence. Hot-wire oscillograms in the near region display large periodic fluctuations with some superimposed turbulence. These may qualitatively be understood by imagining that the flapping causes the instantaneous velocity profile at a given x to oscillate with frequency f about the x-axis. Fig.1 sketches the extreme positions of such instantaneous profiles at three stations. At given x, the time difference between the occurrences of the extreme profiles is T/2 (T = 1/f). A hot-wire on the axis at section 1 sees the maximum velocity during the whole cycle - abstraction being made of the relatively small turbulence - because of the overlapping of the two extreme profiles. At section 2 and 3, however, a hot-wire on the axis sees the maximum velocity twice during a cycle at times T/4 and 3T/4, the beginning of the cycle for fixed x being chosen to coincide with the upper extreme position, and a minimum velocity twice at time 0 and T/2, the minimum being equal to zero at section 3.

From the various shapes of the hot-wire oscillograms it is possible to infer the positions of points C_+ and C_- and thus the flapping angle φ_0. For $\overline{u_o}$ = 32 m/sec it was found that $\varphi_0 = 14^\circ$, also $\varphi_0 = \theta$. This equality may however not be true for other velocities.

In the near region, the pathlines of this unsteady flow are roughly straight lines whereas the streaklines are amplified sinusoids.

Flow Characteristics in the Fully Turbulent Region

With increasing downstream distance the turbulence builds up while the periodic variations are gradually washed out and disappear completely beyond $x/d_0 > 15$. From the forced flapping it is evident that the flapping jet must initially spread faster than the classical two-dimensionnal steady jet. The measurements reveal that the high growth rate persists far downstream.

The plots of the decay of the axial mean velocity $\overline{u_a}$, Fig.2, and of the jet half width $y_{1/2}$, Fig.3, show the strikingly different behavior of the flapping jets as compared to the steady jet whose characteristics are taken from Newman [1]. At $x/d_0 = 50$, for instance, the decay of the velocity on the axis of the flapping jet $\overline{u_0} = 61{,}4$m/sec is twice that, and the half width is four times that of the steady jet. These differences are even considerably larger for the slower flapping jets: the one characterized by $\overline{u_0} = 36{,}3$m/sec sees its velocity $\overline{u_a}$ reduced by a factor of 7 after only 40 d_0, a reduction the steady jet would only achieve after 300 d_0!

Turbulent intensities on the axis of more than 40% have been measured - as compared to the 23% found for the steady case - which may partly account for the high rate of spreading.

Because of the low aspect ratio (=10) of the nozzle and of the high turbulent intensity the flapping jets investigated became threedimensional beyond $x/d_0 \simeq 20$. Since the steady threedimensional axisymmetric jet spreads somewhat slower than the plane jet, this increases the differences in growth rates between flapping and steady jets even more.

Applications

In experimental studies of wind induced vibrations, the properties of flapping jets in the near region could be used to simulate gustiness while their downstream properties could be useful where high turbulent intensities are required.

Acknowledgment

Financial support of the Direction des Recherches et Moyens d'Essais, Ministère d'Etat de la Défense Nationale is gratefully acknowledged.

References

1. Newman, B.G.: Turbulent Jets and Wakes in a Pressure Gradient, in Fluid Mechanics of Internal Flow, G. Sovran, Editor, Elsevier 1967.

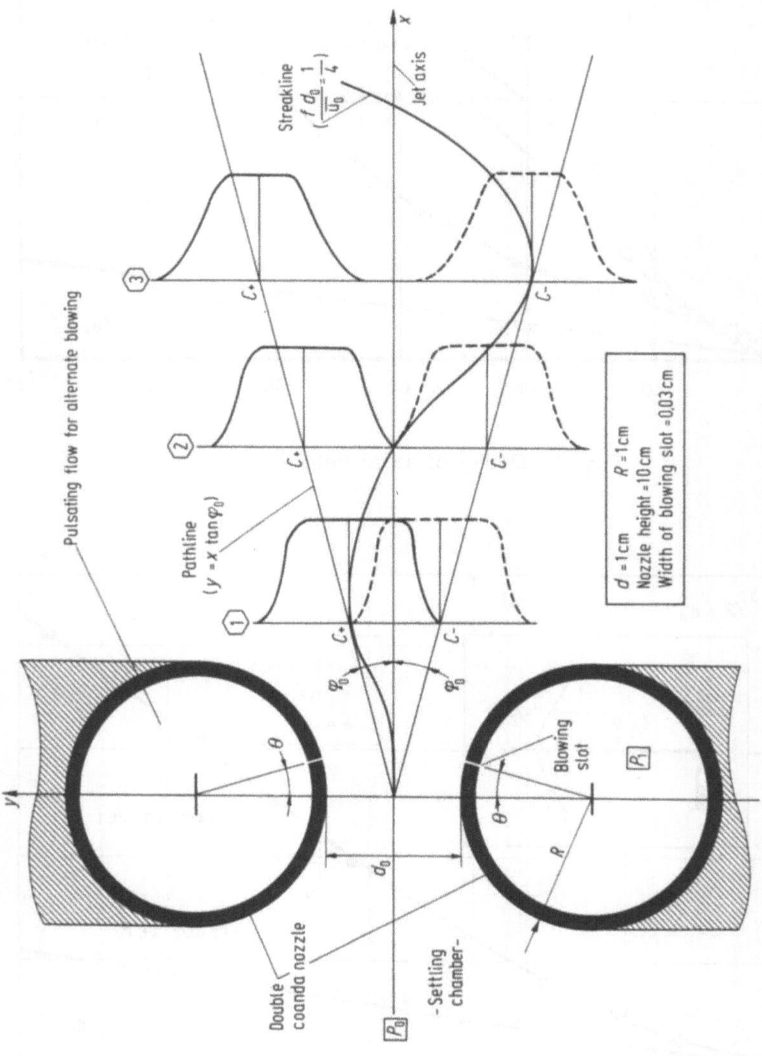

Fig. 1. Nozzle and initial jet development.

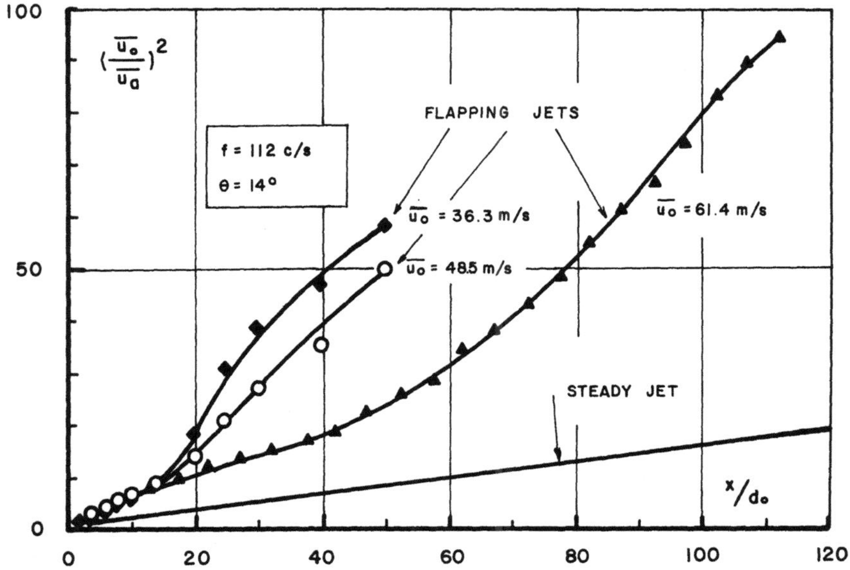

Fig.2. Decay of axial velocity.

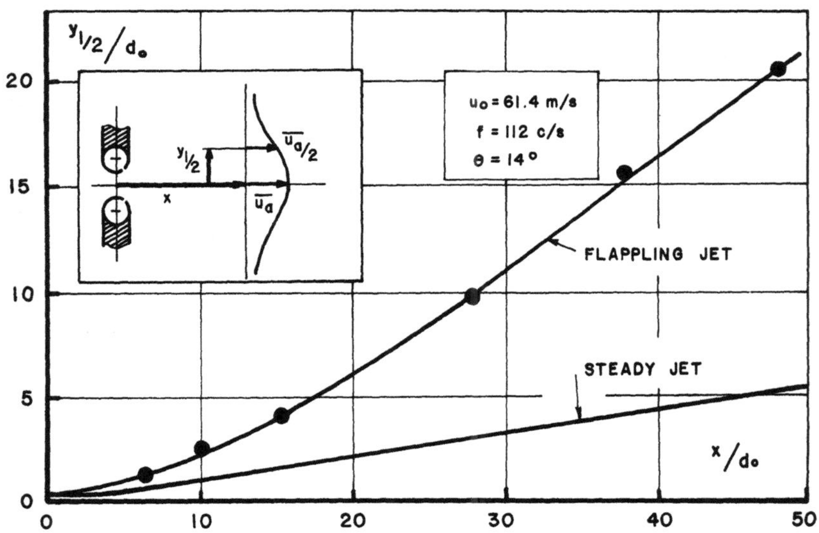

Fig.3. Jet growth.

Oscillations of a Jet in a Rotating Fluid

By
R.M. Curtet, B.V. Martins
Grenoble, France

Summary

In this paper some experimental results on the flow due to a jet placed on or off the axis of a strongly rotating ambient stream, confined in a cylindrical duct [1], are briefly presented.

Experimental Set-Up and Test Conditions

A water jet (Fig.1) issues from a nozzle of diameter $2a = 12$ mm (flow-rate Q_j in l/sec). The swirl is forced on the ambient flow by a system of 24 fixed vanes, followed by a contraction. Mixing between the jet and ambient discharges occurs in a cylindrical chamber of diameter $D = 2R = 216$ mm and of effective length 10D; this duct is terminated by a cross-bar system which suppresses the rotation downstream. The injector is mounted on a circular disk whose axis is excentric with respect to that of the upstream end wall; by turning this disk, the distance of the jet axis to the axis of the chamber can be changed.

During the tests, the jet flow rate varied between 0 and 2 l/sec, and the ambient flow rate from 0 to 5 l/sec. The initial conditions, in the exit section of the jet, are given by dimensionless numbers, respectively related to the impuls Π:

$$\Pi = \int_0^R (p^* + \rho u^2 + \rho \overline{u'^2}) 2\pi r dr \tag{1}$$

and to the angular momentum flux Γ:

$$\Gamma = \int_0^R \rho (uw + \overline{u'w'}) 2\pi r^2 dr, \tag{2}$$

where r is the ordinate with respect to the axis of the duct, $p^* = p + \rho g z$ (piezometric pressure), u and w are the axial and tangential components of the mean-time velocity, u' and w' are velocity fluctuations, and ρ is the density of the fluid.

The dimensionless numbers are:

$$m + \frac{1}{2} = \frac{\pi \Pi R^2}{\rho Q^2}, \tag{3}$$

$$S_d = \frac{2\pi \Gamma R}{\rho Q^2}. \tag{4}$$

m is the "impulse parameter" defined by Craya and Curtet [1]; S_d is the "degree of swirl" and Q the total flow rate in the chamber. The numbers m and S_d characterize the initial conditions; they are invariant with axial distance, if wall friction can be neglected. The geometrical ratio R/a was constant during the tests and equal to 18.

The flow was visualized with small air bubbles (about 0.2 mm in diameter); it was thus possible to study the instantaneous flow patterns.

Jet Located on the Axis of the Chamber

A previous study has shown that, in the absence of the jet, the swirl number S_d in the present set-up was equal to 34 and independent of the flow-rate Q_s ; this swirl is sufficiently high to create an important amount of back-flow on the axis along all the effective length of the chamber: the structure is similar to that of a bicellular vortex [3].

The issuing of the jet against this back-flow produces a very remarkable phenomenon: the potential core of the jet is fixed in space as usually, but beyond point A (Fig.2), at an axial distance of about 8 diameters from the nozzle, the jet is deflected from its axis and is caught by the swirling ambient flow. This results in a circular periodic well-defined motion of the jet, in the same direction as the ambient rotation, about the axis of the chamber. The angular velocity of rotation of the jet is about five times smaller than that of the ambient fluid.

For a discharge ratio $Q_s/Q_j = 2,5$ (Fig.2), the flow pattern shows the following characteristic regions:

The jet, coming from the left, spreads into a plume which, beyond point A, curves and closes itself, forming thus a recirculating eddy; this eddy has a downstream stagnation point B, on the axis of the chamber.

Downstream of point B, is a region BC about one diameter long, characterized by a very high instability of the fluid lines.

Beyond this region, one finds conditions similar to those of the initial swirling ambient flow (bicellular vortex). The air bubbles used for visualization concentrate on the axis and form a string which flows backwards towards the recirculating eddy. It is then entrained by this eddy up to point A, winds around the potential core of the jet, turns around the nozzle up to five diameters upstream of the exit section, before being carried again by the ambient flow. In the unstable region BC, the string of bubbles is intermittently subjected to an intense twisting, followed by a bursting of the filament.

The recirculating region and the back-flowing string together turn about the axis of the chamber, with a well-defined frequency (stable to ± 2% in most cases).

These various aspects are present, but vary in extent with the discharge ratio Q_s/Q_j. The frequency of rotation of the jet varies also, and the experimental points can be plotted in condensed form as shown on Fig.3. On this diagram, $\sqrt{m}$ is along the horizontal axis (m = = impulse parameter, Eq.3) and the Strouhal number D^3f/Q is along the vertical axis (f = frequency in cps). For values of $\sqrt{m}$ smaller than 6, one notices that the experimental points are located near a straight line.

Jet Located off the Conduit Axis

If the nozzle is moved away from the axis of the chamber, the rotation frequency decreases and becomes zero when the distance between the two axes exceeds 5mm, i.e. approximately one jet radius.

References

1. Martins, B.V.: Sur l'écoulement d'un jet dans un fluide en rotation d'intensité élevée. Thèse de Docteur-Ingénieur, Université de Grenoble, février 1972.

2. Craya, A., Curtet, R.: Sur l'évolution d'un jet en espace confiné. C.R.A.S. Acad. des Scienc., Paris, 241, 621-622 (1955).

3. Mc Neil, N.: Sur la structure d'un écoulement tournant en concuit cylindrique avec aspiration axiale. Thèse de Docteur-ingénieur, Université de Grenoble, septembre 1970.

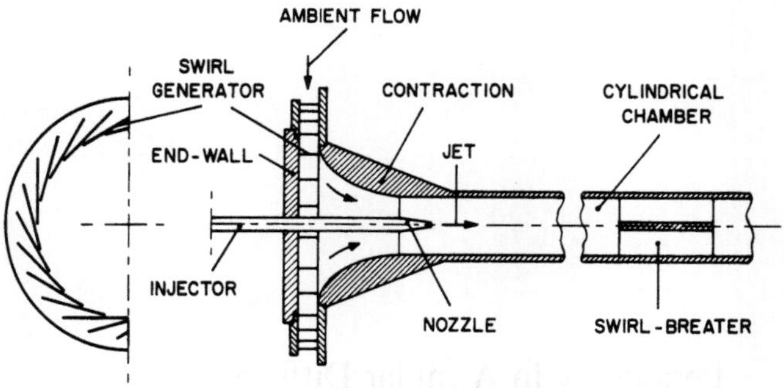

Fig.1. Test apparatus.

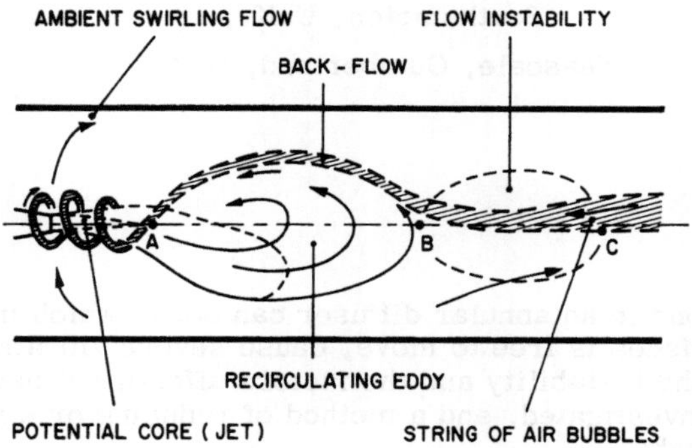

Fig.2. Characteristic regions of the flow.

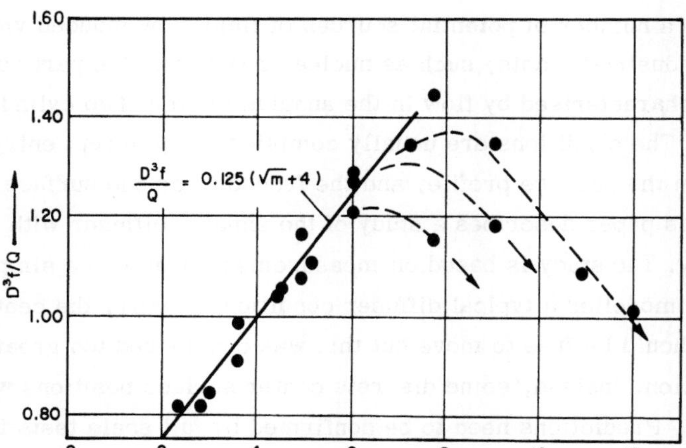

Fig.3. Frequency of the jet oscillations.

Periodicity in Annular Diffuser Flow

By

F.R. Fricke, P.O.A.L. Davies, M.W. Parkin

Southampton, U.K.,

Seascale, Cumberland, U.K.

Summary

Flow oscillations in an annular diffuser can occur which may, if either of the surfaces is free to move, cause severe vibration problems. The nature of the instability and the factors affecting it have been experimentally investigated, and a method of reducing or eliminating it has been sought.

Introduction

There are a number of potential sources of fluid-flow induced vibrations in industrial plants, such as nuclear reactors. One particular group is characterised by flow in the annulus between two cylindrical surfaces. The conditions are usually complicated by abrupt entry, changes in the passage profile, and the freedom for one surface to move. This paper describes a study of the annular diffuser with abrupt entry. The study is based on measurements made on a simple rig which modelled a typical diffuser condition. Ideally, the center surface should be free to move but this was considered too great a complication. Instead, some discrete center surface positions were examined. Predictions need to be confirmed by full scale tests in a real situation.

There is a paucity of published papers on the subject of annular diffusers which are of relevance to the present work, most work being concerned with pressure recovery rather than stability. The published work is of little interest for two further reasons. The area ratio of relevance used in the present work is outside the range considered elsewhere. A comparison with work by Howard et al is shown in Fig.2. This shows the hopelessness of trying to improve the diffuser flow by decreasing the diffuser angle. Work by Howlett shows that the situation is aggravated by center surface eccentricity. He found that this increased the tendency of the diffuser to stall.

Although the results presented here are not conclusive because of experimental-rig inadequacies, it seems that the periodic flow in the diffuser, and the resulting forces on the center surface, is connected with the localised separated flows in the diffuser. It is possible to reduce this instability by placing circumferential discontinuities downstream of the diffuser, or increasing the diffuser angle.

The Flow Process

From the observations and measurements a picture of the flow process has been formed [1], illustrated in Fig.1, in which four principal components are identified. They are: the approaching axial flow (1), which is deflected towards the region of maximum clearance as it enters the eccentric annulus; two localised separated flows (2) formed within the diffuser section; two associated spiralling vortices (3) formed off the leading edge of the centre body (these later break down, the so-called "bursting" process); and finally recirculation (4), as the flow re-establishes itself in the rather more symmetrical downstream annulus.

The above picture has been built up as follows. Starting with the sharp diffuser (e.g. $\emptyset = 30^{\circ}$) separation and an associated reverse flow would be anticipated. The static pressure distribution over the outer surface (the diffuser) revealed the low and high pressure regions characteristic of the separation and re-attachment of two such flows, the implied surface direction of the flow being consistent with the situation. The point of interest is that there are two such separated flows, not one as occurs in the conical diffuser.

The influence of the pair of spiralling vortices seen in dyeline visualisation studies is apparent also in surface flow visualisation. On the center body, flow lines downstream of the localised pressure maxima, deflect towards the line of minimum clearance, whereas the flow direction nearby, is towards the line of maximum clearance. There are thus seen to be two spiralling vortices, symmetrically disposed about the plane of eccentricity of the annulus and rotating in an opposite sense to each other. Each spiral vortex is associated with a diffuser separation vortex which lies in a radial plane drawn through the spiral vortex. Lambourne and Bryer noted separated flow beneath spiralling vortices, and it is thought that this had triggered the two separated flows in the diffuser of one as normally expected.

The recirculating flow is apparent in surface-flow visualisation and was noted in dyeline visualisation studies. It is believed that this is the normally observed stall region for large angled diffusers.

The Instability Mechanism

It seems most likely that the spiralling vortices are responsible for the flow oscillation which causes the center body to vibrate, and that it is the bursting of these vortices which is particularly responsible.

This bursting was observed in dyeline flow visualisation, and there is some other supporting evidence. For example, the localised deflected flow observed on the surface of the center body makes an angle of 40° with the local mean flow direction. This indicates a helix angle of 50°. Ludwieg found for flow between two rotating annular cylinders that the flow becomes unstable if the helix angle of a streamline is less than 42°. Hummel has found experimentally that part of the flow at a cross section immediately upstream of a vortex breakdown over a delta wing is close to this condition. Vortex bursting will occur where a low total pressure is combined with an adverse pressure gradient. Whilst total-head measurements were not successful, there is certainly an adverse pressure gradient.

Dynamic pressure measurements, when cross correlated as in Fig. 3, show that the pressure disturbance frequency in the plane of eccentricity, is twice as great as that perpendicular to the plane eccentricity. This suggests two sources of disturbance, one each side of the plane of eccentricity, as in the vortex street formed by the two spiralling vortices.

Several actual mechanisms are postulated. If one of the spiralling vortices is stronger than the other, the bursting will produce a larger pressure rise on that side of the center body, and consequently the flow resistance there will increase. This in turn will switch the flow to the opposite side. The process will continue to build up until the oscillation reaches a steady level because of the energy dissipation of the system. Alternatively it could be axial or circumferential oscillation of the bursting position, or a combination of both which causes the fluctuating forces.

Stabilisation

Two forms of stabiliser were tested, they were strakes or radial vanes, and circumferential rings. Of the two, the circumferential rings were the more effective when placed in the optimum axial position. One basis of judgment was by means of load measurements on the center body. Referring to the results shown in Fig. 4 the improvement of the rings can be seen. It should also be noted that the 30°-diffuser result is more stable than that from the 2°-diffuser. Both observations support the axial oscillation of the vortex-bursting position as the instability mechanism, the three illustrating a progressively deterministic system.

Conclusion

The mechanism of the instability has been shown, fairly conclusively, to be linked with the two localised separated flows. Two possible methods of stabilisation are suggested, (a) the use of circumferential fences downstream of the diffuser, or (b) the use of diffuser angles of greater than 30° (included angles greater than 60°). Dynamic testing, with a free center body, is being carried out at the present time to test the validity of the present work.

Reference

1. Fricke, F.R.: Periodicity in Annular Diffuser Flow. Nuc. Sci. and Engng. 48 (1972). This publication includes a complete list of references.

Acknowledgements

The authors wish to thank the joint sponsors for supporting this work, and for their helpful comments and criticisms: United Kingdom Atomic Energy Authority, The Nuclear Power Group Ltd, and British Nuclear Design and Construction Ltd.

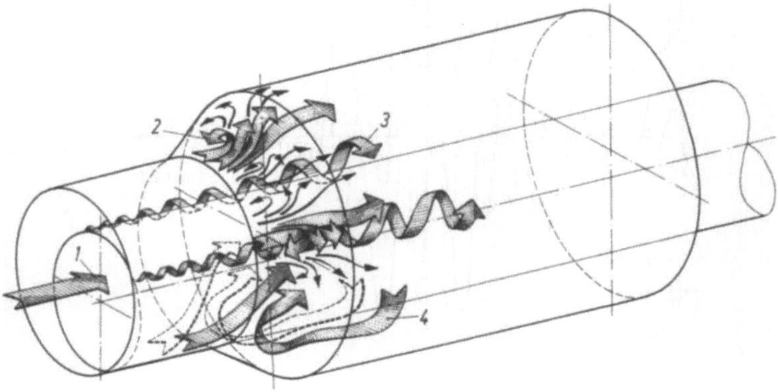

Fig.1. Flow phenomena in the eccentric annular diffuser.

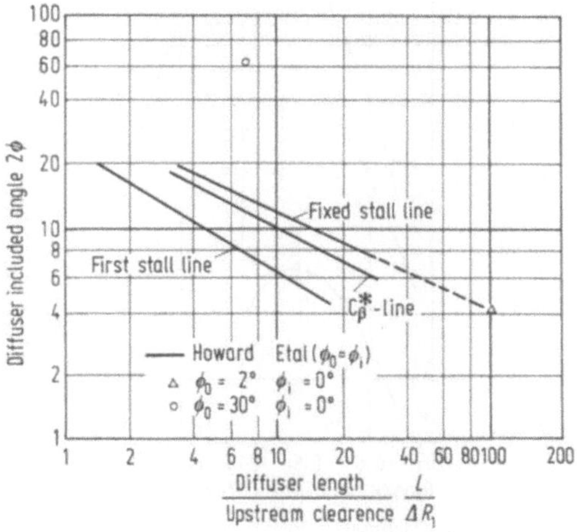

Fig.2. Flow regimes in annular diffusers.

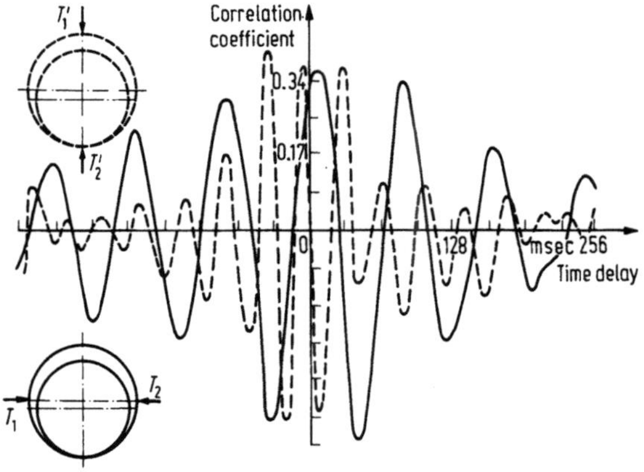

Fig. 3. Normalized pressure cross correlation, 2° diffuser.

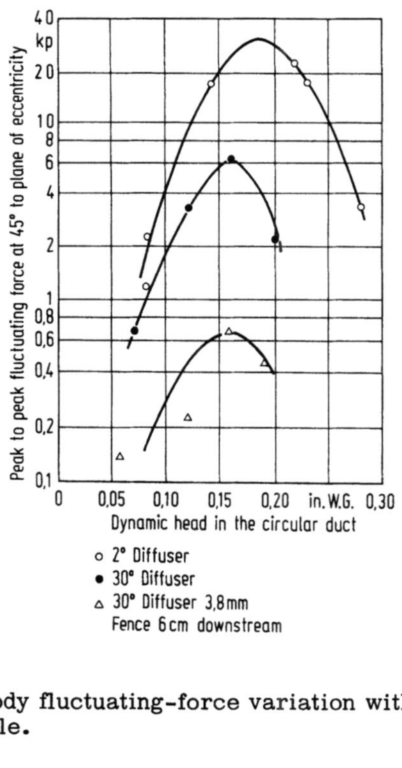

Fig. 4. Center body fluctuating-force variation with velocity and diffuser angle.

On a Global Hydraulic Instability in the Vessel of a Pressurized-Water Reactor [1]

By

A. Daubert, L. Pugnet, F. Boulot

Chatou, France

Introduction

The serious damage which occurred on the internal structures of the SENA and SELNI pressurized-water reactors shows the importance of hydroelastic problems which occur in this type of reactor. After those damages were observed, the study of the hydroelastic problems was started at the "Etudes et Recherches d'Electricité de France" with the aim of obtaining a better knowledge of the phenomena involved and thus avoiding these accidents in the projected plants, particularly those being built at Tihange and Fessenheim. Eventually, these studies could lead to the optimisation of the design of this type of reactor.

The study is limited to the annular space between the reactor vessel and the core-barrel, in which is found another barrel: the thermal shield (Fig. 1). In this communication are presented the results of the first part of the study which deals with the global hydraulic in-

---

[1] For a definition of the various parts of this general study, see: "Analyse des principales instabilités hydrauliques et des problèmes hydroélastiques dans les réacteurs à eau pressurisée - Programme d'études - par Mm. Daubert, Pugnet, Warluzel, Barrouillet - Comité Technique de la S.H.F. - Mars 1971).

stabilities without interaction with the structures and which has been carried out in a hydraulic model of scale 1/10.

Physical Description of the Phenomenon

The visual observation of the model under different working conditions has shown the possible existence of a hydraulic instability phenomenon in the annular space which is between the thermal shield and the reactor vessel, this instability occurring only when there is a thermal shield. During the instability, two great eddies are observed under the inlet nozzle and in the external annular zone which is found between the thermal shield and the impermeable vessel (Fig.2). These generally unstable eddies are formed near the rim of the shield from where they break loose more or less regularly causing an oscillation in the flow with a frequency of 1 Hz.

The formation of eddies is linked to the existence of an ascending flow under the inlet nozzle in the annular zone between the shield and the vessel. This ascending current is probably due to the presence of the shield and the existence of a great depression caused by the high velocities on the fluid paths of small radius of curvature.

From the spectral analysis of the data of the fluctuating pressures and velocities it has been possible to characterise the phenomenon by 1,2,3 or 4 "humps" of relatively small frequency range. The cross spectral analysis of pressure data recorded at two arbitrary points in the model showed that the correlation length of the phenomenon is of the order of the dimensions of the model and that, on the path of the eddy, the difference in phase keeps growing from the upper part of the shield.

For every characteristic frequency of the phenomenon, a map of the mean-square pressure fluctuations and of the dephasing has been drawn from the measurements obtained at a number of points in the annular space. The influence of the Reynolds number on the instability has been studied by varying the flow in the model. As an example, the Figs.3 give - in nondimensional form and on logarithmic scales -

the fundamental frequency and the corresponding pressure spectral density as a function of the Reynolds number for a single loop working configuration.

On account of the dispersion of the experimental points it is observed that, within the range which has been studied (relative Re-variation of about 3), the Reynolds number does not seem to be a parameter; and this would lead to a law of proportionality of the frequency to the inflow and of the pressure variations to the square of the inflow.

Influence of the Geometric Parameters

Instabilities have not been observed visually or by pressure measurements in the absence of the thermal shield. As a consequence, our investigations were directed toward the study of the thermal shield.

The experimental runs have shown a very marked influence of the number of loops in the working configuration, whatever shield was used. With one loop, the eddies of large dimensions can be stable or unstable and can cause important pressure oscillations, with amplitudes which may reach 1 m of water column and periods of about 10 seconds. With two loops, the phenomenon is less evident, less important and even absent for certain shields. With three loops, no eddies have been observed.

By the utilisation of thermal shields of different geometrical configurations, we observed that the mean diameter is an important parameter and, particularly, that there is a value of this parameter for which the instability is maximum. The influence of the thickness of the shield is less evident. The runs with eccentric or inclined shields have shown that, for a one loop working configuration, the most important parameter is the position of the shield near the inlet nozzle of the working loop in the annular space between the reactor vessel and the core barrel.

In particular, instabilities were found in the one- or two-loop configurations of two reactors which are being built; these instabilities could bring about alternating forces (a couple of tons on the thermal shield and the core barrel).

Other geometrical parameters were studied but not found of importance; for example, we studied the influence of a vertical wedging-up of the shield, the inversion of the flow in the lower plenum and the presence of outlet nozzles in the flow.

Conclusions

The study has shown the existence of global instabilities of the flow in the model which resulted in pressure fluctuations and alternating loads on the structure. The main cause of this phenomenon is the presence of a thermal shield. Among all the parameters investigated, it was observed that the position of the thermal shield in the annular zone between the reactor vessel and the core barrel has a great influence on the existence of instabilities and the generation of substantial alternating loads.

The second phase of the studies which has been undertaken concerns an eventual coupling between the flow and the movement or the vibration of the internal structural elements of the reactor.

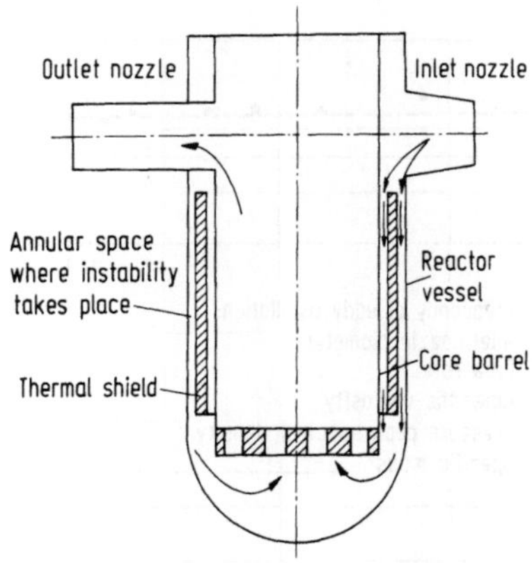

Fig. 1. Scheme of a pressurized water reactor.

Fig. 2a-e. Evolution of the flow under an inlet nozzle in a case of instability.

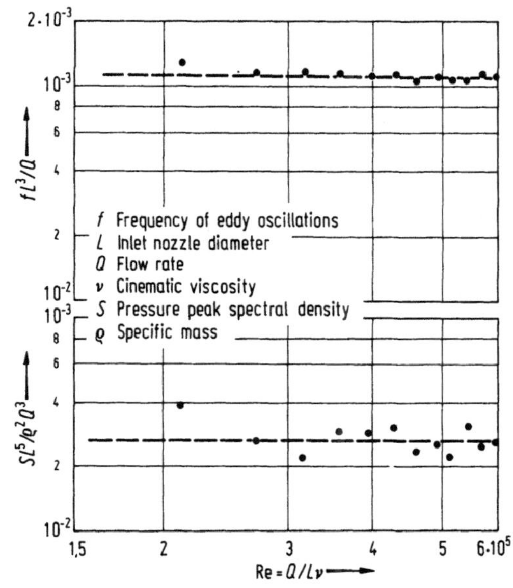

Fig. 3. Effect of the Reynolds number on the frequency and amplitude of a peak in the pressure spectral density.

TECHNICAL SESSION B

"Mathematical Models of Flow-Induced Vibrations"

Chairman: A.G. Davenport, Canada

Co-Chairman: O.H. Schiele, Germany

GENERAL LECTURE

Mathematical Models of Flow-Induced Vibrations of Bluff Bodies

By

G.V. Parkinson

Vancouver, Canada

Summary

Several forms of flow-induced vibration of bodies are discussed briefly, and two, vortex-induced vibration and galloping of long bluff cylinders, are selected for detailed consideration. The importance of the size and shape of the cross-section afterbody and of the details of vortex formation are emphasized. Several semi-empirical mathematical models of the body-oscillation dynamics are examined for both forms of oscillation. Some approaches to improved models compatible with the known characteristics of the separated flows and of the nonlinear oscillations are considered.

Introduction

An elastic structure exposed to a fluid flow may vibrate under the action of the flow for a variety of causes. If the incident flow is oscillatory, either in an organized form, as with the vortex street from an upstream body [1,2], or in the form of random turbulence [3,4] the structure will develop an oscillatory response. If the structure is streamlined, like an airfoil, and the incident flow is steady, the elastic characteristics in two or more degrees of freedom may permit the structure to extract energy from the flow so as to develop a catastrophic flutter [5]. These forms of flow-induced vibration will be considered in other papers presented at this Symposium.

In this paper, certain forms of vibration will be considered arising from another type of flow configuration, one in which the incident flow is steady in the mean, although perhaps turbulent, and the body shape is bluff, so that a broad wake is produced. The two shear layers that separate from the body and bound this wake are the source of the effects that characterize these forms of flow-induced vibration. One important effect, arising from the instability of the shear layers, is their universal tendency to roll up and form discrete vortices alternately from the two layers, so that an oscillatory pressure loading is produced on the generating body, and the various phenomena of vortex-induced vibration result. Another important effect can arise when the body shape is such that a small transverse or torsional motion of the body causes the shear layers to lie sufficiently asymmetrically with respect to the downstream body surfaces. This creates a pressure loading on the body which forces it in the direction of its initial small motion, and galloping instability results.

Certain special forms of flow-induced vibration of bluff bodies will not be examined in detail in this paper although some of them will be treated by others at this Symposium. If the cross-section of the body in the plane of the incident flow and of the transverse vibration is long in the streamwise direction, the initially separated flow will reattach to the downstream body surfaces, and the pressure loading may have enough similarity to that for an airfoil section that a combined plunging and twisting flutter oscillation can develop. This can occur for suspension bridge decks, and the relevant mathematical models of the oscillation phenomena are based on those for airplane wing flutter, with empirical modifications arising from the partly separated nature of the flow [6,7].

With a multi-component bluff structure, such as a transmission line conductor bundle, the wake of an upstream element may impinge on a downstream element to create pressure loadings on it, unstable to small disturbances in two degrees of freedom, and wake galloping results [1,8]. A stranded conductor may also gallop in a flow approaching at a small angle to the normal from the conductor span. In this case the instability results from asymmetric flow separation caused by the different surfaces presented to the flow by the helically wound strands [9].

Two forms of vortex-induced vibration not examined here are the phenomena of breathing oscillation, or ovalling of the open top of a cylindrical stack in a transverse flow [10], and the important problem of tube vibration in transverse flow through tube banks, often occurring in heat exchangers [11].

In the following sections, attention will be concentrated on the transverse vortex-induced or galloping oscillations of long bodies of bluff cross-section in flows normal to their span. Although all such phenomena are to some extent three-dimensional, and some observance of this will be made in the paper, the mathematical models investigated will represent essentially two-dimensional configurations of body and flow. Before examining mathematical models of the oscillations, it is important to understand some physical phenomena and parameters, and these are discussed briefly in the following section. Then models of vortex-induced and galloping oscillations of bodies are considered in subsequent sections, and in a final section some approaches to improved models that would be consistent with the observed phenomena of the flows and of the body oscillations are discussed.

Relevant Physical Parameters

The most important physical parameter of a two-dimensional body subject to vortex-induced or galloping oscillation is the size and shape of its afterbody, the part of the cross-section downstream of the separation points. The pressure loading that causes either form of oscillation is principally on the afterbody surface, and a section with a very short afterbody will not oscillate. Thus, a cylinder of semicircular, or D-, section will exhibit both vortex-induced oscillation and, if given a sufficient initial transverse velocity, galloping, when the flat face is upstream and normal to the incident flow, so that the flow separates at the sharp edges and the semicircular cylinder forms the afterbody. If, however, the flat face is downstream, the flow will separate at the edges, but now there is no afterbody, and the cylinder is completely stable, exhibiting neither vortex-induced oscillation nor galloping [12].

A convenient shape for a systematic study of the effect of afterbody length is the rectangular section, as shown in Fig. 1. The three curves shown summarize a large amount of data, taken in the Reynolds number range $10^4 < Re < 10^5$, for stationary and galloping cylinders of rectangular section. For the stationary cylinders, curves are given based on several investigations [12 to 17] of sectional drag coefficient

$$C_D = \frac{D}{\rho V^2 h/2}$$

and Strouhal number

$$S = \frac{f_v h}{V},$$

where f_v is the vortex formation frequency in one shear layer. For the galloping cylinder, the curve from two investigations [12,13] is the maximum ratio of dimensionless transverse galloping amplitude $\overline{Y}_s$ from rest to dimensionless flow velocity U, both defined later in the paper.

The curves indicate that as the afterbody length d increases, flow conditions change continuously and gradually except in two ranges of d/h, from 0.5 to 1.0, and from 2.5 to 3.0. For the first range, Bearman and Trueman [15] argue that increasing d reduces the size of the base 'cavity' from which entrainment of fluid takes place by the shear layer forming a discrete vortex, and the base pressure is accordingly lowered with an accompanying drag increase. This trend cannot continue indefinitely because a sufficiently large d will begin to interfere with the inward-curving shear layer and force a change in the process. Thus C_D reaches a maximum at about $d/h = 0.62$. Apparently this beginning of interference between the shear layer and the afterbody downstream corner has a significant effect on the pressure loading on the afterbody sides, because Brooks [12] and Smith [13] found that for $d/h < 0.75$, elastically mounted rectangular cylinders were hard oscillators, requiring a considerable initial transverse velocity $\dot{y}$ before galloping would occur, whereas for $0.75 \leqslant d/h \leqslant 3.0$, they were soft oscillators, unstable at rest to small transverse disturbances.

Bearman and Trueman [15] suggest that the interference caused by the longer afterbodies forces the vortex formation further downstream, and this raises the base pressure, reducing C_D, and also makes the shear layers more diffuse prior to vortex formation which, following Gerrard [18], would cause a decrease in f_v and provide an explanation for the observed drop in S.

The other abrupt changes, occurring for $2.5 < d/h < 3.0$, clearly correlate with reattachment of the shear layers on the afterbody sides, with final separation occurring at the downstream corners. Cylinders with $d/h > 3.0$ are stable with respect to transverse galloping because reattached shear layers provide a pressure loading on the afterbody sides that opposes small transverse disturbances. It is interesting to note that in the range of d/h where C_D changes rapidly, S changes only gradually, because the principal effect is on the base pressure, not the shear layer, whereas in the range where S changes rapidly, the opposite is true.

The above discussion indicates that the galloping characteristics of bluff cylinders depend on the interaction between the separated shear layers and the sides of the section afterbody, with conditions causing reattachment on the sides playing an important role. Galloping occurs approximately at a natural frequency f_n of the elastic system, and at incident flow velosicities for which f_v is appreciably higher than f_n. Vortex-induced oscillation also depends on the pressure loading produced on the afterbody sides by the action of the shear layers, but here the details of the discrete vortex formation process and the effect of the cylinder oscillation on this process dominate the relationships.

It is well known that the total circulation Γ_s shed from one side of a stationary bluff cylinder section during a cycle of vortex formation is nearly twice the circulation Γ_o in the discrete vortex. In Ref. [18] Gerrard explains this discrepancy by a model in which some circulation from one shear layer is transferred across to the other one, cancelling a part of it, and vice versa, and another small amount is transferred into the recirculation and entrainment region behind the cylinder base. Wood [19] has performed experiments providing verification of this model, and giving $\varepsilon = \Gamma_o/\Gamma_s = 0.66$, in respectable agreement with the results of a potential flow study by Abernathy and Kronauer [20] of two parallel rows of point vortices given a sinusoidal disturbance. They found a value $\varepsilon = 0.61$.

When the cylinder oscillates with appreciable amplitude under the pressure loading produced by these vortices, the details of vortex formation change, and when the incident flow velocity is in a range where the vortex frequency f_v approaches a natural frequency f_n of the cylinder elastic system, the two frequencies lock in to a common value f_c, close to f_n, and higher pressure loadings cause larger vibration amplitudes in a form of nonlinear resonance.

Models of Vortex-Induced Vibration

General Principles

A rigid two-dimensional bluff body of mass m per unit length, mounted on linear springs giving undamped natural circular frequency ω_n in transverse oscillation, and with viscous-type structural damping a fraction β of critical damping, obeys the differential equation for transverse displacement y,

$$m\ddot{y} + 2\beta m\omega_n\dot{y} + m\omega_n^2 y = F_y = C_y \frac{\rho}{2} V^2 h \qquad (1)$$

under the action of periodic fluid dynamic force F_y, (see inset of Fig. 1) subsequently expressed in terms of coefficient C_y. If Eq.(1) is divided by $m\omega_n^{\ 2}h$, it becomes dimensionless in the form

$$\overset{\circ\circ}{Y} + 2\beta\overset{\circ}{Y} + Y = C_y n U^2,$$

$$\text{where } Y = \frac{y}{h}, \quad \tau = \omega_n t, \quad U = \frac{V}{\omega_n h}, \quad n = \frac{\rho h^2}{2m}, \qquad (2)$$

and ($^\circ$) now denotes differentiation with respect to τ. Equation (2) represents the mechanical model of the vortex-induced oscillator (and indeed the galloping oscillator) used by most investigators of the oscillation phenomena. Where the various studies differ is in the treatment of the fluid dynamic force coefficient C_y.

Here attention will be concentrated on examples in which the vortex-induced oscillation amplitude $\overline{y}$ is an appreciable fraction of the bluff cylinder transverse dimension h, and for these, as mentioned in the previous section, there is an interdependence between excitation and response, which Naudascher [21] and Toebes [22] have called fluid-elastic interaction. Thus the amplitude $\overline{C}_y$ of the periodic fluid dynamic force coefficient depends upon the amplitude $\overline{Y} = \overline{y}/h$ of the cylinder displacement. This fact, and the frequency lock-in mentioned in the previous section, suggest, and a growing body of experimental evidence confirms, that these are nonlinear oscillation phenomena. Before examining some proposed models, it is therefore appropriate to quote Minorsky [23] to the effect that in the theoretical study of nonlinear oscillation very little progress could be accomplished without a continuous guidance by experimental evidence.

From such experimental evidence it is clear that in large-amplitude steady-state vortex-induced oscillation, the force excitation and cylinder displacement response oscillate at the same circular frequency ω_c, close to ω_n, and both have nearly sinusoidal waveforms with little amplitude modulation. The excitation leads the response by phase angle Φ. Accordingly, the expressions

$$Y = \overline{Y}_s \sin \frac{\omega_c}{\omega_n}\tau , \quad C_y = \overline{C}_{y_s} \sin\left(\frac{\omega_c}{\omega_n}\tau + \Phi\right), \tag{3}$$

where $(\ )_s$ refers to steady-state oscillations, can be substituted in Eq. (2), and the resulting coefficients of $\sin(\omega_c/\omega_n)\tau$, $\cos(\omega_c/\omega_n)\tau$ are equated to zero to give the two equations,

$$\left(1 - \frac{\omega_c^2}{\omega_n^2}\right)\overline{Y}_s = nU^2\overline{C}_{y_s} \cos\Phi , \tag{4}$$

$$2\beta \frac{\omega_c}{\omega_n}\overline{Y}_s = nU^2\overline{C}_{y_s} \sin\Phi . \tag{5}$$

Useful information can be obtained by order-of-magnitude examination of Eqs. (4) and (5). Typically, $\overline{Y}_s$ and $\overline{C}_{ys}$ are $0(1)$, and since the frequency lock-in range of flow velocities includes the velocity for which $f_v = f_n$:

$$U = \frac{V}{2\pi f_n h} = 0\left(\frac{1}{2\pi S}\right) = 0(1).$$

For air flow, in which most of the experiments used for comparison with theories discussed below have been carried out, mass parameter n is $0(10^{-3})$ usually, so Eq. (4) confirms that $\omega_c \doteq \omega_n$. For water flow, n is $0(1)$, so ω_c can be appreciably different from ω_n. In Eq. (5), all of the quantities can be measured, fairly accurately, so the equation can be used to check the consistency of the measurements.

Three of the theoretical models to be discussed below [24, 25, 26] make use of comparisons with results from wind tunnel experiments on circular cylinders carried out under the author's supervision [27, 28], so as an example Eq. [5] will be applied to some measurements made by Feng [28]. Feng used the same circular cylinder model as Ferguson [27], and much of the same instrumentation, but with new pressure-measuring and correlating instrumentation, he was able to make better measurements of the phase angle Φ than Ferguson had been able to. In addition he determined $\overline{C}_{ys}$ from measured pressure distributions, and explored the occurrence of oscillation hysteresis. Figure 2 shows Feng's measured variation with U of $\overline{Y}_s$, $\overline{C}_{ys}$, Φ, ω_c/ω_n, ω_v/ω_n, and ω_{vs}/ω_n, where ω_{vs} is the vortex-shedding frequency when the cylinder is experiencing steady-state oscillations.

Consider the measured values (denoted by A) at $U = 0.97$, close to the conditions for maximum cylinder amplitude, when the wind speed is increased from a low value with the cylinder oscillating (the lower branch of the amplitude curve would be reached if the cylinder was released from rest at that wind speed). The values are $\overline{Y}_s = 0.47$, $\overline{C}_{ys} = 1.90$, $\Phi = 50^\circ$, and since the accuracy of the measurements doesn't warrant greater precision in the following calculation,

$\omega_c/\omega_n \doteq 1.0$. Equation (5) can be used to estimate the structural damping coefficient β, using Feng's measured value of mass parameter $n = 0.0026$

$$\beta = \frac{0.0026(0.97)^2 1.90 \sin 50^\circ}{2(0.47)} = 0.0038 .$$

Feng's measured value of β was 0.0010, obtained by observing the still-air decay of free oscillation of a streamlined bar, of the same mass as the circular cylinder, mounted in its place. It is not surprising that the effective structural damping during vortex-induced oscillation of the cylinder is higher than the value determined by the above free-oscillation test because of the effect of the drag load on the cylinder transmitted to the air-bearing mounting system, and the effect of the ambient vibration level of the wind tunnel structure to which the cylinder mounting system is attached, when the tunnel is operating.

It is clear from the above that large-amplitude vortex-induced oscillations require very low damping in the cylinder elastic system. There is no other damping mechanism present. Equation (5) represents a balance between the structural (or elastic system) damping and the component of aerodynamic excitation in phase with the vibration velocity. Some investigators refer to 'aerodynamic damping' during vortex-induced oscillation as though it were an additional effect. In fact, all aerodynamic influence on the cylinder motion is included in the proper specification of $\bar{C}_{ys}$, which represents the integrated effect of the surface pressure loading by the flowing fluid on the vibrating cylinder (shear loading is two orders of magnitude smaller, and can be neglected).

Oscillator Models

Some recently proposed mathematical models of vortex-induced oscillation will now be examined. di Silvio [24] essentially uses Eq. (2) under the assumption that it governs a forced oscillation at frequency ω_{v_s}, which is determined by the assumption that Roshko's wake Strouhal number [29]

$$S^* = \frac{\omega_{v_s} b}{2\pi k V} = \text{const} \tag{6}$$

holds with its stationary-cylinder value under conditions of vortex-induced oscillation of the generating bluff body. In Eq. (6), b is the wake width between the separated shear layers (idealized by Roshko as free streamlines) or the lateral separation of the two rows of wake vortices, and k is the ratio of separation velocity to free stream velocity - di Silvio assumes that k = 1.4, independent of bluff body shape or motion, and further assumes that the wake width is dependent on vibration amplitude, $\overline{Y}_s$ and phase ψ with respect to the time of origin of a growing wake vortex, through the relation

$$\frac{b}{h} = M(1 + 2\overline{Y}_s \sin\psi) , \tag{7}$$

where M is the "bluffness ratio" for the body-wake system, taken empirically as 1.15 for the circular cylinder. Next he assumes (improperly) that the exciting force F_y is given in terms of ρ, V, and wake vortex circulation Γ_o by the Kutta-Joukowsky law [29], derived for steady potential flow past an isolated body with circulation but in the absence of other flow disturbances. To determine Γ_o, he uses Roshko's [14] estimate of the fraction ε of the vorticity originally present in a separation shear layer that appears in the circulation of the wake vortices from that layer, and this leads to the formula for $\overline{C}_{ys}$

$$\overline{C}_{y_s} = \frac{\pi k \varepsilon b}{2S^* h} , \tag{8}$$

The sum of Φ and ψ is a constant, determined by matching experimental values at maximum amplitude. This condition and Eqs. (4) to (8), can then be used to find compatible values for given U of $\overline{Y}_s$, ω_{vs}/ω_n, and Φ, where $\omega_c = \omega_{v_s}$ in Eqs. (4) and (5).

Di Silvio's results, when compared with Ferguson's [27] measurements, show a superficial resemblance to them (peak values of $\overline{Y}_s$

were made to agree in the solution process), and the concept of a variable vortex-wake width, dependent on cylinder amplitude and vortex-origin time may have merit (different forms of this concept have been proposed previously [30, 31, 32]), but the theory cannot be considered satisfactory. Too many empirical assumptions, some demonstrably incorrect, are needed, and the results are in any case inaccurate. For example, di Silvio's values give $\overline{C}_{ys} = 8.4$ for conditions corresponding to Ferguson's [27] maximum cylinder amplitude, more than four times the actual value (see Fig. 2), and his amplitude matching method required him to replace the actual structural damping β, of about the value calculated above from Feng's [28] measurements, by an enormously larger "aerodynamic damping" β of 0.046.

Halle [25] also proposes an oscillator model in which the effect of cylinder oscillation on vortex frequency is accounted for by assuming, like di Silvio, that Strouhal number remains constant (Halle uses S, rather than S*), so that frequency is inversely proportional to an effective shear layer width, empirically dependent on amplitude and phase of cylinder oscillation. Halle does not speculate further about the nature of the fluid motions and forces, but replaces the fluid by a second mechanical oscillator coupled to the cylinder by an equivalent viscous damping force proportional to the relative velocity of this fluid oscillator to the cylinder. (He also includes a random fluid dynamic exciting force on the cylinder, to represent incident and wake turbulence, but its effect is negligible over the lock-in range of flow velocities). This mutual damping force represents the actual pressure loading on the cylinder caused by the wake vortices, and its amplitude, and therefore $\overline{C}_{ys}$, is inversely proportional to a factor Q_g which is chosen empirically to provide a good fit to experimental data.

Halle obtains a variation of $\overline{Y}_s$ with U over the lock-in range for a circular cylinder that agrees quite well with the same measurements (Ferguson) used by di Silvio, but as Halle concedes in the paper, Q_g was chosen judiciously and, in addition, the theoretical and experimental peak amplitudes were matched. This means essentially that the width and height of the Y_s - U curve were made to agree with the experimental data, so the general agreement is not surprising. Moreover the numerical values from Halle's solution correspond to $\bar{C}_{ys} = 0.20$ at $(\bar{Y}_s)_{max}$, almost an order of magnitude too small. Halle does mention that a larger value of Q_g would correspond to a narrower $\bar{Y}_s$ - U curve bending over to the right, and therefore perhaps corresponding qualitatively to the conditions for the oscillation hysteresis of Fig. 2. However, this would correspond to a still lower, and therefore more unrealistic, value of $\bar{C}_{ys}$.

Hartlen and Currie [26] also simulate the fluid dynamics of vortex-induced oscillation by the introduction of a fluid oscillator, but in a different approach than that of Halle. Their cylinder motion is governed by Eq. (2), but instead of introducing empirical specifications of C_y and the dependence of vortex frequency on cylinder motion, they make a case for regarding C_y as a soft nonlinear oscillator, coupled to the cylinder motion over the lock-in range of flow velocities by a linear dependence on cylinder velocity. They choose the simplest oscillator of this class, one studied in another context by van der Pol (see Ref. 23, p. 439):

$$\mathring{\mathring{C}}_y - A\Omega_0 \mathring{C}_y + \frac{B}{\Omega_0} \mathring{C}_y^3 + \Omega_0^2 C_y = D\mathring{Y} , \tag{9}$$

where $\Omega_0 = 2\pi SU = \omega_v/\omega_n$, and A, B, D are constants. Although Eq. (9) was chosen for its simplicity, it is nevertheless capable of representing many of the observed effects of wake vortices. If the cylinder is restrained from moving, so that the right side of Eq. (9) vanishes, the equation represents the oscillating pressure force on a stationary bluff cylinder. Elementary theory of weakly nonlinear equations (see the section on galloping models) then gives the result (which of course was designed into Eq. (9)) that C_y oscillates at the stationary Strouhal circular frequency ω_v, and that the amplitude

$$\bar{C}_{y_0} = \sqrt{\frac{4A}{3B}} . \tag{10}$$

Experimental values of $\bar{C}_{y_0}$ can be used to determine the ratio A/B for insertion in the analysis of the corresponding problem with the cylinder oscillating. This leaves two of the three constants, say A and D, to be determined.

If the cylinder is oscillated mechanically, with a specified amplitude and frequency, as in the experiments of Smirnov and Pavlihina [33], Bishop and Hassan [34], Jones [35], and Tanaka and Takahara [36], only Eq. (9) is needed, since $\mathring{Y}$ is known and Minorsky shows ([23], p. 439) that when the impressed frequency ω_c is sufficiently near ω_v, the solution of Eq. (9) predicts the lock-in of ω_v to ω_c. Hartlen and Currie apply this solution to a comparison with Jones' [35] data on $\overline{C}_{y_s}$ as a function of ω_c/ω_v and $\overline{Y}_s$, with good qualitative results.

The most difficult test of the theory, however, is for vortex-induced oscillation of a cylinder, requiring a solution of Eqs. (2) and (9). Hartlen and Currie concentrate their attention on steady-state oscillations in the lock-in range, so that Eqs. (3), (4), and (5) apply. Equations (3) are substituted in Eq. (9), and coefficients of $\sin\Omega\tau$, $\cos\Omega\tau$ are equated to zero, where $\Omega = \omega_c/\omega_n$. (The third harmonic terms arising from $\mathring{C}_y^{\,3}$ are dropped as unimportant). The two resulting equations are put in terms of the unknowns $\overline{Y}_s$ and Ω through the use of Eqs. (4) and (5), and $\overline{Y}_s$ can then be eliminated from them to give a final equation for frequency ratio Ω

$$\Omega_0 = \frac{\Omega}{\sqrt{1 - \dfrac{nD\beta\Omega^2}{2\pi^2 S^2[(1-\Omega^2)^2 + 4\beta^2\Omega^2]}}} \quad . \qquad (11)$$

From Eq. (11) it can be seen that $\Omega \leqslant \Omega_0$. With Ω given by Eq. (11), $\overline{Y}_s$, $\overline{C}_{y_s}$, and Φ can be solved for as functions of flow velocity parameter Ω_0 from the other equations, and Hartlen and Currie have done this using values of n, β, $\overline{C}_{y_s}$, and S corresponding to Ferguson's [27] data for the circular cylinder. Parameters A and D are chosen to provide the best fit to the data. These theoretical solutions are shown in Fig. 3, and they can be seen to give good qualitative and fair quantitative agreement with the lower branches of Feng's curves in Fig. 2. (These lower branches represent values for steady-state oscillations reached from an initial condition with the cylinder at rest at the given wind speed, as in Ferguson's measurements).

The theoretical solutions from Eqs. (2) and (9) of Hartlen and Currie do not display all of the observed features of vortex-induced oscillations, such as the oscillation hysteresis of Fig. 2 and the fact, observed by Feng [28] for the D-section, that its lock-in range of flow velocities is displaced to the left so that $\Omega \geqslant \Omega_0$, unlike the solution given by Eq. (11). Nevertheless, it seems probable that both of these features could be reproduced in a modified version of Eq. (9), with a somewhat more elaborate nonlinearity and/or coupling term, and the method appears to have considerable merit.

Another approach to simulation of fluid dynamic effects in vortex-induced oscillation by the use of a mechanical oscillator to represent the fluid is taken by Nakamura [37]. He adapts Birkhoff's [38] model of the "dead air region" behind the bluff cylinder as a torsional oscillator, whose linear spring is provided by the aerodynamic restoring torque created by an angular displacement of this dead air region. This torque is given by a formula equivalent to assuming that quasi-steady thin airfoil theory can be applied to the dead air region, except that the effective lift force is apparently considered to act at "midchord" of the region. Nakamura shows that reasonable choices for the dimensions of this dead air region lead to a prediction of the correct stationary-cylinder Strouhal frequency ω_v as the undamped natural frequency of this oscillator. When the cylinder is in transverse motion, he assumes it to be a linear undamped oscillator, so that it is governed by Eq. (2) with $\beta = 0$. For C_y, he appeals to the relationship observed experimentally [39, 40] between magnus-effect lift and wake angular displacement for a rotating circular cylinder, and for the term coupling the equation for the torsional fluid oscillator to the cylinder, he draws an analogy with the governing equation for a compound pendulum whose hinge is oscillating horizontally. This leads to a pair of linear coupled equations for Y and the angular displacement of the dead air region. Conventional flutter analysis reveals a binary flutter in a limited range of flow velocity. However, the absence of nonlinearity and structural damping, and the considerable number of empirical assumptions made about the aerodynamics, make the model rather unsatisfactory, and in fact the flutter frequency variation with flow velocity obtained from it is unlike the lock-in behavior actually observed.

Another model in which the vortex frequency is made a function of cylinder amplitude is proposed by Kawamura [41]. Like di Silvio and Halle, he uses an empirical relationship, but he does not consider any phase effect, and makes ω_{v_s} a quadratic polynomial in $\overline{Y}_s$. He takes Eq. (2) to represent forced oscillation at the amplitude-dependent frequency $\omega_c = \omega_{v_s}$, and assumes $\overline{C}_{y_s}$ is constant, independent of cylinder oscillation. He obtains the usual formula for a linear forced oscillator giving $\overline{Y}_s$ formally as a function of ω_c, and then substitutes the polynomial in $\overline{Y}_s$ for ω_c in the formula. There are numerous errors in the paper, and insufficient data are given to permit a check of the calculated curves of $\overline{Y}_s$ vs ω_v/ω_n (proportional to flow velocity, since S = const), but the qualitative effect of the nonlinear dependence of ω_c on $\overline{Y}_s$ is to cause the resonant peaks to lean to the right, giving conditions corresponding to the sort of oscillation hysteresis of Fig. 2. However, for the values of parameters Kawamura uses, his $\overline{Y}_s$ values seem much too high.

Flow Field Models

None of the mathematical models of vortex-induced oscillation considered so far included any analysis of the flow field. Ideally one would solve the time-dependent Navier-Stokes equations in the presence of the oscillating bluff cylinder, the flow separation and vortex formation would emerge naturally from the solution, and the pressure and shear loading on the cylinder surface would provide the forcing function for the coupled motion of the cylinder, still given by Eq. (2). Numerical solutions of the Navier-Stokes equations have been achieved for the flow fields of stationary bluff cylinders, and recently Jordan and Fromm [42] have produced calculations of drag, lift, torque, pressure distribution, separation angle movement, and Strouhal number at three Reynolds numbers in the vortex-shedding range for a stationary circular cylinder in a uniform incident flow. Their results agree well with experimental data except at the highest Reynolds number (1000), where somewhat larger discrepancies occur, partly because of the difficulty of modelling the thin cylinder boundary layer upstream of separation, and partly because of experimental three-dimensionality of the flow.

To make any progress with an analytical model of the flow field, it is necessary to abandon the full Navier-Stokes equations for something simpler, and since one can argue, with a somewhat dubious eye cast on the wake, that vorticity is significant only in the cylinder boundary layer, the thin shear layers separating from it, and the resulting cores of discrete vortices, many have proposed potential-flow models for bluff bodies. Of these, free streamline models have proved useful for the calculation of time-average properties such as cylinder mean drag and pressure distribution, but they are not of interest here. Instead, the relevant models simulate the actual shear layers and vortices by potential vortices, in which the vorticity is concentrated in the core singularity. The original vortex street model of von Kármán [43] continues to form a basis for investigations [11, 44], although Kármán's stability criterion of a vortex spacing ratio of 0.281 is no longer generally accepted, because of a considerable body of evidence that vortex streets, even in potential flow models [20], form with a range of spacing ratios, and Bearman [44] offers evidence in support of an alternative stability criterion by Kronauer [45].

In any model simulating the actual wake vortex system by discrete potential vortices, such as the one shown in Fig. 4, two important decisions to be made are the strength of circulation Γ_0 to be assigned to the vortex, and the method of creating it, and relating it to the flow separation at some point s. Gerrard [46], in applying potential flow methods to the problem of vortex-induced lift on a stationary circular cylinder, makes use of arguments developed in an earlier paper [18] in empirically determining Γ_0, and creates each vortex at a control surface downstream of s, but fairly near it, where the ability of the formation point to oscillate as part of a feedback mechanism [45] is essential. The basic potential flow is the unseparated flow model given by uniform flow plus a doublet at the origin, and the wake vortices are paired with internal images at their inverse points, as in Fig. 4, to satisfy the surface boundary condition. The computation proceeds by allowing each vortex to move over short time intervals with the velocity calculated for the flow at the vortex position at the beginning of each interval. The relevant field dynamical equation is the unsteady form of Bernoulli's equation,

$$\frac{\partial\varphi}{\partial t} + \frac{p}{\rho} + \frac{u^2 + v^2}{2} = \frac{p_\infty}{\rho} + \frac{V^2}{2} , \qquad (12)$$

where p is pressure, φ is velocity potential, and u, v are cartesian fluid velocity components at any point. This can be integrated to produce general formulas for force and moment on a body in a two-dimensional potential flow, such as the generalized Blasius equations (see Ref. [29]), and Gerrard determines the force F_y on the cylinder in this way. He obtains values for $\bar{C}_y$, S, and other parameters of the vortex wake in reasonable agreement with experimental values. Sarpkaya [47] applies similar methods to the symmetric problem of impulsive flow about a circular cylinder. A principal difference of his method from Gerrard's is his use of an infinite number of feeding vortex sheets supplying each new discrete vortex in a zone centered at the actual separation point s (Fig. 4). Sarpkaya obtains good agreement with experimental data for the calculation of the time build-up of drag on the cylinder.

An interesting but rather cryptic paper by Shioiri [48] links potential-flow field methods for the vortex wake of a circular cylinder with the vortex frequency lock-in produced by forced oscillation of the cylinder over a range of cylinder frequencies. The wake is composed of two infinite sheets of discrete vortices (applied as though they were semi-infinite) which experience an antisymmetric periodic disturbance, varying as in the usual linear stability analysis as

$$e^{i(\omega_v t - \lambda x)} ,$$

where λ is a complex wave number. The effect of the final nonlinear growth process for the disturbances is included empirically, and the feedback of oscillatory disturbance to the separation points s (taken at the transverse diameter, with k = 1.4) is calculated for each vortex and its image as in Fig. 4 (because of the difficulty of the integration over the semi-infinite length of the vortex sheet, an approximation to the integrand is used). Relating this feedback to the oscillatory perturbation of the vorticity shedding at s leads to an equation for the self-sustained oscillation of the feedback system. In this way, Shioiri calculates S = 0.244 and 0.256, for the stationary cylinder, between 20 and 30 % high.

When the cylinder is in forced oscillation at frequency ω_c, Shioiri includes an effect of the cylinder motion on the disturbances to the vortex sheets and, by a modification of the previous analysis, calculates conditions for stable lock-in of the vortex frequency to the cylinder frequency. Figure 5 shows Shioiri's calculated values of the minimum $\bar{Y}_s$ needed to produce lock-in at different frequency ratios ω_c/ω_v. He finds quite good agreement with the experimental values of Tanaka and Takahara [36] and Koopmann [49].

The only flow-field model for a bluff cylinder in vortex-induced oscillation of which the author is aware has been under development by Ujihara [50] for a circular cylinder. His methods are similar to those described in connection with Refs. [46] and [47], except that this discrete vortex strength is determined by what he refers to as a generalized Kutta condition, in which the shear layer vorticity is assumed to be released at points of velocity maxima on the cylinder surface, and by a vortex injection frequency empirically correlated with cylinder Reynolds number. The elastic system for the cylinder is linear, and the natural frequency is taken to be the Strouhal frequency for the stationary cylinder. Preliminary results from the computer program showed a rapid build-up of resonant cylinder oscillation, and values were qualitatively encouraging, although further development was required.

Models of Galloping

General Principles

With vortex-induced vibrations, some details of the phenomena are still interpreted differently by different investigators, and different models of the vibration result, none completely satisfactory. With galloping, no major disagreements appear to exist, and although there have been quite a few studies of galloping, it will be necessary to describe only one model in detail to give an adequate picture.

The name of den Hartog [51] is usually associated with the quasi-static force criterion for galloping instability, and Sisto [52] has applied the nonlinear methods like the ones described here to the stall flutter of airfoils. Scruton [53] has determined galloping amplitudes of bluff cylinders from empirical force data by a graphical method, and Novak, in a sequence of papers [54, 55, 56], has extended the nonlinear model for two dimensional flow described here to continuous elastic systems, such as cantilevered towers, and turbulent shear flows. He has also proposed a "universal response curve" which would permit the prediction of galloping characteristics of a long body of specified bluff cross section from a dynamic wind tunnel test of a scale model, without the necessity of static force measurements. The form of theoretical model presented here was developed by the author [57] and correlated with the results of experiments by Brooks [12], Smith [13], Santosham [58], (who also contributed to the theory), and Laneville [59]. Another application has been made by Slater [60], under the direction of V.J. Modi.

The theory is a two dimensional quasi-static model, in which it is assumed that the bluff section, at an instant during the galloping cycle, (see inset of Fig.1) experiences the same force F_y that it would if held stationary at the angle of attack α and incident flow velocity V_{rel} arising from the transverse cylinder velocity $\mathring{y}$. The validity of this assumption requires that the reduced frequency of galloping be very small, or equivalently, that $U \gg 1$. Santosham [58] has found good agreement between theory and experiment for a rectangular section of $d/h = 2$ if $U \mathring{>} 10$. This requirement also effectively removes the phenomenon of vortex-induced resonant vibration from the analysis of galloping, since for vortex-induced vibration, $U = 0(1/2\pi S)$, and for bluff sections

$$0.05 \mathring{<} S \mathring{<} 0.50 .$$

Analysis of the dynamics of a bluff section capable of galloping in the flow velocity range for vortex resonance would be considerably more complex.

With the usual linear elastic system for the model, Eq. (2) can again be employed, with C_y now given by static force measurements as a function of a. The discussion of the section on physical parameters indicated that C_y results primarily from a pressure loading on the section afterbody caused by interaction between the separated shear layers and the afterbody sides, and no theory capable of predicting it is yet available. The experimental variation of C_y with α (and therefore with $\tan\alpha$) can be approximated by a polynomial in

$$\tan\alpha = \frac{\mathring{Y}}{V} = \frac{\mathring{Y}}{U}$$

such as the 7th degree polynomial used for the square section [57] in Fig. 6, for which the measurements were made in a uniform, low turbulence air flow at a Reynolds number of $2.2(10)^4$:

$$C_y = A_1\left(\frac{\mathring{Y}}{U}\right) - A_3\left(\frac{\mathring{Y}}{U}\right)^3 + A_5\left(\frac{\mathring{Y}}{U}\right)^5 - A_7\left(\frac{\mathring{Y}}{U}\right)^7, \qquad (13)$$

where A_1, A_3, A_5, A_7 are positive constants. We will follow the argument of Ref. [57] through for the square section. If Eq. (13) is substituted in Eq. (2), the result can be written:

$$\overset{\circ\circ}{Y} + Y = nA_1\left\{\left(U - \frac{2\beta}{nA_1}\right)\mathring{Y} - \left(\frac{A_3}{A_1U}\right)\mathring{Y}^3 + \left(\frac{A_5}{A_1U^3}\right)\mathring{Y}^5 - \left(\frac{A_7}{A_1U^5}\right)\mathring{Y}^7\right\}$$

$$= \mu f(\mathring{Y}), \qquad \mu = nA_1 \ll 1 \quad \text{for air flow} . \qquad (14)$$

Eq. (14) has the form of a weakly nonlinear autonomous differential equation capable of solution by the first approximation method of Krylov and Bogoliubov (Ref.[23], Ch.14).

For $\mu = o$, the solution is the familiar

$$Y = \bar{Y}\cos\tau, \qquad \mathring{Y} = -\bar{Y}\sin\tau, \qquad (15)$$

where $\bar{Y}$ is constant. The method assumes that for $\mu \neq o$, but small, the solution can be obtained as a series in powers of μ:

$$Y = \bar{Y}\cos\tau + \mu Y_1(\bar{Y}_1\tau) + \mu^2 Y_2(\bar{Y}_1\tau) + \dots, \qquad (16)$$

where $\bar{Y}$ is now a slowly varying function of τ. (In general, there is also a slowly varying phase angle, but this vanishes in the first approximation of Eq. (14)). In most applications, as here, only the first approximation is needed to give an accurate estimate of the oscillation phenomena. In the first approximation, Eqs. (15) are retained, but $\overline{Y}$ is considered a function of τ. If Eq. (14) is multiplied by $\mathring{Y}$, the result can be written

$$\frac{1}{2}\frac{d}{d\tau}(Y^2 + \mathring{Y}^2) = \mu f(\mathring{Y}) \cdot \mathring{Y} \qquad (17)$$

or, using Eqs. (15)

$$\frac{1}{2}\frac{d\overline{Y}^2}{d\tau} = -\mu f(-\overline{Y}\sin\tau)\overline{Y}\sin\tau\,. \qquad (18)$$

It can be seen from Eq. (18) that $d\overline{Y}^2/d\tau$ is very small, so that the change in $\overline{Y}$ is negligible over one oscillation cycle. Accordingly, it will be satisfactory to replace the right side of Eq. (18) by its average over one cycle, thereby eliminating harmonics:

$$\frac{d\overline{Y}^2}{d\tau} = -\mu/\pi \int_0^{2\pi} f(-\overline{Y}\sin\nu)\overline{Y}\sin\nu\, d\nu\,. \qquad (19)$$

If the function f from Eq. (14) is inserted in Eq. (19), the integration yields

$$\frac{d\overline{Y}^2}{d\tau} = nA_1\left\{\left(U - \frac{2\beta}{nA_1}\right)\overline{Y}^2 - \frac{3}{4}\left(\frac{A_3}{A_1U}\right)\overline{Y}^4 + \frac{5}{8}\left(\frac{A_5}{A_1U^3}\right)\overline{Y}^6 - \frac{35}{64}\left(\frac{A_7}{A_1U^5}\right)\overline{Y}^8\right\}. \qquad (20)$$

Eq. (20) completely determines the galloping behavior of the bluff cylinder of square section in two dimensional uniform air flow. It can be integrated directly, but a preliminary qualitative analysis of its form and its relationship to the behavior of the system in the phase plane $(Y, \mathring{Y})$ may be useful, and could suggest to the interested reader some possibilities for nonlinear oscillator models of the vortex-induced systems of the previous section. The form of Eq. (20) is

$$\frac{dR}{d\tau} = a_1R - a_3R^2 + a_5R^3 - a_1R^4 = F(R)\,, \qquad (21)$$

where $R = \overline{Y}^2$ and a_1, a_3, a_5, a_7 are functions of the system parameters. Stationary oscillations correspond to $dR/d\tau = 0$, and are therefore given by $R = 0$, and by the real positive roots of the cubic

$$a_1 - a_3R + a_5R^2 - a_7R^3 = 0 . \tag{22}$$

In the phase plane (Fig.7), $R = 0$, the initial position of equilibrium, is a singular point at the origin called a focus. The positive roots R_i of Eq. (22) define trajectories in the phase plane called limit cycles, here concentric circles of radius $\overline{Y}_{s_i} = \sqrt{R_i}$. The polar angle ϑ in the phase plane is given by

$$\tan\vartheta = \frac{\mathring{Y}}{Y} = - \tan\tau , \tag{23}$$

so that

$$\frac{d\vartheta}{d\tau} = - 1 , \tag{24}$$

and a stationary oscillation of unit circular frequency (real time circular frequency ω_n) and amplitude $\overline{Y}$ is represented in the phase plane by a point rotating clockwise with unit angular velocity on a circle of radius $\overline{Y}_{s_i}$.

The stability of equilibrium and of limit cycles is determined by investigating the tendency of the oscillator to return to the original focus or limit cycle after a small displacement of δR. The pertinent function is therefore $d\delta R/d\tau$, and from Eq. (21) this is given by

$$\left.\frac{d\delta R}{d\tau}\right|_{R=R_i} = F'(R_i)\delta R , \tag{25}$$

where the prime indicates differentiation with respect to R. It follows that the sign of $F'(R_i)$ determines the stability of the limit cycle, which is unstable, stable, or neutrally stable according as $F'(R_i)$ is positive, negative, or zero. Here

$$F'(R) = a_1 - 2a_3R + 3a_5R^2 - 4a_7R^3 . \tag{26}$$

The stability of the focus at $R = 0$ is determined by the sign of

$$F'(0) = a_1 = nA_1\left(U - \frac{2\beta}{nA_1}\right) \qquad (27)$$

from Eqs. (26) and (20). If β is negligibly small, the sign is that of A_1 which is positive for the square section, showing it to be unstable at rest at any wind speed in the absence of system damping. Now $A_1 = dC_y/d\alpha|_{\alpha=0}$ and this criterion for instability is equivalent to the one proposed by Glauert [61] for auto-rotation of a stalled aerofoil and later by Den Hartog [51] for the galloping of electric transmission lines. If $\beta > 0$, the square section is unstable only for

$$U > U_0 = \frac{2\beta}{nA_1}, \qquad (28)$$

so that galloping can be eliminated below a given wind speed by sufficiently increasing the system damping.

If Eq. (28) is satisfied and $a_1 > 0$, then it is clear that there must be at least one limit cycle corresponding to a positive root R_1 of Eq. (22), since a_3, a_5, a_7 are all positive from Eqs. (20) and (13), and the left side of Eq. (22) will therefore be negative for large R. This limit cycle may be the only one, in which case it is stable, since the slopes of $F(R)$ at successive real roots will alternate in sign, unless a double root intervenes. The phase plane then shows (Fig. 7a) an unstable focus encircled by a stable limit cycle, a system with soft self-excitation designated $\mathcal{US}$, following Minorsky [23], Ch. 7. The corresponding graph of $F(R)$ is also shown in Fig. 7a. Under these conditions the system oscillates spontaneously from rest with increasing amplitude until a stationary oscillation of amplitude $\overline{Y}_{s_1} = \sqrt{R_1}$ is attained.

Another possibilty is that all three roots of Eq. (22), R_1, R_2, R_3 are real and positive, say $R_3 > R_2 > R_1$. With $a_1 > 0 (U > U_0)$ this means that R_1 and R_3 represent stable limit cycles separated by an unstable limit cycle, corresponding to R_2. This system is designated $\mathcal{U}\mathcal{S}\mathcal{U}\mathcal{S}$, and the phase plane and graph of $F(R)$ are shown in Fig. 7b. The transient oscillations, not shown, appear on the phase planes as tightly coiled spirals away from the focus and unstable limit cycle, and towards the stable limit cycles. Inspection of Fig. 7b suggests the remaining possible systems. R_2 and R_3 could coalesce to a double root, giving a system designated $\mathcal{U}\mathcal{S}\,(\mathcal{U}\mathcal{S})$, intermediate between those of Figs. 7a and 7b. Alternatively, R_1 and R_2 could coalesce to a double root, giving a system $\mathcal{U}\,(\mathcal{S}\mathcal{U})\mathcal{S}$, intermediate between that of Fig. 7b and a final possible system in which R_3 is the only real positive root. The phase plane for this system would be like that of Fig. 7a.

For a square prism with given mass and damping parameters, n and β, the choice among these possibilities depends on the reduced wind speed U as parameter. This is an example of bifurcation theory in non-linear oscillations. The coefficients a_1, a_3, a_5, a_7 all depend on U, and as it passes through critical values, the phase portrait of the system changes from one of the forms described above to another.

The dependence of the system on the parameter would commonly be shown on an (R,U) diagram. Here it is convenient to use a diagram giving stationary amplitude plotted against wind speed directly in the form $\overline{Y}_s/U_0$ vs U/U_0 (Fig.8). The theoretical curves predict that the system has oscillation hysteresis. At values of $U < U_0$, $\overline{Y}_s = 0$ and the phase portrait is a stable focus. Small oscillations die out and the cylinder remains at rest. At $U = U_0$ the system forks into a form whose phase portrait is like Fig. 7a, $\mathcal{U}\mathcal{S}$, with the amplitude of the stable limit cycle, $\overline{Y}_{s_1}$, increasing with U, as shown by the curve 012 in Fig. 8. Between U_1 and U_2, however, the phase portrait has become that of Fig. 7b, $\mathcal{U}\mathcal{S}\mathcal{U}\mathcal{S}$, with the intermediate forms $\mathcal{U}\mathcal{S}(\mathcal{U}\mathcal{S})$ at U_1, and $\mathcal{U}(\mathcal{S}\mathcal{U})\mathcal{S}$ at U_2. Between U_0 and U_2 the lower limit cycle $\overline{Y}_{s_1}$ is reached from rest. For $U > U_2$ the form becomes $\mathcal{U}\mathcal{S}$ with the single stable limit cycle of amplitude $\overline{Y}_{s_3}$ reached from rest. If the wind speed is then decreased to a value between U_1 and U_2 while the prism is still oscillating, the upper limit cycle is the one reached, until for $U < U_1$ the amplitude drops to that of the lower cycle again. Thus, the path 012343510 containing the hysteresis loop 12351 is traversed on the $(\overline{Y},U)$ diagram. The unstable limit cycle in the range $U_1 < U < U_2$ discriminates between the stationary oscillations on the basis of initial amplitude. If the initial amplitude is less than $\overline{Y}_{s_2}$, the stationary amplitude reached is $\overline{Y}_{s_1}$. If it is greater than $\overline{Y}_{s_2}$, the stationary amplitude is $\overline{Y}_{s_3}$.

Two other important characteristics of the $(\overline{Y}_s,U)$ variation can be seen by equating the left side of Eq. (20) to zero, and dividing through by $nA_1U\overline{Y}_s^2$

$$\left(1-\frac{2\beta}{nA_1U}\right)-\left(\frac{3A_3}{4A_1}\right)\left(\frac{\overline{Y}_s}{U}\right)^2+\left(\frac{5A_5}{8A_1}\right)\left(\frac{\overline{Y}_s}{U}\right)^4-\left(\frac{35A_7}{64A_1}\right)\left(\frac{\overline{Y}_s}{U}\right)^6=0. \quad (29)$$

It is already known that one stationary amplitude $\overline{Y}_{s_3}$ exists for $U > U_2$. Eq. (29) shows that as $U \to \infty$, $\overline{Y}_s/U$ becomes independent of U, so that the $(\overline{Y}_s,U)$ curve becomes asymptotic to a line through the origin. Moreover, it can be seen from Eq. (29) that a curve of $(nA_1/2\beta)\overline{Y}_s$ or $\overline{Y}_s/U_0$ plotted against $(nA_1/2\beta)U$ or U/U_0 would be independent of n and β so that the theoretical curve of Fig. 8 should be universal for the square section, and experimental values of $\overline{Y}_s$ and U for oscillations of square prisms conducted at various levels of n and β should collapse on this single curve when plotted in this form.

The experimental points shown in Fig. 8 were obtained by Smith [13] for a square-section cylinder with $n = 4.3(10)^{-4}$ and four different values of β ranging from 0.0011 to 0.0037. The incident air flow was uniform and of low turbulence level, and test conditions were closely two dimensional. The data show excellent agreement with the theoretical predictions. The oscillation hysteresis is accurately predicted, and the collapse of data points to the single curve is demonstrated. The poorest agreement is for the lowest value of β, and the reason is the proximity of the vortex resonance range of wind speeds, as indicated by the vertical mark with the appropriate data symbol (•) on the abscissa, denoting the condition $U = 1/2\pi S$ plotted for that value of U/U_0.

It remains to consider the transient oscillations quantitatively by integrating Eq. (20). With some rearrangement of terms, it can be integrated to

$$-\Lambda\tau = \int_{z_0}^{z} \frac{dz}{z(z - z_1)(z - z_2)(z - z_3)} , \tag{30}$$

where

$$\Lambda = \frac{35nA_7U}{64} , \quad z = \left(\frac{\overline{Y}}{U}\right)^2 , \quad z_0 = \left(\frac{\overline{Y}_0}{U}\right)^2 , \quad z_i = \left(\frac{\overline{Y}_{s_i}}{U}\right)^2 , \quad i = 1,2,3 .$$

$\overline{Y}_0$ is the initial amplutide and $\overline{Y}_{s_i}$ are the amplitudes of the limit cycles discussed above. For $U_1 < U < U_2$ the z_i are all real and the integral is

$$-\Lambda\tau = \ln\left\{\left(\frac{z}{z_0}\right)^{\frac{-1}{z_1z_2z_3}}\left(\frac{z - z_1}{z_0 - z_1}\right)^{Z_{1,23}}\left(\frac{z - z_2}{z_0 - z_2}\right)^{Z_{2,13}}\left(\frac{z - z_3}{z_0 - z_3}\right)^{Z_{3,12}}\right\}, \tag{31}$$

where

$$Z_{i,jk} = \frac{1}{z_i(z_i - z_j)(z_i - z_k)} .$$

For $U_0 < U < U_1$, or $U > U_2$, one of the z_i, say z_3, is real and the other two are complex conjugates

$$Z_{1,2} = p \pm iq . \tag{32}$$

The integral then takes the form

$$-\Lambda\tau = \ln\left\{\left(\frac{z}{z_0}\right)^{\frac{-1}{z_3(p^2+q^2)}}\left(\frac{z-z_3}{z_0-z_3}\right)^{\frac{1}{z_3\left[(z_3-p)^2+q^2\right]}}\right.$$

$$\left.\left(\frac{(z-p)^2+q^2}{(z_0-p)^2+q^2}\right)^{\frac{z_3-2p}{2(p^2+q^2)\left[(p-z_3)^2+q^2\right]}}\right\} \tag{33}$$

$$+\frac{\left(p^2-z_3p-q^2\right)}{q(p^2+q^2)\left[(p-z_3)^2+q^2\right]}\left\{\tan^{-1}\left(\frac{z-p}{q}\right)-\tan^{-1}\left(\frac{z_0-p}{q}\right)\right\}.$$

Equations (31) and (33) give the form of the amplitude-time variation of the prism oscillations, and permit an estimate of the time to build up to a stationary oscillation. This must be defined arbitrarily as, for example, the time to build up from $\overline{Y}_0$ to $0.99\overline{Y}_{s_i}$, since the theory gives an asymptotic approach to the focus and limit cycles.

These equations for the transient behavior are also shown in Ref. [57] to produce very good agreement with wind tunnel measurements of the transient build up of galloping.

This example, and the results of other applications in Refs. [54] to [60], show that the quasi-static theory of galloping reviewed in this section gives an accurate model of the important phenomena. In the next section, two of these other applications are described, one for turbulent air flow and one for water flow.

Applications of Galloping Theory

The engineering importance of galloping arises from its occurrence in the natural wind, which is highly turbulent. The investigation of the effects of turbulence in the incident flow on galloping is therefore of interest, and a reasonable starting point for an investigation in the wind tunnel is the introduction of homogeneous isotropic turbulence into the flow incident on a bluff cylinder of constant cross section. The mean flow conditions therefore remain closely two dimensional, and the theory of the last section can again be applied, under the condition that the experimental static variation of C_y with α used in the theory is measured in the same turbulent wind in which the galloping is observed. Ref. [59] describes some early results from such an investigation. It is found that the effect of turbulence integral scale is small, within the range tested, but that there is a consistent effect of turbulence intensity, such that increasing intensity makes a soft oscillator weaker, and even stable for a sufficient intensity, while a hard oscillator tends to become soft. In all cases, the qualitative galloping characteristics predicted by the theory from the $C_y - \alpha$ curves were verified by the actual galloping behavior, and respectable quantitative agreement was also obtained, as shown in Fig. 9 for cylinders of square section and rectangular section with $d/h = 2$. These are both soft oscillators in smooth flow, and the figure shows theoretical curves and experimental points for turbulence intensities of 6.7 %, 9.0 %, and 12.0 %. Both sections become progressively weaker oscillators with increasing intensity, and the rectangular section in fact becomes completely stable in turbulence of 12.0 % intensity. Novak [56] obtains very similar results for finite cantilevered towers of rectangular section in turbulent shear flow.

All of the above behavior is consistent with the remarks in the section on physical parameters about the importance for galloping characteristics of shear layer reattachment on the afterbody sides, if increasing turbulence intensity is considered to cause reattachment on the windward side at a lower value of α for a given sectional shape. Some flow visualization experiments by Laneville, as yet unpublished, seem to confirm this speculation, which is also supported by the results of Vickery [62].

If the assumptions on which the quasi-static theory of galloping is based are valid, so that the flow past the cylinder at any instant during the galloping cycle is equivalent to the flow past the stationary cylinder at the effective angle of attack, then the vortex shedding frequency during galloping should be that for the equivalent stationary cylinder. Since S does not change much for a stationary cylinder over the small α-range typical of a galloping cycle ($-17^\circ < \alpha < +17^\circ$ for the square section), approximately the value of S for the stationary cylinder at $\alpha = 0$ should be measured during galloping. The author has verified this by making hot wire measurements of vortex shedding frequency behind a galloping square section cylinder, and found $S = 0.135$, the same value measured for the stationary cylinder, for $\mathring{U} \gtrsim 2(1/2\pi S)$. Under these conditions, Eq. (2) could be written

$$\mathring{\mathring{Y}} + Y = \mu f(\mathring{Y}) + nU^2\overline{C}_{y_v} \sin 2\pi SU\tau , \tag{34}$$

where the first term on the right side is the same as in Eq. (14), representing the galloping excitation and system damping, and the second term represents the wake vortex excitation at the stationary Strouhal frequency, assumed to be a large irrational multiple of the system frequency. $\overline{C}_{y_v}$ is now used to denote the amplitude of the vortex force coefficient, and it is assumed constant. (This last assumption is rather more dubious, since Vickery [62] shows a considerable variation of $\overline{C}_{y_v}$ with α for a square section).

In air flow, because $n = 0(10)^{-3}$, this direct vortex excitation has negligible effect, as the good agreement shown in the last section between experiment and a theory neglecting it testifies. (The galloping excitation also has very little effect in one cycle, but the nonlinear instability causes a slow build-up to a limit cycle over a large number of cycles). However, in water flow, n is much larger, and Eq. (34) suggests the interesting possibility of asynchronous quenching of the galloping, or autoperiodic oscillation, by the response to the vortex force, the heteroperiodic oscillation (see Ref.[23], Ch.24).

Santosham [58] has considered this possibility, again using the first approximation methods of Krylov and Bogoliubov, and developed the theory for square and rectangular sections. The method is to introduce

$$Y = X + Y_v \sin 2\pi SU\tau , \tag{35}$$

where

$$Y_v = \frac{nU^2\overline{C}_{y_v}}{I - (2\pi SU)^2} \tag{36}$$

the solution of Eq. (34) if $\mu = 0$. On substitution of Eq. (35) into Eq. (34), it becomes

$$\overset{\circ\circ}{X} + X = \mu f\{\overset{\circ}{X} + 2\pi SUY_v \cos 2\pi SU\tau\} . \tag{37}$$

Again the harmonic form of solution for $n = 0$ is assumed:

$$X = \overline{Y}\cos\tau, \qquad \overset{\circ}{X} = -\overline{Y}\sin\tau \tag{38}$$

and on multiplication of Eq. (37) by $\overset{\circ}{X}$, using Eqs. (38), it becomes

$$\frac{d\overline{Y}^2}{d\tau} = -2\mu\overline{Y}\sin\tau f\{-\overline{Y}\sin\tau + 2\pi SUY_v \cos 2\pi SU\tau\} . \tag{39}$$

The argument following Eq. (18) can be used again, and the right side of Eq. (39) averaged over one cycle. Since $f(\overset{\circ}{Y})$ is a polynomial of degree N, it can be put in the form

$$f\{\overset{\circ}{X} + 2\pi SUY_v \cos 2\pi SU\tau\} = f_0(\overset{\circ}{X}) + \sum_{l=1}^{N} f_l(\overset{\circ}{X}) \cos 2\pi SUl\tau . \tag{40}$$

Averaging Eq. (39) over a cycle, using Eq. (40), it becomes

$$\frac{d\overline{Y}^2}{d\tau} = -\frac{\mu}{\pi}\int_0^{2\pi} \overline{Y}\sin\nu f_0(-\overline{Y}\sin\nu)d\nu . \tag{41}$$

The analysis following Eq. (19) can now be repeated, with f_0 replacing f, and the existence and stability of stationary galloping oscillations investigated.

In air flow, Y_v is negligible away from resonance, so $Y = X$, and the result of the last section is obtained. In water flow it is possible to devise an elastic system for which $\mu = nA_1 \stackrel{\circ}{=} 0.1$, so that the requirements of the first approximation method for a weakly nonlinear system are not too seriously violated, and yet Y_v is large enough to be significant. Santosham's results for the square section, using the same f as in Eq. (14), and taking $\mu = 0.13$, $n = 0.05$, $\overline{C}_{y_v} = 1.8$ (using the value for $\alpha = o$ from Ref. [62]), indicate that $\overline{Y}_s$ would be reduced about 30 % at large values of U by the asynchronous quenching effect of the vortices. It would be of interest to seek experimental verification of this theory. Santosham has made some measurements of the galloping of a cantilevered cylinder of square section in a water tunnel, but not enough data were obtained for definite conclusions to be drawn. One difficulty in comparing theory and experiment is the need to include an estimate of the apparent mass of the water with the elastic system mass.

Some Possible Approaches

Vortex-Induced Vibrations as Nonlinear Oscillators

With vortex-induced vibrations, it is obvious that considerable progress must be made before the phenomena are fully understood, and certainly before they are adequately modelled dynamically and fluid dynamically. Since most of the observed effects are typical of nonlinear oscillations, it would appear to be a valuable aid to understanding to seek mathematical models that would explain the effects in terms of the known properties of nonlinear oscillators, quite independently of efforts to describe adequately the flow field that gives rise to the effects. This approach has been rewarding for galloping, and Hartlen and Currie have made a promising beginning in Ref.[26] for vortex-induced vibrations.

Thus by using a linear oscillator (left side of Eq. (2)) coupled to a simple nonlinear oscillator (left side of Eq. (9)) they have demonstrated a system capable of mutual synchronization, and closely representative of the lock-in observed experimentally. Figure 10 shows $\Omega = \omega_c/\omega_n$ or ω_{v_s}/ω_n as a function of $\Omega_0 = 2\pi SU$ as given by Eq. (11) for $n = 0.003$, $\beta = 0.002$, and two values of coupling parameter D. When $D = 1$, the right-hand curve is obtained. The solid and dashed sections of the curve are intended to correspond to conditions for stable and unstable limit cycles of the fluid oscillator, following the notation of Figs. 7 and 8, so that one might expect a frequency hysteresis 123, 456, 21. In fact, the experimental variation for the circular cylinder shows the vortex frequency ω_{v_s} following the trend

123, 45, and shows the mutual synchronization in the range 23, but the trend 46 is not observed, the return jump of ω_{v_s} occurring at or near 4, and ω_c does not jump at all, remaining near ω_n.

Nevertheless, the model predicts frequency behavior encouragingly like that of the circular cylinder. Further, if D = - 1, Eq. (11) shows $\Omega \geqslant \Omega_0$, and the left hand curve of Fig. 10 is obtained, from which one might expect the hysteresis loop 789, 11 12 11 10, 87. Actual behavior of ω_{v_s} for the D-section cylinder [28] seems to follow the part 12 11 10, 8 7 in both directions, and shows the mutual synchronization over the range 10 11. A study of Eqs. (2) and (9) with a more elaborate coupling term in Eq. (9) might clarify some of these discrepancies.

The actual oscillation hysteresis loops observed by Feng for the circular cylinder (Fig.2) occur in the middle of the lock-in range, so that $\omega_{v_s} = \omega_c \doteq \omega_n$. This suggests that it might be useful to think qualitatively of C_y as an autonomous nonlinear oscillator, to see if the behaviour of $\overline{C}_y$ in Fig. 2 can be identified. (It is worthwhile to note that the jump condition originates in the fluid system, and therefore with C_y, and not in the cylinder elastic system. This can be seen in Fig. 23b of Ref. [28] (also reproduced as Fig. 3b of Parkinson [63]), an oscilloscope photograph of fluctuating surface pressure at the transverse diameter, proportional to $\overline{C}_y$, and of cylinder amplitude, during the jump corresponding to 34 in Fig. 11. $\overline{C}_y$ is seen to decrease abruptly, in less than 1 second, and this is followed by a more gradual decrease of $\overline{Y}$ over several seconds.) Fig. 11 which shows the variation of $\overline{C}_{y_s}$ with U over the lock-in range of Fig. 2, indicates that in these terms the curves represent another example of the bifurcations of section 4 in nonlinear oscillations. The hysteresis loop is 123, 456, 21. From U_0 to U_1, the system is of the type $\mathcal{U}\mathcal{S}$, with oscillation building up from an unstable focus to a stable limit cycle. U_1 is a bifurcation point, and the system becomes of type $\mathcal{U}\mathcal{S}\mathcal{U}\mathcal{S}$, in the range from U_1 to U_2, with stable limit cycles 23 and 64 separated by unstable limit cycle 63. U_2 is again a bifurcation point, and the system again becomes of type $\mathcal{U}\mathcal{S}$ in the range from U_2 to U_3.

Although it is difficult to make general statements about coupled nonlinear oscillators, it seems likely that a higher order polynomial for the nonlinearity in Eq. (9) would be capable of generating solutions for $\overline{C}_{y_s}$ like Fig. 11 (and, of course, for $\overline{Y}_s$ and Φ through the solution of Eqs. (2) and the modified Eq. (9)).

Another puzzling feature of vortex-induced vibrations that would merit investigation in terms of properties of nonlinear oscillators is the relatively large-amplitude oscillation of the lightly damped circular cylinder persisting at flow velocities beyond the lock-in range. Thus, Feng's results in Fig. 2 show $\overline{Y}_s$ decreasing from 0.25 to negligible values as U increases from 1.1, the end of the lock-in range, to about 1.3. The vortex frequency, and therefore the frequency of pressure loading, has reverted to ω_v, the stationary Strouhal value, but the cylinder continues to oscillate at $\omega_c \doteq \omega_n$ so it is not a simple forced vibration at the exciting frequency.

It might be regarded as the response of a lightly-damped system to random turbulent excitation, which also happens to contain a dominant frequency ω_v, except that the values of $\overline{Y}_s$ are too large, and completely unmodulated, whereas the pressure loading amplitude has a large random modulation. Fig. 12 is an accurate copy of Fig. 37e from Ferguson [27], an oscilloscope photograph corresponding to these conditions, in which the upper trace shows Y oscillating at 9.0 cps, while the lower trace shows fluctuating surface pressure at the transverse diameter oscillating at 13.0 cps. The extremes of pressure amplitude modulation are greater than those evident in the few cycles of Fig. 12, and in fact have a ratio of maximum to minimum of about 10.

Perhaps these effects could be analysed as an example of asynchronous excitation (Ref.[23], Ch.24), a phenomenon common enough in electrical circuits, but rare in mechanical systems. In these terms, a heteroperiodic oscillation (frequency ω_v) impressed on the system, releases an autoperiodic oscillation (frequency ω_c), where ω_v/ω_c is irrational, and the system would be a hard oscillator in the absence of the heteroperiodic oscillation. Such a model would presumably require the simulation of some fluid dynamic effect (in addition to the pressure loading at frequency ω_v) to make the system equivalent to a hard autonomous nonlinear oscillator of a suitable type.

Flow Field Models

As was indicated in the previous section on flow field models, potential flow models of wake vortex systems can be informative and realistic. However, although they are simpler and, in some ways, more useful than the corresponding numerical solutions of the Navier-Stokes equations, potential flow models like those of Gerrard [46], Sarpkaya [47], and Ujihara [50] are still rather too complex for ready evaluation of the effects they display. Unfortunately, a reduction in complexity seems to require an increase in empiricism, so there is no guarantee of an increase in useful information from a simpler model. Nevertheless, there is probably something to be gained by exploring the possibilities, for models of vortex-induced vibration, of improved versions of a simple model suggested by B. Etkin and applied by D.M. McGregor [64] to the problem of the time-dependent flow field near a stationary circular cylinder in a uniform incident flow.

In Etkin's model, a single wake potential vortex like that of Fig. 4 is placed on the wake center-line (x-axis), and it and its image are given the harmonic variation of circulation

$$\Gamma = \Gamma_0 \sin \omega_v t . \tag{42}$$

The vortex, located at $x = c$, does not move with the flow, and the time dependence enters only through the variation of circulation. If the flow is analyzed in the complex plane $z = x + iy$ its complex potential is

$$F(z) = \varphi + i\psi = V\left(z + \frac{h^2}{4z}\right) + \frac{i\Gamma}{2\pi} \ln(z - c) - \frac{i\Gamma}{2\pi} \ln\left(z - \frac{h^2}{4c}\right) \tag{43}$$

and the complex velocity is

$$w(z) = u - iv = V\left(1 - \frac{h^2}{4z^2}\right) + \frac{i\Gamma}{2\pi(z - c)} - \frac{i\Gamma}{2\pi(z - h^2/4c)} . \tag{44}$$

The pressure at any point can be obtained from Eq. (12), and the transverse force coefficient C_y can be obtained from the unsteady form of Blasius' equation:

$$\frac{\rho}{2} V^2 h(C_x - iC_y) = \frac{i\rho}{2} \oint w^2(z)dz + i\rho \oint \frac{\partial \varphi}{\partial t} d\bar{z} . \qquad (45)$$

The result can be written

$$C_y = -\left(\frac{2\Gamma_0}{Vh}\right)\left(1 - \frac{h^2}{4c^2}\right) \sin \omega_v t - \left(\frac{2\Gamma_0}{Vh}\right) \pi S \left(1 - \frac{h}{2c}\right) \cos \omega_v t . \qquad (46)$$

This should be compared with the commonly used, but incorrect, formula for C_y, based on the Kutta-Joukowsky Law:

$$C_y = -\left(\frac{2\Gamma_0}{Vh}\right) \sin \omega_v t . \qquad (47)$$

The differences arise from the time-dependent term in Eq. (45), and from the proximity of the wake vortex. Of course, Eqs. (43) or (44) do not represent a "correct" model of the flow either, but at least they lead to results consistent with the assumptions made. Etkin's model displays some of the features of a wake vortex system in the presence of the generating cylinder, but does not include any simulation of the alternate vortex generation from the two shear layers.

An improvement therefore might be to use two stationary wake vortices (and their images) symmetrically located on either side of the wake center line at $z = c \pm ib/2$, and with harmonic circulations 180° out of phase:

$$\Gamma_{U,L} = \frac{1}{2} \Gamma_0 (1 \pm \cos \omega_v t) . \qquad (48)$$

The extensions of Eqs. (43) and (44) are obvious, and again Eqs. (12) and (45) can be used to determine the pressure and C_y. Madderom [65] has used this model to calculate the distribution on the cylinder surface of the component of fluctuating pressure oscillating at the fundamental Strouhal frequency ω_v, for various values of $2\Gamma_0/(Vh)$, b/h, c/h. In

Fig. 13 one such calculation (with $2\Gamma_0/(Vh) = 0.53$, $b/h = 0.93$, $c/h = 2.0$) is compared with experimental values by McGregor [64] and Gerrard [66], at Reynolds numbers of $4.4(10)^4$ and $6.3(10)^4$ respectively, and with the corresponding calculation from Etkin's model of Eqs. (43) and (44) (with $2\Gamma_0/(Vh) = 0.62$, $c/h = 1.0$). The ordinate in each case is the RMS values of the pressure coefficient

$$C_p' = \frac{p'(\text{rms})}{\rho V^2/2} .$$

Experimental values of C_p' tend to increase with Reynolds number in the range of Fig. 13, so it is natural that Gerrard's values are somewhat higher than McGregor's. Nevertheless, there is considerable scatter in such measurements, and Madderom's curve can be considered to follow the trend of the data quite respectably, better than the curve for Etkin's one-vortex model.

What is learned from these models? Certainly there is considerable empiricism put into them. Parameters ω_v and b/h were given their measured values, based on $S = 0.2$ and observations by Ferguson [27]. The quantity c/h in Madderom's calculation, and $2\Gamma_0/(Vh)$ in both calculations, were chosen to give good agreement with the C_p' data. It is interesting to note that, in order to produce this agreement, the vortices used were considerably weaker than in reality, and in the case of the two-vortex model were placed further downstream. Another point to note is the phase difference between C_y on the cylinder, as given by Eq. (46), and the vortex formation cycle, as simulated by Eq. (42).

The chief motivation for suggesting extending the 1-or 2-vortex type of model to problems of vortex-induced vibration is their demonstrated capability of producing realistic pressure loadings on the cylinder within the framework of a tractable analytic method. More empiricism would have to be introduced, to account for the effects of the cylinder motion on the vortex system, but useful information could probably be gained. Perhaps such a study could converge on the parallel study of vortex-induced vibrations as nonlinear oscillators, suggested in the previous section, and lead to a valuable combined result.

Finally, what if anything, should be done about a flow field model for galloping? Here the need is not so great, because of the success of the quasi-static model, requiring loading data only for the equivalent stationary cylinder. However, this data can only be obtained experimentally at present, so a theory providing it would be worthwhile. Just as the circular cylinder serves as the prototype for vortex-induced vibration studies, so the rectangular cylinder should be the prototype for galloping studies, because of its geometric simplicity.

The pressure loading that causes galloping of a rectangular section is on the two sides of the afterbody (see inset of Fig. 1), and this is what the static theory would need to predict. On the windward side, as α is increased from zero, the pressure loading is that of a separated flow which eventually reattaches at the trailing edge and forms a separation bubble of decreasing size as α is increased further. On the leeward side, the flow remains separated throughout the positive α range. The pressure levels at the two sides are linked through the base pressure variation, and it in turn depends on the general properties of the near wake, including the vortex formation. The problem is obviously complex, but reasonable approximations can probably be made for the pressure distributions (constant suction on the leeward side, constant suction over the front part, with a linear increase in pressure over the rear part, on the windward side at a given α), and valuable insight into the parameters that would constrain any theoretical model can be gained from Tani [67] and Roshko [68, 69].

References

1. Cooper, K.R., Wardlaw, R.L.: Aeroelastic Instabilities in Wakes. Proc. 3rd Int. Conf. Wind Effects on Buildings and Structures, Tokyo, IV, 1-1 to 1-9 (1971).

2. Mair, W.A., Maull, D.J.: Aerodynamic Behaviour of Bodies in the Wakes of Other Bodies. Phil. Trans. Roy. Soc. London A 269, 425-437 (1971).

3. Davenport, A.G.: The Response of Six Building Shapes to Turbulent Wind. Phil. Trans. Roy. Soc. London A, 269, 385-394 (1971).

4. Bearman, P.W.: N.P.L. Aero Report 1296 (1969).

5. Theodorsen, T.: General Theory of Aerodynamic Instability and the Mechanism of Flutter. NACA Report 496 (1935).

6. Scanlan, R.H.: Studies of Suspension Bridge Deck Flutter Instability, AIAA Paper 69-744, CASI/AIAA Subsonic Aerodynamics Meeting, Ottawa (1969).

7. Smith, I.P.: The Aeroelastic Stability of the Severn Suspension Bridge. N.P.L. Aero Report 1105 (1964).

8. Simpson, A.: Stability of Subconductors of Smooth Circular Cross-Section. Proc. Inst. Elect. Engrs. 117, 4, 741-750 (April 1970).

9. Davis, D.A., Richards, D.J., Scriven, R.A.: Investigation of Conductor Oscillation on the 275 kV Crossing over the Rivers Severn and Wye. Proc. Inst. Elect. Engrs. 110, 1, 205 (1963).

10. Johns, D.J., Allwood, R.J.: Wind Induced Ovalling Oscillations of Circular Cylindrical Shell Structures such as Chimneys. Symp. Wind Effects on Buildings and Structures, Loughborough 2, 28.1.-28.17. (1968).

11. Chen, Y.N.: Fluctuating Lift Forces of the von Kármán Vortex Streets on Single Circular Cylinders and in Tube Bundles, Part 1 - The Vortex Street Geometry of the Single Circular Cylinder, Part 2 - Lift Forces of Single Cylinders, Part 3 - Lift Forces in Tube Bundles. ASME Papers 71-Vibr. - 11, 12, 13 (1971).

12. Brooks, N.P.H.: Experimental Investigation of the Aeroelastic Instability of Bluff Two-Dimensional Cylinders. M.A.Sc. Thesis, U. British Columbia, 1960.

13. Smith, J.D.: An Experimental Study of the Aeroelastic Instability of Rectangular Cylinders. M.A.Sc. Thesis, U. British Columbia, 1962.

14. Roshko, A.: On the Drag and Shedding Frequency of Two-Dimensional Bluff Bodies. NACA TN 3169 (1954).

15. Bearman, P.W., Trueman, D.M.: An Investigation of the Flow around Rectangular Cylinders. Imperial College Aero Report 71-15 (1971).

16. Nakaguchi, H., Hashimoto, K., Muto, S.: An Experimental Study in Aerodynamic Drag of Rectangular Cylinders. J. Japan Soc. Aero and Space Sciences 16, 168 (1968).

17. Bostock, B.R., Mair, W.A.: Pressure Distributions and Forces on Rectangular and D-Shaped Cylinders. Aero Quarterly, 1-6 (February 1972).

18. Gerrard, J.H.: The Mechanics of the Formation Region of Vortices Behind Bluff Bodies. J. Fluid Mech. 25, 401-413 (1966).

19. Wood, C.J.: An Examination of Vortex Street Formation by Particle Tracers, Euromech 17, Cambridge 1970. (See Mair, W.A., Maull, D.J.: Bluff Bodies and Vortex Shedding - a Report on Euromech 17. J. Fluid Mech. 45, 209-224 (1971)).

20. Abernathy, F.H., Kronauer, R.E.: The Formation of Vortex Streets. J. Fluid Mech. 13, 1-20 (1962).

21. Naudascher, E.: From Flow Instability to Flow-Induced Excitation. Proc. ASCE, Hy 4, 15-40 (July 1967).

22. Toebes, G.H.: Fluidelastic Features of Flow around Cylinders. Proc. Int. Res. Sem. Wind Effects on Building and Structures, Ottawa 1967.

23. Minorsky, N.: Nonlinear Oscillations, van Nostrand 1962, p. 436.

24. di Silvio, G.: Self-Controlled Vibration of Cylinder in Fluid Stream. Proc. ASCE, EM 2, 347-361 (April 1969).

25. Halle, H.: Proposed Analytical Model for the Cross-Flow-Induced Vibrations of a Circular Cylinder. Proc. Conf. Flow-Induced Vibrations in Reactor System Components, Argonne Lab., May 1970, pp. 248-269.

26. Hartlen, R.T., Currie, I.G.: Lift-Oscillator Model of Vortex-Induced Vibration. Proc. ASCE, EM 5, 577-591 (October 1970).

27. Ferguson, N.: The Measurement of Wake and Surface Effects in the Subcritical Flow Past a Circular Cylinder at Rest and in Vortex-Excited Oscillation. M.A.Sc. Thesis, U. British Columbia, 1965.

28. Feng, C.C.: The Measurement of Vortex Induced Effects in Flow Past Stationary and Oscillating Circular and D-Section Cylinders. M.A.Sc. Thesis, U. British Columbia, 1968.

29. Milne-Thomson, L.M.: Theoretical Hydrodynamics, 5th ed., McMillan 1968, p. 200.

30. Landweber, L.: Flow About a Pair of Adjacent, Parallel Cylinders Normal to a Stream: Theoretical Analysis. Taylor Model Basin Report 485 (1942).

31. Steinman, D.B.: Problems of Aerodynamic and Hydrodynamic Stability. Bull. 31, Iowa U. Studies in Engrg. (1946).

32. Templin, J.: Private Communication, 1968.

33. Smirnov, L.P., Pavlihina, M.A.: Vortical Traces for Flow around Vibrating Cylinders. Presented at the I.M.A. Communications Meeting, Moscow, October 1957.

34. Bishop, R.E.D., Hassan, A.Y.: The Lift and Drag Forces on a Circular Cylinder Oscillating in a Flowing Fluid. Proc. Roy. Soc. A, 277 (1964).

35. Jones, G.W.: Unsteady Lift Forces Generated by Vortex Shedding about a Large, Stationary and Oscillating Cylinder at High Reynolds Number. Symp. Unsteady Flow, ASME, Philadelphia, May 1968.

36. Tanaka, H., Takahara, S.: Unsteady Air Forces Acting on a Vibrating Circular Cylinder. Proc. 19th Japan National Congress App. Mech., 1969.

37. Nakamura, Y.: Vortex Excitation of a Circular Cylinder Treated as a Binary Flutter. Reports of Res. Inst. App. Mech., Kyushu U., XVII, 59 (1969).

38. Birkhoff, G.: Formation of Vortex Streets. J. App. Phys. 24, 1, 98 (1953).

39. Swanson, W.M.: The Magnus Effect: A Summary of Investigations to Date. J. Basic Engrg., Trans. ASME D 83, 461 (1961).

40. Tietjens, O.G.: Applied Hydro- and Aero-Mechanics, Dover, 1957, p. 286.

41. Kawamura, S.: A Design Criterion of Wind Loads of High Steel Stacks. Proc. 3rd Int. Conf. Wind Effects on Buildings and Structures, Tokyo, 1971, IV, 707-716.

42. Jordan, S.K., Fromm, J.E.: Oscillatory Drag, Lift, and Torque on a Circular Cylinder in a Uniform Flow. Phys. Fluids 15, 3, 371-376 (March 1972).

43. von Kármán, T.: Über den Mechanismus des Widerstands, den ein bewegter Körper in einer Flüssigkeit erfährt. Göttinger Nachrichten, Math. Phys. Kl. 509-19 (1911).

44. Bearman, P.W.: On Vortex Street Wakes. J. Fluid Mech. 28, 4, 625-641 (1967).

45. Kronauer, R.E.: Predicting Eddy Frequency in Separated Wakes. Presented at IUTAM Symp. on Concentrated Vortex Motions in Fluids, U. of Michigan, Ann Arbor, July 1964.

46. Gerrard, J.H.: Numerical Computation of the Magnitude and Frequency of the Lift on a Circular Cylinder. Phil. Trans. Roy. Soc. A 261, 137-162 (January 1967).

47. Sarpkaya, T.: An Analytical Study of Separated Flow About Circular Cylinders. ASME Symp. Unsteady Flow, 1968, Paper 68-FE-15.

48. Shioiri, J.: Synchronization Phenomenon in Vortex Shedding and its Role in Vortex Induced Oscillation of Structures. Proc. 3rd Int. Conf. Wind Effects on Buildings and Structures, Tokyo, 1971, IV, 677-686.

49. Koopmann, G.H.: The Vortex Wakes of Vibrating Cylinders at Low Reynolds Numbers. J. Fluid Mech., 28, 3, 501-12 (1967).

50. Ujihara, B.H.: Hydroelastic Analysis of a Circular Cylinder. North American Rockwell Space Division Rep. SD68-996, 1968.

51. den Hartog, J.P.: Transmission Line Vibration due to Sleet. Trans. AIEE 49, 444 (1930).

52. Sisto, F.: Stall Flutter in Cascades. J. Aero. Sci. 20, 9, 598-604 (September 1953).

53. Scruton, C.: The Use of Wind Tunnels in Industrial Aerodynamic Research. N.P.L. Aero Report 411 (1960).

54. Novak, M.: Aeroelastic Galloping of Prismatic Bodies. Proc. ASCE, EM 1, 115-142 (February 1969).

55. Novak, M., Davenport, A.G.: Aeroelastic Instability of Prisms in Turbulent Flow. Proc. ASCE, EMI, 17-39 (February 1970).

56. Novak, M.: Galloping Oscillations of Prismatic Structures. Proc. ASCE, EM 1, 27-46 (February 1972).

57. Parkinson, G.V., Smith, J.D.: The Square Prism as an Aeroelastic Nonlinear Oscillator. QJMAM XVII, 2, 225-239 (May 1964).

58. Santosham, T.V.: Force Measurements on Bluff Cylinders and Aeroelastic Galloping of a Rectangular Cylinder. M.A.Sc. Thesis, U. British Columbia, 1966.

59. Laneville, A., Parkinson, G.V.: Effects of Turbulence on Galloping of Bluff Cylinders. Proc. 3rd Int. Conf. Wind Effects on Buildings and Structures, Tokyo, 1971, IV, 787-798.

60. Slater, J.E.: Aeroelastic Instability of a Structural Angle Section. Ph. D. Thesis, U. British Columbia, 1969.

61. Glauert, H.: The Rotation of an Aerofoil about a Fixed Axis. R. and M. 595 (1919).

62. Vickery, B.J.: Fluctuating Lift and Drag on a Long Cylinder of Square Cross-Section in a Smooth and in a Turbulent Stream. NPL Aero Report, 1146 (1965).

63. Parkinson, G.V.: Wind-Induced Instability of Structures. Phil. Trans. Roy. Soc. London A, 269, 395-409 (1971).

64. McGregor, D.M.: An Experimental Investigation of the Oscillating Pressures on a Circular Cylinder in a Uniform Stream. UTIAS TN 14 (1957).

65. Madderom, P.: Two-Vortex Potential Model for Turbulent Flow Past Circular Cylinders. UBCME Dept. Unpublished Note, April 1966.

66. Gerrard, J.H.: An Experimental Investigation of the Oscillating Lift and Drag of a Circular Cylinder Shedding Turbulent Vortices. J. Fluid Mech., 11, 244-256 (1961).

67. Tani, I.: Low-Speed Flows Involving Bubble Separations. Progress in Aero Sci., 5, 70-103 (1964).

68. Roshko, A., Lau, J.C.: Some Observations on Transition and Reattachment of a Free Shear Layer in Incompressible Flow. Proc. Heat Trsfr. and Fluid Mechn. Inst., Stanford, 1965, pp. 157-167.

69. Roshko, A.: A Review of Concepts in Separated Flow. Proc. 1st Canadian Congress Appl. Mech. 3, 1967, pp. 81-115.

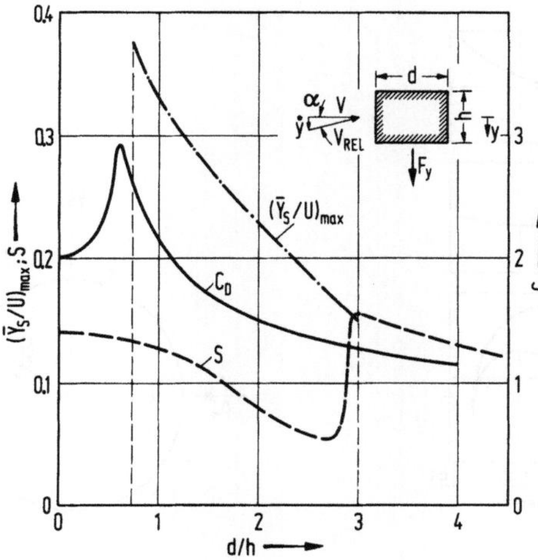

Fig.1. Effects of rectangular section afterbody length [12 to 17].

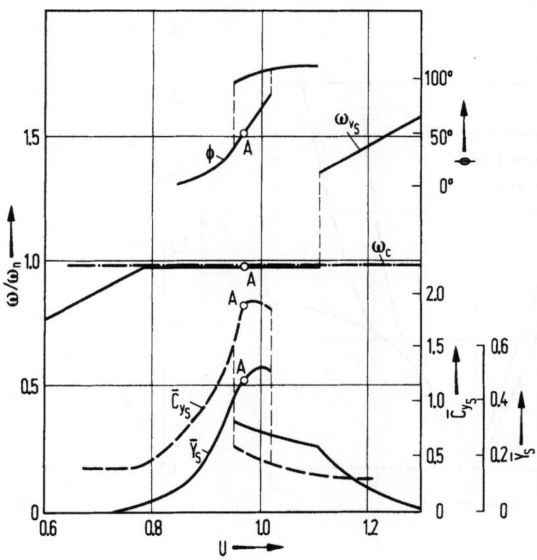

Fig.2. Experimental results for vortex-induced vibration of circular cylinder [28].

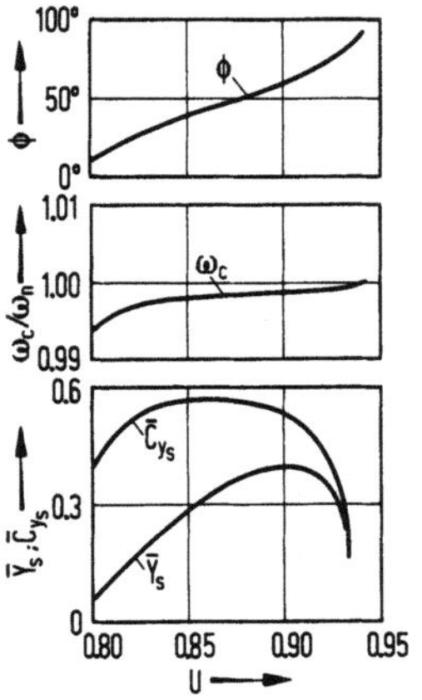

Fig. 3. Theoretical predictions for vortex-induced vibration of circular cylinder [26].

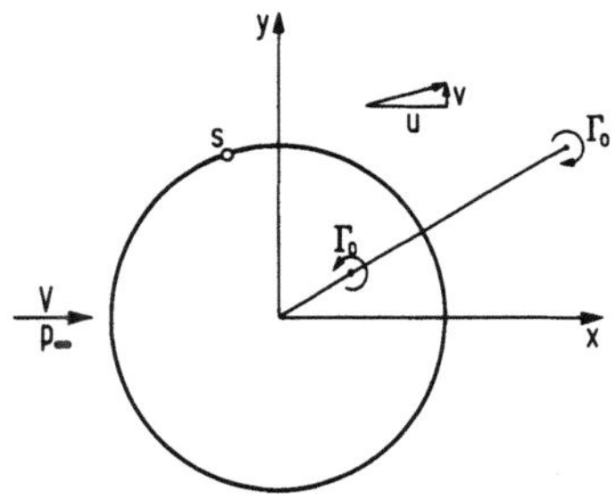

Fig. 4. Potential flow modelling of wake vortex system.

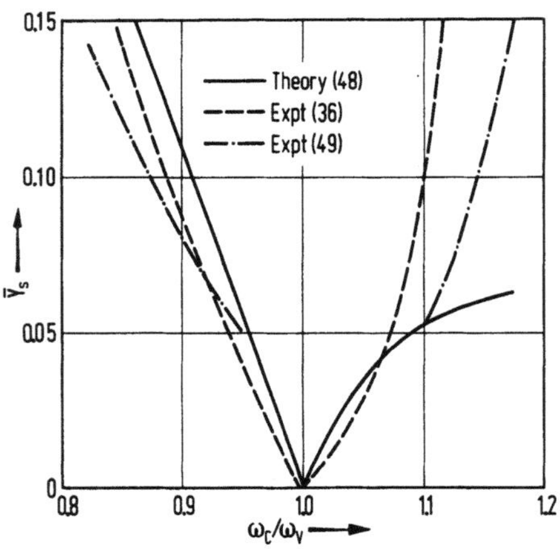

Fig. 5. Minimum cylinder amplitude for synchronization [48].

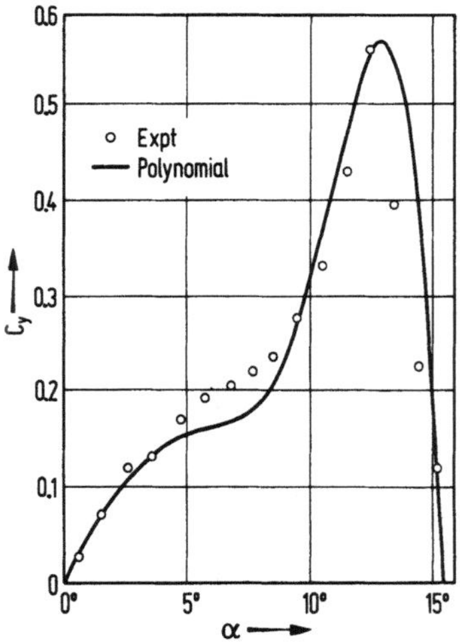

Fig. 6. Transverse force vs angle of attack for square cylinder [57].

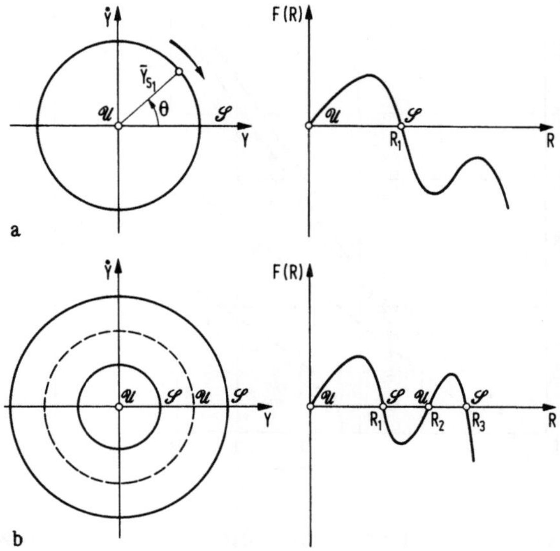

Fig.7. Phase planes and corresponding functions F(R) [57].

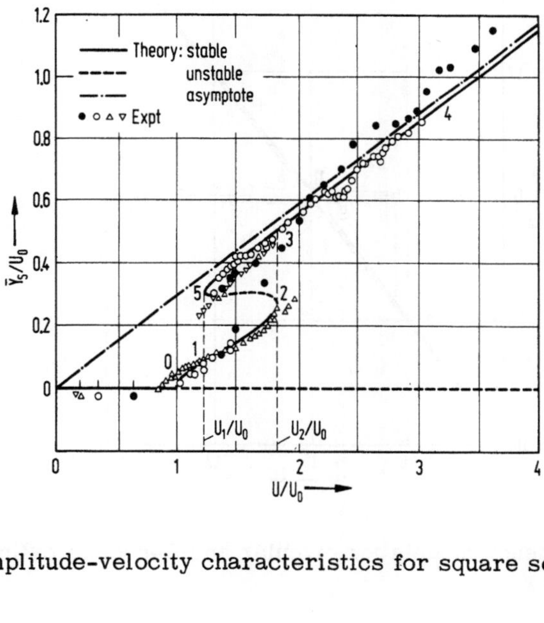

Fig.8. Amplitude-velocity characteristics for square section [57].

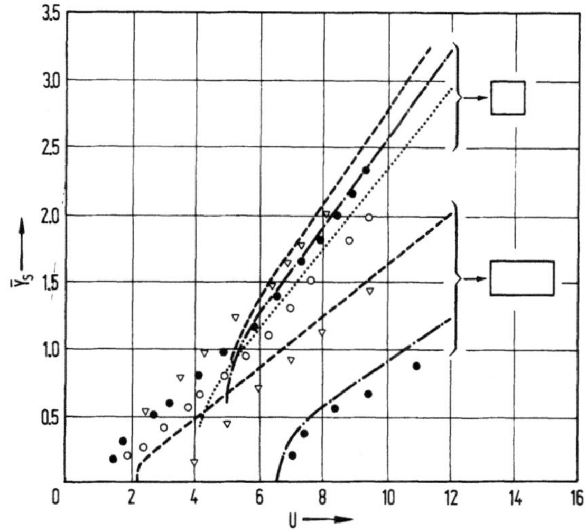

Fig.9. Amplitude-velocity characteristics in turbulent flow [59].

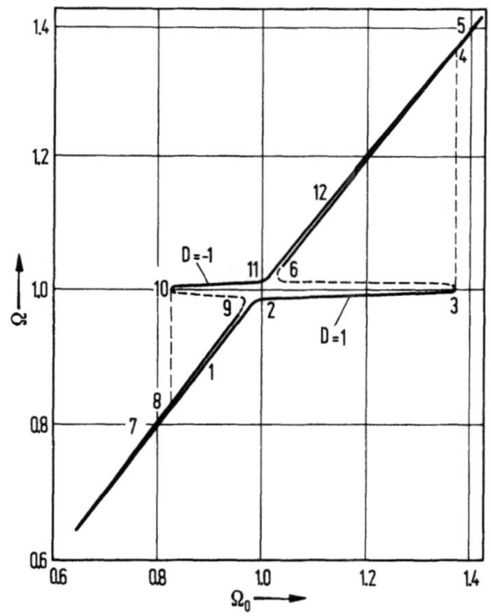

Fig.10. Frequency models for circular and D-section cylinders.

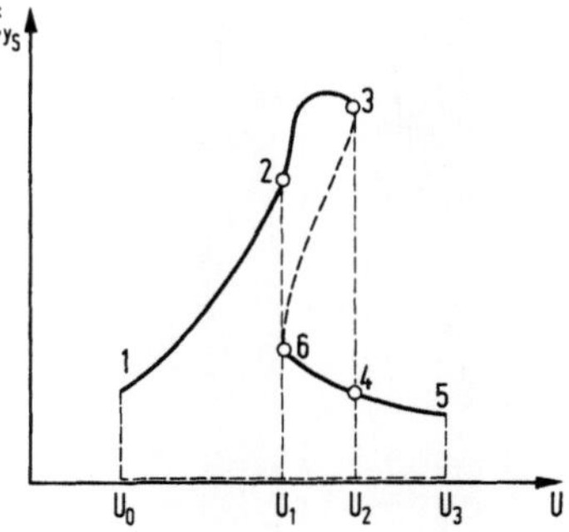

Fig. 11. Vortex-induced force on circular cylinder as nonlinear oscillator.

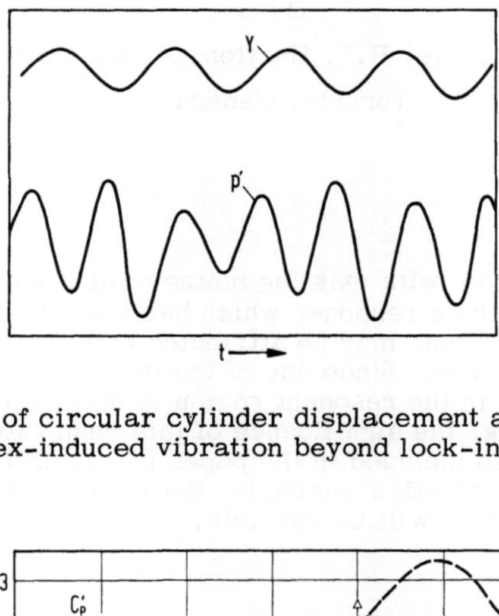

Fig. 12. Traces of circular cylinder displacement and surface pressure in vortex-induced vibration beyond lock-in range [27].

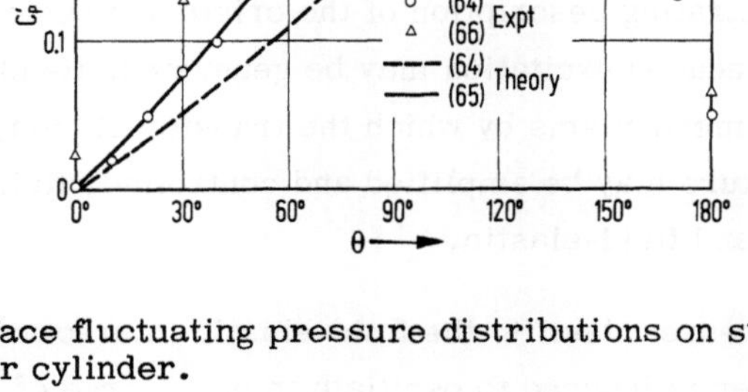

Fig. 13. Surface fluctuating pressure distributions on stationary circular cylinder.

CONTRIBUTIONS

The Response of Circular Cylinders to Vortex Shedding

By

I.G. Currie, R.T. Hartlen, W.W. Martin

Toronto, Canada

Summary

It is shown, consistent with existing mathematical models, that the double-valued amplitude response which has been observed for vortex-excited cylinder motions, may be attributed to a small nonlinearity in the mechanical spring. Since one of the two possible amplitudes which are observed in the resonant region is much larger than that for the linear spring, the significance of this result to practical situations is great. Also included in the paper is a stability diagram which predicts, by analytical methods, the range of wind speeds over which a given structure will be unstable.

Introduction

Flow-induced excitation is a consequence of energy transfer from a fluid to a structure around which the fluid is flowing. Naudascher [1] provided a penetrating description of the origin and basic mechanisms by which flow-induced excitation may be generated. He classified three feedback mechanisms by which the transfer of energy from a fluid to a structure may be amplified and controlled; fluid-dynamic, fluid-resonant and fluid-elastic.

This paper addresses itself to the fluid-elastic interaction in which a circular cylinder is induced to oscillate in one degree of freedom by vortex shedding. The equations governing such a situation are the dynamic equation of motion for the cylinder and the Navier-Stokes equations. This system of equations has proved to be mathematically intractible so that mathematical models have been sought which produce cylinder responses which match experimental observations.

This paper discusses existing mathematical models which attempt to match the experimental observations of physical systems as described above and to extend these results to new situations. An explanation is presented for the apparent discrepancy between these mathematical models and the experimental observation of a double-valued response in the cylinder motion. A stability diagram, of the type introduced by Scruton [2], is also presented and the role of aerodynamic damping is discussed.

Review of Existing Mathematical Models

At least five mathematical models for the response of cylinders to vortex shedding have recently appeared in the literature. One of these, due to Di Silvio [3], is based on the observation that the lift and drag of the cylinder and the vortex shedding frequency for a stationary cylinder may be related to the wake width. In the dynamic model, the wake width is assumed to be controlled by the position of the edge of the cylinder at the so-called "vortex-origin time" which, in turn, is controlled by the cylinder motion. The gross-parameter equations which result from this assumption yield, over a range of damping values, three possible amplitudes of vibration for a given wind speed. Two of these solutions increase in amplitude without limit as the wind speed increases while the third solution decreases in amplitude. There appears to be no theoretical criterion, however, for the system to jump from one solution to another as does the physical system.

A second model, due to Nakamura [4], is based on Birkhoff's oscillator model for the dead air region behind a circular cylinder. The magnitude of the lift force acting on the cylinder is taken to be proportional to the angular displacement of the dead air region with a constant empirical factor of proportionality being employed. The equations of motion are then taken to be the equation of an undamped linear oscillator, which is forced due to a Magnus effect, and a similar equation governing the angular displacement of the dead air region, which is based on Birkhoff's model. Solutions to these equations showed that a binary response was possible over a limited range of flow velocities near the resonant velocity.

Halle [5], in his model, considered the fluid to be a one-degree of freedom linear oscillator which is driven by a sinusoidally varying lift force which is in phase with the cylinder displacement. A damping term is included in the fluid system and also in the equation of motion for the cylinder. The system is assumed to be energized by white noise which is characterized by a displacement spectral density function which varies as the fourth power of the flow velocity. The lift force which drives the fluid is assumed to vary as the square of the velocity.

The results obtained by Halle from his model showed a "lock-in" frequency region, and good qualitative agreement with experimental values of the phase angle between the lift force and the cylinder motion, vibration amplitude, and cylinder frequency. It was also observed that by adjusting the magnitude of the fluid-cylinder damping factor, the amplitude response curve could be made to bend over causing an "overhang" which can lead to a double-valued response. The author indicated that he would pursue this possibility.

A fourth mathematical model, introduced by two of the authors [6], utilizes a self-excited oscillatory equation to describe the instantaneous lift coefficient which acts on the cylinder. This equation, for which the Van der Pol equation was chosen, together with the dynamic equation of motion for the cylinder, yields a pair of coupled ordinary differential equations which govern the cylinder displacement and lift coefficient. These equations yield three possible solutions for the vibration frequency and amplitude of the cylinder over a range of speeds near the critical. However, at some upper limit of the wind speed, two of these frequencies become a complex pair so that a jump to the third solution is mandatory.

Another model, due to Kawamura [7], utilizes the equation of motion of a linear, damped oscillator in which the lift force on the cylinder is assumed to vary sinusoidally with time. From experimental data, the vibration frequency is expressed as a polynomial in the dimensionless amplitude of the cylinder motion. The resulting equation is then solved for the amplitude of vibration and phase lag between the cylinder oscillations and those of the lift force. For small values of the damping, the amplitude response predicted by this system is skewed in the vicinity of the resonant frequency.

Justification of the Wake-Oscillator Model

The wake oscillator concept is the basis of the mathematical model presented in Ref. [6]. The pair of equations which result from this concept are reproduced here for discussion and future reference.

$$x_r'' + 2\zeta x_r' + x_r = a\omega_0^2 C_L , \tag{1}$$

$$C_L'' - \alpha\omega_0 C_L' + \frac{\gamma}{\omega_0} (C_L')^3 + \omega_0^2 C_L = b x_r' . \tag{2}$$

Equation (1) is the dynamic equation of motion for the cylinder and is exact for viscous damping and a linear restoring spring. The external force which drives the cylinder is written in terms of the instantaneous lift coefficient C_L which, in turn, is supposed to satisfy the Van der Pol equation in which the lift coefficient is linearly affected by the cylinder velocity. In these two equations, the dependent variables are the dimensionless cylinder displacement x_r and the instantaneous lift coefficient C_L while the independent variable is the dimensionless time, derivatives with respect to which are denoted by primes. The parameters in Eq. (1) are the damping factor ζ, the mass parameter a and the velocity parameter ω_0, all of which will be known for a given cylinder immersed in a given flow field. The parameters in Eq. (2) are the Van der Pol coefficients α and γ and the interaction parameter b. The ratio α/γ and the order of magnitude of α may be estimated from the steady state and transient behavior of the fluctuating lift, respectively; b is selected to fit experimental response data.

The justification for Eq. (2) is that it exhibits the correct behavior of the lift force on a stationary cylinder and the solutions to Eqs. (1) and (2) agree generally with experimental observations. Since Eq. (2) was originally formulated, some further general support for the wake-oscillator concept has materialized.

Local velocity fluctuations near a vibrating cylinder are consistent with the wake-oscillator concept. Toebes [8] measured the velocity fluctuations in the boundary layer and in the near flow field of an externally driven circular cylinder. Among the conclusions of this work are the following;

(1) "... the unsteady flow field might be likened to a non-linear oscillator that can be excited both at subharmonic and superharmonic frequencies".

(2) Cylinder oscillation such that $f_r = 1.0$ associates with much stronger vortices than any other f_r ratio.

(3) The amount of periodic shifting of the two separation lines on the cylinder ... increases strongly under conditions of forced vortex formation.

The motion of the separation point on a vibrating cylinder is also consistent with the wake-oscillator concept. Mei and Currie [9] made such measurements on a circular cylinder which was performing aeolian vibrations. The amplitude and phase of the motion of the separation point around the surface of the cylinder, relative to the mean position of approximately 82°, were determined as functions of the cylinder amplitude and frequency. The motion of the separation point was observed to have a frequency-dependent behavior, characterized primarily by an amplitude maximum at $V/fd \simeq (1.15)^{-1}$, and a varying phase angle relative to the cylinder motion.

Surface pressure measurements also support the wake-oscillator model. The model predicts that the lift coefficient, and therefore the magnitude of the local pressure fluctuations, peaks at a lower flow velocity than does the cylinder vibration amplitude. This prediction was found to be in agreement with the experimental observations of Ferguson and Parkinson [10].

Some recent experiments on externally driven cylinders also agree with the results obtained from the wake-oscillator model. Yano and Tokahara [11] reported experimental results for an externally driven cylinder and these results exhibit the same behavior as the model predictions for the externally driven case.

The foregoing experimental results further support the idea that there is not only a mechanical oscillator in the system but also a lift oscillator. That is, these experiments may be interpreted in terms of an oscillating lift coefficient whose instantaneous magnitude is dependent upon the cylinder motion. This type of reasoning is the basis of Eq. (2) which we refer to as the "wake-oscillator model".

Double-Amplitude Response

The solutions to Eqs. (1) and (2) can be made to fit all of the experimentally observed results, with respect to frequency and amplitude of vibration, except the double-valued amplitude response. For very light mechanical damping, Parkinson, Feng and Ferguson [11] have observed that over a certain range of wind speeds a cylinder may vibrate with one of two amplitudes. The failure of the wake-oscillator model, and indeed all other mathematical models, to reproduce this type of response may be regarded as a failure of these models and it may be taken as an indication that the physics of the situation is far more complex than the models take into account. However, there are other possible explanations for the discrepancy, one of which will be investigated here.

It is well known that nonlinear mechanical oscillators exhibit a double-amplitude response. It is therefore proposed to investigate the consequence of a nonlinearity in the mechanical oscillator as described by Eq. (1). The most obvious nonlinearity which has not been accounted for in Eq. (1) is the aerodynamic damping which will exist when the cylinder is in motion. Indeed, the fact that the double-amplitude response is observed only when the mechanical damping is very light makes this mechanism appear most attractive. However, a study of the order of magnitude of the various forces which are represented by Eq. (1) shows that, under the nearly-resonant conditions in which the double-valued amplitude is observed, the inertia and spring forces are about three orders of magnitude greater than the damping force and the external force. It therefore seems highly improbable that a nonlinearity in the damping, a relatively small force, could explain the observed phenomenon.

A second nonlinearity which may occur in the mechanical system is the restoring spring force characteristic. As is well known, a nonlinear spring produces distortion of the frequency-response curve, particularly in the vicinity of the resonant frequency. The fact that in the experiments the frequency at which the double amplitude is observed is less than the natural frequency suggests that the spring in question is of the "soft" variety.

Figure 1 shows the amplitude versus frequency response of a soft-spring oscillator for various magnitudes of the external force. It is known from experiments that as the flow velocity increases the vibration frequency increases slightly but remains near the natural frequency. The amplitude of vibration also increases and the value of the lift coefficient at first increases. The cylinder therefore follows a path as indicated in Fig. 1 between the points A and B. At the point B, the lift coefficient must decrease if the amplitude is to continue to increase so that the path of the cylinder is from B to C as shown. That this in fact happens is well known from experiments; the pressure coefficient is observed to reach a maximum before the peak amplitude is reached. Finally, at point C, the cylinder amplitude cannot change continuously for a continuous change in lift coefficient. The alternative is for the system to jump to a similar force level at a much lower amplitude, as indicated by the points C and D.

The foregoing explanation of the double-amplitude phenomenon is compatible with the observation that it is only with very light damping that such a response occurs. Since the magnitude of the external force is small compared to the spring force, a nonlinearity of say 1% will produce a negligible curvature in the amplitude versus frequency response for a heavily damped oscillator, but for very light damping this nonlinearity will be pronounced near the resonant frequency.

The qualitative explanation of the double-amplitude response which is given here has been tested quantitatively. This was done by modifying Eq. (1) as follows.

$$x_r'' + 2\zeta x_r' + x_r(1 - \varepsilon x_r^2) = a\omega_0^2 C_L . \tag{3}$$

That is, the spring force has been made nonlinear by adding a softening term which is proportional to the cube of the amplitude of vibration. The result of solving Eqs. (3) and (2) in the resonant region is shown in Fig. 2 in the form of an amplitude versus wind velocity graph. Results are shown for nonlinearities of 1%, 2% and $2\frac{1}{2}$%. There are three solutions for the amplitude of vibration for any given value of the nonlinearity ε. One of these solutions corresponds to that for the linear spring while the other two are due to the nonlinearity. Of these two additional solutions, the one corresponding to the lower amplitude is unstable so that the oscillation will assume one of the other solutions. In the case studied, the larger-amplitude solution due to the nonlinearity is about twice the amplitude which corresponds to the linear spring.

The conclusion which may be drawn from the foregoing is that a small nonlinearity in the restoring spring characteristic may produce a double-amplitude response if the damping is sufficiently light. The amplitude of vibration for the nonlinear case may be significantly higher than for the linear case. Since structures such as bridge components, tall buildings, transmission towers, etc., are comprised of complex elements, it is likely that the restoring force is a nonlinear function of deflection for these structures. In such cases the nonlinear effects discussed may be amplified so that in order to reduce the probability of increased amplitude of vibration, it is particularly important that the mechanical damping should be as high as possible.

Stability Diagram

With respect to aeolian vibrations, the questions facing designers are; through what range of wind speeds will the structure be subjected to forced oscillations and what will be the amplitudes of vibration? The designer normally has a specific structure in mind and can calculate or estimate the natural frequencies and the magnitude of the inherent damping. With this type of information available it is possible to answer the first of the above questions by use of a stability diagram of the type introduced by Scruton [2].

Using Eqs. (1) and (2) it is possible to construct such a diagram from analytical considerations. The method of solving these two equations has been described in Ref. [6] and will not be repeated here. It is sufficient to recall that of the parameters which appear in Eqs. (1) and (2), two of them, α and b, are indeterminate from theoretical considerations. These two parameters are determined by fitting the calculated results to established experimental results.

In order to construct the stability diagram one must have a set of vibration amplitude versus wind speed curves available and, having decided on a lower limit of vibration amplitude below which the system may be classified as stable, the wind speed corresponding to this stability limit may be determined for various levels of mechanical damping.

The amplitude versus wind speed curves obtained from Eqs. (1) and (2) are shown in Fig. 3 for various values of the mechanical damping ζ. Also shown on Fig. 3 are the experimental results of Feng [13]. The ordinate in Fig. 3 is the cylinder amplitude, nondimensionalized by its diameter D. The abscissa is the dimensionless wind speed $\omega_0 = SV/f_n D$ where S is the Strouhal number, V is the wind speed and f_n is the natural frequency of the cylinder. The two parameters which may be determined from physical considerations are the mass parameter $a = \rho D^2 L/(8\pi^2 S^2 M)$ and γ. Here ρ is the fluid density, L is the length of the cylinder and M is its mass. The parameter γ is one of the Van der Pol parameters; it is defined by Eq. (2) and is determined, in terms of α, by the value of $C_{LO} = (4\alpha/3\gamma)^{\frac{1}{2}}$. The undetermined parameters are α and b.

The values of a and C_{LO} which were used in the calculations correspond to Feng's experiments. In this way the calculated results may be compared directly with the experimental results. The values of α and b were determined by trial and error to give the best fit to the experimental curves. In establishing this fit, the larger of the two vibration amplitudes which are observed at low values of the damping was ignored since we are postulating that this phenomenon is due to a slight nonlinearity in the spring.

In addition to the mechanical damping, a form of aerodynamic damping was included in the calculations which lead to Fig. 3. The case for the inclusion of aerodynamic damping and the details of the magnitude and form of this damping, as used to construct Fig. 3, are given in the Appendix. It should be noted that the quantity ζ refers to the mechanical damping only and that the aerodynamic damping as described in the Appendix has been added to the quoted values of the mechanical damping.

Having established Fig. 3 it is possible to construct a stability diagram once an amplitude limit has been chosen. Here we define the stability boundary as the wind speed which, for a given value of the mechanical damping, gives a dimensionless cylinder amplitude x/D of 0.02. Then for all amplitudes greater than this value the system will be classified as unstable.

Having defined a stability limit, the value of the wind speed which corresponds to this limit may be read off Fig. 3 for the various values of the mechanical damping. The resulting plot is shown in Fig. 4. Also shown on Fig. 4 is the curve which identifies the wind speed which gives the maximum amplitude for a particular value of the mechanical damping. The lower stability boundary follows from Fig. 3 in the manner described above, but the upper stability limit is obtained differently. The solution obtained from Eqs. (1) and (2) ceases to exist at some wind speed just above the value which gives the maximum amplitude. The termination of the synchronous solution is signified by the roots of the frequency equation becoming complex and the physical interpretation of this is that the solution jumps to a different branch of the frequency diagram. That is, the "lock-in" ceases and the frequency returns to a value close to the Strouhal frequency. Although the mathematical model does predict termination of the synchronous solution, it does not define an absolute upper stability boundary since appreciable amplitudes persist at higher velocities in a non-synchronous mode. Therefore, the physical observation of Feng [13] that the peak amplitude occurs at about the middle of the unstable wind speed range has been used. That is, the upper stability boundary has been placed at the same distance above the peak amplitude location as the lower stability boundary is below it.

In constructing Fig. 4, the well established wind speed and dimensionless mechanical damping co-ordinates have been used. In the latter quantity, δ_m is the logarithmic decrement of the mechanical damping. The values of the various parameters which were used to construct Fig. 4 are identified thereon. However, it should be noted that the result is universally valid provided aeolian vibrations are taking place at a frequency which is close to the natural frequency of the structure.

Appendix

Aerodynamic Damping

When a cylinder moves in a fluid, some form of damping force will act on the cylinder due to the presence of the fluid. This damping force will be particularly important when the mechanical damping is light. In order to obtain an estimate of the effect of aerodynamic damping, consider the idealized situation in which the wake aligns itself, at each instant in time, behind the fluid velocity vector relative to the cylinder. Then the aerodynamic force F_a which acts along the line of the cylinder motion will be

$$F_a = \frac{1}{2}\rho\,(V^2 + \dot{x}^2)\,DL\,C_D \frac{\dot{x}}{\sqrt{V^2 + \dot{x}^2}} \,,$$

in which x is the dimensional cylinder velocity and C_D is the drag coefficient of the cylinder. If the cylinder velocity is small compared to the flow velocity, that is if $\dot{x}/V \ll 1$, the above expression may be expanded in terms of $\dot{x}/V$. The leading term in such an expansion is

$$F_a = \frac{1}{2}\rho VDL\, C_D\, \dot{x}.$$

This shows that, to the first order in the velocity ratio, the aerodynamic damping behaves like a linear, viscous damper. Introducing an aerodynamic damping factor ζ_a, analogous to the mechanical damping factor ζ, then

$$2\zeta_a = a\omega_0\,, \tag{4}$$

where the parameter a was introduced in Eq. (1). In deriving Eq. (4), the drag coefficient C_D has been taken as unity.

Although the assumptions which led to Eq. (4) represent a gross oversimplification of the physical situation, the result does give an order of magnitude indication of the damping coefficient and its effect. It will be seen that for very light mechanical damping, the aerodynamic damping will be of the same order of magnitude. Furthermore, Eqs. (1) and (2) have been solved both with and without aerodynamic damping for comparison purposes. The resulting cylinder amplitude and lift coefficient for no aerodynamic damping cannot be made to fit the experimental curves, particularly for low mechanical damping. Furthermore, the solutions for x_r and C_L are double-valued in the absence of any aerodynamic damping. For these reasons we believe that aerodynamic damping is very important and should be taken into account. However, the exact nature of such damping requires further investigation.

References

1. Naudascher, E.: Proc. ASCE, HY 4 (July 1967).
2. Scruton, C.: Proc. Int. Conf. on Wind Effects on Buildings and Structures, Teddington, June 1963.
3. Di Silvio, G.: Proc. ASCE, EM2, 347-361 (April 1969).
4. Nakamura, Y.: Reports of Research Institute for Applied Mechanics, Kyushu University, XVII, No. 59 (1969).

5. Halle, H.: Proc. Conf. on Flow-Induced Vibrations in Reactor System Components, Argonne National Laboratory, Report ANL-7685, May 1970.

6. Hartlen, R.T., Currie, I.G.: Proc. ASCE, EM5, 577-591 (October 1970).

7. Kawamura, S.: Proc. Third International Conference on Wind Effects on Buildings and Structures, Tokyo, September 1971.

8. Toebes, G.: ASME J. of Basic Eng. 91, No. 3, 493-505 (September 1969).

9. Mei, V.C., Currie, I.G.: Physics of Fluids 12, No. 11, 2248-2254 (1969).

10. Ferguson, N., Parkinson, G.V.: ASME J. of Eng. for Industry 89, 831-838 (November 1967).

11. Yano, T., Takahara, S.: Proc. Third International Conference on Wind Effects on Buildings and Structures, Tokyo, September 1971.

12. Parkinson, G.V., Feng, C.C., Ferguson, N.: Proc. Symposium on Wind Effects on Buildings and Structures, Loughborough, April 1968.

13. Feng, C.C.: M.A.Sc. Thesis, Department of Mechanical Eng., University of British Columbia, October 1968.

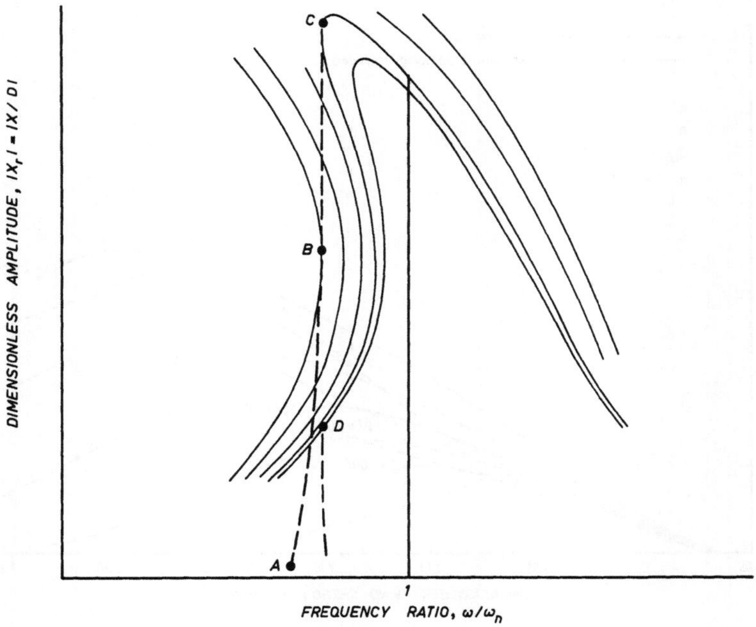

Fig.1. Amplitude response of damped forced oscillator with soft spring showing path of vortex-excited cylinder.

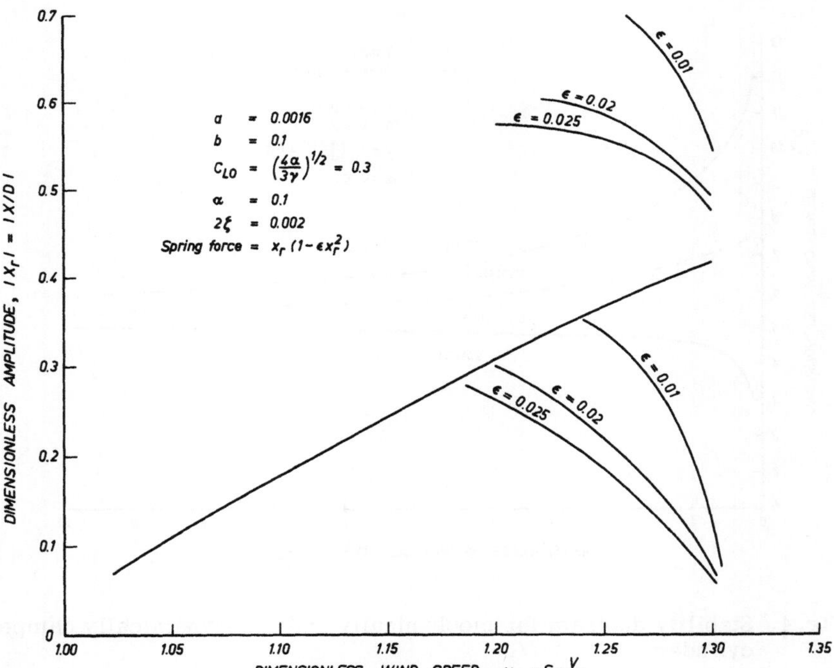

Fig.2. Amplitude versus wind speed for slightly nonlinear spring.

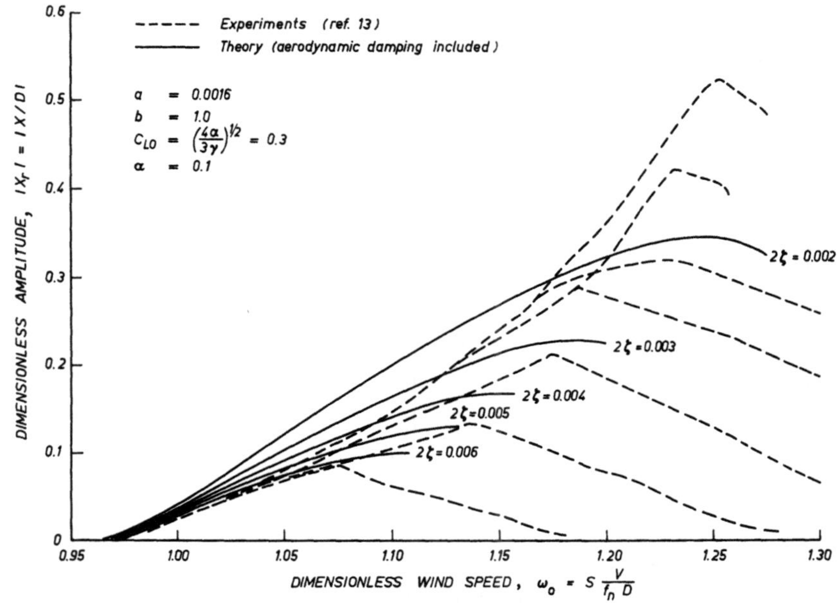

Fig. 3. Amplitude versus wind speed for various values of mechanical damping.

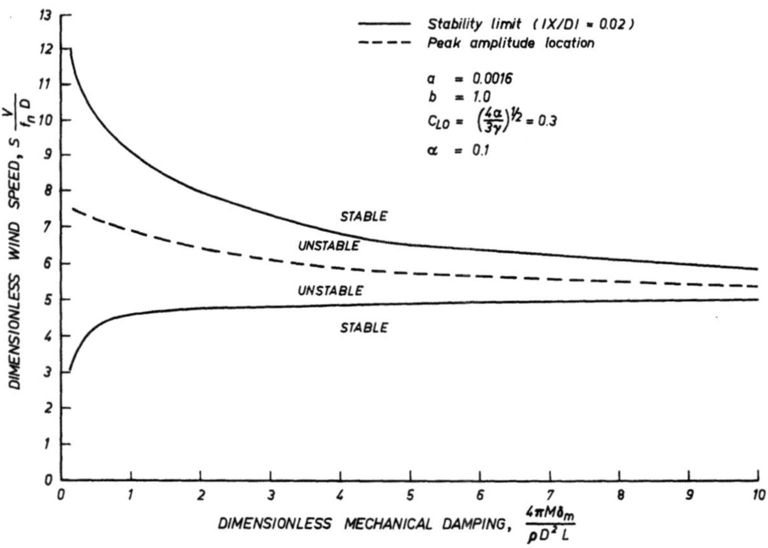

Fig. 4. Stability diagram for mechanically and aerodynamically damped cylinder.

Wake Swing and Vortex Shedding in a Cross Flow Past a Single Circular Cylinder

By

Yian-nian Chen

Winterthur, Switzerland

Summary

The real Kármán vortex street will be simulated by an ideal, semi-infinitely long rectilinear vortex street. By means of this model, the lift force exerted by the vortex on the cylinder can be calculated. It is equal to the fluid momentum flux through the front end of the vortex street, namely $L_{max} = \rho u \Gamma$, which is confirmed as the maximum lift force ever measured by a great number of investigators, as given in a previous paper [1].

Since the vortex needs a certain distance s from the cylinder for its formation, the reaction force of the cylinder on the flow will form a torque with respect to the front end of the vortex street. This torque will act on the vortex field next to the cylinder and force it to perform a rotational movement, which is known as the swing of the near wake in the literature. As the moment of the inertia force arising with the movement can be calculated as the fluctuating moment of the fluid-momentum content in the field of the vortex associated with its formation, a dynamical equation can be established between this moment and the torque formed by L_{max} and s. From this equation, the length of the vortex formation zone can be determined. These theoretical results correspond reasonably well to the experimental values given by Schaeffer/Eskinazi and Bloor [4,5].

By means of the moment of the inertia force mentioned above and that of the mass inertia of the corresponding vortex field with its rotating angular displacement, a wake number fh^2/Γ can be derived, which should remain constant in spite of the variation of the Reynolds number. The theory is confirmed by the experimental results compiled in a previous paper [2], according to which the dimensionless expression fh^2/Γ is equal to a value of 0.165 for an ideal flow up to a Reynolds number of 10^7, the highest value measured hitherto. Thus, the fluid model used in this paper can be considered to reflect reasonably well the behavior of the real vortex street as far as the force and the swing movement incorporated in it are concerned.

Introduction

A mathematical model for the real vortex street formed behind a single circular cylinder in a cross flow will be established by means of an ideal, infinitely long, rectilinear vortex street cut at a certain section simulating the existence of the tripping cylinder. The instantaneous lift force exerted by the vortex street on the cylinder will therefore be equal to the reaction to the momentum flux through the section mentioned. Since the momentum content in this semi-infinitely long vortex street is always in equilibrium, no consideration needs to be given to it. The amplitude of the lift force can be derived to yield the expression $L_{max} = \rho u \Gamma$. However, the moment of the fluid momentum content in it is no longer always in equilibrium, as will be shown in a corresponding calculation.

The formation of the vortex street begins at some distance s from the tripping cylinder of diameter d, so that a moment will be generated by the lift force with respect to this distance. The reaction to this moment is the torque exerted by the cylinder on the vortex street, forcing the vortex field next to the cylinder of the real vortex street to perform a rotational movement. The fluctuation of the moment of the fluid momentum content per unit time will act as a moment of the inertia force, resisting this periodical, rotational movement, known as the swing of the near wake.

By equating the forcing torque sL_{max} to the moment of the inertia force $M_{t,max} = \alpha\rho\omega\Gamma(\ell/2\pi)^2$, the length of the vortex formation zone can be derived, namely $s/d = M_{t,max}/L_{max}d = (\alpha/2\pi)(\ell/d)(U_\infty-u)/u$, with α = constant. Using the data given in a previous paper [2], this dimensionless length can be determined as a function of the Reynolds number.

On the other hand, the moment of the inertia force is equal to the moment of the mass inertia Θ of the corresponding vortex field times its angular accerlation $\beta_{max} = \omega^2 \varphi_{max}$. Since Θ can be set proportional to $\rho h \ell^3$ and φ_{max} can be set proportional to h/ℓ, a new wake number fh^2/Γ can be deduced by equating the expression $\omega^2 \varphi_{max} \Theta$ to that $M_{t,max}$. This dimensionless number should be independent of the Reynolds number according to the theory. The values of it will be computed using the data given by the writer elsewhere [2].

Since the theoretical results derived for the lift force L, the length of the vortex formation zone s, and the constancy of the wake number fh^2/Γ are confirmed by the evaluation of the experimental investigations published in the literature, the flow model introduced in this paper for the real vortex street is proved to be adequate for predicting the force and the wake swing accompanying the vortex shedding. Especially the new wake number fh^2/Γ established from the theory may yield a useful tool for characterizing the vortex street because of its universal nature - being equal to a value of 0.165 for a flow with low turbulence level and a tripping cylinder with smooth surface up to a Reynolds number of 10^7, the highest value ever measured.

A Mathematical Model for the Vortex Street

The ideal, inviscid, infinitely long vortex street with rectilinear staggered vortices in the two rows according to the Kármán model will be considered in a coordinate system moving with the translation velocity u of the vortices. The vortex street will then appear stationary to the observer. Since nothing changes in it with time, it is in an equilibrium condition.

If this ideal vortex street is cut through a section perpendicularly to the axis, this equilibrium condition will be destroyed. A force and a torque are needed to overcome the internal adhesion resistance, which is just corresponding to the internal stress of a mechanical beam with-standing a shearing force. Let us consider only one of these semi-infinitely long vortex streets, for example the right-hand portion in Fig. 1. The fluid momentum content for the y-component velocity v_y, $J_y = \iint \rho v_y \, dx \, dy$, vanishes in every field of one wavelength to the right of the free end, as pointed out in a previous paper [1].

Thus the fluid momentum content will always vanish in the whole ideal, semi-infinitely long vortex street. Since the pressure on this front end acts only in the x direction, the force required for the cutting is equal to the reaction to the y-component of the momentum flux through this section alone, i.e., $S_y = \int \rho v_x v_y dy$. As this momentum flux varies with the position of the cutting, the force will vary with this correspondingly. The momentum flux will be a maximum for a cutting section just midway between two adjacent vortices of the different rows, namely with an amount of $(S_y)_{max} = \rho u \Gamma$ (u = translation velocity of the vortex and Γ = circulation of a completed vortex), as was shown in a previous paper [1]. Its sign will be positive for the position 1 and negative for the position 1' in Fig.1. However, it will be zero when the section just goes through a vortex center.

Let us imagine that the cutting of the vortex street is performed by means of a circular cylinder. Then, it has to be put in some distance (s) in front of the cutting section to produce a torque in addition to the cutting force. This simulates the need of the vortex for this distance in its generation process. This situation is sketched in Fig.2. Thus, we have constructed a fluid model which should substitute the real vortex street shed by a cylinder. The distance s corresponds then to the length of the formation zone of the vortex. The variation of the flow condition on the free end of the ideal, semi-infinitely long vortex street with the variation of the cutting position simulates the variation of the near wake of the real vortex street during the shedding process. The force exerted by the cylinder on the vortex street will vary between $+\rho u \Gamma$ and $-\rho u \Gamma$ in a periodic function of time, because the variation of the cutting section must be carried out with a constant velocity of $(U_\infty - u)$ to the left direction for simulating the formation of the real near wake. The function of the cylinder, here, is therefore confined to producing the force and torque on the semi-infinitely long vortex street necessary for keeping it in equilibrium. The movement of the cylinder with a relative velocity of $(U_\infty - u)$ determines only the position of the cutting, but will not influence the flow pattern in the existing vortex street. The reaction of this force on the cylinder is nothing less than the fluctuating lift of the vortex street with an amplitude of $L_{max} = \rho u \Gamma$. According to a previous paper [1], this theoretical value corresponds well to the maximum value measured by numerous authors in the interesting Reynolds number range. Thus, the model can be considered to be adequate to predict the force exerted by the real vortex street.

The Moment of the y-Component of the Fluid Momentum Content

In the foregoing section it is stated that the y-component of the fluid momentum content $J_y = \iint \rho v_y dx\, dy$ always vanishes in the ideal, semi-infinitely long vortex street. Contrary to this, the moment of this fluid momentum content is not always zero.

Let us consider first the ideal vortex street with a free end just going midway through two adjacent vortices of the different rows (see Fig. 3). Since the distribution of the flow pattern is mirror-imaged symmetrically about the vertical plane through the vortex center in each field of a half wavelength to the right of the free end, the moment of the y-component of the fluid momentum content will be equal in amount in every field, but change sign from one field to another. Thus, this moment will cancel out its neighbour in the adjacent field and its total value in the whole, ideal, semi-infinitely long vortex street will vanish for the position of the free end chosen in the way mentioned above. In this manner, the moments can be balanced between each two adjacent fields. This condition is shown in Fig. 4a again.

When the vortex street grows in the left direction, a new flow area will join up to the old balanced vortex street. The moment of the new fluid momentum content can no longer be balanced by the old vortex street, this being contrary to the generated fluid momentum content itself. This moment with respect to the border of the old, balanced vortex street can be calculated by using the following equation:

$$M_y = -\int_{x=0}^{x} \int_{y=-\infty}^{y=+\infty} \rho\, v_y\, x\, dx\, dy\,, \tag{1}$$

where
$$v_y = \frac{\Gamma}{2l}\left\{-\frac{\sin\frac{2\pi}{l}\left(x-\frac{\ell}{4}\right)}{\cosh\frac{2\pi}{\ell}\left(y-\frac{h}{2}\right)-\cos\frac{2\pi}{\ell}\left(x-\frac{\ell}{4}\right)} + \frac{\sin\frac{2\pi}{\ell}\left(x+\frac{\ell}{4}\right)}{\cosh\frac{2\pi}{\ell}\left(y+\frac{h}{2}\right)-\cos\frac{2\pi}{\ell}\left(x+\frac{\ell}{4}\right)}\right\} \tag{2}$$

represents the y-velocity component [2,3] with ℓ = longitudinal spacing and h = transverse spacing of the vortices (see Fig. 1).

The absolute value of M_y reaches its maximum, when $-x$ increases to $\ell/4$, whereby the free end just goes through a vortex center, as shown in Fig. 4b. The maximum is negative for this special position (see Fig. 4f).

By further growth of the vortex, we obtain the flow pattern in Fig. 4c for $-x = \ell/2$, where the vortex has just completed its formation. The flow pattern is symmetrical in each field of one half wavelength next to the free end. The balance of the moments of the fluid momentum content within two adjacent fields will be re-established, i.e., $M_y = 0$ for $-x = \ell/2$, see Fig. 4f. The moment of the momentum content for the field growing from instance b to instance c (i.e., the field from $-\ell/4$ to $-x$, see Fig. 4b and 4c, respectively) can therefore be computed in an indirect way, whereby the complementary field (i.e., the field from $-x$ to $-\ell/2$) should be first formed, which is needed as addition to the existing fields to form the new, balanced vortex street (i.e. the condition in Fig. 4c). The moment of the fluid momentum content in this complementary field with respect to the free end of the new, balanced vortex street (see Fig. 4c) is then the required result.

The M_y-curve for $-x = \ell/4$ to $\ell/2$ is therefore symmetrical to that for $-x = 0$ to $\ell/4$ (see Fig. 4f). Likewise, the further course of the curve M_y can be constructed (see Fig. 4f).

The maximum amplitude of curve $M_y : M_{y\,max} = \int_{x=0}^{x=-\ell/4} \int_{y=-\infty}^{y=+\infty} \rho v_y x \, dx \, dy$ can be computed by inserting v_y from Eq. (2) to the following value:

$$M_{y\ max} = \frac{\rho\Gamma}{2\ell} \left\{ \left(\frac{\ell}{2\pi}\right)^3 \times 4\alpha\pi \right\} = \alpha\rho\Gamma \left(\frac{\ell}{2\pi}\right)^2 , \tag{3}$$

where α is a constant, depending on the ratio of ℓ/h. This is shown in Table 1. For Kármán's criterion $\ell/h = 1/0.281 = 3.55$ we obtain $\alpha = 0.555$.

Table 1

ℓ/h	3	3.55	4	4.5	5
α	0.538	0.555	0.555	0.568	0.568

The instantaneous value of the moment of the y-momentum content will be

$$M_y(t) = - M_{y,max} \sin \omega t \tag{4}$$

and its variation in unit time will be

$$M_t = \frac{dM_y(t)}{dt} = -\alpha\rho\omega\Gamma \left(\frac{\ell}{2\pi}\right)^2 \cos \omega t \, . \tag{5}$$

This expression, originating from the variation of the moment of the momentum content of the flow field, accompanying the generation of the new vortex, is nothing but the moment of the inertia force of the corresponding vortex field.

Swing of the Wake as a Rotational Movement of the Vortex Street

The well-known swing phenomenon of the near wake can be imagined as a rotational movement of the free end of the ideal, semi-infinitely long vortex street[1]. The relations between the swing movement and the variables S_y, F_y, sF_y and M_t are shown in Fig. 4a to i. In example (a), a vortex of the upper row has just been completed. The momentum flux S_y through the free end is positive maximum (upward) and the induced lift force F_y on the cylinder is negative maximum (downward), $F_y = -S_y$ (see subfigure g). The moment of the fluid momentum content in the vortex street is zero ($M_y = 0$), and its time derivate $M_t = dM_y/dt$ is negative maximum (rotation direction ↻) (see subfigure f and g). Since the two vectors S_y and F_y are separated by a distance s, a moment sF_y arises. The reaction of this moment, $-sF_y$, acts on the vortex street and forces the wake to rotate. The direction of rotation is also shown in subfigure a.

The symbol Γ_t denotes the instantaneous circulation of the vortex just formed next to the free end (see subfigure h). A counter circulation Γ_c with a counter sign must be induced around the cylinder. This is shown in subfigures a and h. Since the coordinate system is chosen to possess a velocity u (the translation velocity) in a direc-

---

[1] As a matter of fact, the vortex field of the near wake of a circular cylinder is very complicated. The theory proposes that the influence of this near wake, with its swing movement and its vortex field, on the existing vortex street be simulated by a semi-infinitely long ideal vortex street with a free front end, the vortex field distribution in which is uniform everywhere. This ideal vortex street will always remain straight, despite the swing of the real near wake. This means that the influence of the swing movement is supposed to be always balanced by the influence of the variation in the vortex field of the near wake. Thus, the resultant influence of these two factors on the existing real vortex street is simulated in the ideal vortex street model by a growth in its (uniform) length while maintaining a free space s behind the tripping circular cylinder.

The swing of the near wake will therefore not change the straight form of the ideal vortex street. As will be shown later on, this idealisation is justified owing to the reasonable agreement between the theoretical result and the experimental evidence obtained on the circular cylinders. The rotation of the front end of the ideal vortex street shown in Fig. 4a to e therefore only symbolises the swing of the near wake of the real vortex street, and is included purely to simplify the presentation. Thus, we are concerned here not with the swing of the front of the ideal vortex street itself, but with that (the swing) of the real near wake; the influence of which is incorporated in the front end of the ideal vortex street. More details of this concept will be given in the discussions with Dr. Zdravkovich to be found in the appendix.

tion to the left (←), the flow field must be considered to be superimposed with a relative velocity u in a counter direction (→) to the coordinate system. This relative velocity is also shown above the cylinder. The counter circulation Γ_c and this velocity u will generate a force of $\rho u \Gamma_c$ on the cylinder, as shown in subfigure i. It is the lift force F_y (see subfigure g). The lift force will thus always be in an opposite direction to that of the shed vortex generating this force. This theoretical result corresponds exactly to the experimental findings reported by Dreher 1956 ([7] in his Fig. 2) and by Dye 1970 [8]. Thus we reach the same result with a purely physical consideration. In this manner the phase relationship between the quantities S_y, F_y, M_t, sF_y and Γ_t have been established.

The swing direction of the near wake is always on the side where a vortex has just been formed. This direction coincides with that of the flow streaming out of the free end in the ideal model.

As mentioned above, the torque $-sF_y$ will force the near wake to swing. On the other hand, the reaction $-M_t$ to the torque M_t will act on the vortex field also. However, this reaction torque is always in counter-phase with $-sF_y$ and therefore exhibits a resistance to the swing movement (see Fig. 4a and f). The forcing torque $-sF_y$ and the inertia moment $-M_t$ must be in equilibrium, namely

$$sF_y = M_t$$

or

$$s\rho u\Gamma = \alpha\rho\omega\Gamma\left(\frac{\ell}{2\pi}\right)^2$$

and

$$s = \alpha\frac{\omega}{u}\left(\frac{\ell}{2\pi}\right)^2 .$$

By inserting $\omega = 2\pi f = 2\pi(U_\infty - u)/\ell$ it follows

$$s = \alpha\frac{\ell}{2\pi}\,\frac{U_\infty - u}{u} . \tag{6}$$

In this manner, the length s of the vortex formation zone can be related to the longitudinal vortex spacing ℓ and the velocity ratio

$(U_\infty - u)/u$. The above equation can be made dimensionless in the following form:

$$\frac{s}{d} = \frac{\alpha}{2\pi} \frac{\ell}{d} \frac{U_\infty - u}{u} . \qquad (7)$$

Since the quantities ℓ/d and $(U_\infty - u)/u$ are given in a previous paper of the writer [2], the dimensionless vortex formation zone length s/d can be calculated up to a Reynolds number of 10^7. The result is plotted in Fig.5 as curves $(s/d)_{theory}$. Curve 1 denotes the ideal case (polished cylinder in a flow with low turbulence level) and curve 2 denotes the usual case (rough cylinder in a usually encountered flow). Only a few experimentally determined values for the vortex formation zone length are available. The investigation results given by Schaeffer/Eskinazi [4] and Bloor [5] are shown as point groups $(s/d)_{exp}$ in Fig.5. The definitions for this length are quite different by these authors. Schaeffer/Eskinazi defined this length as the distance from the tripping cylinder up to the narrowing neck of the vortex street, behind which the stable range with practically constant street width begins. Bloor, on the other hand, considered the formation region as that of the wake up to the first appearance of the periodic vortex street. Although these two definitions are not quite the same as that length denoted with s in this paper, a great similarity between them is obvious. An exact coincidence between the theoretical curve $(s/d)_{theory}$ and the experimentally determined point group $(s/d)_{exp}$ can, of course, not be claimed. Nevertheless, the theoretical curves show a great agreement with the experimental points. The curves have a minimum at $Re = 400$ and a maximum at $Re = 1$ to 2×10^3. A further minimum appears at about $Re = 2 \times 10^4$ and a further maximum at $Re = 2 \times 10^5$ for the ideal case. In the transcritical Reynolds number range, the vortex formation zone length will be much less ($s/d = 1.35$ at $Re = 2 \times 10^6$ and $s/d = 1.11$ at $Re = 10^7$) according to the theoretical curve 1.

The good agreement between theory and experiment demonstrates the usefulness of the flow model presented in this paper.

The Universal Wake Number

Let Θ denote the moment of inertia of the fluid mass participating in the rotation movement and φ_{max} the maximum angle of rotation. It can be expected that Θ is proportional to $\rho h \ell^3$ and φ_{max} is proportional to the central angle σ of a vortex field with one wavelength (see Fig. 2). From this we obtain $\varphi_{max} \sim \sigma \sim h/\ell$. The maximum angular acceleration will be $\beta_{max} = \omega^2 \varphi_{max}$. Since the expression $\beta_{max}\Theta$ denotes the maximum moment of the inertia force just as $M_{t,max}$ does, these two expressions must be equal to each other, namely

$$M_{t,max} = \beta_{max}\Theta .$$

Inserting the corresponding values into the equation yields

$$\alpha\rho\omega\Gamma\left(\frac{\ell}{2\pi}\right)^2 = \varepsilon\omega^2 \frac{h}{\ell} \rho h \ell^3 ,$$

where ε is a constant. By re-arrangement we obtain

$$\frac{\omega h^2}{\Gamma} = \frac{1}{4\pi^2}\frac{\alpha}{\varepsilon}$$

or

$$\frac{f h^2}{\Gamma} = \frac{1}{8\pi^3}\frac{\alpha}{\varepsilon} .$$

Since ε is set as a proportionality constant denoting the property of the rotation of the near wake of the vortex street, it must be independent of the Reynolds number, as long as the generation mechanism of the vortex street from the cylinder remains comparable. Since, on the other hand, α remains practically constant within the possible range of the spacing ratio of the vortices (i.e., $\alpha = 0.538$ to 0.555 for $\ell/h = 3$ to 4 respectively, see Table 1), the expression fh^2/Γ must be a constant too, i.e., independent of the Reynolds number if the condition mentioned above is fulfilled. This theory is completely confirmed by the evaluation of the experimental results given in a previous paper of the writer [2]. Figure 6 shows the investigation result as curves fh^2/Γ No. 1 and 2. This reveals that the expression fh^2/Γ remains constant with a value of about 0.175 in the whole Reynolds number range from $Re = 10^2$ to 10^7 for an ideal flow with a low turbulence level and for a tripping cylinder with a smooth surface (curve No. 1). This quantity varies only in the low Re-range from a value of 0.180 at $Re = 40$ to a value of 0.168 at $Re = 100$. Since the generation of the vortex street in this low Re-range originates from the wave-like movement of the wake far downstream, thus differing from the generation mechanism for the Reynolds numbers above this Re-range, this lying in the rolling of the free-shear layer directly separated from the cylinder surface. The higher values of fh^2/Γ in this low Re-range are therefore intelligible [2]. The dimensionless street width h/d calculated by using the data given in the quoted paper [2], is also plotted in Fig.6, just to show its large variations in the three Reynolds number ranges (the sub, super- and transcritical ones), in contrast to the constancy of the quantity fh^2/Γ in these ranges. Some experimental results for h/d (Schaeffer/Eskinazi, Bloor/Gerrard, Berger 1964 and Wille/Timme 1957) are included also. It is to be noticed, that only the widths of the vortex street corresponding to its stable region, are shown [9,10,11].

Thus, the quantity fh^2/Γ with its value of 0.165 is a universal number for a normal vortex street - with the cylinder as the direct, vortex shedding source in an ideal flow - being independent of the behavior of the vortex, i.e., whether the vortex is laminar or highly turbulent. If the flow is not ideal, either with a high turbulence level or with a rough tripping cylinder surface, this universal number will increase to 0.172 (see curve 2) [2]. The quantity $c = fh^2/\Gamma$ will be called "the universal wake number" of the vortex street. This criterion says that in the vortex street, the geometry, the vortex strength, and the shedding frequency always remain in a firm relationship with fh^2/Γ = constant, owing to the swing of the wake as a rotational system, coupled with the shedding of the vortex.

Thus, the fluid model introduced in this paper for the vortex street proves to be adequate for describing the behavior of the vortex street concerning the force and the swing movement incorporated in it. The derivation of the universal wake number fh^2/Γ by means of this model seems especially beneficial for the further investigation of the vortex street.

References

1. Chen, Y.N.: ASME-Vibrations Conference, Toronto, September 8-10, 1971, Paper 71-Vibr. 12.

2. Chen, Y.N.: ASME-Vibrations Conference, Toronto, September 8-10, 1971, Paper 71-Vibr. 11.

3. Kochin, N.E., Kibel, I.A., Roze, N.V.: Theoretical Hydromechanics, New York - London - Sydney: Interscience Publishers, John Wiley 1964, pp. 197-245.

4. Schaeffer, J.W., Eskinazi, S.: J. of Fluid Mech. 6, 241-260 (1959).

5. Bloor, M.S.: J. of Fluid Mech. 19, 290-304 (1964).

6. Chen, Y.N.: Proc. of the ASME-Symposium on "Flow-Induced Vibration in Heat-Exchangers", 1970, pp. 85-88.

7. Drescher, H.: Z. Flugwiss. 4, 17-21 (1956).

8. Dye, R.C.F.: J. of Fluid Mech. 45, 213 (1971).

9. Bloor, M.S., Gerrard, J.H.: Proc. Royal Society, London A, 294, 319-342 (1966).

10. Berger, E.: Z. Flugwiss. 12, 41-59 (1964).

11. Wille, R., Timme, A.: Jahrbuch der Schiffbautechnischen Gesellschaft 1957, pp. 215-221.

Fig. 1

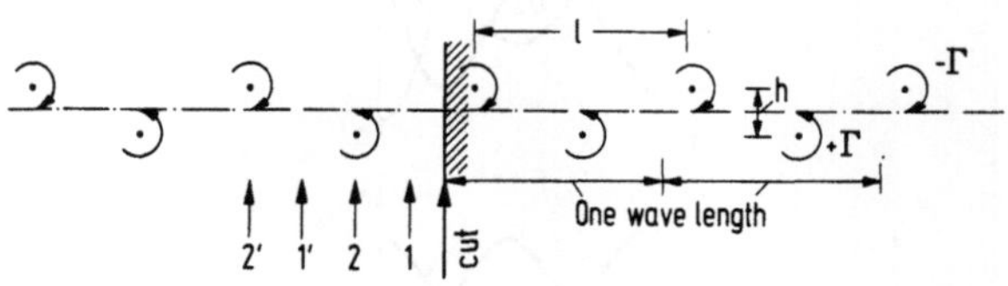

Fig. 2

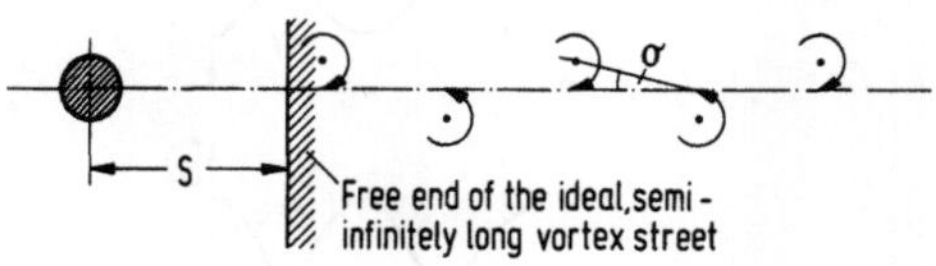

Fig. 3

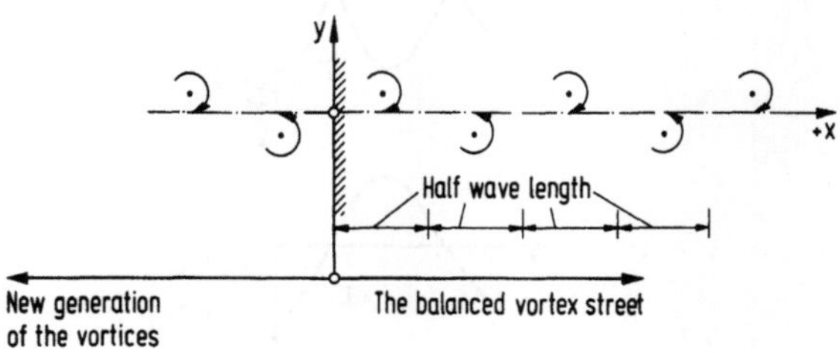

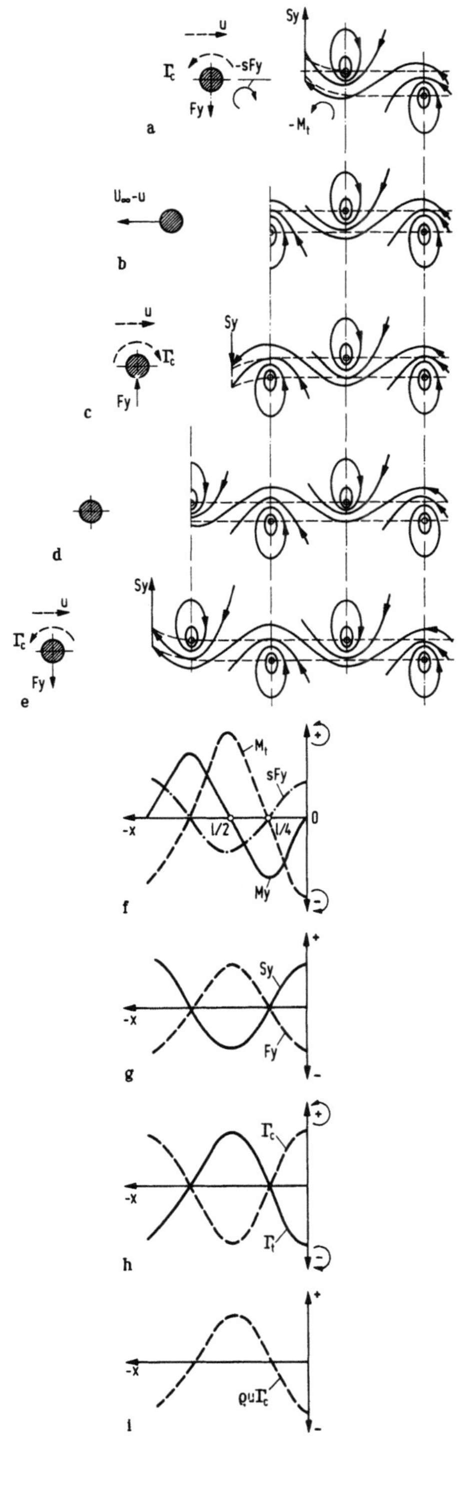
u
Sy
Γ_c
-sFy
Fy
a
$-M_t$
U_∞-u
b
u
Sy
Γ_c
Fy
c
d
Sy
u
Γ_c
Fy
e
M_t
sFy
-x
l/2
l/4
0
My
f
Sy
-x
Fy
g
Γ_c
-x
Γ_t
h
-x
$\rho u \Gamma_c$
i

Fig.4

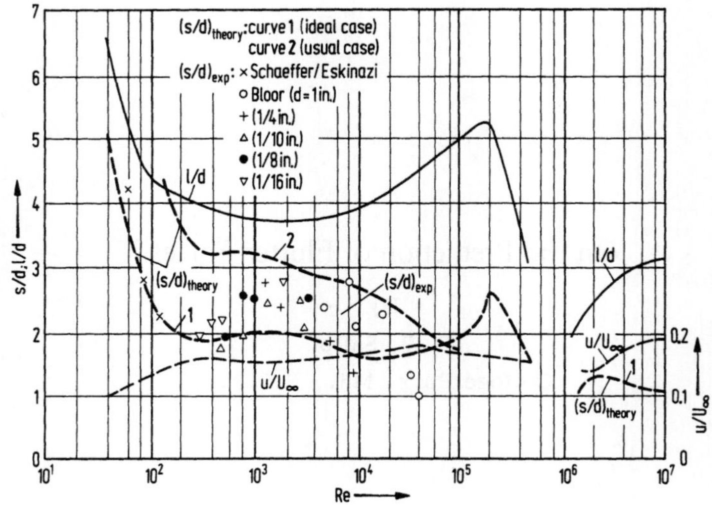
(s/d)theory: curve 1 (ideal case)
curve 2 (usual case)
(s/d)exp: × Schaeffer/Eskinazi
○ Bloor (d = 1 in.)
+ (1/4 in.)
△ (1/10 in.)
● (1/8 in.)
▽ (1/16 in.)
l/d
(s/d)theory
(s/d)exp
u/U∞
s/d; l/d
u/U∞
Re

Fig. 5

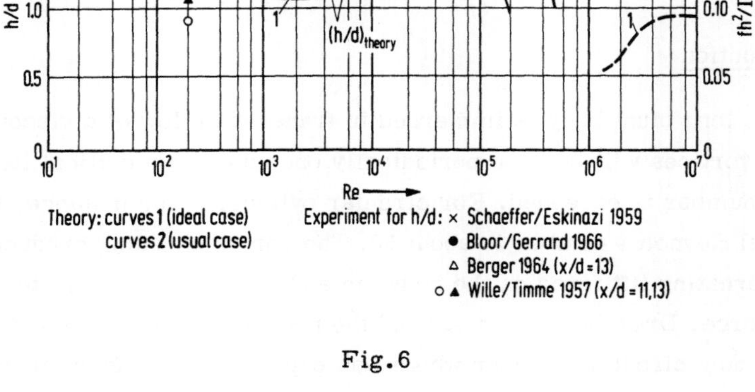
fh²/Γ
(h/d)theory
h/d
fh²/Γ
Re
Theory: curves 1 (ideal case)
curves 2 (usual case)
Experiment for h/d: × Schaeffer/Eskinazi 1959
● Bloor/Gerrard 1966
△ Berger 1964 (x/d = 13)
○ ▲ Wille/Timme 1957 (x/d = 11,13)

Fig. 6

On the Prediction of Flutter Forces

By

D.W. Sallet

College Park, Md., U.S.A.

Summary

The prediction of the fluctuating lift forces which occur on cylindrical bodies in cross flow is discussed. Equations are developed which permit the estimation of the maximum lift coefficient for a bluff cylindrical body exposed to uniform approach flow. By combining the results of a potential flow analysis with experimentally found values for the Strouhal number and the coefficient of drag, the lift coefficient can be expressed as a function of the Reynolds number. It is found that the lift forces are inversely proportional to the frequency of vortex shedding, after certain simplifying assumptions are introduced.

Introduction

When a long blunt body is immersed in transverse flow of a viscous fluid, vortices will be shed periodically once a certain critical Reynolds number is exceeded. For circular cylinders, for instance, this critical Reynolds number is about 50. The vortex shedding produces an alternating lift force on the body, in addition to the nearly steady drag force. Drescher [1] measured the non-steady pressure distribution on a circular cylinder which was exposed to a uniform approach flow. From these pressure measurements and the results from observations of the wake of the cylinder, Drescher related the vortex shedding to the fluctuating lift as shown in Fig. 1.

The fluctuating lift forces may initiate vibrations which are usually detrimental to the engineering structures as structural failure may occur. The now classic report by Meier-Windhorst [2] and the more recent extensive work by Toebes (e.g., Ref. [3]) describe and analyse the response of the cylinder and the flow field around it when the cylinder is elastically supported [2] or when the cylinder is externally excited [3]. Meier-Windhorst's investigation was preceded by a lucid discussion on the nature of flow induced vibrations by Thoma [4]. Although a fluctuating wake is not a necessary requirement for aeroelastic instability of non-circular cross-sections, the instability of bodies with circular cross sections is due in part to the unsteady wake as explained by several investigators, e.g., Steinman [5] or Parkinson [6]. The question now arises, whether the geometry of the nonsteady wake can yield an estimate of the nonsteady lift forces acting on the body.

Estimation of Lift by Momentum Analysis of the Vortex Street

Let the complex potential of the flow field be denoted by

$$w(z) = \varphi(x,y) + i\Psi(x,y)\,, \tag{1}$$

where φ is the velocity potential and Ψ is the stream function. Then the complex velocity is

$$\frac{dw}{dz} = u - iv\,, \tag{2}$$

where u is the velocity component in the x-direction, and v is the velocity component in the y-direction, and z equals $x + iy$. The complex potential of the staggered, infinitely long double row of vortices shown in Fig.2 is:

$$w(z) = i\frac{\Gamma}{2\pi}\ln\frac{\sin\frac{\pi}{l}(z - z_0)}{\sin\frac{\pi}{l}(z + z_0)}\,, \tag{3}$$

where $z_0 = +\frac{l}{4} + i\frac{h}{2}$ and where Γ is the circulation of one vortex.

The stream function of this vortex street is

$$\Psi = -\frac{\Gamma}{4\pi} \ln \left(\sin^2\left[\left(x+\frac{1}{4}\right)\frac{\pi}{l}\right]\cosh^2\left[\left(y+\frac{h}{2}\right)\frac{\pi}{l}\right] + \right.$$
$$\left. + \cos^2\left[\left(x+\frac{1}{4}\right)\frac{\pi}{l}\right]\sinh^2\left[\left(y+\frac{h}{2}\right)\frac{\pi}{l}\right]\right) \Big/$$
$$\left(\sin^2\left[\left(x-\frac{1}{4}\right)\frac{\pi}{l}\right]\cosh^2\left[\left(y-\frac{h}{2}\right)\frac{\pi}{l}\right] + \right.$$
$$\left. + \cos^2\left[\left(x-\frac{1}{4}\right)\frac{\pi}{l}\right]\sinh^2\left[\left(y-\frac{h}{2}\right)\frac{\pi}{l}\right]\right) . \tag{4}$$

The equation of the stream line separating the flow fields associated with the upper and lower vortex rows then becomes:

$$\sin\left(\frac{2\pi x}{l}\right) + \sinh\left(\frac{2\pi y}{l}\right)\sinh\left(\frac{\pi h}{l}\right) = 0 . \tag{5}$$

Von Kármán [7] showed that the condition for stability of the above vortex street is

$$\sinh\left(\frac{\pi h}{l}\right) = 1 . \tag{6}$$

The equation of the separatrix for the Kármán vortex street is

$$\sin\left(\frac{2\pi x}{l}\right) + \sinh\left(\frac{2\pi y}{l}\right) = 0 . \tag{7}$$

Equation (7) shows that the flow field due to the vortices has the spatial period l as was naturally expected. The momentum principle indicates that a blunt body creating such a flow field will have a fluctuating lift in addition to a fluctuating drag force. In view of Eq. (7) it is plausible to assume a sinusoidal function for the lift coefficient which has a spatial period of l. This assumption is supported by the results of a numerical study by Dawson and Marcus [8]. Figure 3 shows the fluctuating lift coefficient of a circular cylinder at a Reynolds number of 100 (private communications to the author by Mr. Dawson). The slight assymmetrical time history of the numerically obtained $C_L(t)$ is not due to plotting errors. The coefficient of lift is defined by Eq. (28) below. A relation between the spatial periods of the sinusoidal and hyperbolic functions and the time period of the computed function is given in the following.

The momentum principle which von Kármán applied in his classical paper [7] to deduct the average drag acting on the body which produces a vortex street can be employed to estimate the maximum lift acting on the blunt body. The momentum principle simply states, that the resultant force acting on the control volume is equal to the rate of increase of linear momentum within the control volume plus the net efflux of linear momentum from the control volume. Let the control volume ABCD (see Fig. 4) translate with the same velocity as the vortex street, namely u_s. It is seen that due to the uniform motion of the control volume with respect to the fluid at rest fluid enters the control volume with a uniform velocity u_s at the boundary BC, and that the velocity of the body with respect to the control volume is $U - u_s$, where U is the velocity of the cylinder with respect to the fluid. It is also seen that after any successive interval Δt, where

$$\Delta t = \frac{l}{(U - u_s)} , \tag{8}$$

the control volume will contain one more vortex pair.

Before expressing the momentum balance in mathematical form the following two assumptions are made: (1) The complex potential of the infinite vortex street does not differ significantly from that of the semi-infinite street which results from the introduction of the cylinder; (2) the forces acting on the cylinder are only due to the production of the vortex wake.

The momentum equation now expresses the drag force F_x and the lift force F_y acting on a cylinder of unit length as:

$$F_x = -\frac{dM_x}{dt} + \rho \frac{d}{dt}\left[\int_D^A \varphi dy - \int_C^B \varphi dy\right] - \frac{\rho}{2}\int_D^A (u'^2 - v'^2)\, dy \tag{9}$$

and

$$F_y = -\frac{dM_y}{dt} + \rho \frac{d}{dt}\left[\int_B^A \varphi dx - \int_C^D \varphi dx\right] - \rho u_s \Gamma_T - \rho \int_D^A u'v' dy, \quad (10)$$

where $u' = u - u_s$ and $v = v'$, ρ is the density of the fluid, M_x and M_y is the momentum of the fluid inside the control volume in the x- and y-direction, and Γ_T is the total circulation around the control volume. The propagation velocity of the whole street as given by von Kármán [7] is

$$u_s = \frac{\Gamma}{2l} \tanh \frac{\pi h}{l}. \quad (11)$$

Equation (10) is now simplified to

$$F_y = -\frac{dM_y}{dt} - \rho u_s \Gamma_T + \frac{1}{2} \rho u_s \Gamma, \quad \text{if } z_0 = +\frac{l}{4} + i\frac{h}{2}, \quad (12)$$

or

$$F_y = -\frac{dM_y}{dt} - \rho u_s \Gamma_T - \frac{1}{2} \rho u_s \Gamma, \quad \text{if } z_0 = -\frac{l}{4} + i\frac{h}{2}. \quad (13)$$

When the control volume is drawn as in Fig. 4 then Eq. (12) applies and Γ_T equals zero. If the control surface $\overline{AD}$ is drawn midpoint between vortex 1 and 2 or vortex 0 and -1 then Eq. (13) applies and $\Gamma_T = -\Gamma$. It is also noted that as soon as the sign of M_y changes the value of the total circulation also changes. Therefore, Eqs. (12) and (13) become

$$F_y = \pm\left[\frac{dM_y}{dt} - \frac{1}{2} \rho u_s \Gamma\right]. \quad (14)$$

The increase in momentum in the control volume per unit time can be determined by comparing the momentum in the control volume at time t_1 with the momentum content at time $t_1 + \Delta t$. It is noted that during this time interval, the control volume gains one vortex pair, say vortex 2 and 3. The momentum of this isolated vortex pair in the x and y direction is

$$M_x = \rho \Gamma h \tag{15}$$

and
$$M_y = \rho \Gamma \frac{l}{2} \tag{16}$$

respectively. However, any vortex in the wake must be considered as a part of two vortex pairs. In Fig. 4, for instance, vortex 2 is part of the vortex pair 1, 2 and the vortex pair 2, 3. The maximum positive or the maximum negative increase of momentum in the y-direction which can occur during any time interval Δt therefore is

$$M_y = \frac{1}{4} \rho \Gamma l . \tag{17}$$

The peak lift acting on the body per unit length is now readily deduced from Eq. (14):

$$F_y = \frac{1}{4} \rho \Gamma [U - 3u_s] . \tag{18}$$

Expression (18) shows that the lift per unit length acting on a body is proportional to the product of a density, a circulation and a velocity term. The result shows in principle agreement with the Kutta-Joukowsky theorem.

It is desirable to obtain an estimation for the lift in terms of easily measured and readily available experimental data. An expression relating the longitudinal vortex spacing l and the velocity of the vortex street u_s with the experimentally determined coefficient of drag C_D and the experimentally determined Strouhal number S was obtained by Sallet [9] as follows: From Eq. (9), von Kármán's result for the average drag force per unit time length is obtained, namely

$$F_x = \rho \Gamma \frac{h}{l} (U - 2u_s) + \rho \frac{\Gamma^2}{2\pi l} . \tag{19}$$

Equating the frequency of vortex shedding of the above potential flow model to the vortex shedding frequency which is experimentally observed yields

$$\frac{U - u_s}{U} = S \frac{l}{d} , \tag{20}$$

where the Strouhal number S is a function of the Reynolds number. Roshko [10], for instance, found that for circular cylinders

$$S = 0.212\left(1 - \frac{21.2}{Re}\right), \qquad \text{if } 50 < Re < 150\,, \tag{21}$$

$$\text{and} \quad S = 0.212\left(1 - \frac{12.7}{Re}\right), \qquad \text{if } 300 < Re < 2{,}000\,. \tag{22}$$

while for the Reynolds number range from 2,000 to 400,000, the Strouhal number is reported to lie between 0.195 and 0.210 [11]. A further increase in the Reynolds number shows a transition zone, $(4 \times 10^5 < Re < 3.5 \times 10^6)$ where the Strouhal number has been observed to lie between 0.21 and 0.46. Vortex shedding occurs with a Strouhal number of 0.27 [12] for Reynolds numbers larger than 3.5×10^6. Defining the coefficient of drag C_D in the customary way, namely

$$C_D = \frac{2F_x}{\rho U^2 d} \tag{23}$$

and introducing Eqs. (11), (19) and (20) into expression (23) yields

$$C_D = \frac{4}{d}\left[h\left(1 - \frac{Sl}{d}\right)\left(\frac{2Sl}{d} - 1\right)\coth\frac{\pi h}{l} + \frac{l}{\pi}\left(1 - \frac{Sl}{d}\right)^2 \coth^2\frac{\pi h}{l}\right]. \tag{24}$$

If the vortex street obeys von Kármán's stability criterion, then Eq. (24) becomes

$$C_D = 4\sqrt{2}\,\frac{l}{d}\left[1 - \frac{Sl}{d}\right]\left[\frac{h}{l} - 2\left(1 - \frac{Sl}{d}\right)\left(\frac{h}{l} - \frac{1}{\pi\sqrt{2}}\right)\right] \tag{25}$$

or, since $\cosh\frac{\pi h}{l}$ equals $\sqrt{2}$ in this case,

$$S^2\left(\frac{l}{d}\right)^3 + 0.529\,S\left(\frac{l}{d}\right)^2 - 1.529\,\frac{l}{d} + 1.593\,C_D = 0\,. \tag{26}$$

Equations (11), (18) and (20) now combine to yield an expression for the lift coefficient C_L,

$$C_L = \frac{l}{d}\left(1 - \frac{Sl}{d}\right)\left(3\,\frac{Sl}{d} - 2\right)\coth\frac{\pi h}{l}\,, \tag{27}$$

where C_L is defined as

$$C_L = \frac{2F_y}{\rho U^2 d} \,. \tag{28}$$

Just as for Eq. (24), the derivation of Eq. (27) does not assume the vortex street to obey von Kármán's stability criterion. If von Kármán's stability condition is satisfied, then Eq. (27) further simplifies to

$$C_L = \sqrt{2}\,\frac{1}{d}\left[1 - \frac{Sl}{d}\right]\left[3\,\frac{Sl}{d} - 2\right] . \tag{29}$$

The fluctuating lift coefficient $C_L(t)$ may now simply be approximated by

$$C_L(t) = C_L \sin \frac{2\pi US}{d}\, t \,. \tag{30}$$

Equation (29) gives the maximum lift coefficient of a blunt cylinder in terms of the longitudinal vortex spacing l, the Strouhal number S and the width of the body d. Equation (25) relates the coefficient of drag of a blunt cylinder with vortex spacing l, the Strouhal number S and the width of the body d. Since the coefficient of drag C_D and the Strouhal number S as functions of the Reynolds number for a particular blunt cylinder such as circular cylinders, wedges of flat plates are readily available in the literature, the lift coefficient of a blunt cylinder can now be estimated. Should the vortex spacing ratio be known, then the use of Eqs. (24) and (27) is preferable to the use of Eqs. (25) and (29) as the derivation of the former equations require one assumption less.

The results shown above differ significantly from earlier estimates of the lift, such as those by Ruedy [13], Steinman [5], Schwabe [14] and Chen [15]. While the former three investigators obtained results which differ from the above presented equations in functional dependence as well as in numerically evaluated predictions for C_L, Chen obtained a prediction equation which, when numerically evaluated, will yield similar results if the relations given in the paragraph second to the next are used.

Estimation of Lift by an Extension of Lagally's Theorem

Sarpkaya [16] extended Lagally's theorem and obtained an expression for the forces experienced by a circular cylinder due to moving and growing vortices in time dependent flow. For the lift force per unit length Sarpkaya's theorem states that

$$F_y = -\rho \sum_{k=1}^{m} \Gamma_k (U - u_k + u_{ki}) - \rho \sum_{k=1}^{m} p_{ki} \frac{\delta \Gamma_k}{\delta t}, \tag{31}$$

where $$u_{ki} = \frac{\left(\frac{d}{2}\right)^2}{\left(p_k^2 + q_k^2\right)^2} \left[\left(q_k^2 - p_k^2\right) u_k - 2p_k q_k v_k \right], \tag{32}$$

and $$q_{ki} = \frac{q_k \left(\frac{d}{2}\right)^2}{p_k^2 + q_k^2}. \tag{33}$$

In the above equations, p_k and q_k are the coordinates of the center of the kth vortex having the circulation Γ_k and velocity components u_k, v_k. The additional index i refers to the image vortex, inside the circular streamline which represents the cylinder. Led in part by the results of the computer study simulating viscous flow around a circular cylinder [8] the following assumptions are made: (1) The vorticity which results in the formation of concentrated vortices is created near the upstream face of the cylinder; (2) the vorticity creation is steady and is not dependent upon the vortex shedding; (3) the vorticity produced per unit time near the left and right half of the cylinder results in the formation of the same number of vortices of equal strength but of opposite rotation; (4) while one vortex is shed, the other vortex, next in line for shedding, stays attached to the cylinder.

With the above assumptions Eq. (31) becomes

$$L = -\rho \sum_{k=1}^{m} \Gamma_k (U - u_k + u_{ki}). \tag{34}$$

The individual concentrated vortices which shed will join the vortex street. By the time such a vortex is in its proper position in the street its coordinates are such, that according to (32) and (33) $u_{ki} \ll u_k$. However, once a vortex has left the cylinder, there will be a net circulation about the cylinder due to the vortex next in line for shedding. The resulting lift per unit length therefore simply is

$$L = \rho \Gamma u_s . \tag{35}$$

With Eqs. (11) and (20), the coefficient of lift can now be obtained:

$$C_L = 4 \frac{l}{d} \left(1 - \frac{Sl}{d}\right)^2 \coth \frac{\pi h}{l} \tag{36}$$

or

$$C_L = 4\sqrt{2} \frac{l}{d} \left(1 - \frac{Sl}{d}\right)^2 , \tag{37}$$

where for the latter equation the vortex spacing ratio is given by Eq. (6).

The Lift Coefficient as a Function of the Strouhal Number

Bearman [17] succeeded in deriving a relationship between the vortex spacing ratio h/l and the velocity ratio u_s/U by applying the Kronauer stability condition,

$$\frac{\partial C_{DS}}{\partial \left(\frac{h}{l}\right)_{u_s/U}} = 0 . \tag{38}$$

Bearman obtained that

$$2 \cosh \frac{\pi h}{l} = \left(\frac{U}{u_s} - 2\right) \sinh \frac{\pi h}{l} \left(\cosh \frac{\pi h}{l} \sinh \frac{\pi h}{l} - \frac{\pi h}{l}\right) . \tag{39}$$

As indicated by Sallet [9], it is found, that

$$\frac{l}{d} = 0.86 \frac{1}{S} \tag{40}$$

if von Kármán's stability criterion, namely Eq. (6) is also satisfied. With Eq. (40) Eq. (29) simplifies to

$$C_L = \frac{0.099}{S} \tag{41}$$

and Eq. (37) simplifies to

$$C_L = \frac{0.095}{S} . \tag{42}$$

Hence the analyses in Sections 2 and 3 give for the here stated assumptions and specializations substantially the same numerical result. The Strouhal number is a function of the Reynolds number for a given bluff cylinder. For circular cylinders, for instance, Roshko's [10] emperically found equations of the Strouhal number as a function of the Reynolds number may be substituted into the last two equations. Using Eq. (41) one obtains

$$C_L = \frac{0.467Re}{Re - 21.2}, \quad \text{if } 50 < Re < 150 , \tag{43}$$

and

$$C_L = \frac{0.467Re}{Re - 12.7}, \quad \text{if } 300 < Re < 2000 . \tag{44}$$

As an approximation, Eqs. (43) and (44) can be extrapolated to cover the Reynolds number range from 150 to 300. Since there is little change of the Strouhal number in the Reynolds number range 2,000 to 400,000, Eq. (44) is also a good approximation for the coefficient of lift of circular cylinders in this domain. For the estimation of lift at still higher Reynolds numbers, Eq. (41) in conjunction with the proper Strouhal number should be used.

The Change of Lift Coefficient due to Transverse Vibrations

The momentum analysis, in general, cannot yield an estimate of the lift when the cylinder is elastically supported, i.e., free to undergo flow induced vibrations. In order to establish qualitatively a trend as to how a transverse vibration will influence the lift forces in the absence of all other effects, such as inertial and added mass effects, the vibrating cylinder of width d is replaced by a fictitious steady cylinder of width $\bar{d}$ (see Fig. 5). The width $\bar{d}$ is equal to the projected distance between the two extreme positions of the vibrating body. Assuming that the vortex formation process and the vortex street properties do not differ from each other in the above described two models, Eqs. (11) and (19) will stay the same except that h and l will now become $\bar{h}$ and $\bar{l}$. Over a certain frequency range the vortex shedding of the vibrating cylinder has been observed to occur at the frequency of its own vibration [6]; Eq. (20) therefore becomes

$$\frac{U - u_s}{U} = k\, S\, \frac{\bar{l}}{\bar{d}} \,. \qquad (45)$$

According to Den Hartog [18], the "locked-in" region in which the vortex shedding takes place at or near the frequency of the vibrating cylinder extends from 0.8 S to 1.2 S, i.e., the values for k will lie between 0.8 and 1.2 for a free vibration. The vortex spacing ratio $\bar{l}/\bar{d}$ can be calculated from Eq. (26) by replacing l/d by $\bar{l}/\bar{d}$ and S by $k \cdot S$. It is seen that for values of k which are larger than 1.0, the vortex spacing ratio $\bar{l}/\bar{d}$ decreases and that for values of k which are smaller than 1.0 $\bar{l}/\bar{d}$ increases. Equation (29) then yields, after replacing l/d by $\bar{l}/\bar{d}$ and S by $k \cdot S$, the lift acting on the vibrating body. The coefficient of lift is seen to decrease for k larger than 1.0 and to increase for k smaller than 1.0. If one further stipulates the validity of Kronauer's stability condition and that l/d is proportional to $\bar{l}/\bar{d}$, it is found that the coefficient of lift is inversely proportional to the product of k and S. Due to the omission of all other effects, which was necessary to replace the vibrating body by a fictitious steady body, an evaluation of any proportionality constants would only yield rough estimates of the lift coefficients. There is, however, some experimental verification for the above indicated trends. A decrease in longitudinal vortex spacing at the upper frequency limit of the free vibration and an increase in vortex spacing at the lower limit was observed by Koopman [19]. Bishop and Hassan [20] found that the lift coefficient increased as the cylinder was oscillating at a frequency lower than the natural Strouhal frequency of the stationary cylinder, and that the lift coefficient decreased (below its value for the stationary cylinder) when the cylinder was vibrating at a higher frequency. As an example, the lift of a vibrating circular cylinder is compared with that of a stationary cylinder. Let the k values be 1.0 no vibration, 0.9 and 1.05. At a Reynolds number of 6,000 the coefficient of drag for a circular cylinder is 1.00 and the Strouhal number is 0.205. For values of $k = 0.9$, 1.0 and 1.05, Eq. (26) yields the values of $\frac{l}{d}$ of 4.60, 4.03 and 3.76, respectively. Equation (29) now predicts coefficients of lift for $k = 0.9$, 1.0 and 1.05 of $C_L = 0.538$, 0.473 and 0.436 respectively. The ratios of C_L at $k = 0.9$ to C_L at $k = 1.0$ and C_L at $k = 1.05$ to C_L at $k = 1.0$, therefore, become 1.12 and 0.93, respectively. Experiments by Bishop and Hassan [20] at approximately the same Reynolds number show these ratios to be 1.21 and 0.80 (extrapolated). The experimental results were found when the ratio of the amplitude of the vibration to the diameter of the cylinder was 0.3.

Discussion

The derivation of the above equations which permit estimating of the lift coefficients is based on two basic assumptions in addition to the series of simplifying assumptions which were stated in the development of the analysis. The two basic assumptions are: (1) A two-dimensional analysis is justified; (2) the staggered potential vortex street sufficiently simulates the real wake.

The first basic assumption is usually justified by requiring a large length to diameter ratio of the vortex shedding cylinder. According to experimental results the coefficient of drag of a cylinder is independent of the length-to-diameter ratio once this ratio exceeds a value of 10. For the above given determination of the lift, however, a two-dimensional analysis is only justified if the vortex shedding and the formation of the vortex street is truly two-dimensional. One can for instance easily envision the possibility, that a vortex is not shed simultaneously all along the cylinder. Gerrard [21] discusses these and other three-dimensional characteristics of the vortex wake.

The second basic assumption invites more comment than can be given here. At the beginning of the analysis, the finite vortices in the actual wake were replaced by potential line vortices. The momentum gained in the control volume and the momentum efflux from the control volume do however depend upon the fluid mass moving with a vortex pair and the velocity profile across the wake, respectively. In addition, for vortices with finite cores, Eq. (11) only approximates the relation between the propagation velocity of the vortex street and its spacing ratio, provided the core of the real vortices is small when compared to the shortest distance between two neighboring vortices. For a complete description of the development of a vortex street in a real fluid, i.e., the size and growth of the vortex cores, the change of the spacing ratio h/l and the change of the propagation velocity of the street with increasing distance from the cylinder, the reader is referred to the article by Wille and Timme [22]. While at higher Reynolds number ranges, there is generally less objection in analyzing viscous flow phenomena with invicid flow theory, such a statement cannot be offered in support of the above developed expressions for the coefficient of lift. All of the prediction equations are seen to be dependent upon the frequency of vortex shedding. At certain higher Reynolds-number ranges, however, the spectra of the Strouhal number suddenly widen and may even exhibit two peaks, as discussed by Bearman [23].

In view of the stated reasons why most of the above assumptions cannot be fully justified, the comparison of the estimated lift with the experimentally measured lift shows surprisingly good agreement (see Fig.6). This agreement may in part be explained by the fact that the relation giving the vortex spacing is in terms of the experimentally obtained coefficient of drag and Strouhal number. It is desirable that the parameters C_D and S, which are substituted in the prediction equations, originate from the same experiments. Since such data are not available in the literature, only the results of the simplified equations as given in section 4 are presented in Fig.6. In all of the above numerical evaluations for the lift coefficient and for the longitudinal vortex spacing, equations were used which included von Kármán's value for h/l. In view of the strong influence which the vortex spacing has upon the predicted values of the coefficient of lift, this ratio should also be considered a variable and experimentally determined as a function of the Reynolds number. The fact that the potential vortex street would then become more unstable for values of h/l other than 0.280549 is of secondary importance, since the purpose of the potential-flow model is to predict the lift forces due to a hypothetical vortex street which is so constructed that it correctly predicts the easily measured drag forces, and not to predict the flow pattern of the real wake.

The inviscid flow analysis is of more value as a hypothetical model which sufficiently explains the nature and magnitude of the lift rather then giving an exact description of the details of the wake. Equations (41) and (42) are in accord with the general conclusion reached from an experimental study by Protos, Goldschmidt and Toebes [24], namely, that the lift force on the steady cylinder varies in inverse proportion to the frequency of the vortex shedding.

References

1. Drescher, H.: Messung der auf querangeströmte Zylinder ausgeübten zeitlich veränderten Drücke. Z. Flugwiss. 4, 17-21 (1956).

2. Meier-Windhorst, A.: Flatterschwingungen von Zylindern im gleichmässigen Flüssigkeitsstrom. Mitteilungen des Hydraulischen Instituts der Technischen Hochschule München, H. 9, 1-29 (1939).

3. Toebes, G.H.: The Unsteady Flow and Wake Near an Oscillating Cylinder. J. of Basic Engng. Trans. ASME. 493-505 (September 1969).

4. Thoma, D.: Die Entstehung des Antriebes bei Flatterschwingungen. Mitteilungen des Hydraulischen Instituts der Technischen Hochschule München, H. 8, 93-98 (1936).

5. Steinman, D.B.: Problems of Aerodynamic and Hydrodynamic Stability. Proceedings of the 3rd Hydraulics Conference, Iowa University, Bulletin No. 31, 136-163 (1947).

6. Parkinson, G.V., Brooks, N.P.H.: On the Aeroelastic Instability of Bluff Cylinders. J. Appl. Mech., Trans. ASME, 252-258 (June 1961).

7. von Kármán, Th., Rubach, H.: Über den Mechanismus des Flüssigkeits- und Luftwiderstandes. Phys. Z. No. 2, 49-59 (Januar 1912).

8. Dawson, Ch., Melvyn, M.: DMC-A Computer Code to Simulate Viscous Flow About Arbitrarily Shaped Bodies. Proc. of the 1970 Heat Transfer and Fluid Mechanics Institute, Stanford: Stanford University Press 1970, pp. 323-338.

9. Sallet, D.W.: On the Spacing of Kármán Vortices. J. Appl. Mech., Trans. ASME, 370-372 (June 1969).

10. Roshko, A.: On the Development of Turbulent Wakes From Vortex Streets. NACA Report 1191.

11. Goldstein, S., ed.: Modern Developments in Fluid Dynamics, 1st ed., Vol. 2, Oxford: Oxford University Press 1938, p. 571.

12. Roshko, A.: Experiments on The Flow Past a Circular Cylinder at Very High Reynolds Number. J. of Fluid Mechanics, Vol. 10, Pt. 3, May 1961, pp. 345-356.

13. Ruedy, R.: Vibrations of Power Lines in a Steady Wind. Canadian Journal of Research 13, 82-92 (October 1935).

14. Schwabe, M.: Über Druckermittlung in der nichtstationären ebenen Strömung. Ing.-Arch. 6, 34-50 (1935).

15. Chen, Y.N.: Fluctuating Lift Forces of the Kármán Vortex Street on Single Circular Cylinders and in Tube Bundles. ASME Paper No. 71-Vibr.-12.

16. Sarpkaya, T.: Lift, Drag, and Added-Mass Coefficients for a Circular Cylinder Immersed in a Time-Dependent Flow. J. Appl. Mech., Trans. ASME, 13-15 (March 1963).

17. Bearman, P.W.: On Vortex Wakes. J. Fluid Mech. 28, 625-641 (1967).

18. Den Hartog, J.P.: Recent Technical Manifestations of Von Kármán's Vortex Wake. Proc. of the National Academy of Sciences 40, 155-157 (1954).

19. Koopmann, G.H.: The Vortex Wakes of Vibrating Cylinders at Low Reynolds Numbers. J. of Fluid Mech. 28, 501-512 (1967).

20. Bishop, R.E.D., Hassan, A.Y.: The Lift and Drag Forces on a Circular Cylinder Oscillating in a Flowing Fluid. Proc. Royal Society, London, Ser. A, 277, 51-75 (1964).

21. Gerrard, J.H.: The Three-Dimensional Structure of the Wake of a Circular Cylinder. J. Fluid Mech. 25, 143-164 (1966).

22. Wille, R., Timme, A.: Über das Verhalten von Wirbelstrassen, Jahrbuch der Schiffbautechnischen Gesellschaft, Vol. 51, 215-221 (1957).

23. Bearman, P.W.: On Vortex Shedding from a Circular Cylinder in the Critical Reynolds Number Regime. J. Fluid Mech. 37, 577-587 (1969).

24. Protos, A., Goldschmidt, V.W., Toebes, G.H.: Hydroelastic Forces on Bluff Cylinders. J. of Basic Engineering. Trans. ASME, 348-386 (September 1968).

25. Bishop, R.E.D., Hassan, A.Y.: The Lift and Drag Forces on a Circular Cylinder in Flowing Fluid. Proc. Royal Soc., London, Ser. A, 277, 32-50 (1963).

26. Fung, Y.C.: Fluctuating Lift and Drag Acting on a Cylinder in a Flow at Supercritical Reynolds Numbers. J. Aerospace Science, 27, 801-814 (November 1960).

27. Goldman, R.L.: Kármán Vortex Forces on the Vanguard Rocket. The Shock and Vibration Bulletin, No. 26, Part 2, 171-179 (December 1958).

28. Jordan, K.S., Fromm, J.E.: Dynamic Interaction between a Circle Moving at Terminal Velocity and the Surrounding Fluid Medium. AIAA Paper No. 72-III, AIAA 10th Aerospace Sciences Meeting, San Diego, Ca., 1-9 (January 1972).

29. Keefe, R.T.: Investigation of the Fluctuating Forces Acting on a Stationary Circular Cylinder in a Subsonic Stream and of the Associated Sound Field. J. Acoustical Soc. of America 34, 1711-1714 (November 1962).

30. McGregor, D.M.: An Experimental Investigation of the Oscillating Pressures on a Circular Cylinder in a Fluid Stream. University of Toronto Institute of Aerophysics, Technical Note 14, June 1957.

31. Phillips, O.M.: The Intensity of Aerolian Tones. J. Fluid Mech. 1, 607-624 (December 1956).

32. Schmidt, L.V.: Measurements of Fluctuating Air Loads on a Circular Cylinder. J. Aircraft 2, 49-55 (1965).

33. Warren, W.F.: An Experimental Investigation of Fluid Forces of an Oscillating Cylinder. Ph. D. Thesis, University of Maryland 1962.

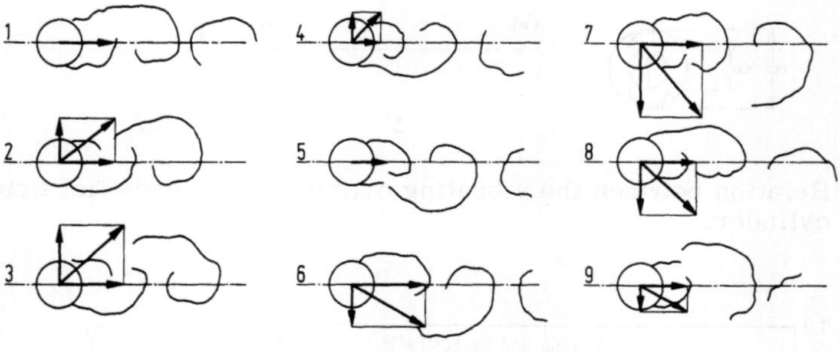

Fig. 1. Fluctuating lift due to uniform approach flow after Drescher.

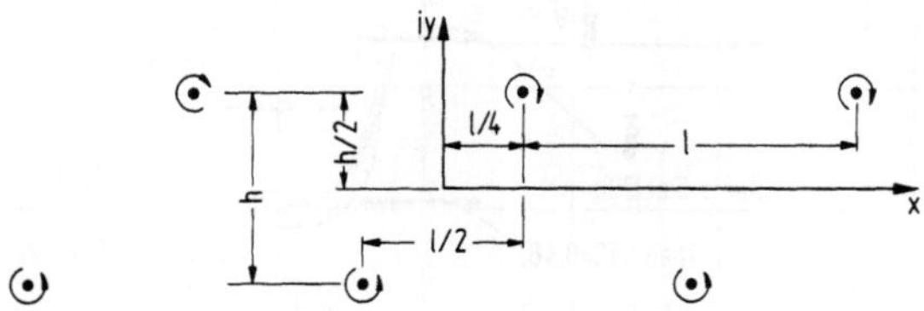

Fig. 2. Vortex street.

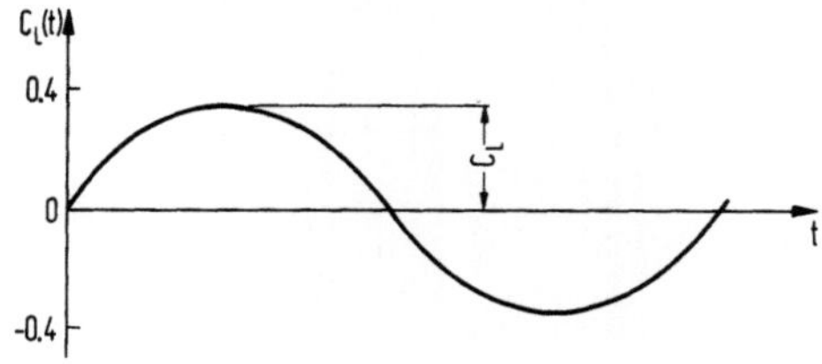

Fig. 3. The coefficient of lift $C_L(t)$ of a circular cylinder.

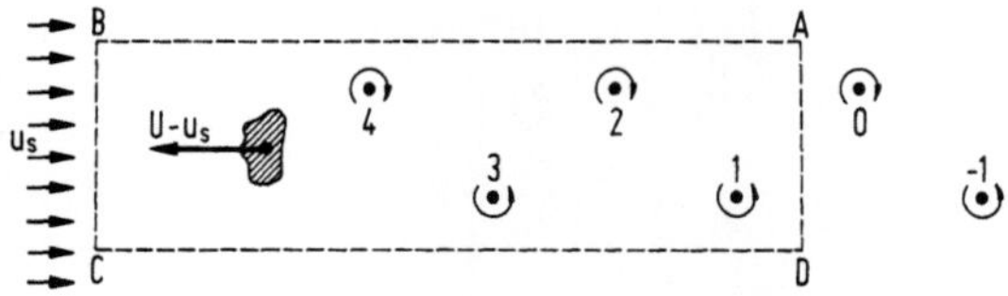

Fig. 4. Control volume.

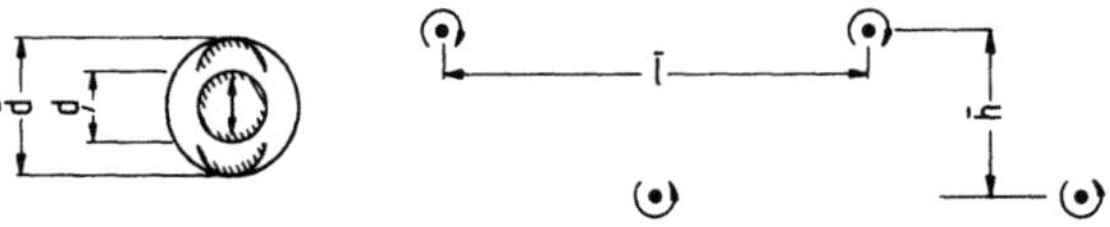

Fig. 5. Relation between the vibrating cylinder and steady fictitious cylinder.

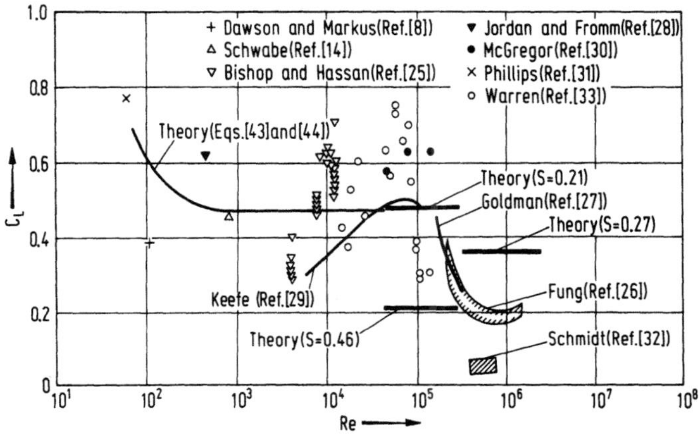

Fig. 6. Experimentally obtained values of C_L and estimated values of C_L (simplified theory) as a function of the Reynolds number.

The Low Frequency Components of Separated Flows

By

J.H. Gerrard

Manchester, U.K.

Summary

A numerical solution of the equations of motion for the incompressible two-dimensional flow down a step at a Reynolds number of 100 produced a basic flow on which to perform numerical experiments. The basic flow was found to execute oscillations spontaneously. The frequency of these corresponded with the most unstable frequency of a free shear layer in parallel flow. Low frequency fluctuations were absent. The basic flow was perturbed in a manner which modelled turbulence in the separated shear layer at high Reynolds numbers. The disturbances were found to produce the same frequencies as those occurring spontaneously and again no low frequency components. Recirculation within the separated region is a possible mechanism for the production of low frequencies. Scrutiny of the development of the disturbances showed that they died out due to the effects of diffusion of vorticity within the low velocity separated region. It is concluded that a low Reynolds number two-dimensional separated flow will be free from low frequency fluctuations. Low frequency components may also be absent at higher Reynolds numbers when the flow is two-dimensional.

Introduction

It has been amply demonstrated that the characteristics of the wakes of bluff bodies are strongly dependent upon, if not entirely governed by, the exchange of vorticity bearing fluid in that part of the wake immediately downstream of the separation points. This part of the wake is the vortex formation region. Besides the vortex shedding or Strouhal frequency other frequencies are detected in the wake to an extent depending upon position in the wake and the Reynolds number. Frequencies very much less than the shedding frequency have been reported at all Reynolds numbers at which vortex shedding occurs. At low Reynolds numbers these oscillations can be regular. The low frequencies, apparent as a modulation of the shedding frequency signal in the developed wake, are predominant in the formation region.

There are a variety of causes of the low frequency fluctuations. They have been shown to be at least partly attributable to three-dimensional effects by Gaster [1] and by the author [2]. A numerical two-dimensional model of the oscillating wake has shown that low frequency oscillation may exist without any three-dimensional effects [3], but this result is inconducive on this point.

The present work is aimed at investigating low frequency fluctuations in a two-dimensional flow by numerical experiments. For this purpose the simplest flow has been chosen, namely the flow over a rearward facing step. This choice has been made because it was found that when vortex shedding from a circular cylinder at high Reynolds number was inhibited by attaching a long splitter plate, the spectral components at other than the shedding frequency were essentially unaltered. It thus appears that the low and high frequency components are decoupled from the Strouhal frequency. In other fields the mutual effect of widely differing scales upon each other has been demonstrated to be strong (see [4]). It may be that fluid injected into the separated region attending the contortions of the shear layer by the transition waves could trigger and maintain low frequency oscillations.

The aim of the numerical computations was to produce a steady flow on which to do numerical experiments. The oscillations which arose spontaneously are found to be as expected from other analytical and experimental work [5,8]. The response of the flow to perturbations has also been investigated. The study is restricted to laminar flow at a Reynolds number based on step height of 100. The frequencies observed are those expected from the instability of the free shear layer. Low frequency oscillations of the separated region are absent. A reason for this is suggested which may also hold at higher Reynolds numbers. The conclusion is thus the negative result that a two-dimensional separated region is not expected to undergo low frequency oscillations.

Experimental Results

The Strouhal frequency detected in the wake of a bluff body which is shedding vortices may be suppressed by the insertion of a long splitter-plate down the centre of the wake. Figure 1b shows this effect. The high frequency, transition wave, component is still present when the splitter plate is attached though it is a little different in character and frequency. In order to see the effect of the plate on the low frequency components a low-pass filter was inserted in the circuit. Figures 1c and d show the low frequency components. In Fig. 1c particularly we see that it is impossible to distinguish by eye which trace was obtained with the splitter in position. This strongly indicates that the low frequency fluctuations are independent of the shedding frequency.

Calculation of the Basic Flow

The two-dimensional flow was determined numerically by a solution of the vorticity and continuity equations for an incompressible fluid by well established methods. The equations were expressed in central finite difference form and a steady state solution sought by an explicit method for the time-wise advance from an assumed initial state. Presentation of the equations and discussion of the method are omitted to economise in space. The method is discussed by the author, for axisymmetric flow in Ref. [6]. At each time step the vorticity, ζ, and the stream function, ψ, are calculated at each interior mesh point.

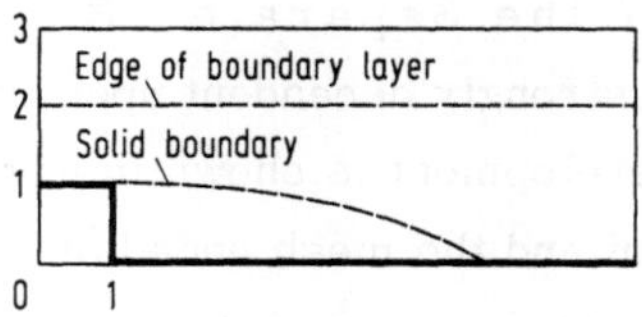

Values on some of the boundaries are then adjusted. The boundary conditions require some comment. The boundaries of the flow are shown in the sketch. On the solid boundaries the stream function, $\psi = 0$ and the value of the vorticity, ζ is extrapolated from the flow. If suffixes indicate number of mesh lengths from the solid surface along a line perpendicular to the surface, then

$$\zeta_0 = 6\psi_1/h^2 - 3.5\,\zeta_1 + 2\,\zeta_2 - 0.5\,\zeta_3$$

where h is the mesh length perpendicular to the surface. At the entry, ψ and ζ are independent of time: between 1 and 2 step heights their values correspond to a Blasius boundary layer, between 2 and 3 step height to a uniform stream of speed U. At the right hand side of the flow ψ and ζ are equated to the values they have one mesh length to the left. On the top boundary ψ is linearly extrapolated from the two adjacent mesh points immediately below and ζ is kept constant at zero. This upper boundary condition is almost equivalent to a free surface above which there is fluid of zero viscosity.

The initial conditions were arbitrarily chosen. The Blasius boundary layer continued unaltered along the step, separated from the corner and reattached to the lower wall. There was no flow in the separated region. The flow eventually became independent of these initial conditions. The computer program was arranged to expand the field of flow when the point of reattachment moved out of the field of calculation.

After the initial period during which the mesh length in the free stream direction was changed the mesh lengths were 0.1 s and the time step 0.05 s/U, where s is the step height and U is the free stream speed. A change to 1/5 of this time step made little difference.

The Development of the Flow from the Assumed Initial Conditions

(a) Development of the Separated Region. The first stages of development are strongly dependent upon the arbitrary initial conditions. The later development is shown in Figs. 2a to f. The mesh length remained constant and the mesh ended at 7 step heights (s) downstream of the step for times greater than 23 s/U. Figure 2b indicates the magnitude of the reversed flux in the separated region. The region grew larger with time and the reversed flux increased to a maximum value considerably larger than the final 'steady' value. This was followed by a relaxation and oscillation of all quantities considered. The relaxation was associated with a shedding of some of the separated-region circulation. It would be interesting to know whether this occurs in a real flow started from rest. During this period the contours of ψ = -0.01 are shown superimposed in Fig. 2d the development is like a tadpole losing its tail whilst the head remains of fixed size and position.

All the quantities considered continue to oscillate with slowly decreasing amplitude. The frequencies of oscillation are discussed later.

(b) Development Outside the Separated Region. Small oscillations in the stream function and vorticity were detected in the early stages of development in the region above the separated shear layer. They are seen in the developed phase in Fig. 3. The wave motion is seen to be concentrated near to the "free" surface. If the streamline indicated and the top of the field were coincident the top of the flow would be truly a free surface. Gradients of vorticity in the vertical direction are small in this region.

It is considered that these waves are associated with the free surface. They have been observed by Mattingley [7] above the oscillating wake of a cylinder when it was very close to the free surface in a water channel. It was observed in the numerical experiments that these waves were modulated with a frequency which was related to the oscillatory motions near to the wall.

Eventually the waves broke at the downstream end of the flow. The computing was terminated by the excessively large vorticity values generated in this process. The process is not unlike transition to turbulence. It would be interesting to see whether it occurred or could be made to occur in the absence of the free surface.

(c) The Final State of the Separated Region. The oscillations of the separated regions are small enough for us to refer to a final steady state. To the topic of low frequency fluctuations the representation of Fig. 4 is the most apposite. This shows time lines and particle paths within the separated region. Three particle lines initially radiating in straight lines from the centre of rotation were followed. The numbers on the curves indicate the distance in step heights travelled by the free stream since the particles left their initial positions. Reference to Fig. 3 shows that vorticity is neither concentrated nor uniformly distributed over this region. It is remarkable that the motion is very like solid body rotation in the top half of the region and not very different from an irrotational vortex flow (velocity inversely proportional to distance from the centre) in the lower half.

The oscillations discussed in the previous sections and below have periodic times less than 15 s/U. If we seek a mechanism for lower frequency fluctuations in the rotation within the separated region we see from Fig. 4 that this relates to a period in excess of 60 s/U. It is also apparent that disturbances will rotate in widely differing times depending upon their distance from the centre of rotation which would indicate a broad spectrum of oscillations due to this cause. It was not possible in the numerical work to study the flow for the very long times required. On the ICL 1906 computer in use at Manchester the computation occupied about 5 minutes of mill time for a lapse of 10 s/U. The disruption of the surface waves in this model eventually terminated the computing after a total time from the initiation of the flow of 55 s/U. Conclusions however can be drawn from tracing round their path in the recirculating flow the history of disturbances specifically introduced.

The Effect of Disturbances

The basic flow at a Reynolds number of 100 remains laminar. Flows of practical interest are at least partly turbulent. To model turbulent eruptions from the separated shear layer into the separated region, the vorticity distribution was disturbed near the edge of the separated layer ($\psi \simeq 0$). At this low Reynolds number, the effect of the disturbance diffuses as well as being convected. Only to some extent does this model the turbulent diffusion at higher Reynolds numbers. The velocity in the separated region is less than one tenth of the free stream speed. The effect of various patterns of vorticity distribution on the position of reattachment is shown in Fig. 5. The separation streamline and the positions of the disturbances are shown on the left of the figure. The first four disturbances have regions (of 9 mesh points forming a square) where the vorticity is elevated or reduced by 0.2. In this region the basic flow gradient of vorticity is about 0.2 per 0.1 s change in distance. The delay in onset of the third disturbance at the reattachment point indicates a speed of the disturbance of about half the free stream speed. The second and fourth cases show that oppositely signed excursions of reattachment position are produced by oppositely signed disturbances. With certain distributions of disturbance one might imagine that an amplified effect may be produced. The effect of one random distribution of vorticity with the same maximum amplitude (0.2) is shown. The amplitude and wavelength are about the same as those caused by the other disturbances. Though the excursions of the reattachment point are small, the oscillations of velocity nearby are about 10% of the local velocity.

The Observed Frequencies and Absence of Low Frequencies

Many of the oscillations computed have about the same period. From Fig. 2 we deduce a non-dimensional period T U/s of 9 to 17 from the oscillations of reattachment and of 13 from the minimum stream function. The disturbances (Fig. 5) produce a predominant period of about 10 s/U. The surface waves had a wavelength of 0.85 s and a speed of 0.6 U and thus produce a period of 1.4 s/U. These waves were modulated with a wavelength of about 3 s and a modulation speed of 0.24 U corresponding to a period of 12.5 s/U.

Betchov and Criminale [8] consider the stability of a free shear layer with a velocity profile $U = U_0 \tanh y/H$ which has a width of 5H. The wave number of maximum instability (non-dimensionalised with H) is 0.45. In our case 5H = 2s: the velocity profiles over most of the flow approximate to tanh curves of width 2 s. Also our free stream speed corresponds to $2U_0$. Taking the velocity of the disturbances to be half the free stream speed results in a most unstable period of 11.2 s/U. This we expect to correspond to the so called transition wave period. It certainly agrees with the observed periods of all but the surface waves. It is also in agreement with the dominant frequency observed in the separated flow downstream of a leaf gate by Narayanan [5] at high Reynolds numbers.

Let us now turn to the lower frequencies associated with rotation of the separated flow. These we have conjectured (at a Reynolds number of 100) will have periods of 60 s/U and above. This is the order of magnitude of the period expected from studies of oscillating wakes of bluff bodies. Here the equivalent shedding period is about $5 \times 2s/U$ and the low frequencies several times this value and above. Such low frequencies have not been observed in this numerical work. It may appear that this is because we have not followed the development for a sufficiently long time. A detailed mapping of the disturbance vorticity field reveals that the disturbance effects die out. When a disturbance is introduced several effects follow. Initially there is the kinematic effect due to the presence of the boundary. Introduction of positive vorticity close to a wall result in the creation of negative vorticity at the wall. Both convection and diffusion of vorticity are important and the resulting effect depends on their relative importance. The local Reynolds number in the separated region is very small and vorticity is diffused faster than it is convected. Diffusion is, in fact, so important that the vorticity disturbance is reduced to about 1% of its initial value in times of the order of 1 s/U. It is also destroyed by inter-diffusion of the disturbance and its oppositely signed wall effect. We can conclude with some certainty that the separated region at low Reynolds number will have no low frequency fluctuations if the flow is strictly two-dimensional. At higher Reynolds number, not studied here, convection becomes more important. Experiment suggests [9] that in the free wake (without a splitter plate) the velocities in the region close behind the body are extremely low even at high Reynolds numbers.

Conclusions

The numerical studies of the development of the incompressible two-dimensional flow down a step have revealed oscillations in the flow.

The only frequency remaining after the flow has developed (ignoring the high frequency free-surface waves) is attributable to instability of the free shear layer and corresponds to what have been termed transition waves. The same frequencies were produced when the flow was disturbed. The absence of low frequency fluctuations at this Reynolds number of 100 is attributed to the annihilation of vorticity disturbances in the separated region where the speed is very low and diffusion consequently important. It is suggested that even at higher Reynolds numbers low frequency fluctuations will be absent in a separated region when it is two-dimensional.

References

1. Gaster, M.: Vortex Shedding from Slender Cones at Low Reynolds Number. J. Fluid Mech. 38, 565 (1969).

2. Gerrard, J.H.: The Three-Dimensional Structure of the Wake of a Circular Cylinder. J. Fluid Mech. 25, 143 (1966).

3. Gerrard, J.H.: Numerical Computation of the Magnitude and Frequency of the Lift on a Circular Cylinder. Phil. Trans. Roy. Soc. A 261, 137 (1967).

4. Mollo-Christensen, E.: Physics of Turbulent Flow. AIAA J. 9, 1217 (1971).

5. Narayanan, R., Reynolds, A.J.: The Reattaching Flow Downstream of a Leaf Gate. J. ASCE. Hydraulics Division (May 1972).

6. Gerrard, J.H.: The Stability of Unsteady Axisymmetric Incompressible Flow Close to a Piston. Part I, Numerical Analysis. J. Fluid Mech. 50, 625 (1971).

7. Mattingley, G.E.: An Experimental Study of the Three Dimensionality of the Flow around a Circular Cylinder. Univ. of Maryland Tech. Note BN-295 (1962).

8. Betchov, R., Criminale, W.O.: Stability of Parallel Flows, Academic Press 1967.

9. Nielsen, K.T.W.: Vortex Formation in a Two-Dimensional Periodic Wake. Ph. D. Thesis Oxford, Engineering Science, 1970.

a 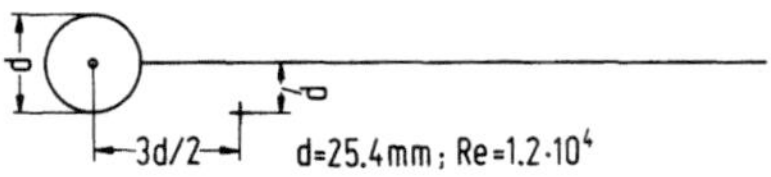

Circular cylinder with splitter plate and hot wire position. (Shedding frequency 50 Hz; Transition wave frequency ca. 600 Hz; Predominant low frequency component ca. 18 Hz.)

b 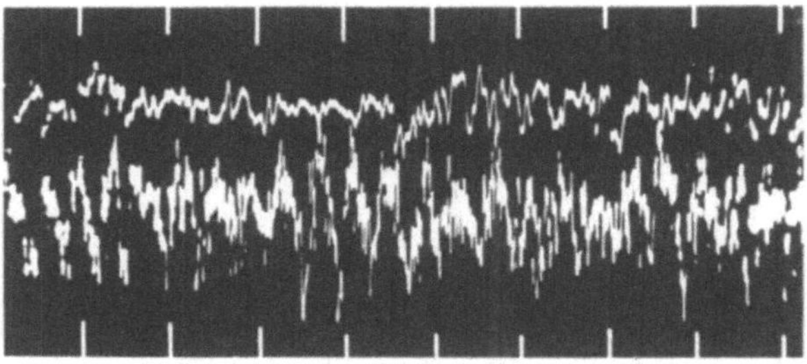

Time base 50 msec/cm.

c 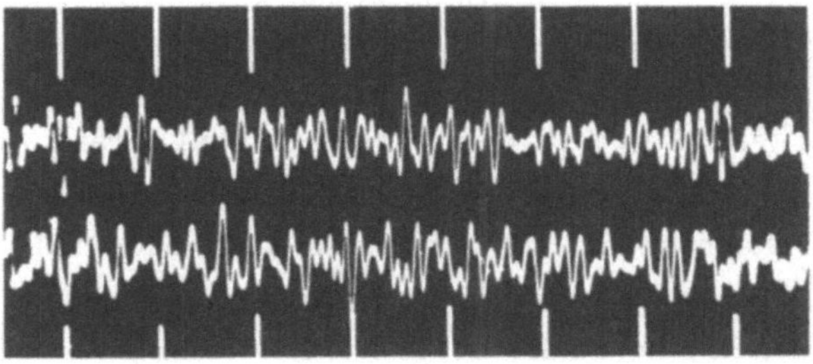

Time base 0.5 sec/cm (low pass filter with 20 Hz cut off).

d 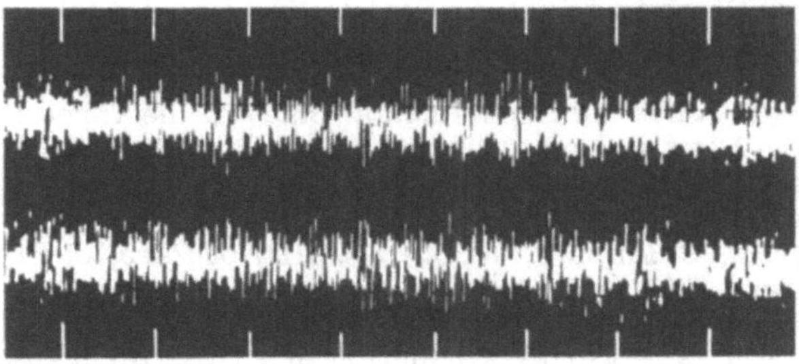

Time base 2 sec/cm (low pass filter with 20 Hz cut off).

Fig.1. Oscillograms from hot wire. (Top trace with splitter plate; lower trace cylinder alone.)

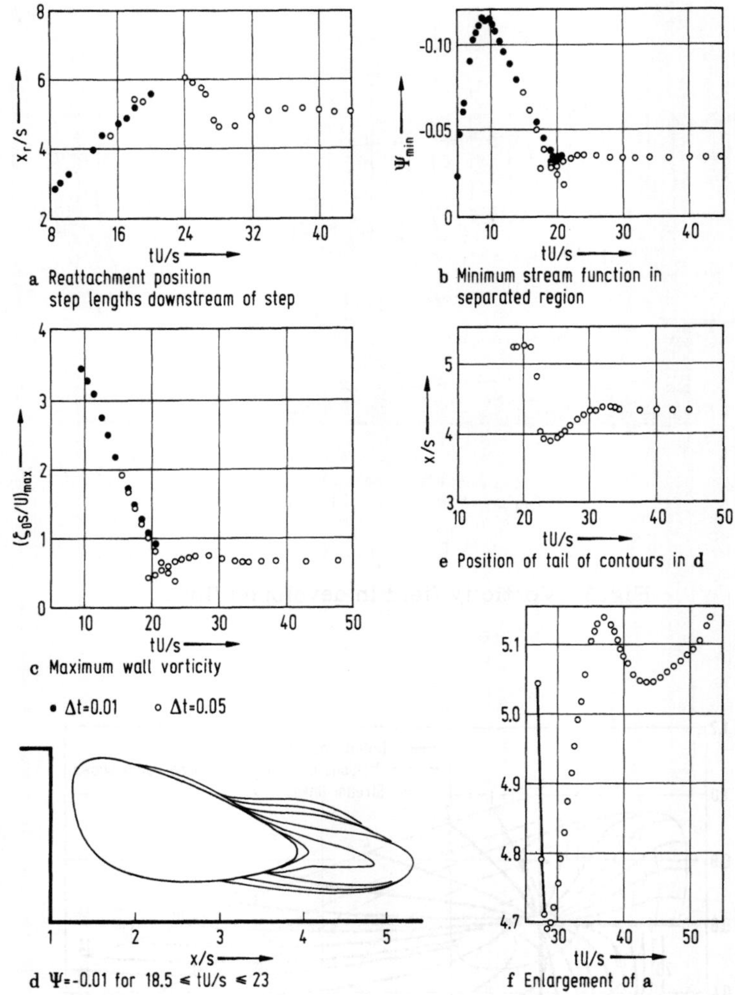

Fig. 2a-f. Development of the flow down a step.

(a) Reattachment position x_r in step lengths downstream of step.

(b) Minimum stream function in separated region.

(c) Maximum wall vorticity.

(d) $\psi = -0.01$ for $18.5 \leqslant tU/s \leqslant 23$.

(e) Position of tail of contours in (d).

(f) Enlargement of plot (a).

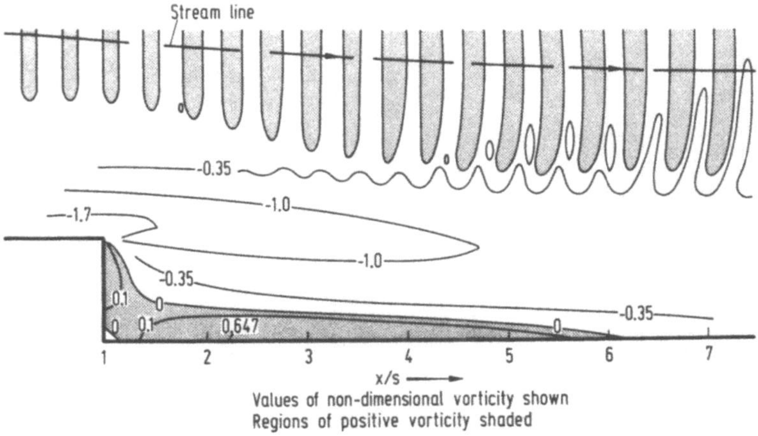

Fig. 3. Vorticity field in developed flow.

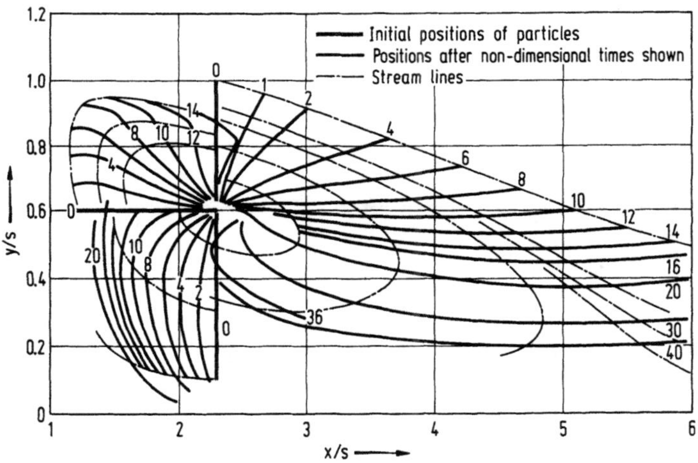

Fig. 4. The developed separated region.

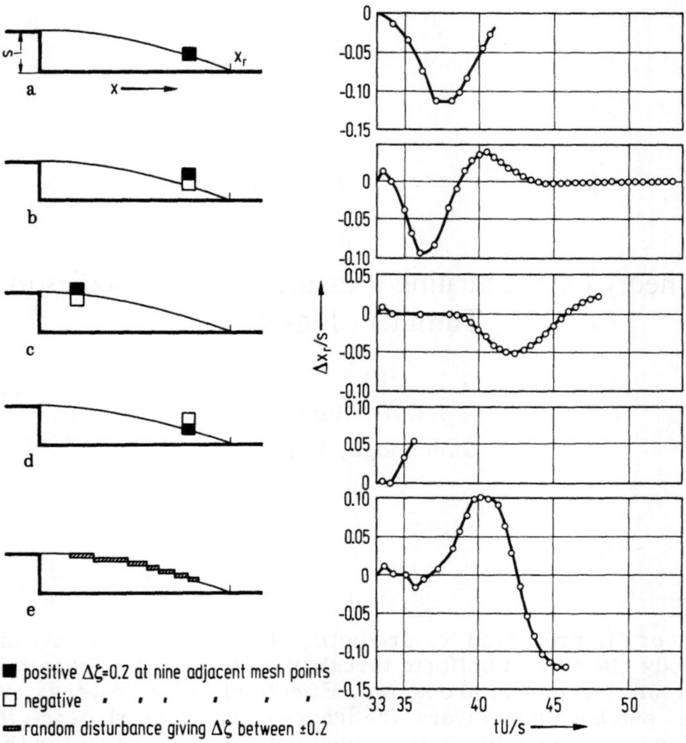

Fig. 5. Effect of vorticity disturbances on the point of reattachment.

A Theory for Fluctuating Pressures on Bluff Bodies in Turbulent Flows

By

J.C.R. Hunt

Cambridge, U.K.

Summary

A new theory is presented for predicting the turbulent velocity around a bluff body placed in a uniform turbulent flow when the properties of the upstream turbulence are known. Effects of the wake (e.g., vortex shedding) and boundary layers are ignored, and the turbulence is weak enough for the mean velocity to be unaffected by it. The linear theory enables spectra and correlations of velocity and pressure on the body to be calculated. These results are used to examine the formulae for fluctuating pressures and loads and their physical interpretations, as developed by Davenport and Vickery. The theory confirms experimental results of Bearman that these formulae are invalid when the scale of the incident turbulence is comparable with the size of the body. New formulae are proposed and, more important, new concepts are developed to describe these complicated flows.

Introduction

The aim of this paper is to demonstrate the inadequacy of the present theory and to provide a new theory for calculating the fluctuating pressures and loads on bluff bodies caused by the pressure of turbulence in the incident flow.

The present basis developed by Davenport and Vickery [6] for calculating fluctuating pressures and loads on bluff bodies in turbulence is as follows: Firstly, it is assumed that the pressure p^* at a point

(x',y') is a function of the upstream velocity u_∞ and rate of change of velocity far upstream at $(-\infty, y')$, so that

$$p^*(t) = \frac{1}{2}\rho u_\infty^2(t)\, a^2 C_p + \rho a^3 (du_\infty/dt) C_m(x',y'). \qquad (1)$$

where ρ is the density and $2a$ is the diameter of the body (Fig. 1). C_p and C_m can be found by potential flow theory, only if the upstream velocity is assumed to be constant over a region large compared with a. In fact Vickery [10] assumed Eq. (1) to be valid even if u_∞ does vary over a distance $0(a)$. The second step in the calculation has been to calculate the load on the body from the knowledge of the correlation coefficients of the upstream turbulence. This step, introduced by Vickery [10], is of course valid only if Eq. (1) is also. Bearman [3] showed experimentally that when the integral scale of the incident turbulence L_x is of the order of a, this calculation procedure is very inaccurate. By examining the theory of turbulent flow round a bluff cylindrical body we show why this procedure is, in general, incorrect. Of necessity the analysis cannot be presented here in full, but it will be shortly.

Analysis and Results

To find the fluctuating pressures, the turbulent velocity distribution must first be determined. We first divide the flow into three regions, the wake (W), the boundary layers (B), and the external flow (E), and then assume that the turbulent velocity in these three regions is uncorrelated. (This has been shown experimentally by D.G. Petty.) To the first approximation, the fluctuating velocity and pressure in the wake are unaffected by the incident turbulence and therefore can be assumed to be given experimentally. Our object is to calculate the fluctuating velocity and pressure in region (E) where analysis is possible, using the inviscid Navier-Stokes equations.

Our analysis for region (E) is a generalisation of the "rapid distortion" theory developed by Batchelor and Proudman [1] for the effect of wind tunnel contractions on turbulence intensity. The essential physical idea is that, if the turbulence is distorted rapidly enough by the presence of a body or by changes in the mean velocity, then each wavenumber (or eddy) is distorted separately before it can exchange energy non-linearly with other wavenumbers. The assumptions required by this linear analysis are that

$$u'_\infty/\bar{u}_\infty \ll 1, \quad \bar{u}_\infty a/\nu \gg 1, \quad u'_\infty a/(L_x \bar{u}_\infty) \ll 1, \quad L_x \gg \delta, \tag{2}$$

where ν is the kinematic viscosity, u'_∞ is the r m s turbulent velocity, $\bar{u}$ the mean velocity, and L_x the integral scale in the upstream flow. δ is some typical boundary layer width. Both these conditions are adequately satisfied for wind flow round buildings.

From Eq. (2) we assume $\bar{u}_i$, the mean velocity, to be unaffected by the turbulence to the first order and therefore to be irrotational. In addition, we ignore the wake in order to simplify the analysis near the upstream face of the body (in this case a circular cylinder). Then, having normalised all the variables in terms of $\bar{u}_\infty$ and a, the first step is to calculate the change in vorticity of a fluid element produced by the mean flow, $\bar{u}_i$. Since $D\omega_i/Dt \propto Dl_i/Dt$, where l_i is the vector of a fluid element ([2] p. 267), by calculating $l_i(t)$, ω_i can be found, and visualised (see Fig. 2). In fact, $\omega_i(x,y,z,t)$ is found in terms of $\omega_{\infty j}(x, y-\Delta_y, z, t-\Delta_t)$, the vorticity the same fluid element would have had in the absence of the body. If $\Psi(x,y)$ is the stream function, where $\bar{u}_1 = -\partial\Psi/\partial y$, $\bar{u}_2 = +\partial\Psi/\partial x$, then $\Delta_y(x,y) = \Psi(x,y) + y$. Also $\Delta_t(x,y) = \int_{-\infty}^{x} (1/\bar{u}_1(x,y) - 1)\,dx$, the time for a fluid particle to travel to (x,y) less the time it would have taken in the absence of the body.

Then
$$\omega_i(x,y,z,t) = \gamma_{ij}(x,y)\,\omega_{\infty j}(x, y-\Delta_y, z, t-\Delta_t), \tag{3}$$

where

$$\gamma_{ij} = \begin{bmatrix} \bar{u}_1 & -\partial\Delta_t/\partial x & 0 \\ \bar{u}_2 & 1+\partial\Delta_t/\partial y & 0 \\ 0 & 0 & 1 \end{bmatrix}, \quad \Delta_t = \frac{1}{\rho}\left\{(1-\rho^2/2)\left[K(\rho^2) \mp F\left(\theta-\tfrac{\pi}{2}, \rho^2\right)\right] - \left[E(\rho^2) \mp E\left(\theta-\tfrac{\pi}{2}, \rho^2\right)\right]\right\}, \quad \theta \gtrless \pi/2$$

$$\rho^2 = 4/(4+\Psi^2),$$

$F\left(\theta-\frac{\pi}{2},\rho^2\right)$, $E\left(\theta-\frac{\pi}{2},\rho^2\right)$, and $K(\rho^2)$, $E(\rho^2)$ are elliptic integrals and complete elliptic integrals of the first and second kind, respectively. (This expression for Δ_t is only valid for a circular cylinder.) Thus, given the vorticity field of the incident turbulence (derived from the velocity field), using Eq. (3), the vorticity can be calculated everywhere. The second step is to calculate the turbulent velocity u_i from ω_i, which by definition must satisfy

$$\text{(a)}\ \varepsilon_{ijk}\,\partial u_k/\partial x_j = \omega_i\,, \quad \text{(b)}\ \partial u_i/\partial x_i = 0\,, \tag{4}$$

assuming incompressible flow. (ε_{ijk} is the alternating tensor.)

The first boundary condition, assuming that $L_x \gg \delta$, is that at the boundary between (E) and (B)

$$u_i n_i = 0 \tag{5}$$

on the body upstream of (W), where n_i is a vector normal to a body. The second follows from the definition of $u_{\infty i}$;

$$u_i \to u_{\infty i} \quad \text{as } (x^2 + y^2) \to \infty, \ x < 0\,. \tag{6}$$

The fluctuating pressure $p^{*\prime} = p'/(\rho\,\bar{u}_\infty u'_\infty)$ is found from u_i by solving

$$\partial p'/\partial x_i = -(\bar{u}_j\,\partial u_i/\partial x_j + u_j\partial\bar{u}_i/\partial x_j + \partial u_i/\partial t)\,. \tag{7}$$

Note that it follows from Eq. (2), or Taylor's hypothesis, that as $x \to -\infty$, $\partial p'/\partial x_i = 0$. For many diameters downstream of the body, neither $\bar{u}_i$ nor u_i revert to their upstream value. However, since potential flow calculations, assuming no wake, give reasonable approximations to the mean velocity near the upstream face of the body, we shall assume that the wake does not exist and that the mean velocity is everywhere given by potential flow. We expect that this should give an adequate approximation to the turbulence near and fluctuating pressure on the upstream face. Thus $\bar{u}_i$ is given by the potential flow velocity distribution, and Eq. (5) is assumed valid over the entire surface of the body.

Since the upstream turbulence is defined in terms of its spectra, the velocity has to be Fourier analysed as follows. Upstream

$$\begin{Bmatrix} u_{\infty i} \\ \omega_{\infty i} \end{Bmatrix}(x,y,z,t) = \iiint_{-\infty}^{\infty} e^{i(\varkappa_1 x+\varkappa_2 y+\varkappa_3 z-\varkappa_1 t)} \begin{Bmatrix} S_{\infty i} \\ s_{\infty i} \end{Bmatrix}(\varkappa_1 \varkappa_2 \varkappa_3) d\varkappa_1 d\varkappa_2 d\varkappa_3 . \tag{8}$$

Note the implication of Taylor's hypothesis.

Near the body the flow is inhomogeneous in x and y, so that

$$\begin{Bmatrix} u_i \\ p' \end{Bmatrix}(x,y,z,t) = \iint_{-\infty}^{\infty} e^{i(\varkappa_3 z-\varkappa_1 t)} \begin{Bmatrix} \hat{S}_i \\ \hat{W} \end{Bmatrix}(x,y;\varkappa_1,\varkappa_3) d\varkappa_1 d\varkappa_3 . \tag{9}$$

To find u_i and p' in terms of $u_{\infty i}$, we express $\hat{S}_i$ and $\hat{W}$ in terms of $S_{\infty i}$ as follows

$$\begin{Bmatrix} \hat{S}_i \\ \hat{W} \end{Bmatrix} = \int_{-\infty}^{\infty} \sum_n \begin{Bmatrix} M_{in} \\ Q_n \end{Bmatrix}(x,y;\varkappa_1,\varkappa_2,\varkappa_3)\, S_{\infty_n}(\varkappa_1 \varkappa_2 \varkappa_3) d\varkappa_2 .$$

From M_{in}, Q_n, the one-dimensional or power density spectra for velocity and pressure Θ_{ij}, Θ_p near the body can be expressed in terms of the three-dimensional spectrum of the upstream turbulence $\Phi_{\infty_{nm}}$:

$$\begin{Bmatrix} \Theta_{ij} \\ \Theta_p \end{Bmatrix}(\varkappa_1;x,y) = \iint_{-\infty}^{\infty} \sum_n \sum_m \begin{Bmatrix} M_{in}\, M^*_{jm} \\ Q_n\, Q^*_m \end{Bmatrix} \Phi_{\infty_{nm}}(\varkappa_1 \varkappa_2 \varkappa_3) d\varkappa_2 d\varkappa_3 , \tag{10}$$

where

$$\Phi_{\infty_{nm}} = \frac{1}{(2\pi)^3} \iiint_{-\infty}^{\infty} \overline{u_{\infty_n}(x,y,z,t) u_{\infty_m}(x,y+r_y,z+r_z,t+\tau)} \cdot$$

$$\cdot \exp\left\{ -i[\varkappa_2 r_y+\varkappa_3 r_z-\varkappa_1 \tau] \right\} dr_y\, dr_z\, dz ,$$

$$\begin{Bmatrix} \Theta_{ij} \\ \Theta_p \end{Bmatrix}(\varkappa_1;x,y) = \frac{1}{2\pi} \int_{-\infty}^{\infty} \begin{Bmatrix} \overline{u_i(x,y,z,t)\, u_j(x,y,z,t+\tau)} \\ \overline{p(x,y,z,t)\, p(x,y,z,t+\tau)} \end{Bmatrix} e^{i\varkappa_1 \tau} d\tau .$$

To provide practical examples, we assume $\Theta_{\infty_{11}}$ is given by von Kármán's spectrum

$$\Theta_{\infty_{11}} = c_1 / \left[c_2 + \varkappa_1^2 \right]^{5/6} .$$

Then, if we further assume that the upstream turbulence is isotropic,

$$\Phi_{\infty_{nm}} = \frac{A \left[k^2 \delta_{nm} - \varkappa_n \varkappa_m \right]}{\left[c_2 + k^2 \right]^{17/6}} , \tag{11}$$

where

$$A = 55 c_1 / (36\pi) , \quad k^2 = \varkappa_1^2 + \varkappa_2^2 + \varkappa_3^2 ,$$

$$c_1 = g_1 (a/L_x)^{2/3} , \quad c_2 = g_2 (a/L_x)^2 , \quad g_1 = 0.1955 , \ g_2 = 0.577 .$$

To find M_{in}, it is convenient to express the difference between u_i and $u_{\infty i}$ in terms of a potential Φ and stream function Ψ_i as:

$$u_i = - \partial\Phi/\partial x_i + \varepsilon_{ijk} \partial\Psi_k/\partial x_j + u_{\infty i} . \tag{12}$$

Substituting u_i into Eq. (4), and taking Ψ_k such that $\partial\Psi_k/\partial x_k = 0$, it follows that

$$\partial^2 \Psi_i / \partial x_k \partial x_k = -(\omega_i - \omega_{\infty i}) , \quad \partial^2 \Phi / \partial x_k \partial x_k = 0 , \tag{13}$$

subject to the boundary conditions (for the circular cylinder)

$$\left. \begin{array}{ll} \text{at} & r = 1 , \ n_i \varepsilon_{ijk} \partial\Psi_k/\partial x_j = 0 , \ n_i \partial\Phi/\partial x_i = n_i u_{\infty i} , \\ \text{as} & (x^2 + y^2) \to \infty , \ |\varepsilon_{ijk} \partial\Psi_k/\partial x_j| \to 0 , \ |\partial\Phi/\partial x_i| \to 0 . \end{array} \right\} \tag{14}$$

If Φ and Ψ_i are expressed in terms of their Fourier transforms as

$$\left\{ \begin{matrix} \Phi \\ \Psi_i \end{matrix} \right\} = \iiint_{-\infty}^{\infty} e^{i(-\varkappa_1 t + \varkappa_3 z)} \left\{ \begin{matrix} \beta_j(\varkappa_1 \varkappa_2 \varkappa_3 ; x, y) S_{\infty j} \\ \alpha_{ij}(\varkappa_1 \varkappa_2 \varkappa_3 ; x, y) s_{\infty j} \end{matrix} \right\} d\varkappa_1 d\varkappa_2 d\varkappa_3 \tag{15}$$

then the required solution for M_{il} can be found from β_j and α_{ij} since, using the relation

$$s_{\infty_i} = i\,\varepsilon_{ijk}\,\varkappa_j\,S_{\infty_k},$$

$$M_{il} = M_{il}^{(d)} + M_{il}^{(s)} + M_{il}^{(\infty)}, \tag{16}$$

where

$$M_{il}^{(d)} = i\,a_{ij}\,\varepsilon_{jkl}\,\varkappa_k, \quad a_{ij} = \left(\frac{\partial\alpha_{3j}}{\partial y} - i\varkappa_3\,\alpha_{2j},\ \frac{-\partial\alpha_{3j}}{\partial x} + i\varkappa_3\,\alpha_{1j},\ \frac{\partial\alpha_{2j}}{\partial x} - \frac{\partial\alpha_{1j}}{\partial y}\right)$$

$$M_{il}^{(s)} = (-\partial\beta_1/\partial x,\ -\partial\beta_1/\partial y,\ -i\varkappa_3\,\beta_1), \quad M_{il}^{(\infty)} = \delta_{il}\,e^{i(\varkappa_1 x + \varkappa_2 y)}.$$

Substituting (15) into (13) gives the actual equations which must be solved for the entire wavenumber space, namely

$$\left(\frac{\partial^2}{\partial x^2} + \frac{\partial^2}{\partial y^2} - \varkappa_3^2\right)\alpha_{ij}(\varkappa_1, \varkappa_2, \varkappa_3; x, y) = -\Omega_{ij} \text{ for } i = 1,2,3;\ j = 1,2,3, \tag{17}$$

where

$$\Omega_{ij}(\varkappa_1, \varkappa_2; x, y) = \left(\gamma_{ij}\,e^{i(\varkappa_1\Delta_t - \varkappa_2\Delta_y)} - \delta_{ij}\right) e^{i[\varkappa_1 x + \varkappa_2 y]}$$

and

$$\left(\frac{\partial^2}{\partial x^2} + \frac{\partial^2}{\partial y^2} - \varkappa_3^2\right)\beta_j(\varkappa_1, \varkappa_2, \varkappa_3; x, y) = 0 \tag{18}$$

subject to:

$$\left.\begin{aligned} &\frac{\partial\alpha_{ij}}{\partial x_i} = \lambda\,\varkappa_j, \quad \infty > r \geqslant 1, \\ &n_i\,\varepsilon_{ijk}\,\frac{\partial\alpha_{kl}}{\partial x_j} = \mu\,\varkappa_l \text{ at } r = 1, \quad |\alpha_{kl}| \to 0 \text{ as } r \to \infty \end{aligned}\right\} \tag{19}$$

and

$$n_i\,\frac{\partial\beta_j}{\partial x_i} = n_i\,\delta_{ij}\,e^{i(\varkappa_1 x + \varkappa_2 y)} \quad \text{at } r = 1, \quad |\beta_j| \to 0 \text{ as } r \to \infty. \tag{20}$$

(λ, μ are scalars in wave-number space and are used because from Eq. (4b): $\varkappa_1 S_{\infty 1} = 0$.) Analytical solutions to Eqs. (17) and (18) have been obtained as asymptotic expansions in k $\left(=\sqrt{\varkappa_1^2+\varkappa_2^2+\varkappa_3^2}\right)$ or k^{-1} as $k \to 0$ or $k \to \infty$, respectively. For values of k of 0(1), computer solutions are necessary, of which only a few have been obtained, because to calculate M_{il} at a sufficient number of values of $\varkappa_i$ to compute spectra is too lengthy. We are forced to concentrate on calculating spectra for large-scale and small-scale turbulence, corresponding to $k \to 0$, $k \to \infty$.

In the limit as $k \to 0$, it follows from Eq. (16) that $M_{il}^{(d)} = 0$, from the solution to Eq. (18) that

$$M_{il} = \begin{bmatrix} 1-(x^2-y^2)/(x^2+y^2)^2 & -2xy/(x^2+y^2)^2 & 0 \\ -2xy/(x^2+y^2)^2 & 1+(x^2+y^2)/(x^2+y^2)^2 & 0 \\ 0 & 0 & 1 \end{bmatrix}, \quad (21)$$

which is the "quasi-steady" value and comes from the solution of Eq. (18), and the boundary condition (19). However, when $k \to \infty$ the more difficult Eq. (17) has to be solved, which is done using a Green's function method.

$$\alpha_{ij}(\varkappa_1\varkappa_2\varkappa_3;x,y) = \frac{1}{4\pi}\iint_{-\infty}^{\infty}\Omega_{ij}(x',y')K_0\left(\varkappa_3\sqrt{(x-x')^2+(y-y')^2}\right)dx'dy', \quad (22)$$

where $K_0(\;)$ is the zero order modified Bessel function. Expanding $\Omega_{ij}(x',y')$ about the point (x,y), if $(r-1) \gg k^{-1}$ and if

$$\chi^2 = \left\{\left[\varkappa_1\left(\frac{\partial\Delta_t}{\partial x}+1\right)-\varkappa_2\frac{\partial\Delta_y}{\partial x}\right]^2+\left[\varkappa_1\frac{\partial\Delta_t}{\partial y}+\varkappa_2\left(1-\frac{\partial\Delta_y}{\partial y}\right)\right]^2+\varkappa_3^2\right\},$$

$$\alpha_{ij} = \frac{\gamma_{ij}}{\chi^2}\exp\{i[\varkappa_1(x+\Delta_T)+\varkappa_2(y-\Delta_y)]\}-\frac{\delta_{ij}}{k^2}\exp\{i(\varkappa_1 x+\varkappa_2 y)\},$$

whence $M_{il}^{(d)}$ is calculated using Eq. (16). The solution to Eq. (18) shows that $M_{il}^{(s)}$ is $0(e^{-k(r-1)})$, so that when $(r-1) \gg k^{-1}$; $M_{il} = M_{il}^{(d)} + M_{il}^{(\infty)}$. When $(r-1) \ll k^{-1}$, then a rather careful evaluation of Eq. (22) is required because $\Delta_t \to \infty$ and $\gamma_{22} \to \infty$ as $r \to 1$, where the vortex lines accumulate and vortex stretching becomes singular. The analysis of this region is particularly important for calculating the surface pressure.

The pressure tensor Q_n can be calculated from Eq. (7) using our results for M_{il}. Since $p' = \text{constant} = 0$, say, as $x \to -\infty$, it follows that for the important case of stagnation pressure,

$$Q_n(-1,0) = i\varkappa_1 \int_{-\infty}^{-1} \left(M_{1n}^{(d)} + M_{1n}^{(s)}\right)(x,0)dx + M_{1n}^{(\infty)}(-1,0) . \quad (23)$$

Then around the surface $Q_n(r=1,\theta) = Q_n(r=1,\theta=\pi) - i\varkappa_1 \int_\theta^\pi \tilde{M}_{2n}(r=1,\theta')d\theta' + 2\sin\theta\tilde{M}_{2n}(\theta)$, where $\tilde{M}_{2n} = M_{2n}\cos\theta - M_{1n}\sin\theta$. The asymptotic values at the body's surface of $Q_n\,(r=1,\theta)$ are found from the results for M_{il} to be:

as $k \to 0$: $Q_n = (1-4\sin^2\theta,\ 2\sin^2\theta,\ 0)$ to zero order, $2\pi \geqslant \theta > 0$ (24)

$$Q_n = \left(1-2i\varkappa_1-\varkappa_1^2[0.92-\ln|\varkappa_{3/2}|]/2,\ \varkappa_1\varkappa_2(1.08+\ln|\varkappa_{3/2}|)/2,0\right)$$

at $\theta = \pi$, to second order, (25)

as $k \to \infty$, $$Q_n = \frac{i\varkappa_1}{2\chi'} e^{i\varkappa_1[\ln|\chi'/2|-1+\ln 2/2]} \Gamma(-i\varkappa_1/2) e^{-\varkappa_1\varphi/2} (-\varkappa_3, 0, \varkappa_1), \quad (26)$$

where $\chi' = \left[\varkappa_3^2 + 9\varkappa_1^2/16\right]^{1/2}$, $\varphi = \tan^{-1}(3\varkappa_1/(4|\varkappa_3|))$.

From these values of M_{il} and Q_n can be found many of the most useful spectra Θ_{ij} and Θ_p, and the co-variances $\overline{u_i u_j}$ and $\overline{p}^2$. In Fig.3, $\hat{\Theta}_{11}(\hat{\varkappa}_1) = \Theta_{11}/(a/L_x)$ is plotted against $\hat{\varkappa}_1 = \varkappa_1/(a/L_x)$ for the velocity u_1 on $y = 0$ to show up the enormous effect of the turbulence scale on the turbulence near the body. It is also interesting to note that the greatest amplification occurs for small scale turbulence at low frequencies. In fact we find that as $\xi \to 0$, $a/L_x \to \infty$, $\hat{\Theta}_{11}(\varkappa_1 = 0) \propto 1/\xi^2$ if $r - 1 = \xi$ and $(\xi a/L_x) \to \infty$, but $\hat{\Theta}_{11} \propto (a/L_x)^{5/3} \xi^{1/3}$ as $(\xi a/L_x) \to 0$. In Fig. 4, the theoretical results for $\sqrt{\left(\overline{u_1^2}\right)}/u'_\infty$ are shown for $(a/L_x) \to 0$, $(a/L_x) \to \infty$, along with some experimental results of Dr. Petty which bear out the theoretically predicted effects of (a/L_x) [8]. Rather different results occur in the spectra of u_2, which are amplified when $a/L_x \to 0$, and in general diminish when $a/L_x \to \infty$.

The explanation for these large differences in behaviour between large and small scale turbulence is simple:

There are two ways in which an obstacle affects the turbulence flowing over it; firstly, it acts as a "source" of turbulence in order that $u_i n_i = 0$, and, secondly, by changing $\overline{u}_i$ it distorts the turbulence. Our analysis shows why the first effect dominates when $k \to 0$ or $a/L_x \to 0$, and the second when $k \to \infty$ or $a/L_x \to \infty$. Then physical arguments using potential flow and vortex stretching show why we find the particular results shown in Figs. 3 and 4.

From Eqs. (10), (11), and (24) to (26), we obtain the following results for and $\overline{p}^2$, shown graphically in Fig. 5:

$$\left.\begin{array}{l} a/L_x \to 0 \\ r = 1,\ 2\pi \geqslant \theta > 0 \end{array}\right\} \left\{\begin{array}{l} \hat{\Theta}_p(\hat{\varkappa}_1) \\ \overline{p^2} \end{array}\right\} = \left\{\begin{array}{l} \Theta_{\infty 11}(\hat{\varkappa}_1) \\ 1 \end{array}\right\} (5 - 4\cos 2\theta) , \quad (27)$$

If $(a/L_x) \ll 1$, $\varkappa_1 \ll 1$, $\theta = \pi$,

$$\hat{\Theta}_p(\hat{\varkappa}_1) = g_1 \frac{\left[1 + (a/L_x)^2 \hat{\varkappa}_1^2 \left[2 \cdot 71 + (1/4) \ln\left\{(a/L_x)^2 \left(g_2 + \hat{\varkappa}_1^2\right)\right\}\right]\right]}{\left(g_2 + \hat{\varkappa}_1^2\right)^{5/6}} . \quad (28)$$

If a/L_x is arbitrary, $\varkappa_1 \gg 1$; $r = 1$, $\theta = \pi$,

$$\hat{\Theta}_p(\hat{\varkappa}_1) \sim 1.31\, e^{-(\pi/2)(a/L_x)\hat{\varkappa}_1} (a/L_x)^{-2/3} \hat{\varkappa}_1^{-7/3} . \quad (29)$$

Note that as $(a/L_x) \to 0$ $\hat{\Theta}_p$ and $\overline{p}^2$ increase 9 times from $\theta = \pi$ to $\theta = \pi/2$ because near $\theta = \pi/2$ the fluctuating pressure is affected by u_{∞_1} and u_{∞_2}. This result (27) is a rigorous demonstration of what can be obtained directly by a quasi-steady analysis. In that case, the experimentally determined values of the mean pressure $\overline{p^*(\theta)}$ can be used, whence the general quasi-steady result for the fluctuating pressure $p'(\theta)$ (due to incident turbulence) is, in dimensional variables, $\overline{(p^{*\prime}(\theta))^2} = 4\left[(\overline{p(\theta)})^2\,\overline{\left(u'_{\infty_1}\right)^2} + (\partial\overline{p}/\partial\theta)^2\,\overline{\left(u'_{\infty_2}\right)^2}\right] / (\overline{u}_\infty)^2$. This result due to Armitt (unpublished) is being compared with full scale measurements of $\overline{(p')^2}$ on a tall chimney for which $(a/L) \simeq 0.1$. (Tunstall, to be published.)

The result for $\hat{\Theta}_p(\hat{\kappa}_1)$ when κ_1 is small but finite, which corresponds to the spectrum of Davenport's equation (1), is significant in three ways. (a) The first order term is $0\left(\kappa_1^2 \ln|\kappa_1|\right)$ not $0\left(\kappa_1^2\right)$, since the spatial variation of the incident turbulence in the z-direction is as important as the du_∞/dt (added mass) term of Eq. (1). (b) The term in $0\left(\kappa_1^2\right)$ is not only produced by the added-mass effect, but also by the spatial variations of u_∞, and by effects of u_{∞_2} (from the term Q_2 in Eq. (25)). Both the latter effects were neglected by Davenport [7], Vickery [10], and Bearman [4]. (c) The theory provides a criterion for the validity of the expression (28), i.e., $\kappa_1 \ll 1$. Thus, for reasonable accuracy, Eq. (28) should only be used for $\kappa_1 \lesssim 0.3$ as in Fig. 5. When $\kappa_1 \gg 1$, whatever the turbulence scale $\hat{\Theta}_p$ is given by Eq. (29), a most unusual result in that the spectrum decays exponentially. The reason is that, since $p(r=1, \theta=\pi)$ involves the integral (Eq. (23)) of u_1 along the stagnation line and since, on account of the piling up of vortex lines, $u_1 \sim e^{-i\kappa_1 \ln \xi}$, it follows that the net contribution to this integral is very small. In fact $\hat{\Theta}_p(\hat{\kappa}_1)$ does not decay exponentially as $\hat{\kappa}_1 \to \infty$ because non-linear effects of order $(u'_\infty/\overline{U})$ contribute terms which are small compared with $\hat{\Theta}_p(\hat{\kappa}_1 = 0)$ but large compared to $e^{-\kappa_1}$. When $a/L \to \infty$, $\hat{\Theta}_p(\hat{\kappa}_1)$ can be calculated for all $\hat{\kappa}_1 > 0$, but not analytically. Figure 5 shows the results which demonstrate how reducing the scale of incident turbulence reduces $\hat{\Theta}_p$ for all $\hat{\kappa}_1 > 0$ and thence $\overline{p}^2$.

There are still a number of immediately outstanding problems, all of which should be soluble:

(a) The effect of the higher order terms on $\hat{\Theta}_p(\hat{\varkappa}_1)$ as $\hat{\varkappa}_1 \to \infty$; (b) the evaluation of $\overline{(\partial p'/\partial\theta)^2}$ round the circumference, in order to investigate thoroughly the theory of Taylor [9] for turbulent transition of the boundary layer; (c) the extension of the theory to three-dimensional bodies. The effects of shear, anisotropy, and inhomogeneity in the incident turbulent flow may yield to analysis in the long term.

Acknowledgements

I am greatly indebted to many colleagues who have contributed so much to this research, most notably J. Armitt, R.A. Scriven, P.W. Bearman, H.K. Moffatt and G.K. Batchelor. J.C. Mumford has given valuable assistance with the computing.

References

1. Batchelor, G.K., Proudman, I.: Quart. J. Mech. Appl. Math. 7, 83 (1954).
2. Batchelor, G.K.: An Introduction to Fluid Dynamics, Cambridge University Press 1967.
3. Bearman, P.W.: N.P.L. Aero Rep. 1296 (1969).
4. Bearman, P.W.: J. Fluid Mech. 46, 177 (1971).
5. Bearman, P.W.: Agard Conf. Turbulent Shear Flows (1971).
6. Davenport, A.G.: Phil. Trans. Roy. Soc. A. 269, 385 (1971).
7. Davenport, A.G.: Proc. Inst. Civ. Engrs. 19, 449 (1961).
8. Hunt, J.C.R.: Phil. Trans. Roy. Soc. A. 269, 457 (1971).
9. Taylor, G.I.: Proc. Roy. Soc. 161, 307 (1936).
10. Vickery, B.J.: N.P.L. Aero Rep. 1143 (1965).

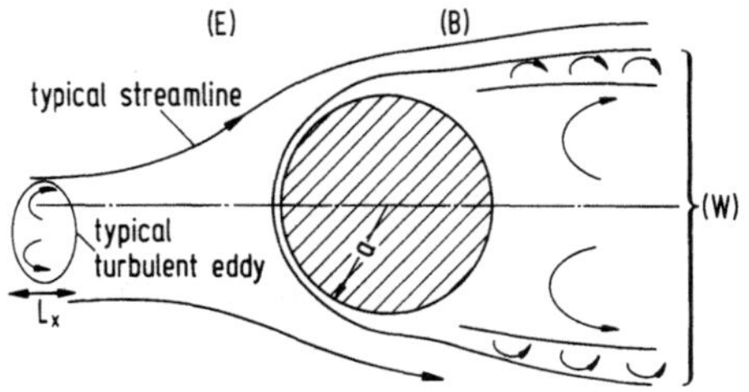

Fig.1. The regions of flow round a bluff body, showing the relevant length scales of the body and of the incident turbulence.

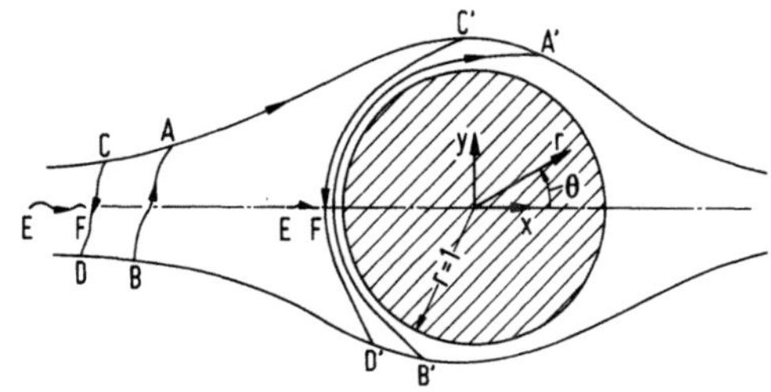

Fig.2. The idealised problem of an inviscid, weakly turbulent flow over a circular cylinder, showing how the mean flow stretches, compresses, and "piles up" typical vortex lines which lie between AB, CD, EF at time $t = t_1$, and between A'B', C'D', E'F' at time $t = t_2$.

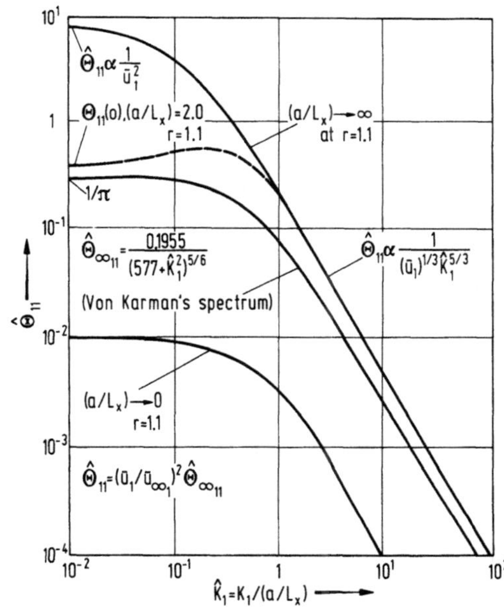

Fig.3. One-dimensional velocity spectrum on the stagnation line of a circular cylinder, for large and small scale turbulence.

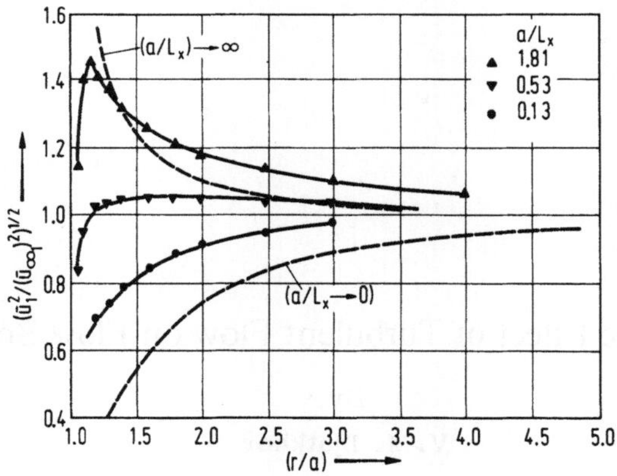

Fig. 4. The change in r m s turbulent velocity along the stagnation line showing the effect of the scale of turbulence. Dashed lines are theoretical. The experimental points were obtained by Dr. Petty.

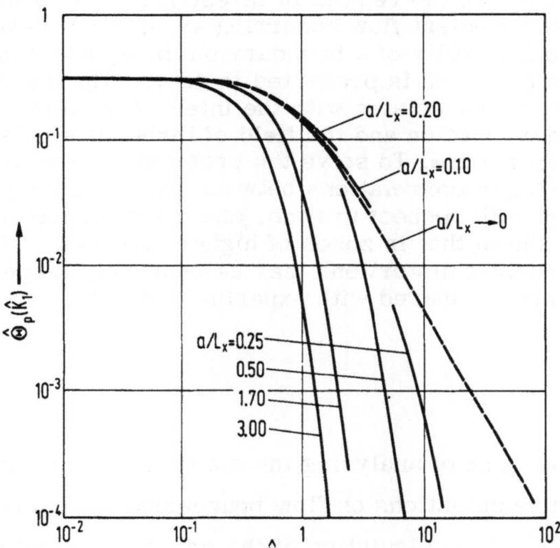

Fig. 5. Pressure spectra at the stagnation point for various scales of turbulence calculated by means of the asymptotic theories.

Hydrodynamic Effect of Turbulent Flow on Flow Boundaries

By

V.M. Ljatkher

Moscow, USSR

Summary

The paper summarizes the results of investigations on wall-pressure pulsations due to turbulent flow occurring at different flow schemes. The effect of the flexibility of a boundary on the spectral characteristics of pressure pulsation is presented in detail. The distortion of pressure pulsations associated with the interaction of the flow field due to the boundary motion and the field of turbulence pulsations is studied for the first time. To solve the problem, it was necessary to calculate the triple correlations between the pressure pulsations, their derivatives with respect to time, and the shear stress on the boundary. It is shown that in zones of higher turbulence, this "substantially hydroelastic distortion" may be relatively large. The theoretical results are compared with experimental data.

Introduction

The general procedure of analysing the statistical characteristics of turbulent pressure pulsations on flow boundaries is described in the writer's book [1]. The calculation of the pressure pulsations is based on the solution of Poisson's equation (or wave equation for a compressible fluid) in case the right-hand part is known. This part comprises the kinematic flow characteristics or the so-called "kinematic function".

The solution of particular problems associated with "two-dimensional" turbulence and the kinematic function, linearized relative to fluctuation components of velocity, has been presented elsewhere [2,3,4,5]. The comparison of laboratory and field results [3,4,6] demonstrates that such a schematization permits in many cases a fair description of the low and medium frequency spectrum of pressure pulsations. In accordance with the theory, the relative intensity of pressure pulsation acting on a wall at large Froude numbers turned out to be considerably lower than at small ones.

However, consideration of a "two-dimensional" turbulence is inadequate in many cases, and by utilization of the linearized kinematic function as is mentioned in Ref. [1], it is impossible to evaluate the high-frequency spectrum of pressure pulsations and to clarify the mechanism of distortion of boundary pressure pulsations resulting from the interaction of the initial turbulence field and the currents induced by forced vibrations of the boundary. At the present, problems of "three-dimensional" turbulence effects on the flow boundary are being considered as well. The main results of the study are mentioned in Ref. [6] and a more extensive survey is to be found in [1,7].

The present paper mainly deals with the problem of the effect of processes in the fluid film close to a flexible wall on the characteristics of pressure pulsations acting on that wall. The simplest aspect of this problem associated with the role of the "added mass" has been discussed more than once [1,8]. In this paper, principal attention is paid to the more complicated (and, in many cases, the most important) aspect of the problem, i.e. the evaluation of the interaction between the turbulent field and the flow field caused by the vibrations of the flow boundary, or of the "substantially hydroelastic distortion" as the writer terms it [1]. Experiments proving the significance of such interaction were first published by the writer in 1965 [9].

The interaction between the turbulence-fluctuation field and the flow field induced by the flow-boundary vibrations might be one of the principal causes for the existence of the "Kramer effect" of drag reduction along a flexible wall. Up to now, this effect has been studied without proper consideration of the turbulence-fluctuation field.

Problem Formulation and General Solution

We shall compare the current near the rigid wall (instantaneous velocity $\tilde{U}_i$, pressure $\tilde{P}$) with the current near the compliant wall ($\tilde{\tilde{U}}_i$, $\tilde{\tilde{P}}$), the deformation of which are assumed to be small. The distortion brought about by the effect of compliance is evaluated by the quantities $\underset{\sim}{U}_i$, $\underset{\sim}{P}$:

$$\underset{\sim}{U}_i = \tilde{\tilde{U}}_i - \tilde{U}_i \, ; \quad \underset{\sim}{P} = \tilde{\tilde{P}} - \tilde{P} \, . \tag{1}$$

The currents near the compliant and the rigid walls are described by the Navier-Stokes and the continuity equations. Using Eq. (1), we obtain

$$\frac{\partial \underset{\sim}{U}_i}{\partial t} + \sum_{j=1}^{3} \underset{\sim}{U}_j \frac{\partial \underset{\sim}{U}_i}{\partial x_j} + \sum_{j=1}^{3} \tilde{U}_j \frac{\partial \underset{\sim}{U}_i}{\partial x_j} + \sum_{j=1}^{3} \underset{\sim}{U}_i \frac{\partial \tilde{U}_i}{\partial x_j} = -\frac{1}{\rho} \frac{\partial \underset{\sim}{P}}{\partial x_i} + \nu \sum_{j=1}^{3} \frac{\partial^2 \underset{\sim}{U}_i}{\partial x_j^2} \, , \tag{2}$$

$$\sum_{j=1}^{3} \frac{\partial \underset{\sim}{U}_j}{\partial x_j} = 0 \, . \tag{3}$$

Along a smooth boundary $x_2 = \eta(x_1, x_3, t)$ the condition of continuity is fulfilled

$$\tilde{\tilde{U}}_1 \frac{\partial \eta}{\partial x_1} - \tilde{\tilde{U}}_2 + \tilde{\tilde{U}}_3 \frac{\partial \eta}{\partial x_3} + \frac{\partial \eta}{\partial t} = 0 \, . \tag{4}$$

When the adhesion and imperviousness of the rigid boundary ($\tilde{U}_j = 0$) and the smallness of the flexible-boundary deformations ($\underset{\sim}{U}_j \, \partial\eta/\partial x_j \simeq 0$, $j = 1,3$) are taken into account, the condition (4) takes the form $\underset{\sim}{U}_2 = \partial\eta/\partial t$ (Eq. 5). We do not apply any restrictions to the quantities $\underset{\sim}{U}_j$ ($j = 1,3$) on the boundary; this permits various statements concerning the disturbance problem (with or without the adhesion of the disturbed fluid to the walls).

The "induced" pressure $\underset{\sim}{P}$ will be introduced as the sum total of the pressure P_1, occurring on the vibrating compliant boundary in the still fluid or in a flow with the constant non-disturbed velocity

$\partial\tilde{U}_i/\partial x_j\big|_1 = 0$, and the pressure P connected with the velocity gradient in the initial flow $\underset{\sim}{P} = P_1 + P$ (Eq. 6).

In problems of vibrations of structural elements in fluids, the pressure P_1 has been studied as the most important manifestation of hydroelastic interaction up to now [1,6,10]. It was shown that this pressure is proportional to a local acceleration of a point on the boundary:

$$P_1 = \mu \frac{\partial \underset{\sim}{U}_2}{\partial t}\Bigg|_{x_2=0} . \tag{7}$$

The dimensional proportionality factor μ being the "virtual-mass coefficient".

According to the writer's terminology [1], the pressure P characterizes the substantially hydroelastic effect to interaction between turbulence flow and the flow boundary. As will be shown below, in some cases at higher turbulence intensities, the influence of the component "P" is larger than that of the component "P_1".

Taking into account the conditions on the rigid wall for the initial current and the determinations of the components P and P_1, the system of Eqs. (2) and (3), taken for the points situated just on the vibrating wall ($x_2 = 0$), now becomes

$$-\frac{\partial P}{\partial x_1} = \frac{1}{\nu}\underset{\sim}{U}_2\tilde{\tau}_1 \; ; \quad -\frac{\partial P}{\partial x_3} = \frac{1}{\nu}\underset{\sim}{U}_2\tilde{\tau}_3 , \tag{8}$$

where $\tau_j = \nu\rho\partial\tilde{U}_j/\partial x_2$ are the instantaneous values of the components of the shear stress on the rigid wall ($j = 1,3$).

Neglecting secondary effects, we shall assume[1] the wall-shift velocity to be described by the linear operator L which depends upon the initial pressure $\tilde{P}$:

$$\underset{\sim}{U}_2 = \frac{\partial}{\partial t} L\{\tilde{P}\} . \tag{9}$$

[1] The substitution of the value of the total pressure $\tilde{\tilde{P}}$ into Eq. (9) instead of $\tilde{P}$ in cases of a rather rigid wall does not change results qualitatively. The quantitative effect of this refinement is presently under study.

With the use of Eq. (9), Eq. (8) is transformed into a two-dimensional Poisson's equation, the solution of which is obtained with the aid of Green's function. Let us consider less complicated particular solutions stemming from Eqs. (1), (6), and (8). It follows from (1) that the correlation function of pressure pulsations on a vibrating wall can be represented in the following form:

$$R_{\approx{p}} = \langle \approx{p}\approx{p}' \rangle = R_{\tilde{p}} + \left[R_{\tilde{p}p_1} + R_{p_1\tilde{p}} + R_{\tilde{p}p} + R_{p\tilde{p}} \right] + \left\{ R_{p_1} + R_{p} + R_{p_1 p} + R_{pp_1} \right\}. \quad (10)$$

The last bracket on the right-hand side comprises the second-order terms relative to the amplitude of the wall vibration. The first bracket, as one may see, combines the correlations of the different components of the "induced" and the initial pressure field.

In accordance with Eqs. (7) and (9) - taking into consideration the linearity of the operators L - one can calculate the mathematical expectation and derive the following formula for the "distortion" of the correlation function of the initial pressure due to the "virtual-mass effect".

$$R_{\tilde{p}p_1} = \langle \tilde{p}(x_1, x_3, t) p_1(x_1', x_3', t') \rangle = \mu \frac{\partial^2}{(\partial t')^2} L' \{ R_{\tilde{p}}(x_j, x_j', t, t') \}. \quad (11)$$

Here L' denotes the operator L applied to suboperator function in the points (x_j', t').

Multiplying the right and the left-hand parts of Eq. (8) by $\tilde{P}(x_j', t')$, calculating the mathematical expectation, and again making use of properties of rearrangement of linear operators, we get the following formulae for determining mutual correlation functions of the "substantially hydroelastic addition" P and the initial pressure $\tilde{P}$:

Along the flow $(x_3 = x_3')$:

$$R_{p\tilde{p}} = \langle p(x_1, x_3, t) \tilde{p}(x_1', x_3', t') \rangle =$$

$$= -\frac{1}{\nu} \frac{\partial}{\partial t''} \int_{-\infty}^{x_1} \langle L\{ \tilde{p}(\xi_1, x_3, t'') \} \tilde{\tau}_1(\xi_1, x_3, t) \tilde{p}(x_1', x_3, t') \rangle d\xi_1 \Big|_{t''=t} . \quad (12)$$

Across the flow ($x_1 = x_1'$):

$$R_{p\tilde{p}} = \langle p(x_1,x_3,t)\tilde{p}(x_1,x_3',t')\rangle =$$

$$= -\frac{1}{\nu}\frac{\partial}{\partial t''}\int_{-\infty}^{x_3}\langle L\{\tilde{p}(x_1,\xi_3,t'')\}\tilde{\tau}_3(x_1,\xi_3,t)\tilde{p}(x_1,x_3',t')\rangle d\xi_3\Big|_{t''=t} . \quad (13)$$

The formulae (11) to (13) show that the "virtual-mass effect" is easily evaluated with the correlation theory. The "substantially hydroelastic effect" on the contrary can be evaluated only with the knowledge of the triple correlations of the pulsations of pressure and shear stress in the turbulent flow along the rigid wall.

The theoretical investigation can be considered completed in the scope of the formulated problem, if the relation between the pressure pulsation and the shear stress on the rigid boundary of the turbulent flow is revealed. This relation seems not always to be unique. Detailed research on the problem possesses independent significance and is beyond the scope of this paper. Below, an approximate relation between the pulsations of shear and normal stresses is presented; an important merit of it is its simplicity and qualitative conformity with the available experimental data. For the present, we shall divert from these results and consider some special cases of Eqs. (10) to (13). Let us assume that the local shift of the boundary points is proportional to the local pressure (Vincler's hypothesis): $L = -1/C_z$ (Eq. 14), where C_z is the coefficient of subgrade reaction.

For the case of ($x_1 = x_1'$), we get from Eq. (12) the changes in autocorrelation function of pressure pulsations in a statistically stationary process:

$$R_{p\tilde{p}}(\tau) = \langle p(t)\tilde{p}(t+\tau)\rangle = \frac{1}{\nu}\int_{-\infty}^{x_1}\frac{1}{C_z}\langle\frac{\partial\tilde{p}}{\partial t}(\xi_1,t)\tilde{\tau}_1(\xi_1,t)\tilde{p}(x_1,t+\tau)\rangle d\xi_1 . \quad (15)$$

Assume that condition (14) is fulfilled in a certain zone $x_0 - r \leqslant x \leqslant$ $\leqslant x_0 + r$, beyond which the boundary can be considered to be rigid

(the effect of distortion of results by the compliant pressure pickup). Taking into account that $C_z = \infty$ at $\xi_1 < x_0 - r$, we find from Eq. (15) for the center ($x = x_0$) of the diaphragm of a pick-up with small diameter:

$$R_{p\tilde{p}} = \frac{1}{\nu C_z} \int_{x_0-r}^{x_0} \langle \frac{\partial \tilde{p}}{\partial t}(\xi_1, t) \tilde{\tau}_1(\xi_1, t) \tilde{p}(x_1, t+\tau) \rangle d\xi_1 \simeq$$

$$\simeq \frac{r}{\nu C_z} \langle \frac{\partial p_0}{\partial t}(t) \tau_0(t) p_0(t+\tau) \rangle . \qquad (16)$$

Here the index "0" characterizes a stress in the point of the rigid boundary. From Eq. (16) it is seen, that if, for instance, one neglects the variation with time of the shear stress ($\tau_0 \simeq$ const), then substantially hydroelastic distortion of pressure pulsations will always cause a small increase of the dispersion of pressure pulsations. Indeed, in this case

$$R_{p\tilde{p}} \simeq \frac{r}{\nu C_z} \tau_0 < \frac{\partial p_0}{\partial t}(t) p_0(t+\tau) > = 0 \quad \text{at} \quad \tau = 0$$

and the dispersion change will be determined by the curly bracket in Eq. (10), where in case of an adequate pick-up rigidity the positive dispersion of the "induced" pressure will be of main significance. Its value can be easily calculated using the first of Eqs. (8) within the scope of the correlation (or spectral) theory and without utilization of triple correlations. The same situation prevails to the changing pressure pulsations on a rigid wall, which is situated upstream (or downstream) from the zone with elastic insert, if its distance from the measuring point is about equal to or larger than the integral scale of the pressure pulsations. In this case, the square bracket in Eq. (10), which is proportional to the first order of the amplitude of the boundary vibration, will transform to zero, and all distortions will be described by the expression in the curly bracket. The corresponding experimental data are presented in Ref. [11].

Pulsation of Shear Stress on a Rigid Wall

Equations for the fluctuating components of velocity immediately at the wall can be written as follows [1]:

$$\frac{\partial u_i}{\partial t} + \sum_{j=1}^{3} u_j \frac{\partial U_i}{\partial x_j} + \sum_{j=1}^{3} U_j \frac{\partial u_i}{\partial x_j} + \frac{1}{\rho}\frac{\partial p}{\partial x_i} - \nu \sum_{j=1}^{3} \frac{\partial^2 u_i}{\partial x_j^2} = 0\,; \quad \sum_{j=1}^{3} \frac{\partial u_j}{\partial x_j} = 0\,. \tag{17}$$

Here, the direction x_1 is chosen along the main stream, the capital letters denote averaged velocity components, small letters denote fluctuating components. In Eq. (14), the nonlinear terms, which are insignificant in the immediate proximity to the wall, are omitted.

To determine the shear stresses on the wall $\tau_j = \rho\nu\partial u_j/\partial x_2$, Eq. (17) can be transformed as follows

$$\begin{aligned} \frac{1}{\nu}\frac{\partial \tau_1}{\partial t} - 2\frac{\partial^2 \tau_1}{\partial x_1^2} - \frac{\partial^2 \tau_3}{\partial x_1 \partial x_3} - \frac{\partial^2 \tau_1}{\partial x_3^2} &= \frac{\partial^2 \tau_1}{\partial x_2^2}\,, \\ \frac{1}{\nu}\frac{\partial \tau_3}{\partial t} - 2\frac{\partial^2 \tau_3}{\partial x_3^2} - \frac{\partial^2 \tau_1}{\partial x_1 \partial x_3} - \frac{\partial^2 \tau_3}{\partial x_2^2} &= \frac{\partial^2 \tau_3}{\partial x_2^2}\,. \end{aligned} \tag{18}$$

The right-hand parts of this system have not yet been determined and are to be obtained from the solution of the problem for a certain neighborhood of the wall. The coefficients in Eq. (17) including components of average velocities are considered to be known functions which change slowly along coordinates parallel to the boundary.

The conditions on the boundary ($x_2 = 0$) are $u_j = 0$, $\partial u_2/\partial x_2 = 0$ (Eq. 19). Furthermore, the characteristics of pressure pulsations on the boundary are to be considered preset $x_2 = 0$, $p = p_0(x_1, x_3, t)$. The motion is considered to be statistically stationary.

To solve Eqs. (17), the boundary conditions are insufficient. In Ref. [12,13], the condition of solution limitation "in infinity" is used. In the present case, this condition is not logical, because the pulsation amplitudes at a sufficiently large distance from the wall can be present on account of the influence of the nonlinear terms not included in Eqs. (17).

It seems to be more logical to prescribe fluctuations of velocity components parallel to the wall at some distance from the wall

$$x_2 = \delta, \quad u_1 = {}^0u_1 (x_1, x_3, t) ; \quad u_3 = {}^0u_3 (x_1, x_3, t) . \qquad (20)$$

The problem is then to find relations between pressure pulsations and fluctuations of the wall shear stress for different spectral forms of velocity pulsations on the outer boundary of the above mentioned layer with thickness δ. This formulation of the problem is close to that of the non-stationary laminar boundary layer [14,15].

The solution of this problem turns out to be rather cumbersome even with the help of computer methods. Using the assumption of small velocity-gradient fluctuations near the wall, results can be obtained which agree with experimental data. The outcome of such a procedure for a two-dimensional turbulence is published in Ref. [16]. A modified solution referring to three-dimensional pulsations and containing more logical formulations of the mentioned assumption about the fluctuations near the wall is given below.

Let us consider the flow to be uniform in the longitudinal and lateral directions. Then

$$U_j = 0, \ (j = 2,3) ; \quad \frac{\partial U_i}{\partial x_j} = 0 \ (i \neq 1 \ \text{or} \ j = 1) ; \quad \frac{\partial U_1}{\partial x_3} = C(x_2) . \qquad (21)$$

The latter condition corresponds to the particular case when the transverse gradient is constant across the flow over distances of the order of the greatest pulsation scale. The other important but simple spatial case is obtained when considering the flow in which large transverse velocity gradients are concentrated within a narrow layer of displacement perpendicular to the wall.

Taking into consideration the conditions of uniformity (21), we shall use the three-dimensional Fourier transformation (by coordinates x_1, x_3, and time t) to the Eqs. (17).

$$i\omega u_1^* + i\varkappa_1 U_1 u_1^* + \frac{\partial U_1}{\partial x_2} u_2^* + \frac{\partial U_1}{\partial x_3} u_3^* + i\varkappa_1 p^* + \frac{1}{Re}\left(\varkappa_1^2 + \varkappa_3^2 - \frac{d^2}{dx_2^2}\right) u_1^* = 0\,, \tag{22}$$

$$i\omega u_2^* + i\varkappa_1 U_1 u_2^* + \frac{dp^*}{dx_2} + \frac{1}{Re}\left(\varkappa_1^2 + \varkappa_3^2 - \frac{d^2}{dx_2^2}\right) u_2^* = 0\,, \tag{23}$$

$$i\omega u_3^* + i\varkappa_1 U_1 u_3^* + i\varkappa_3 p^* + \frac{1}{Re}\left(\varkappa_1^2 + \varkappa_3^2 - \frac{d^2}{dx_2^2}\right) u_3^* = 0\,, \tag{24}$$

$$i\varkappa_1 u_1^* + \frac{du_2^*}{dx_2} + i\varkappa_3 u_3^* = 0\,. \tag{25}$$

The boundary conditions for the "images" of the pulsations (variables noted with *) remain similar to the original ones (19).

Here and below, all variables reduce to dimensionless form with the help of the typical length scale L_0, the density ρ_0, and the velocity U_0; $Re = U_0 L_0/\nu$ is the characteristic Reynolds number, and $\varkappa_1$, $\varkappa_3$ are wave numbers in the direction of the x_1 and x_3 axes. When deriving Eqs. (22) to (25), it is taken into account, in particular, that the relation between the "images" of the derivative and the function becomes $(\partial z/\partial x_j)^* = i\varkappa_j z^*$ with $j = 1, 3$.

Using the equation of continuity (25) for determining u_3 and Eq. (24) for obtaining p^*, we get the system of two equations with unknown images of longitudinal and lateral pulsations (u_1^* and u_2^*):

$$i(\omega + \varkappa_1 U_1) u_1^* + u_2^* \frac{\partial U_1}{\partial x_2} + \frac{1}{\varkappa_3}\left(-\varkappa_1 u_1^* + i\frac{du_2^*}{dx_2}\right) \cdot \frac{\partial U_1}{\partial x_3} =$$

$$= -\frac{i\varkappa_1}{\varkappa_3^2}\left\{\varkappa_1\left[\omega + \varkappa_1 U_1 - i\frac{\varkappa_1^2 + \varkappa_3^2}{Re}\right] u_1^* - \left[i(\omega + \varkappa_1 U_1) + \frac{\varkappa_1^2 + \varkappa_3^2}{Re}\right]\frac{du_2^*}{dx_2}\right\} -$$

$$- \frac{\varkappa_1^2 + \varkappa_3^2}{Re} u_1^* + \alpha_1(x_2) \tag{26}$$

$$i(\omega + \varkappa_1 U_1) u_2^* = -\frac{1}{\varkappa_3^2}\left\{\varkappa_1^2 U_1 \frac{du_1^*}{dx_2} - i\varkappa_1 U_1 \frac{d^2 u_2^*}{dx_2^2}\right\} - \frac{\varkappa_1^2 + \varkappa_3^2}{Re} u_2^* + \alpha_2(x_2)\,.$$

Here the functions $\alpha_1(x_2)$ and $\alpha_2(x_2)$ combine the terms of corresponding equations each of which is not to be zero directly at the wall.

$$\alpha_1(x_2) = -\frac{i\varkappa_1}{\varkappa_3^2}\left(\frac{i\varkappa_1}{Re}\frac{d^2u_1^*}{dx_2^2}+\frac{1}{Re}\frac{d^3u_2^*}{dx_2^3}\right)+\frac{1}{Re}\frac{d^2u_1^*}{dx_2^2} = 0 \quad \text{at} \quad x_2 = 0\,, \tag{27}$$

$$\alpha_2(x_2) = -\frac{1}{\varkappa_3^2}\left\{\left[\varkappa_1\omega-\frac{i\varkappa_1}{Re}\left(\varkappa_1^2+\varkappa_3^2\right)\right]Re\cdot\tau^*+\frac{i\varkappa_1}{Re}\frac{d^3u_1^*}{dx_2^3}-\right.$$

$$\left.-\left[i\omega+\frac{1}{Re}\left(\varkappa_1^2+\varkappa_3^2\right)\right]\frac{d^2u_2^*}{dx_2^2}+\frac{1}{Re}\frac{d^4u_2^*}{dx_2^4}\right\}+\frac{1}{Re}\frac{d^2u_2^*}{dx_2^2} = 0 \quad \text{at} \quad x_2 = 0\,. \tag{28}$$

The functions $\alpha_1(x_2)$ and $\alpha_2(x_2)$ contain derivatives of x_2 of the highest order. In the viscous sublayer, the fluctuating motions directly near the wall should be very small. Therewith the functions $\alpha_1(x_2)$ and $\alpha_2(x_2)$ at $x_2 \to 0$ can be linearized, that is

$$\frac{d^2\alpha_1}{dx_2^2} = \frac{d^2\alpha_2}{dx_2^2} = 0\,. \tag{29}$$

Differentiating Eq. (26) twice and taking into consideration Eq. (29) at $x_2 \to 0$, we find:

$$\frac{d^2u_1^*}{dy^2}\left[i\omega+i\frac{\varkappa_1^2}{\varkappa_3^2}\omega+\frac{\varkappa_1^2}{\varkappa_3^2}\left(\varkappa_1^2+\varkappa_3^2\right)+\left(\varkappa_1^2+\varkappa_3^2\right)\right]+\tau_1^*\left[i\varkappa_1 2-\frac{2C\varkappa_1}{\varkappa_3}+\frac{2i\varkappa_1^3}{\varkappa_3^2}\right]+$$

$$+\frac{d^2u_2^*}{dy^2}\left[1+\frac{2iC}{\varkappa_3}+\frac{2\varkappa_1^2}{\varkappa_3^2}\right]-\frac{i\varkappa_1}{\varkappa_3^2}\left(i\omega+\varkappa_1^2+\varkappa_3^2\right)\frac{d^3u_3^*}{dy^3} = 0\,, \tag{30}$$

$$\left(i\omega+\varkappa_1^2+\varkappa_3^2\right)\frac{d^2u_2^*}{dy^2}+\frac{2\varkappa_1^2}{\varkappa_3^2}\frac{d^2u_1^*}{dy^2}-i\frac{2\varkappa_1}{\varkappa_3^2}\frac{d^3u_2^*}{dy^3} = 0\,. \tag{31}$$

Here scales of velocity and length are chosen in such a way that $Re = 1$ and $\partial U_1/\partial y\big|_{y=0} = 1$, where $y = x_2/L$ (Eq. 32). This can always be done, if the value $\partial U_1/\partial x_2$ proportional to the average value of the local shear stress on the wall is not equal to zero.

From the exact ratio (27) we obtain, at $y = 0$,

$$U_0 = U_* = \sqrt{\nu \left| \frac{\partial U_1}{\partial x_2} \right|}, \quad L_0 = \frac{\nu}{U_*}, \quad \frac{d^2 u_1^*}{dy^2} = \frac{i\varkappa_1}{\varkappa_1^2 + \varkappa_3^2} \frac{d^3 u_2^*}{dy^3}. \tag{33}$$

Substituting Eq. (33) into Eqs. (30) and (31), we obtain the following approximate results corresponding to the accepted hypothesis (29)

$$\frac{d^2 u_2^*}{dy^2} = -2\tau_1^* \frac{\varkappa_1\left(i\varkappa_3^2 - C_0\varkappa_3 + i\varkappa_1^2\right)}{\varkappa_3^2 + i2C_0\varkappa_3 + 2\varkappa_1^2} \quad \text{at} \quad y = 0, \tag{34}$$

$$\frac{d^3 u_2^*}{dy^3} = -\tau_1^* \frac{\left(\varkappa_1^2 + \varkappa_3^2\right)\left(i\omega + \varkappa_1^2 + \varkappa_3^2\right)\left(\varkappa_3^2 + \varkappa_1^2 + iC_0\varkappa_3\right)}{\varkappa_3^2 + i2C_0\varkappa_3 + 2\varkappa_1^2}. \tag{35}$$

Here, $C_0 = CL_0/U_0$ is a dimensionless constant determining the velocity gradient.

From Eqs. (24) and (25), the formula for the pressure "image" is obtained which at the flow boundary, considering the exact ratio (33), is expressed as

$$p_0^* = \frac{1}{\varkappa_1^2 + \varkappa_3^2} \frac{d^3 u_2^*}{dy^3}. \tag{36}$$

By comparing Eqs. (35) and (36), the approximate relation between the pressure pulsation "images" and the wall shear stress can be found:

$$p_0^* = -\tau_1^* \frac{\left(i\omega + \varkappa_1^2 + \varkappa_3^2\right)\left(\varkappa_1^2 + \varkappa_3^2 + iC_0\varkappa_3\right)}{\varkappa_3^2 + i2C_0\varkappa_3 + 2\varkappa_1^2}. \tag{37}$$

Let us note a particular case of two-dimensional turbulence in the viscous sublayer when $\varkappa_3 = 0$

$$p_0^* = -\tau_1^* \frac{i\omega + \varkappa_1^2}{2} \; . \tag{37a}$$

In this case the covariance between pressure and shear stress is

$$\langle p\tau \rangle = \iint_{-\infty}^{\infty} \langle p_0^* \bar{\tau}_1^* \rangle \, d\omega d\varkappa_1 = -\frac{1}{2} \int_{-\infty}^{\infty} S_\tau \varkappa_1^2 d\varkappa_1 < 0 \, .$$

The relation between pressure spectra and shear stress has also a very simple form:

$$S_{p_0}(\omega, \varkappa_1) = \langle p_0^* \bar{p}_0^* \rangle = S_\tau \frac{\varkappa_1^4 + \omega^2}{4} \; . \tag{38}$$

With $C = 0$ in Eq. (37), we obtain an important relationship for flow with a uniform distribution of mean velocity over the width:

$$p_0^* = -\tau_1^* \frac{\left(i\omega + \varkappa_1^2 + \varkappa_3^2\right)\left(\varkappa_1^2 + \varkappa_3^2\right)}{\varkappa_3^2 + 2\varkappa_1^2} \; . \tag{39}$$

From Eqs. (37) to (39) it is seen that in relation to the pressure spectrum near the wall, the spectrum of shear stress is shifted into the direction of lower frequencies.

This conclusion relating to two-dimensional turbulence was obtained by the writer in 1968 [16].

Recently, the results of direct spectral measurements of shear stress on a uniform flow boundary were published by a group of Novosibirsk researchers [17]. In particular, it was stated that already at values of dimensionless circular frequency $\omega = \tilde{\omega}\nu/U_*^2 = 0.1 \div 0.3$, the spectral density of shear stress is 100 to 500 times smaller than the values of spectral density at $\omega \to 0$ ($\omega < 10^{-3}$).

On the contrary, the proper treatment of results of measurements of pressure pulsation published, for example in [18], show that the spectral density of pressure pulsations is not yet diminished for those frequency values where shear stress pulsations are very low.

Taking into account that nonzero values of low and average frequency components of space-time spectra of pressure pulsations are concentrated along the surface $\omega/\varkappa_1 = -U_k/U_*$; $\varkappa_1 \lesssim \varkappa_3 < 10\,\varkappa_1$ (Eq. 40), where U_k is the velocity of disturbance drift close to the velocity in the middle of the flow [1,2,18], it can be stated that the following estimates are valid in the frequency range corresponding to the relatively large values of spectra of shear stress and pressure:

$$\varkappa_1^2 \simeq \omega^2 \left(\frac{U_*}{U_k}\right)^2 \ll \omega\,; \quad \varkappa_3^2 < \omega\,. \tag{41}$$

With Eq. (41), Eq. (39) can be simplified, and the following approximate relation between the dimensionless images of pressure pulsations and shear stress on the boundary is obtained:

$$p_0^* \approx -\tau_1^*\left(i\omega + \varkappa_1^2 + \varkappa_3^2\right) \approx -\tau_1^*\left(i\omega + \varkappa_3^2\right) \approx -\tau_1^*\, i\omega\,. \tag{42}$$

Returning to original and dimensional variables, but maintaining the previous symbols, we can write:

$$\tilde{p} \approx -\frac{\partial \tilde{\tau}_1}{\partial t}\,\frac{\nu}{U_*^2}\,. \tag{43}$$

Figure 1 presents a comparison between the pressure pulsation spectrum obtained from relations (42) and (43) with the aid of measurements of shear stress spectra and the data of direct measurements of pressure pulsation spectra. The comparison confirms the acceptability of the developed theory.

Examples of Estimates of Hydroelastic Distortion of Pressure Pulsations

Now, let us analyze the formulae (10) to (16) considering Eqs. (42) and (43) for the simplest problem of "substantially hydroelastic distortion", i.e., for the recording of pressures in highly turbulent flow with the help of a compliant pick-up. Changes of pressure-pulsation dispersion estimated by the formula (16) at $\tau = 0$, taking into account Eq. (42), can be written as:

$$\langle p\tilde{p}\rangle = \frac{r\delta_1}{\delta_*^2}(\tilde{p}')^2 \cdot A . \tag{44}$$

Here

$$A = -\frac{1}{(\tilde{p}')^3}\left[\langle\frac{\partial\tilde{p}}{\partial t}(t)\tilde{p}(t)\int_{t_0}^{t}\tilde{p}(t_1)dt_1\rangle + \frac{\nu}{U_*^2}\langle\frac{\partial\tilde{p}}{\partial t}(t)\tilde{p}(t)\tilde{\tau}_1(t_0)\rangle\right], \tag{45}$$

$\delta_1 = \tilde{p}'/C_z$ is the displacement of the pick-up membrane, produced by the pressure $\tilde{p}'$, and $\delta_* = \nu/U_*$ is the thickness of the viscous boundary layer (close to values of "displacement thickness" in the theory of laminar boundary layer [14]).

From Eq. (45) it can be seen that the values of coefficient A, and therefore the effect of load distortion, depends on

$$T(\tau) = \frac{1}{(\tilde{p}')^3}\langle\frac{\partial\tilde{p}}{\partial t}(t)\tilde{p}(t)\tilde{p}(t+\tau)\rangle . \tag{46}$$

Figure 2 shows the function $T(\tau)$, obtained from data of field measurements of pressure pulsations on the spillway of the Bratsk hydroelectric station[1]. This same function determines the effect of hydroelastic distortion even under more severe conditions. At larger values of t, the coefficient A must approach a constant that depends on flow properties (because of the stationarity of the pressure pulsations and the shear stress). Defining

$\frac{\partial p}{\partial t}(t)\tilde{p}(t) = (\tilde{p}')^2\varphi_1(t)$ and $\frac{1}{\tilde{p}'}\int_{t_0}^{t}\tilde{p}(t_1)dt_1 = \varphi_2(t)$, we can write at $t \to \infty$

---

[1] The flow was close to uniform. The measurements were carried out by L.A. Goncharov and V.M. Semenkov who kindly presented their preliminary results to the writer. The treatment of the results on the computer was performed according to the program and under the guidance of N.V. Halturina.

$$A = -\langle \varphi_1(t) \cdot \varphi_2(t) \rangle = -\int_{-\infty}^{\infty} S_{\varphi_1\varphi_2}(\omega)d\omega , \tag{47}$$

where $S_{\varphi_1\varphi_2} = \langle \varphi_1^* \overline{\varphi}_2^* \rangle = \overline{\Pi} \langle \varphi_1^* \overline{p}_0^* \rangle = \overline{\Pi} S_T(\omega)$,

$$S_T(\omega) = \frac{1}{2\pi} \int_{-\infty}^{\infty} T(\tau) \exp - i\omega\tau d\tau ;$$

and $\overline{\Pi}$ is the transfer function complexly conjugated with the function Π which transforms the image p_0^* into the image φ_2^*. In this case

$$\varphi_2^* = \Pi p_0^* ; \quad \Pi = \frac{1}{i\omega} ; \quad \overline{\Pi} = -\frac{1}{i\omega} . \tag{48}$$

Substituting Eq. (48) into (47), we obtain

$$A = -\frac{1}{2\pi} \int_{-\infty}^{\infty} T(\tau) \int_{-\infty}^{\infty} \frac{i}{\omega} \exp - i\omega\tau d\omega d\tau = -\frac{1}{\pi} \int_{-\infty}^{\infty} T(\tau) \left[\int_0^{\infty} \frac{\sin \omega\tau}{\omega\tau} d(\omega\tau) \right] d\tau =$$

$$= -\frac{1}{2} \int_{-\infty}^{\infty} T(\tau) d\tau . \tag{49}$$

Integration of the function $T(\tau)$, shown in Fig. 2, yields $\int_{-\infty}^{\infty} T(\tau)d\tau =$ $= 0{,}0102$. Thus, in the example under consideration, the relative error in the estimate of pressure-pulsation dispersion accounted for by the substantially hydroelastic effects is

$$\frac{2\langle p\tilde{p}\rangle}{(\tilde{p}')^2} = -0{,}0102 \frac{r\delta_1}{\delta_*^2} \tag{50}$$

This error can be compared with the error which is included due to the effect of the "virtual mass" (Eq. 7). As in this case $\underset{\sim}{U}_2 = (\partial\tilde{p}/\partial t)/C_z$, then with regard to the known properties of stationary random functions [19, p. 64], one can find

$$\langle p_1\tilde{p}\rangle = -\frac{\rho\mu_1 r}{C_z} \langle \frac{\partial^2 \tilde{p}}{\partial t^2} \tilde{p} \rangle = \frac{\rho\mu_1 r}{C_z} \langle \left(\frac{\partial\tilde{p}}{\partial t} \right)^2 \rangle \tag{51}$$

and $2\langle p_1\tilde{p}\rangle/(\tilde{p}')^2 = 2\mu_1\rho r/(C_z\lambda_p^2)$, where λ_p is the Taylor time scale of pressure pulsation and μ_1 is the dimensionless coefficient of "virtual mass" ($\mu_1 \simeq 0.67$). As it is seen from a comparison of Eqs. (50) and (51), the deviations in pressure dispersion produced by inertial effects and by substantially hydroelastic distortion are opposite in sign.

The ratio of these errors in the given example is

$$\left|\frac{\langle p\tilde{p}\rangle}{\langle p_1\tilde{p}\rangle}\right| = \frac{0,01}{1,34}\,\frac{\tilde{p}'\lambda_p^2 U_*^2}{\rho\nu^2} = B^2 . \qquad (52)$$

The magnitude of B^2 depends upon the structure of the pressure pulsations near the boundary. For instance, in a uniform flow [2,18] $\tilde{p}' = 3\rho U_*^2$; $\lambda_p \simeq 0,28\ \delta^*/U_\infty$ and the ratio B^2 can be of the order of $2.5 \cdot 10^{-6}\,Re^{5/4}$. These hydroelastic effects can be rather large. (For the special example of pressure pulsations on the spillway of the Bratsk hydroelectric station, $B \simeq 30$.)

The role played by hydroelastic effects is especially important within the zone of flow separation where the functions of pressure-pulsation distribution can appreciably differ from the normal law, where the pressure pulsations are intense, and the thickness of the wall layer is small.

References

1. Ljatkher, V.M.: Turbulence in Hydraulic Structures, Moscow: Energia Publishers 1968.

2. Ljatkher, V.M.: Pressure Pulsation on the Boundary of a Uniform Turbulent Flow. Izvestia AN USSR, Mechanika Zhidkosti i Gaza, No. 5 (1967).

3. Ljatkher, V.M.: Pressure Pulsation Acting on the Bottom of the Turbulent Flow Downstream of the Projection, Trudy Coordinatsionnych Soveshchany po Gidrotechniks, Vyp. 52, Leningrad: Energia Publishers 1969.

4. Ljatkher, V.M., Smirnov, L.V.: Pressure Pulsations on the Boundary of Turbulent Flow, Sb. Turbulentnye Techenia, Moscow: Nauka Publishers 1970.

5. Ljatkher, V.M.: Pressure Pulsation on Boundaries of Sharply Non-Uniform Cavitating Flow, Proc. XIII Congr. IAHR, V.5-I, Japan, 1969.

6. Ljatkher, V.M., Skladnev, M.F., Sheinin, I.S.: Investigations into the Field of Dynamic Hydroelastisity of Hydraulic Structures. Gidrotekhnicheskoe Stroitelstvo No. 8 (1971).

7. Miniovich, I.Y., Pernick, A.D., Petrovsky, V.S.: Hydrodynamic Sources of Sound, Leningrad: Sudostrojenie Publishers 1972.

8. Basin, A.M., Korotkin, A.I., Kozlov, L.F.: Management of the Ship Boundary Layer, Leningrad: Sudostrojenie Publishers 1968.

9. Ljatkher, V.M.: Hydroelasticity Effects in Hydraulic Structures, Proc. XI Congr. IAHR, Report 4-6, Leningrad 1965.

10. Moshkov, L.V.: Electrical Analogy in the Problems of Vibrations of Structural Elements in Fluid, Leningrad: Energia Publishers 1967.

11. McCormick, M.E., Riply, T.C.: On Turbulence-Induced Vibrations of a Thin Ribbon with Velocity Dependent Damping. Trans. ASME, Ser. E, No. 4 (1970).

12. Gukhman, A.A., Kadar, B.A.: The Structure of Turbulence in the Viscous Sublayer near a Flat Wall. Proc. IV Intern. Heat Transfer Conf., Versaille 1970.

13. Sternberg, J.: A Theory for the Viscous Sublayer of a Turbulent Flow. J. Fluid Mech. 13 (1962).

14. Schlichting, H.: Grenzschicht-Theorie, Karlsruhe: G. Braun 1964.

15. Pushkareva, I.V.: Laminar Flow Boundary Layer on the Flat Plate in Oscillating Flow. J. Prikladnaya Mechanika i Teoreticheskaya Physika, No. 4 (1968).

16. Ljatkher, V.M.: Turbulent Pulsations in Viscous Sublayer. Doklady AN USSR 180, No. 2 (1968).

17. Kutateladze, S.S., Nakorjakov, V.E., Burdukov, A.P., Guvakov, A.I.: Spectral Density of Friction Pulsations in Turbulent Near-the Wall Current, Doklady AN USSR 196, No. 5 (1971).

18. Bljudze, Yu.G., Dokuchaev, O.N.: Measurement of Velocity Pulsations and Pressure in Turbulent Flow Boundary Layers. Izvestija AN USSR, Mechanike Zhidkosti i Gaza 5 (1969).

19. Sveshnikov, A.A.: Applied Methods of Theory of Random Functions, Moscow: Nauka Publishers 1968.

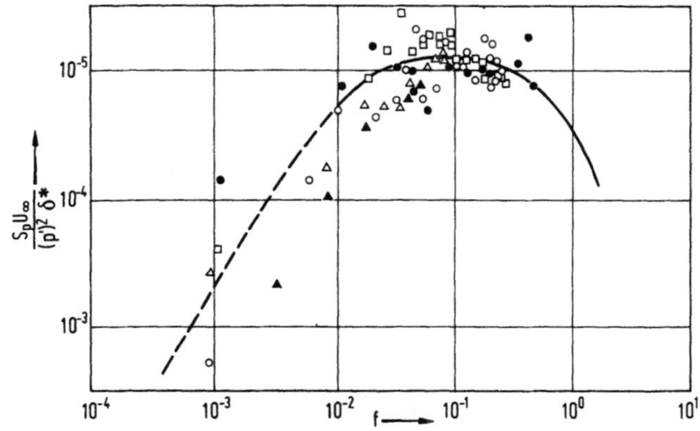

Fig. 1. Comparison between the shape of the pressure-pulsation spectrum calculated according to Eq. (42) on the basis of spectral measurements for shear stress on the wall [17] and the data of direct measurements of pressure spectra in the boundary layer [18] (heavy line)[1].

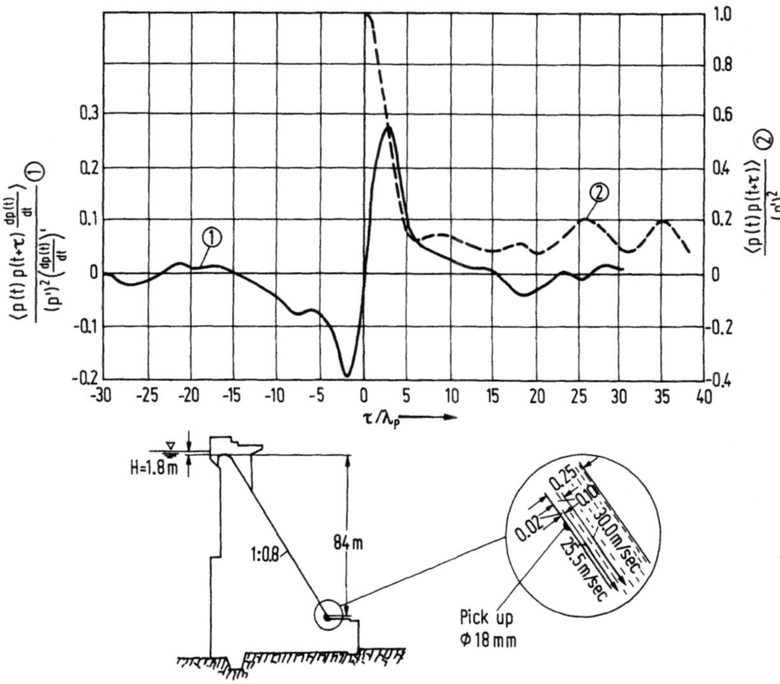

Fig. 2. The double and triple correlations of pressure pulsations on the spillway of the Bratsk hydroelectric station. The flow is aerated. $\lambda_p = p'/(dp/dt)' = 6.88 \cdot 10^{-4}$ sec.

---

[1] In Fig. 1, $f = \bar{\omega}\,\delta^*/(2\pi U_1)$; $\delta^* = \int(1-U/U_\infty)dy$ is the thickness of the displacement boundary layer; and $U_1 = U_\infty$ is the velocity along the channel axis. In the shear stress measurements [17], the Reynolds number varied from 5900 to 13300.

TECHNICAL SESSION C

"Flow-Induced Vibrations of Hydraulic Structures"

Chairman: H. W. Partenscky, Germany

Co-Chairman: B. Treiber, Germany

GENERAL LECTURE

Field Experiences with Hydraulic Structures

By

Jacob H. Douma

Washington, D. C., USA

Summary

Brief descriptions are presented of field experiences with vibration and cavitation problems which have occurred at a number of dams in the United States. These problems have resulted in structural damage to several types of valves and gates; concrete erosion in sluices, tunnels and stilling basins; and destruction of steel liners. Deficiencies in design and operation are discussed and remedial measures which were taken to repair damage and overcome deficiencies are described.

Summersville Dam

Three 108-in. Howell-Bunger valves are installed at Summersville Dam. The valve bodies and six radial vanes were fabricated from 1-in.-thick steel. A butterfly valve with a vertical axis is located 18.5 ft upstream from each Howell-Bunger valve. The conduits reduce from 11 feet in diameter to 9 ft downstream of the butterfly valves. Under normal operation, the butterfly valves are fully open and flow is regulated with the Howell-Bunger valves (Fig.1).

Operation in 1967 under heads up to 330 ft resulted in excessive vibration and severe damage to the Howell-Bunger valves. On two of the valves nearly all of the vanes were damaged or fractured. One vane was bent so as to nearly close the area between adjacent vanes.

An upstream section of one vane was completely torn out and missing. The third valve, which was operated less frequently and had vanes with rounded instead of champhered leading edges as for the other two valves, was not damaged (Fig.2).

Cavitation damage did not occur on the valves. Undoubtedly, the damage was caused by fatigue failure due to excessive vibration. Although there is no conclusive evidence that the butterfly valves contributed to the failure, the mechanical locking mechanisms of these valves may not have maintained the valve leaf rigidly in the full-open position and the resulting turbulence could have caused excessive vibration of the Howell-Bunger valves.

The valves were repaired by replacing all vanes with new vanes two inches in thickness. The valves have operated satisfactorily since without any damage. Evidently, the thicker vanes have sufficient rigidity to prevent the occurrence of damaging vibration.

Modes of vibration comparable to flapping and twisting can be identified for the vanes of Howell-Bunger valves, but an analysis has not been made to determine a critical flow velocity in terms of vane thickness and discharge. Research is needed to establish the vane thickness required as a function of valve size and flow velocity.

San Antonio Dam

An 84-in. butterfly valve is installed at the upstream end of an 84-in. steel outlet pipe at San Antonio Dam. The valve and steel pipe are located in a 12-ft diameter concrete lined tunnel which extends some 300 ft in length underneath the dam. The butterfly valve serves for emergency closure of the outlet. An 84-in. Howell-Bunger valve, installed at the downstream end of the steel pipe, serves to regulate reservoir releases.

In January 1969, flood releases under heads up to 160 ft were made through the outlet with the Howell-Bunger and butterfly valves fully open. The downstream Howell-Bunger valve operated well, but as the heads increased the butterfly valve experienced considerable vibration.

Distressing movements occurred in the mechanical linkage which held the valve in the open position. As it was necessary to release flood waters, valve operation continued until a failure occurred by the sudden closing of the valve and collapse of the adjacent downstream 30-ft section of steel pipe (Figs.3 and 4).

Inspection revealed no failure of the mechanical linkage which might cause the valve leaf to close suddenly. It was surmised that vibration was of sufficient magnitude to disengage the linkage holding mechanism, causing the valve leaf under high-head flow to close with great force. In addition to the pipe failure, minor damage was done to the valve leaf seat, and the air vent, constructed of light galvanized iron, was collapsed.

Analyses have indicated that on sudden closure of the valve the decrease in discharge resulted in a large air demand which sufficiently decreased pressure in the lightly constructed air vent to cause it to collapse. With the sudden failure of the air vent, insufficient air was supplied to the 84-in. steel pipe causing it to collapse too.

Repairs included providing a more rigid leaf linkage and several 12-in. automatic air valves along the crown of the 84-in. pipe. These repairs have made the system operate satisfactorily. This experience indicates that for high-velocity flow there is a strong tendency for a butterfly leaf to vibrate excessively and that very rigid control systems are needed to prevent sudden closure of the valve leaf.

Mossyrock Dam

Three 20.5-ft diameter penstocks are installed at Moosyrock Dam for three 150,000 kW power generators. Each penstock has a rectangular bellmouth entrance with a service gate located near the downstream end. The roller type service gates, 18 ft 8 in. wide by 34 ft 8 in. high, operate in slots which are installed on the upstream face of the 400-ft high concrete arch dam (Fig.5).

In 1968, during commissioning of the first power unit, the bulkhead gate was removed and the service gate was opened six inches to fill the penstock. During the first filling of one penstock in this manner with the reservoir filled to 176 ft above the gate sill, violent vertical movement of the gate occurred. It appeared that some hydraulic condition had caused the gate to be lifted in its slots in excess of 40 ft from where it fell, causing high impact forces on the gate and sill and damaging their various parts. Marks on the sill indicated at least two heavy impacts. The bottom gate beam was cracked at the joints, the gate seals and their supporting elements were twisted and broken, the hoist pulleys were torn from the gate, the guidelines were damaged, and the sill was creased and severely bent.

Model tests to a 1:26 scale confirmed that when the penstock was full high-velocity flow was directed upward into the gate well, causing the gate to be lifted at least 40 ft in the prototype from where it dropped with great force on the gate sill. The high-velocity jet into the gate well resulted from high reservoir pressure in front of the gate and low penstock pressure in back of the gate. The forces which caused the gate to be lifted were uplift on the beams on the back of the gate, sudden increase in pressure on the gate bottom, rapid loss in submerged weight of the gate, uplift due to trapped air under the gate beams, and tension in the hoist lines due to the gate weight and static lifting friction. These forces greatly reduced when the gate was lifted sufficiently to clear the constriction at the bottom of the gate well, causing the gate to drop rapidly.

This experience indicates that in filling a closed penstock with a large closure gate the area of the gate-well slot opening behind the gate should be considerably greater than the area of gate opening. The model tests indicated that a ratio of 10 to 1 should be used. Instead of filling a penstock with the closure gate, an equalizing piping system, independent of the gate, can be provided so that the gate would not need to be operated until upstream and downstream pressuress are equal.

Amistad Dam

At Amistad Dam, five 14.5-ft diameter penstocks have been constructed for a future powerhouse. Two of the penstocks were designed for making irrigation releases until the powerhouse was completed. Each of these penstocks have two tractor gates, 13 ft wide by 16 ft high, and bulkhead slots at the upstream end. A hydraulically operated slide gate 5 ft 8 in. wide by 10 ft high is installed at the downstream end of one penstock for regulating irrigation releases.

The installation was completed and action was started to place the irrigation regulating outlet in operation. With the downstream slide gate closed and upstream tractor gate open the downstream tractor gate was opened a small amount to fill the penstock and tractor gate-well prior to opening the slide gate. During this operation the tractor gate opened and closed suddenly, damaging the gate frame and bottom seal (Fig.6).

Analysis of the problem indicated that when the penstock became full water under reservoir pressure flowed upward through the slot formed by the back of the tractor gate and the back wall of the gate-well. This flow exerted an uplift force on the gate by jet action on its horizontal web members. Coupled with uplift on the gate bottom, the forces were sufficient to cause uncontrolled upward movement of the gate. As the gate lifted upward, the uplift forces were relieved, causing the gate to fall closed with severe impact.

As previously suggested for Mossyrock Dam, filling of the penstock at Amistad could be accomplished successfully provided that the gate opening area during filling is restricted to 10 percent of the flow area in the gate-well slot behind the gate.

Nacimiento Dam

At the 150-ft high earth Nacimiento Dam a 24-in. outlet pipe is located at the base of the dam for low-flow releases and a high-level outlet is located to discharge into the spillway channel. The latter outlet works consist of two 8-ft by 8-ft conduits through the base of the concrete ogee spillway. Flow is controlled by two slide gates located at the upstream end of these conduits. Each gate is attached by a 3-in. steel stem to an electric motor mounted on the spillway bridge (Fig.7).

In January 1969, the reservoir filled to 8 ft above the spillway crest. Prior to spillway operation, the two outlet gates were opened 6 feet to release flood flows. Except for the noise of high-velocity air flow in the 12-in. round air vent for each gate, the outlets operated satisfactorily. The gates were retained in the partially open position as the spillway went into operation. With several feet of head on the spillway crest an attempt was made to close the gates without success. After the reservoir level was lowered, it was found that the two gates had broken loose from the gate stems and were moved upstream from the gate slots. Considerable damage was sustained by the gates and slots (Fig.8).

Analysis of the problem indicated that when significant flow occurred in the spillway channel the air supply at the downstream end of the outlets was shut off and the small air vents did not supply sufficient air to sustain free flow in the outlet conduits. As a consequence, full conduit flow occurred which produced surging pressures on the backside of the gates. The resulting forces caused severe gate vibration and failure. It is not known whether the gate failures occurred suddenly due to water hammer surge or after a considerable period of operation due to fatigue failure. In any event, unbalanced hydraulic loads occurred in the upstream direction as the gates were pushed upstream.

In addition to repairing the gate frames, stems and slots, an elliptical entrance and a larger air vent was provided for each conduit. Also, it was specified that the gates should not be operated partially open when the spillway is in operation. Based on experience at other dams, these remedial measures should insure satisfactory operation of the high-level outlet.

Arkansas River Dams

The Arkansas River Project has 17 navigation locks and dams. Each dam has a spillway with tainter gates to maintain navigation pool levels and regulate river discharges. The spillways consist of low concrete sills, surmounted by either 50-ft or 60-ft wide tainter gates located within the main river channel (Fig.9).

Vibration problems have been encountered with the gates at some of the dams which caused oscillating movement of gate structural elements; vibration of walkways; humming, fluttering and roaring noises; and, in the more severe cases, a ripply wave pattern developed on the upper pool surface upstream from the gates. Cracks in structural gate members and welded connections were determined to be fatigue failures due to excessive gate vibration.

Model tests revealed that the source of vibrations was the gate lip and bottom seal, particularly when a flexible J-rubber seal was mounted on the bottom horizontal gate member. For these gates, instability of flow and pressures on the gate bottom during small gate openings caused the flow control point to shift over the bottom of the gate and the flexible J-rubber seal to flutter. When the bottom member of the gate was placed normal to the skin plate, vibration was reduced but under some flow conditions fluttering of the J-rubber seal still produced excessive vibration. All vibration was eliminated by a sharp lip consisting of the skin plate and vertical bottom gate member without a J-rubber seal. This design also did not vibrate when a small rectangular rubber strip was securely mounted in back of the sharp lip. Revision of the prototype gate bottoms to simulate a sharp lip has corrected this vibration problem (Fig.10).

Pine Flat Dam

Pine Flat Dam is a 440-ft high concrete gravity dam which has ten outlet conduits located in two tiers. Flow through each 5-ft by 9-ft conduit is controlled by two slide gates placed in tandem. The gate bottom slopes downward on a 45-degree angle to a narrow sharp lip which makes a metal to metal seal when the gate is closed (Fig.11).

In many early dams, slide gate bottoms were made normal to the gate faces for nearly the full gate thickness to attain structural strength. These gates experienced excessive vibration and severe cavitation damage to the gate bottoms, particularly when operated at small gate openings and high heads. Consequently, a maximum head limitation of 75 ft was adopted for their safe operation.

Hydraulic model tests revealed that flow under the flat bottom gate was controlled near the upstream face and unstable low pressures and cavitation occurred downstream therefrom. Further tests indicated that shifting the control to the downstream gate face by using either a vertical or 45-degree lip on the bottom of the gate produced stable pressures on the gate bottom, thereby greatly reducing vibration and eliminating cavitation erosion of the gate. Tests of the 45-degree lip slide gates at Pine Flat Dam under heads up to 380 ft resulted in only slight vibration and there was no indication of cavitation erosion on the gate. On the basis of these tests, slide gates with 45-degree lips are now designed to operate under heads up to 300 ft (Fig.12).

The high-head prototype tests resulted in slight cavitation erosion of the 8-ft sections of stainless steel clad conduit liners downstream of the gates. This resulted from high-velocity vortices which originated in the gate slots, passed under the gate lip and collapsed against the liner. However, these liners are providing adequate protection against this cavitation erosion (Fig.13).

Downstream of the steel liner, excessive erosion occurred on the concrete walls and invert of the test conduit. This was attributed to drag and impact forces created by high-velocity flow on relatively inferior concrete surfaces and not to cavitation erosion. Tests were made with various epoxy and aluminite protective coatings. Both proved effective but the aluminite protection is used because of lower cost and greater ease of application (Fig.14).

Lucky Peak Dam

The outlet works at Lucky Peak Dam, a 266-ft high earth and gravel fill, consist of an intake structure and a 23-ft diameter steel penstock leading to a valve-controlled manifold with a design capacity of 30,000 cfs. Primary regulation of flows is by six slide gates 5 ft 3 in. by 10 ft. Flow downstream of these gates is directed by individual short expanding channels and flip buckets into the river channel (Figs.15 and 16).

The reservoir filled in 1955 to within one foot of the spillway crest, resulting in heads up to 232 ft above the slide gate sills. Reservoir releases were made through one or two gates, on a rotational basis, at openings ranging from 5 to 8 ft. Later this method of operation was revised to release flows through five gates at openings of approximately 2 ft. After two weeks, it was apparent that erosion of the concrete inverts and walls downstream of the gates was occurring with the latter method of operation. Closure of the gates revealed that the eroded areas extended in symmetrical patterns from the gates well into the flip buckets, with maximum erosion being 40 in. in depth (Fig.17).

The erosion appeared to begin just downstream of the gate slots due to impact and cavitation pressures produced by downward deflected jets from the gate slots. The expanding channels downstream of the gates and operation at small gate openings were conducive to further development of cavitation pressures. The deep erosion in the flip buckets resulted from impact pressures and not cavitation.

The damaged areas were repaired by filling with a superior quality concrete and steel liners were installed in the outlet channels. No further damage has occurred since 1956.

Belton Dam

Belton Dam is a 192-ft high earth structure which contains a 22-ft diameter concrete lined outlet tunnel 790 ft in length. The intake structure contains three openings 7 ft wide by 22 ft high which are controlled by three tractor-type service gates operated by cable drum hoists. A 3-ft wide by 5-ft high low flow discharge conduit enters the right gate passage at an acute angle downstream of the service gate (Fig.18).

Operation of the low-flow conduit for several years was entirely satisfactory. Later operation of the large gates for flood control resulted in cavitation damage just downstream of a short section of steel liner at the junction of the low-flow conduit and the flood control gate passage. Similar damage did not occur in the left gate passage which is identical with the right gate passage except for the opening in the side wall made by the low-flow conduit. It was concluded that flow at high velocity in the right flood control passage was affected sufficiently by the low-flow conduit opening to cause cavitation pressures to occur downstream of the steel liner. The problem was corrected by extending the steel liner to cover the area subject to cavitation pressures (Fig.19).

Folsom Dam

Folsom Dam is a 340-ft high concrete gravity structure which contains eight 5 ft wide by 9 ft high conduits in two tiers controlled by tandem slide gates of the same size. The outlet conduits discharge onto the downstream fade of the dam which also serves as the invert for the tainter gate controlled overflow spillway (Fig.20).

In 1958, one year after completion of the dam, a large flood filled the reservoir to a level which required combined outlet and spillway operation for a period of about 10 days. After the reservoir was lowered to below spillway crest, extensive concrete erosion was noted at the exits of the upper tier concuits. The damage, which was characteristic of cavitation erosion, occurred on the side walls of the conduit exit openings and in the downstream face of the dam several feet below the intersection of the conduit inverts and the face of the dam (Figs.21 and 22).

Analysis of the problem indicated that the side wall erosion resulted from cavitation at the conduit exits due to insufficient air supply when the conduits and spillway operated simultaneously. The erosion on the downstream face of the dam below the exits was caused by cavitation due to the abrupt angle at the junction of the conduit inverts and dam face and the inability of air to reach the region below the junction during simultaneous conduit and spillway operation.

This problem was corrected by installing an "eyebrow" on the dam face just above each conduit exit to deflect spillway flows over the exit opening. In addition, it was specified that the conduit gates should be closed during spillway operation to enable ample air from the conduit air vents to reach the exits during spillway operation.

Bull Shoals Dam

Bull Shoals Dam is a 280-ft high concrete gravity structure which contains an overflow spillway surmounted by tainter gates and 16 flood control conduits, 4 ft wide by 9 ft high, in the overflow section. The combined stilling basin for the spillway and conduits, as constructed initially, consisted of a 200-ft long apron containing four steps with ellipsoidal upstream faces, terminated by a 4-ft high stepped end sill. The elliptical steps were designed, based on hydraulic model tests, to prevent cavitation pressures for individual and combination conduit and spillway flows (Fig.23).

During the period July 1951 to October 1952, the flood control conduits were used to regulate downstream flows during filling of the reservoir. The maximum head above the lowest step was about 200 ft. Thereafter, an inspection was made of the stilling basin which revealed that the floor of the basin was damaged, to varying degrees, immediately downstream from 12 of the 16 flood control conduits. The most severe damage occurred just downstream from the first step. In some places, erosion of the concrete reached a maximum depth of 3.5 feet. The appearance of the damaged concrete in some areas indicated that initial erosion was caused by cavitation at or near the tops of the ellipsoidal steps and that thereafter a combination of cavitation and abrasive action caused the damage to progress at an increasing rate (Fig.24).

Pressure cells were embedded in steps which had not been damaged in line with one conduit that had experienced only minor operation. Prototype tests disclosed that pressure fluctuations downstream from the tops of the steps were in the cavitation range, contrary to model predictions. Repairs were made by covering the steps with a sloping concrete apron and raising the end sill several feet to provide additional tailwater required to assure satisfactory hydraulic jump action for the spillway design flood.

Bonneville Dam

Bonneville Dam is a concrete gravity structure containing twelve 50-ft wide by 50-ft high vertical lift gates and six 50-ft wide by 60-ft high vertical lift gates surmounted on a 40-ft high ogee spillway weir. The original stilling basin contained two rows of 6-ft high trapezoidal shaped baffle piers located on a horizontal apron with no end sill. The spillway design discharge is 1,600,000 cfs. Full operation of the spillway began in April 1938. During the first years of operation, the spillway was in constant use because river discharge exceeded power flows (Fig.25).

Unwatering of the south half of the stilling basin in November 1954 revealed extensive damage to the concrete and reinforcing bars of the baffles and the horizontal baffle deck. Maximum erosion occurred on the upstream row of baffles. All corners of these baffles were well rounded, worn and deeply nicked, exposing large aggregate to spillway flows. Reinforcing bars were torn loose as holes 12 to 15 in. deep were eroded in the baffle side walls. Areas up to 2 ft in depth were eroded in the baffle deck downstream of each baffle which resulted from cavitation pressures within eddies at the downstream corners of the baffles. In general, erosion of the downstream row of baffle blocks was similar to that of the upstream blocks but not as extensive in depth and area (Figs.26 and 27).

Previously, several baffle blocks had been modified to test methods of providing protection against cavitation erosion. One block was completely armored with a 3/8-in. thick steel plate anchored with bolts spaced on one foot centers. All of the armor plate was ripped off and deep erosion occurred on the baffle sides. The leading edges and tops of several baffles were rounded to provide stream-lined type baffles which model tests had indicated would be less susceptible to cavitation erosion. These baffles experienced significantly less erosion than the original sharp-cornered baffles, but they did not decrease either the extent or severity of damage to the baffle deck downstream of the baffles (Figs. 28 and 29).

Subsequent model tests have indicated that cavitation pressures could be eliminated by constructing a solid end sill at the location of the downstream row of baffles, streamlining all of the upstream row of baffles, and operating the spillway gates with essentially equal openings at all times to assure maximum tailwater levels for any gate discharge. These repairs coupled with reduced spillway flows due to greater power demands have proven to be successful.

Closing Remarks

The foregoing field experiences with structural vibration and cavitation, the resulting consequences, and remedial measures taken to overcome deficiencies by no means cover all such experiences which have occurred. Cavitation erosion of baffle blocks and outlet conduits has occurred at many dams. Other types of problems have occurred at some dams such as deep erosion of the inclined spillway tunnels at Boulder and Yellowtail dams, caused by irregularities in concrete surfaces subjected to velocities in the order of 150 fps.

The experiences cited, however, indicate conclusively that serious damage results whenever excessive cavitation or vibration occurs during project operation. This damage can be reduced or eliminated only by improved design criteria and construction techniques, more careful operation practices, and the development of materials more resistant to cavitation erosion and fatigue failures.

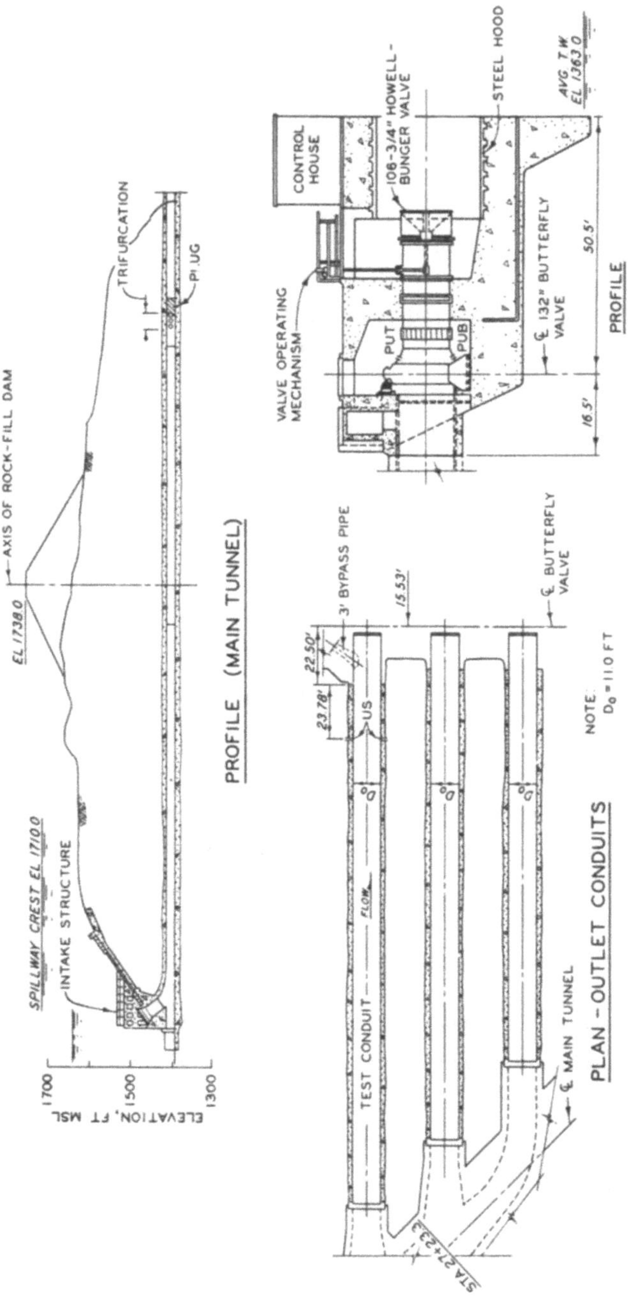

Fig.1. Summersville project.

Fig.2. Damaged Howell-Bunger valve at Summersville Dam.

Fig.3. Downstream view of collapsed 84" steel penstock at San Antonio Dam.

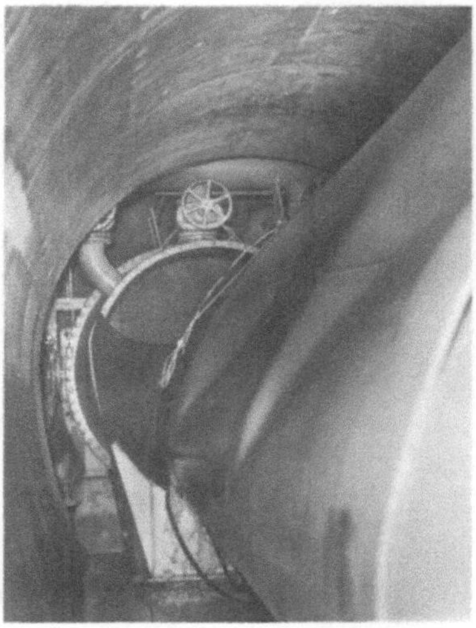

Fig.4. Upstream view of collapsed pipe and closed butterfly valve.

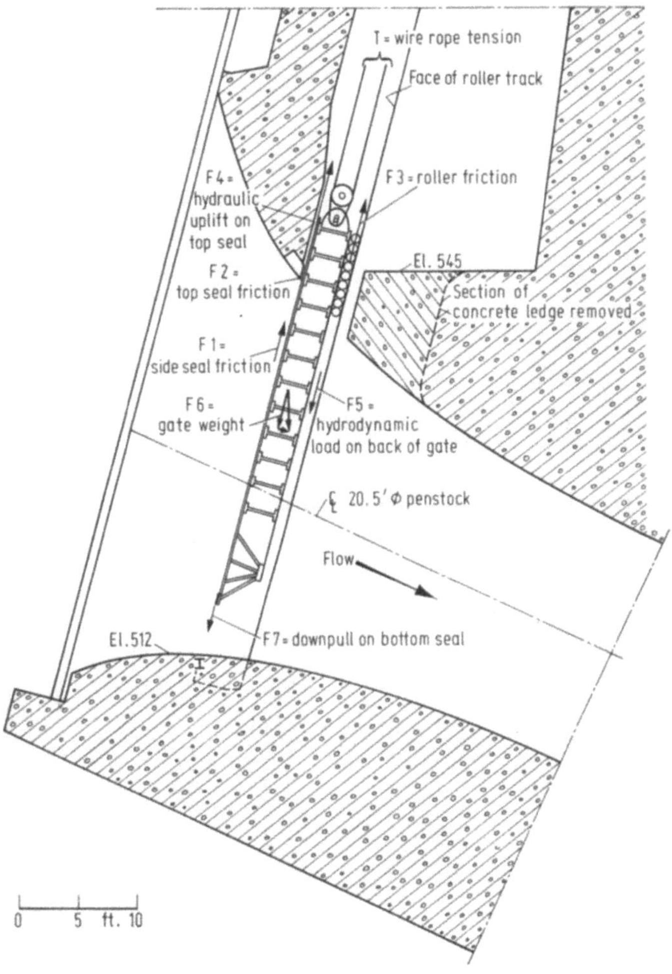

Fig.5. Penstock closure gate at Mossyrock Dam.

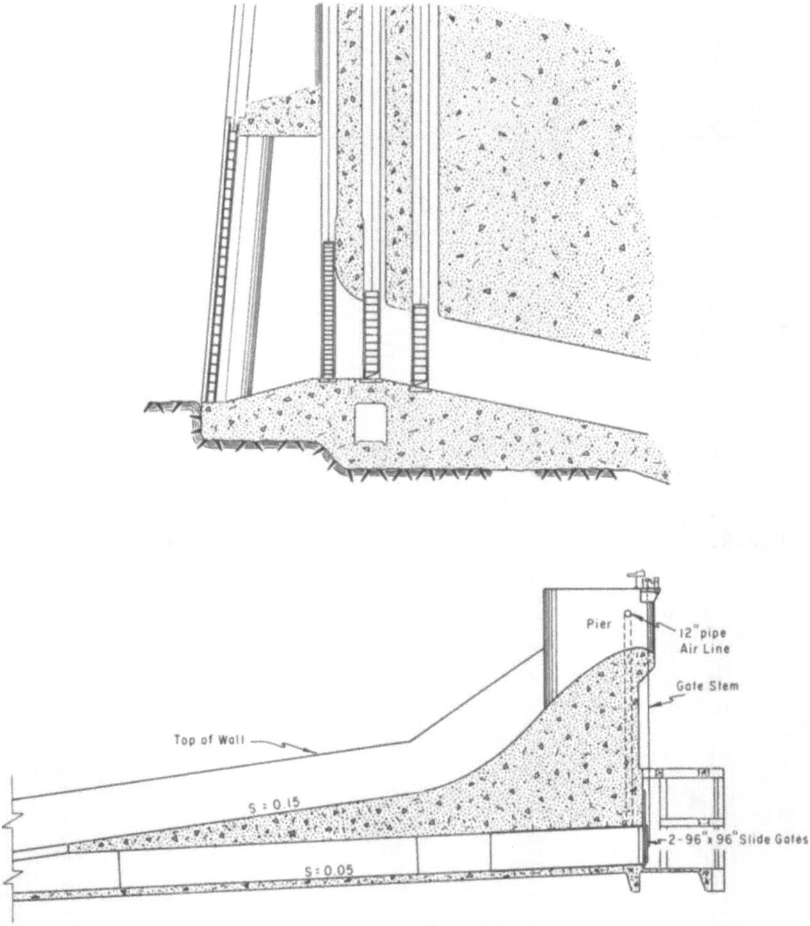

Fig.7. Spillway and outlet at Nacimiento Dam.

Fig.8. Gate failure at Nacimiento Dam.

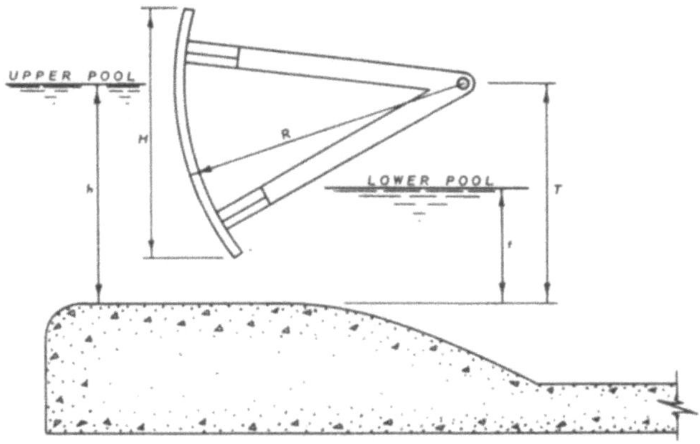

Fig.9. Typical spillway section for Arkansas River Dams.

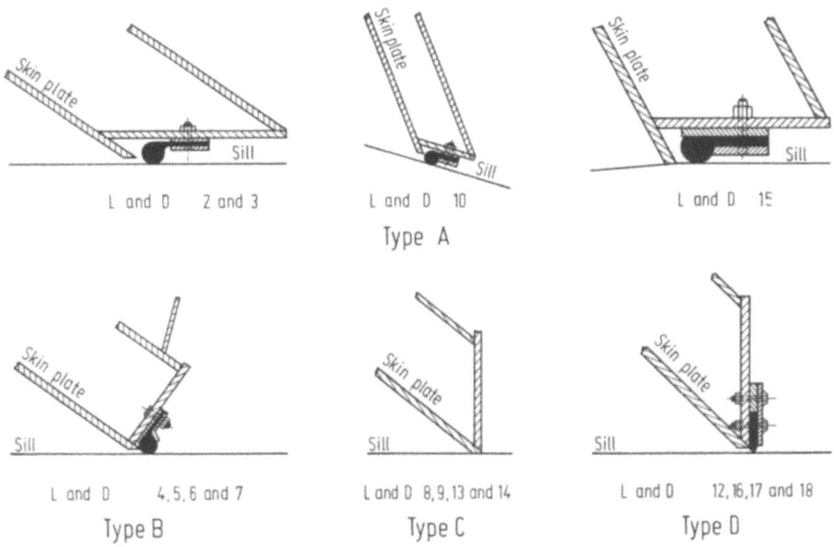

Fig.10. Tainter gate lips for Arkansas River Dams.

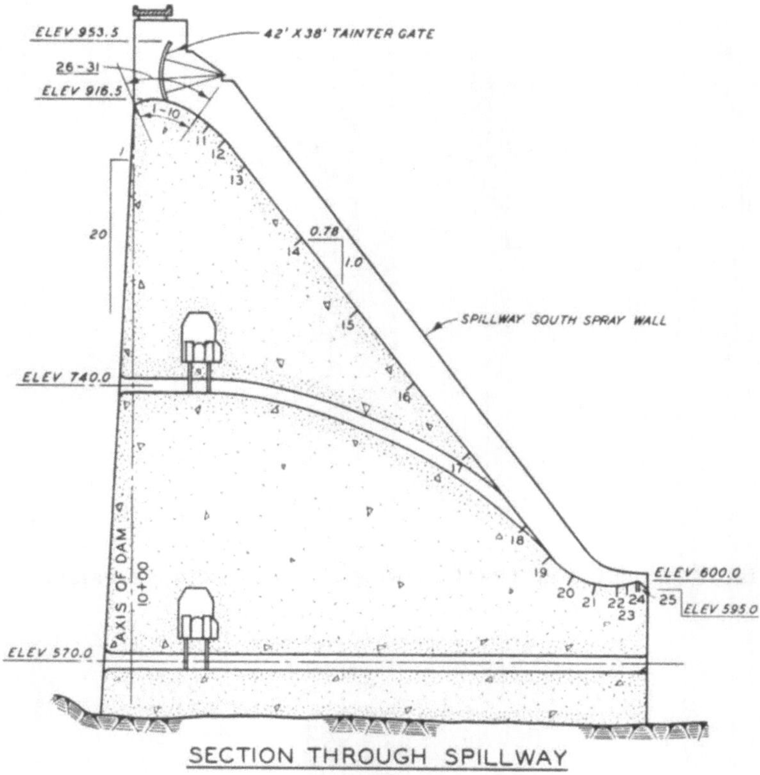

Fig. 11. Pine Flat Dam.

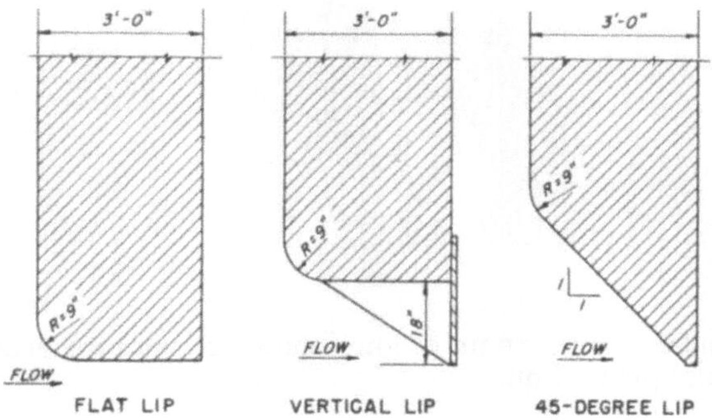

Fig. 12. Slide-gate bottom shapes.

Fig.13. Pine flat gate slot showing slight cavitation erosion.

Fig.14. Conduit downstream of gate showing concrete erosion and epoxy protection.

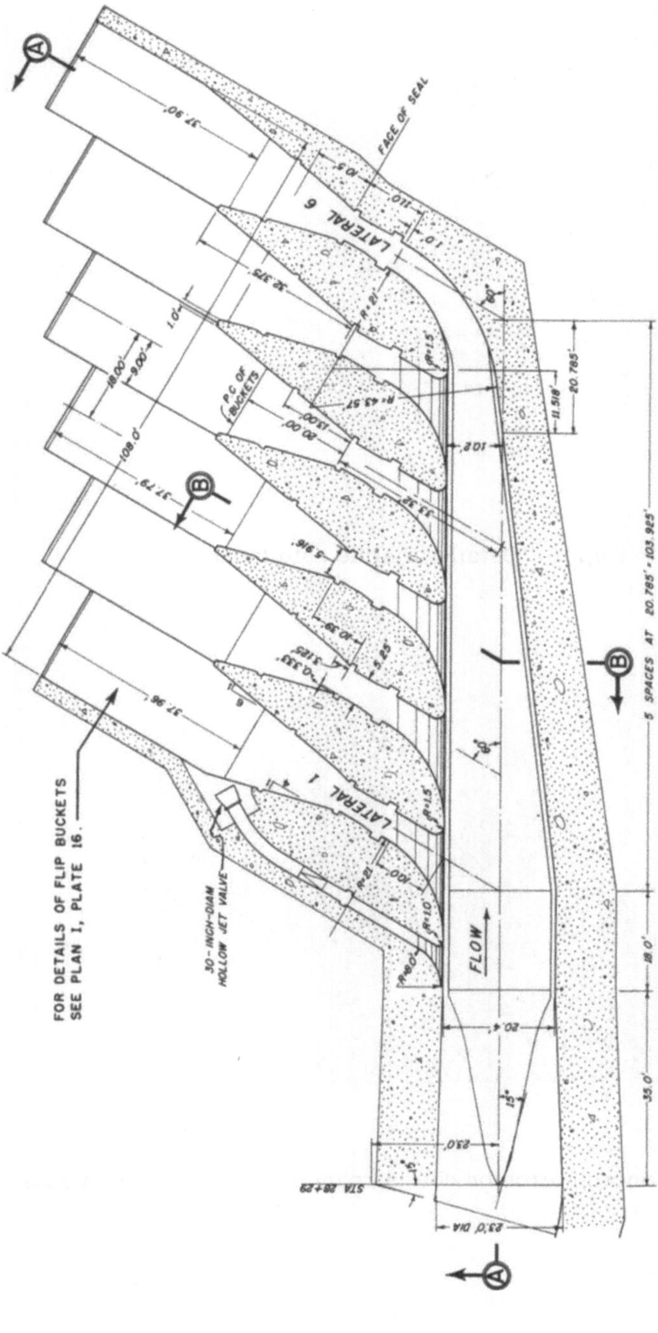

Fig.15. Lucky Peak outlet works.

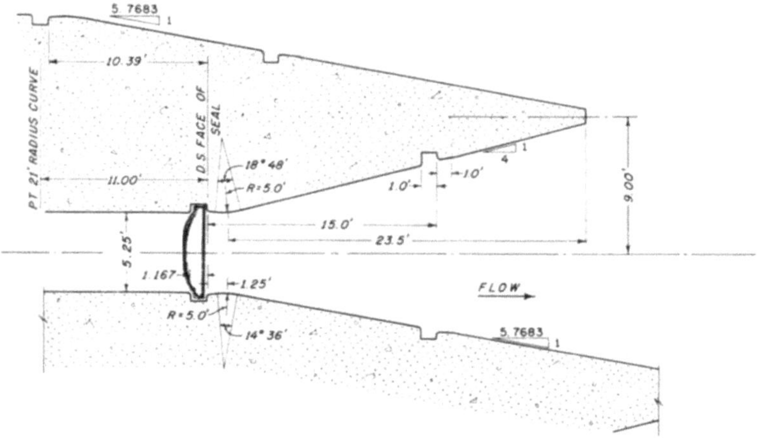

Fig. 16. Details of slide gate location.

Fig. 17. Concrete erosion downstream of lucky peak slide gate.

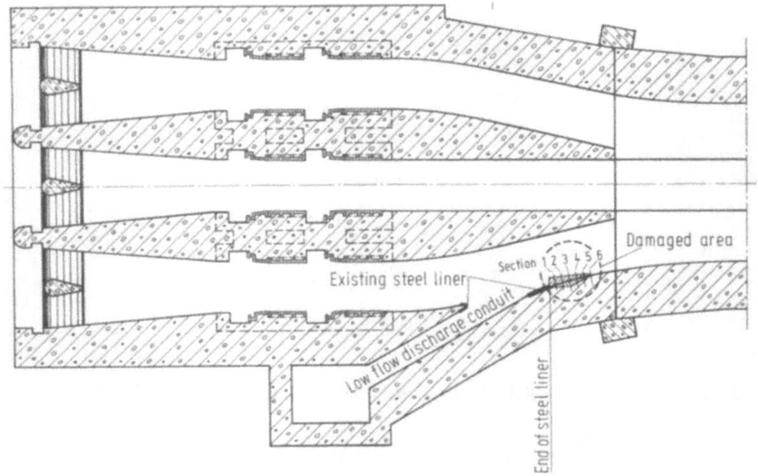

Fig. 18. Belton Dam outlet gates.

Fig. 19. Cavitation erosion downstream of the low-flow conduit at Belton Dam.

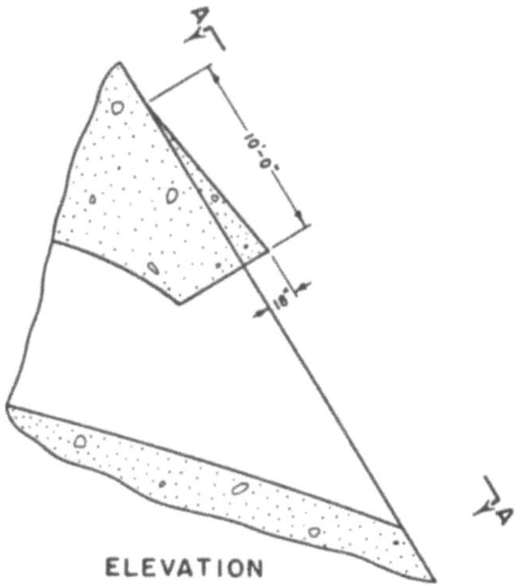

Fig.20. Sluice exit in spillway invert at Folsom Dam.

Fig.21. Cavitation erosion on side wall of sluice exit at Folsom Dam.

Fig.22. Cavitation erosion on spillway invert below sluice exit at Folsom Dam.

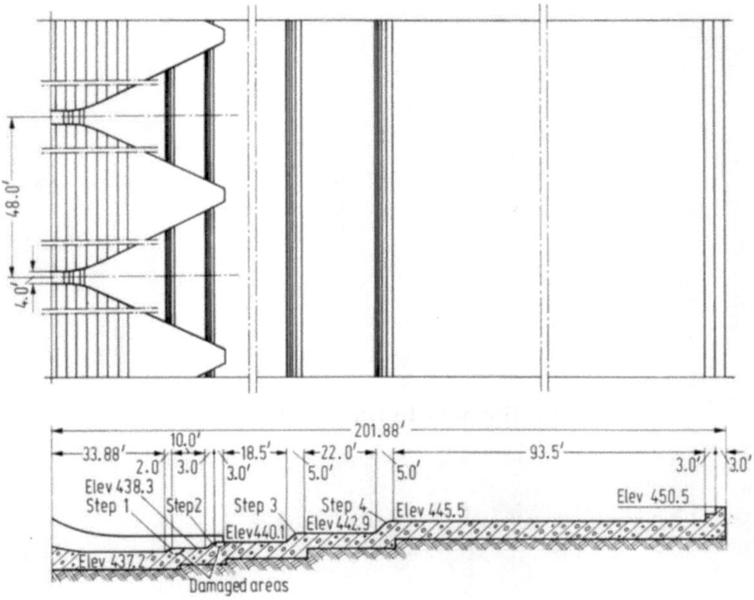

Fig.23. Sluice exits and stilling basin at Bull Shoals Dam.

Fig.24. Cavitation erosion in exit channel at Bull Shoals Dam.

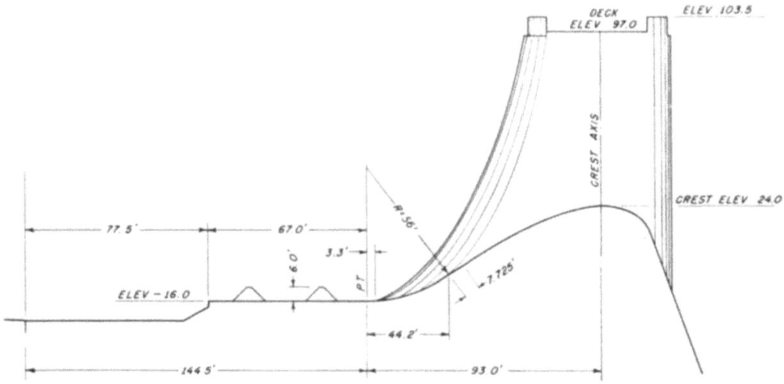

Fig.25. Bonneville Dam spillway.

Fig.26. Cavitation erosion on sides of battles at Bonneville Dam.

Fig.27. Cavitation erosion on apron downstream of battles.

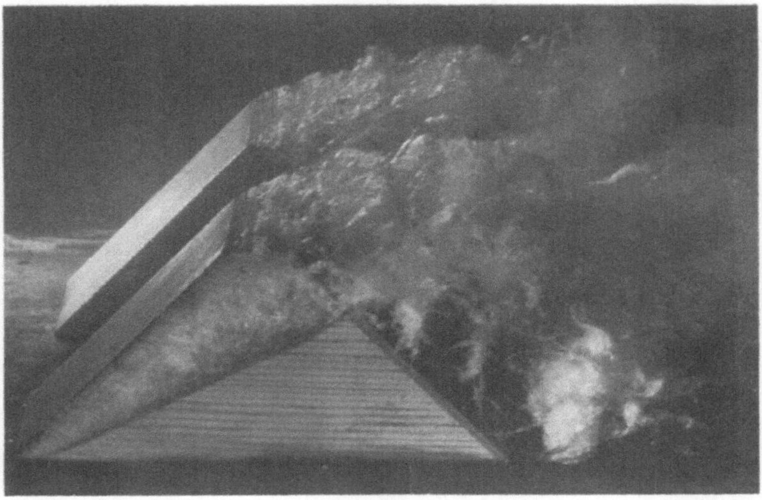

Fig. 28. Cavitation in model of trapezoidal-shaped baffle.

Fig. 29. Baffle from which 3/8-in. steel armor plate was removed.

CONTRIBUTIONS

On Flow-Induced Structural Oscillations

By

M.F. Skladnev, I.S. Sheinin,

Leningrad, U.S.S.R.

Summary

A review of investigations conducted at the B. E. Vedeneev All Union Research Institute of Hydraulic Engineering (VNIIG) is given. Presented are the main research results on hydroelastic interaction between the structure and the flow, mathematical models of flow-induced vibrations, and elaboration of measuring techniques for process parameters in the field of hydroelasticity.

Introduction

Extensive studies in the field of flow-induced structural oscillations are conducted at the VNIIG aimed at:

(1) defining the specific features of dynamic loads exciting structural oscillations in flows; in paricular, study of regularities in the formation of turbulent flows at various hydraulic structural elements and evaluation of their quantitative characteristics;

(2) analysis of phenomena and regularities in the hydroelastic interaction between flow and structure;

(3) consideration of mathematical models of flow-induced vibrations;

(4) development of measuring techniques and simulation of processes in the field of dynamic hydroelasticity;

(5) research on flow-induced vibrations of specific gates and elements of hydraulic structure.

(6) analysis of the effects of different factors (compressibility, viscosity, homogeneity, aeration, waves and ice formation on the free surface, elastic pliability and percolation characteristics of reservoir banks and bottom, deformability of the oscillation member, etc.) in problems on oscillations of structure in a water medium.

The present paper is restricted to the most interesting findings concerned with items 2 and 4, since a short review [1] published recently embraces items 1, 3, and 6, while item 5 is dealt with in a paper by Abelev, A.S. and Dolnikov, L.L. submitted to this Symposium.

Analysis of Hydroelastic Interaction

The most essential result in the analysis of hydroelastic interaction between flow and structure seems to be the solution to a hydroelastic problem for a fairly wide range of cases satisfying the following conditions:

(1) A solution should be known to a static boundary value problem for an elastic body through a set of influence functions (tensor) $f_{hl}(M_0, M)$ equal to the limit of the ratio between the M_0 point displacement along the x_h-axis $(h = 1, 2, 3)$ and the point force $P_{3l}(M)\Delta V$ causing the displacement (the volume forces P_{3l} uniformly distributed over the volume ΔV containing M), the force acting along the x_l-axis $(l = 1,2,3)$, with ΔV tending to zero and $p_{3l}(M)$ = const. The elastic body occupies in space the region V, therefor $M_0 \in V$ and $M \in V$. The boundary surface of the body S contains the surface S_1 over which the elastic body interacts with the fluid.

(2) A solution should be known to a hydrodynamic boundary value problem for a fluid region interacting with the

elastic body. The solution should have the form $p = p(\overline{U}, M)$ expressing the relationship between the hydrodynamic pressure p at the point M of the boundary S_1, and the displacement vector $\overline{U}(M,t)$ of the points of the body at the boundary. In the majority of cases such a relationship may be represented as [2,3]:

$$p = A \sum_{k,n=0}^{\infty} \frac{A_{kn}}{\tau_{kn}} \varphi_{kn}(M) , \qquad (1)$$

where $\varphi_{kn}(M)$ are eigenfunctions of the boundary value problem for the fluid on S_1; A is a constant conditioned by the geometry of the fluid region, oscillation frequencies, fluid density, and other parameters of the hydrodynamic problem; τ_{kn} is the eigenvalue of the boundary value problem for the fluid; A_{kn} are time functions, sign fying expansion coefficients of displacements $\overline{U}$ over S_1 in eigenfunctions φ_{kn}. They have the form

$$A_{kn} = \int_{S_1} \bar{n}(M) \cdot \overline{U}(M,t) \varphi_{kn}(M) dS , \qquad (2)$$

in case the eigenfunctions are othonormalized or differ in a factor independent of M, if they are neither orthogonal nor normalized. Under harmonic oscillations the time function $e^{i\omega t}$ is taken out from under the integral, and $\overline{U}_a(M)$ is substituted for $\overline{U}(M,t)$, the former symbol carrying information only on amplitudes and phases of displacement components at the point M; $\overline{n}(M)$ is the unit vector of the normal at the point M of the surface S.

Let us consider the problem on harmonic oscillations of an elastic body subjected to the volume forces $e^{i\omega t}P_{31}(M)$, the forces $e^{i\omega t}P_{21}(M)$ distributed over the surface S, the forces $e^{i\omega t}P_{11}(M)$ distributed along the L-line, and the forces $e^{i\omega t}P_{01}(M)$ concentrated at isolated points M. The displacement vector of the point M_0 is

$$\bar{U}_a(M_0) = \sum_{h=1}^{3} \bar{i}_h \sum_{l=1}^{3} \int_V \Big[P_{31}(M) + \omega^2\rho(M)\bar{U}_a(M)\cdot\bar{i}_l + P_{21}(M)\delta(S) + P_{11}(M)\delta(L) + P_{01}(M)\delta(M_*) \Big] f_{hl}(M_0,M)dV , \tag{3}$$

where $\delta(S)$, $\delta(L)$, $\delta(M_*)$ are delta functions concentrated on the S-surface on the L-lines and at the points M_* of the three-dimensional space, respectively [4]; $\rho(M)$ being the density of the body at point M and $\bar{i}_h$ the unit vector of the x_h-axis.

Taking into account both the disturbing forces and the hydrodynamic pressure due to oscillations of the body, Eqs. (1,2), and grouping separately the terms depending on $\bar{U}_a$, the following equations can be obtained

$$\bar{U}_a(M_0) = \sum_{h=1}^{3} \bar{i}_h \sum_{l=1}^{3} \Big\{ Q_{hl}(M_0) + \int_V \Big[\omega^2\rho(M)\bar{U}_a(M) + \delta(S_1)\bar{n}(M)p(M) \Big]\cdot \bar{i}_l f_{hl}(M_0,M)dV \Big\}, \tag{4}$$

or

$$\bar{U}_a(M_0) = \sum_{h=1}^{3} \bar{i}_h \sum_{l=1}^{3} \Big\{ Q_{hl}(M_0) + \int_V \Big[\omega^2\rho(M)\bar{U}_a(M) + \delta(S_1)\bar{n}(M)A \sum_{kn=0}^{\infty} \frac{\varphi_{kn}(M)}{\tau_{kn}} \int_{S_1} \bar{n}(M)\cdot\bar{U}_a(M)\varphi_{kn}(M)dS \Big] \times \bar{i}_l f_{hl}(M_0,M)dV \Big\}, \tag{5}$$

where

$$Q_{hl}(M_0) = \int_V \Big[P_{31}(M) + P_{21}(M)\delta(S) + P_{11}(M)\delta(L) + P_{01}(M)\delta(M_*) \Big] f_{hl}(M_0,M)dV . \tag{6}$$

Further, by changing the order of integration and summation and then the order of integration of the second term in curly brackets, Eq. (5), we obtain

$$\bar{U}_a(M_0) = \sum_{h=1}^{3} \bar{i}_h \sum_{l=1}^{3} \left[Q_{hl}(M_0) + \int_V \bar{K}_{hl}(M_0,M) \cdot \bar{U}_a(M)dV \right], \quad (7)$$

where

$$\bar{K}_{hl}(M_0,M) = \omega^2 \rho(M) f_{hl}(M_0,M)\bar{i}_l +$$
$$+ \delta(S_1)A\bar{n}(M) \sum_{k,n=0}^{\infty} \frac{\varphi_{kn}(M)}{\tau_{kn}} \int_{S_1} \bar{i}_l \cdot \bar{n}(M)\varphi_{kn}(M) f_{hl} \cdot (M_0,M)dS . \quad (8)$$

The integral in Eq. (8) is, in the physical sense, the integral influence coefficient of the component along the x_l-axis of the hydrodynamic pressure distributed as the k,n-th eigenform on the displacement component of the point M_0 along the x_h-axis. The integral may be denoted by

$$B_{kn,h,l}(M_0) = \int_{S_1} \bar{i}_l \cdot \bar{n}(M)\varphi_{kn}(M) f_{hl}(M_0,M)dS . \quad (9)$$

Integral equation (7) is sufficient for defining forms and phases of the induced oscillations $\bar{U}_a(M)$. In practice, the calculations reduce to the solution of a linear set of algebraic equations by partitioning the body into n/3 points. Then, designating each component at each point by j (or q), we obtain n^2 influence functions δ_{jq}, n masses m_j, n forces P_j, and n unknown displacements U_j to be determined from the set

$$\left(\omega^2 m_j \delta_{jj} + C_{jj} - 1\right)U_j + \sum_{q=1}^{n}{}' \left(\omega^2 m_q \delta_{jq} + C_{jq}\right)U_q = P_j , \quad (10)$$

$$(j = 1,2,\dots,n),$$

where

$$\left.\begin{aligned} m_j &= \rho(M_j)\Delta V_j; & \delta_{jq} &= f_{hj,lq}(M_j, M_q); \\ U_j &= \bar{i}_j \cdot \bar{U}_a(M_j); & P_j &= \sum_{l=1}^{3} Q_{hj,l}(M_j); \end{aligned}\right\} \tag{11}$$

$$C_{jq} = A\Delta S_q \bar{i}_q \cdot \bar{n}(M_q) \sum_{k,n=0}^{\infty} \frac{B_{kn,hj,lq}(M_j)\varphi_{kn}(M_q)}{\tau_{kn}}; \tag{12}$$

ΔV_j, ΔS_j are finite volumes and areas reduced to a point whose displacement components are designated by j; $\bar{i}_j$ is a unit vector directed similarly to the j-component; the subscript j or q with a scalar value means that the quantity refers to a point possessing a j- or q-component; the same subscript beside the subscript h or l shows that the h- or l-component is taken parallel to the corresponding j- or q-component. The prime beside the summation sign means summation of all terms except the q = j term.

When U_j and therefore $\bar{U}_a(M)$ are determined, the hydrodynamic pressure may be found from Eqs. (1) and (2). The hydrodynamic pressure is known to generally vary with time in phase with the inertia forces of the oscillating body; hence, the ratio of the pressure at each point to the acceleration of the point can be regarded as a certain added mass of the fluid $m_f(M)$ non-uniformly distributed over the surface S_1 [3]:

$$m_f(M) = \frac{p(M)}{\bar{n}(M) \cdot \bar{U}_a(M)\omega^2}. \tag{13}$$

By substituting the expression for the hydrodynamic pressure from Eq. (13) into (4) we have

$$\begin{aligned} U_a(M_0) = \sum_{h=1}^{3} \bar{i}_h \sum_{l=1}^{3} \Big\{ Q_{hl}(M_0) + \omega^2 \int_V [\, \rho(M) + \\ \delta(S_1) m_f(M)\,]\, f_{hl}(M_0, M) \bar{U}_a(M) \cdot \bar{i}_l dV \Big\}, \end{aligned} \tag{14}$$

instread of Eq. (10) the following equation may be written

$$\left[\omega^2(m_j + m_{fj}) - 1\right]\delta_{jj}U_j + \omega^2 \sum_{q=1}^{n}{}' (m_q + m_{fq})\delta_{jq}U_q = P_j, \quad (15)$$

$$(j = 1,2,\dots,n),$$

where

$$m_{fj} = m_f(M_j)\Delta S_j . \quad (16)$$

For plates and bars, Eqs. (4), (5), (7), and (14) assume a scalar form. When considering the oscillations of a vertical plate $x = 0$ of width B and height H at the boundary with a rectangular volume of an ideal incompressible fluid limited from above by the free surface $z = 0$, where waves are neglected, below by the rigid impervious bottom $z = H$, and on the three remaining sides $y = 0$, $y = B$, and $x = C$ by rigid impervious walls, the hydrodynamic pressure on the plate is known [3] to be

$$p = \rho_0\omega^2 \sum_{k,n=0}^{\infty} \frac{\Delta_n A_{kn}}{\sigma_{kn}} \operatorname{cth}\frac{\sigma_{kn}C}{H} \cdot \cos\frac{n\pi}{B}y \cdot \sin\sigma_k z, \quad (17)$$

where ρ is the fluid density,

$$\Delta_n = \begin{cases} \frac{1}{2} & \text{at} \quad n = 0, \\ 1 & \text{at} \quad n = 1,2,3,\dots \end{cases} \quad (18)$$

$$\sigma_k = \frac{(2k+1)\pi}{2H}, \qquad \sigma_{kn} = \sqrt{\sigma_k^2 + \left(\frac{n\pi}{B}\right)^2}, \quad (19)$$

$$A_{kn} = \frac{4}{BH}\int_0^H\int_0^B U_a(y,z)\cos\frac{n\pi}{B}y \cdot \sin\sigma_k z\, dy\, dz . \quad (20)$$

In case the form of the influence function $f(y_0,z_0,y,z)$ of the plate which depends both on boundary conditions and the law of thickness variation is not preset, provided the distributed load $p_H(y,z)e^{i\omega t}$ is supposed to be applied to the plate, one can write instead of Eqs. (3) and (4)

$$U_a(y_0,z_0) = \int_0^H \int_0^B \Big[p_H(y,z) + \omega^2 m(y,z)U(y,z) + p(y,z) \Big] f(y_0,z_0,y,z)\,dydz, \qquad (21)$$

or, including Eqs. (17) to (20), one can write instead of Eq. (7)

$$U_a(y_0,z_0) = P(y_0,z_0) + \int_0^H \int_0^B K(y_0,z_0,y,z)U_a(y,z)\,dydz, \qquad (22)$$

where

$$P(y_0,z_0) = \int_0^H \int_0^B p_H(y,z)f(y_0,z_0,y,z)\,dydz, \qquad (23)$$

$$K(y_0,z_0,y,z) = \omega^2 \Bigg[m(y,z)f(y_0,z_0,y,z) + \frac{4\rho_0}{BH} \sum_{k,n=0}^{\infty} \frac{\Delta_n B_{kn}(y_0,z_0)}{\sigma_{kn}} \operatorname{cth} \frac{\sigma_{kn} C}{H} \cdot \cos \frac{n\pi}{B} y \cdot \sin \sigma_k z \Bigg], \qquad (24)$$

$$B_{kn}(y_0,z_0) = \int_0^H \int_0^B f(y_0,z_0,y,z) \cos \frac{n\pi}{B} y \cdot \sin \sigma_k z\,dydz. \qquad (25)$$

On the other hand, using expression (13) one can write instead of Eq. (14)

$$U_a(y_0,z_0) = P(y_0,z_0) + \omega^2 \int_0^H \int_0^B [m(y,z) + m_f(y,z)]f(y_0,z_0,y,z)U_a(y,z)\,dydz, \qquad (26)$$

which results in Eq. (15) when it is integrated by the finite different method.

Taking the disturbing forces in Eqs. (7), (10) or (22) as equal to zero, the frequencies and forms of free oscillations of the elastic body in the fluid may be estimated.

Knowing the frequencies and forms (both of natural and induced oscillations) of the elastic body, the velocity and pressure fields in the fluid may be determined from the solution to the hydrodynamic problem which is given (see the condition 2, p.2). Thus, a general solution to the hydroelastic problem may be derived.

Since Eqs. (15) and (26) are identical in form with the corresponding equations for a body not immersed in fluid, and essentially differ only in the law of mass distribution, then for a body immersed in fluid all the theorems on frequencies and forms of slight oscillations of the elastic body are valid, i.e. the theorems on the positivity and isolation of roots, on the independence of oscillation eigenforms of the initial conditions, on their orthogonality, on expansibility of any oscillation form in eigenforms, etc.

Further generalizations are as follows:

(1) In case of polyharmonic influences the displacements of the points of the body result from summing up those of each harmonic.

(2) In case of non-stationary phenomena, the displacement spectra at each point may be obtained by representation of the influence in spectral form.

(3) In case one is to take into account damping in the hydroelastic system due to waves on the free surface of the fluid, percolation losses, non-ideal bottom pliability, losses caused by fluid viscosity, and, finally, the internal nonelastic resistance to deformation of the material of the body, then the problem is reduced to the solution of the same equations: either Eq. (7) or (10)

or (22), but with complex kernels or with coefficients received by introducing the complex elasticity modulus into the expressions for influence function and by representing eigenfunctions of the boundary value problem for the fluid as complex functions.

Using the above procedures, oscillations of vertical plates of both constant and linearly varying thickness were calculated under different boundary conditions and for different ratios of plate sides. Various models of fluid were introduced (considering or neglecting vertical nonhomogeneity and compressibility), with waves on the free surface being, ignored, and the other boundaries of the fluid volume considered absolutely rigid and impermeable. Some calculations were carried out even for more complicated cases. In general, influence functions having been calculated for a very large number of structures, and solutions to the hydrodynamic problem in terms of Eq. (1) being available for a great variety of cases [2,3,5,6], the procedures suggested permit to perform calculations for a wide range of hydroelastic problems.

Measuring and Modelling Techniques

A set of instruments for model and in-situ investigations were developed. The set comprises hydrodynamic pressure and vibration pickups, amplifiers, transducers, recording instruments, and devices for reduction of oscillograms or other recordings of oscillatory processes. One of the major problems in experimental studies of fluid-induced oscillations of an elastic body is the impossibility to eliminate the vibrations of the hydrodynamic pressure pickups (hydrophones) submerged in water. With hydrophones being sensitive not only to pressure variation but also to acceleration, the output data on both pressure and acceleration are unreliable.

A pickup which is free from the above disadvantages has been developed at the VNIIG [7] (Fig.1). With the aim in view to exclude the acceleration effect while measuring pressure, and vice versa, its sensing element is designed as a sphere composed of two piezoceramic semi-spheres with identical polarization. The semi-spheres are mounted on an insulating base which serves as a separator and is connected to the walls of the flume. When measuring pressure or acceleration, the unipolar surfaces are commuted in parallel or in series, respectively. The dimensions of the sphere are selected so as to be small enough relative to the fluid wave length, while the average density of the ball inside the sphere should be approximately equal to that of the fluid.

To measure low-frequency (1 to 100 cps) oscillations of underwater hydraulic-structure members, one could utilize only seismograph-type pickups with galvanometric recording, the amplitudes of oscillations to be measured being very small. However, such pickups are known to require periodic zero adjustment, unlocking, and then arresting at each new measurement. Therefore, no watertight design is commercially available.

If such pickups are to be placed inside a watertight unit, it is necessary first to find a solution to the remote control problem for this condition. Attempts to use electric motors for remote zero adjustment and arresting proved a failure because prolongued storage inside the moist atmosphere of the watertight unit, and the impossibility of periodical lubrication put the motor out of operation. A different design principle was suggested at the VNIIG [8]. The vibration pickup is remotely controlled by a special regulating coil, 4, (Fig.2) fixed to the pendulum, 1, and located in the gap of the magnet mounted on the case of the pickup. With the regulating coil de-energized, the pendulum is arrested because the tightness of the spring, 5, is adjusted so that the pendulum is in the lower position for the vertical pickup and at one of the extremes for the horizontal one. When the pickup is in operation, the pendulum can be put in the working position by supplying direct current to the regulating coil, the current being regulated so that without a measured signal the pendulum should point to zero. The signalling system (not shown in Fig.2), which indicates that the pendulum is at zero, comprises a neon glow lamp in a casing, a slotted plate mounted on the pendulum, a diaphragm fixed to the pickup case, and a photo-diod or photo-resistor. The slot in the pendulum coinciding with that in the fixed diaphragm, the photo-resistor is lighted by the neon glow lamp, the current increases and reaches its maximum value when the axes of the slots are in line, which corresponds to the zero-position of the pendulum.

The pickups described above were employed in numerous model and some field studies of structural oscillations in a water medium. Amplifiers, transducers and recording instruments of the commercial type may be used for the pickups. To eliminate manual labour in oscillogram reduction, a series of devices is designed at the VNIIG for a complete automation of the process. The devices have been recently reported [9], therefore their description is not presented herein.

Special Aspects of Flow-Induced Structural Oscillations

The writers have not included in this presentation the relatively well-known phenomena of flutter, divergence, etc. where the secondary velocity and pressure fields in the fluid generated by structural oscillations do not affect the mechanism of the phenomena. Instead, emphasis is placed on the cases when the secondary fields are of primary interest. Naturally, one begins with the simplest case when it is feasible to study separately the primary pressure field in the fluid during its motion relative to the stationary structure and the secondary pressure field induced by structural oscillations. This allows to concentrate on the study of the secondary field, viz. on structural oscillations in a water medium without time-constant velocity components.

The nature of hydrodynamic phenomena due to structural oscillations in a liquid medium appears to be mainly conditioned by the geometry and the character of the boundaries of the fluid region since those are the governing factors affecting the magnitude of the first resonance (acoustical) frequency of the fluid region and the dissipative properties of the oscillatory system formed by the fluid region and its boundaries. Let us dwell upon these phenomena in greater detail. As shown in Ref. [10], with the fluid considered ideal, the relative error in the quantitative evaluation of velocity and pressure amplitudes will not exceed $R^{-1/2}$, where the modified Reynolds number is defined as

$$R = \frac{\omega h^2}{\nu}$$

including instead of the velocity of fluid motion, the product of oscillation frequency ω and the typical dimension of the oscillating element, h. For the majority of practical problems, the error amounts only to percent fractions, therefore fluid viscosity may be neglected.

Energy dissipation due to oscillations can also be attributed to the transfer of energy by progressive waves arising on the free surface during structural oscillations in the fluid. The value of the logarithmic decrement of oscillations is the smaller, the higher the modified Froude number introduced in Ref. [11]

$$Fr = \frac{\omega^2 h}{g}$$

so that at $Fr > 100$, the surface waves my be disregarded in practical calculations.

Another case of energy dissipation is associated with energy absorption by pliable boundaries. In Ref. [12], recommendations are available on the selection of criteria for quantitative evaluation of the effect of such pliability.

Provided all the three types of energy dissipation may be ignored, the structure will be affected solely by the inertia of the fluid, if the oscillation frequency is lower than the first resonance (acoustical) frequency of the fluid region. This inertia effect is responsible for the decrease in natural oscillation frequencies of the structure in the fluid, the decrease being different both for each natural frequency and each form of induced structural oscillations. Therefore, natural oscillation frequencies of structures in fluid may be treated conditionally, viz. if the way of generation of oscillations, the shape of the fluid region and the boundary conditions are indicated.

If the oscillation frequency of the structure is higher than the first resonance frequency of the fluid region, the structure may be subjected both to the inertia and damping effects whose correlation is governed by the geometry of the fluid region and the character of boundary conditions. If the boundary conditions are such that progressive waves may form with beyond-resonance oscillations, it may be definitely stated that the portion of the dissipated energy is equal to the energy portion attributable to those eigenforms whose frequencies are lower than those of the induced oscillations. The remaining energy is related to the inertial components of the hydrodynamic pressure.

Naturally, if one or more of the above types of energy dissipation are to be taken into account, then the portions of energy accounted for by inertial and damping components are redistributed. In specific calculations, those energy portions are estimated proceedings from the fact that the hydrodynamic pressure component varying with time in phase with the internal inertial forces of the structure produces an inertia effect on the structural oscillations, while the hydrodynamic pressure component shifted in phase by $\pi/2$, i.e. varying with time in phase with the non-elastic internal resistance forces of the structure, has a dissipative effect.

References

1. Lyatkher, V.M., Skladnev, M.F., Sheinin, I.S.: Studies in the Field of Dynamical Hydroelasticity of Hydraulic Structures. Gidrotekhnitcheskoye Stroitel'stvo, No. 8 (1971).

2. Kulmatch, P.P.: Hydrodynamics of Hydraulic Structures. Akademia Nauk, SSSR, 1963.

3. Sheinin, I.S.: Vibrations of Hydraulic Structures in Fluid. Handbook on Dynamics of Hydraulic Structures, Pt. 1, Energia, 1967.

4. Gelfand, I.M., Shilov, G.E.: Generalized Functions and Operations on them, Fizmatgiz, 1958.

5. Moshkov, L.V.: Dynamic Interaction between Outlet Works Elements of Hydraulic Structures and the Fluid. Synopsis of Doctor's Thesis, Leningrad, 1971.

6. Shulman, S.G.: Water Pressure due to Seismic Effects on Hydroaulic Structures, Energia, 1970.

7. Sheinin, I.S., Tsuvarev, Yu.V., Noskov, L.D.: Dynamic Pickup. Author's Specification No. 226316, Byulleten' izobretenii, No. 28 (1968).

8. Sheinin, I.S., Noskov, L.D.: Remote Control Vibration Pickup. Autor's Specification No. 176083, Byulleten' izobretenii, No. 21 (1965).

9. Noskov, L.D., Sheinin, I.S.: Devices for Reducing Graphical Realizations of Processes. Sb. Dinamika gidrotekhnitcheskikh sooruzhenii. Trudy koordinatsionnykh soveshchanii po gidrotekhnike. Energia, Vyp. 54 (1970).

10. Sheinin, I.S.: Hydroelastic Problems in the Dynamics of Hydraulic Structures. Izvestia VNIIG 84 (1967).

11. Sheinin, I.S.: On the Effect of Surface Waves due to Vertical Wall Vibrations in Liquid. Izvestia VNIIG 83 (1967).

12. Sheinin, I.S., Voronova, L.S.: Damping Effect at the Bottom on Plate Oscillations in Fluid with Local Pliability of the Bottom (Two-Dimensional Problem). Dinamika gidrotekhnitcheskikh sooruzhenii. Trudy koordinatsionnykh soveshchanii po gidrotekhnike. Energia, Vyp. 54 (1970).

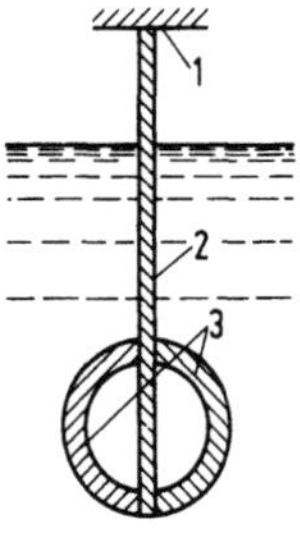

Fig. 1. Dynamic pickup.
1 Flume walls;
2 Pickup base;
3 Piezoceramic semi-spheres.

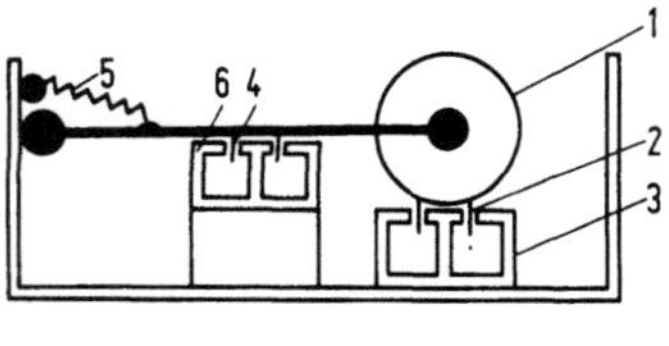

Fig. 2. Remote control vibration pickup with a regulating coil.
1 Pendulum;
2 Measuring coil;
3 Permanent magnet of the measuring system;
4 Regulating coil;
5 Spring;
6 Permanent magnet of the control system.

Experimental Investigations of Self-Excited Vibrations of Vertical-Lift Gates

By

A.S. Abelev, L.L. Dolnikov

Leningrad, U.S.S.R.

Summary

The paper presents the results of experimental and theoretical investigations on the conditions of existence of self-excited vibrations of submerged vertical-lift high-head gates with flow passing both over and under them.

Introduction

Hydraulic gates may be subjected to various modes of self-excited vibrations differing in their physical nature and onset conditions. The mechanism of excitation of these vibrations is governed mainly by the character of the non-stationary hydrodynamic forces dependent on the motion of the gate in the flow. Accordingly, we assume that self-excited vibrations of gates may be divided into two basic categories:

(1) Self-excited vibrations due to the vorticity associated with the separation of the flow from the lower lip of the gate when the eddy formation in the wake past the gate is synchronized with and controlled by the structural vibrations (the eddy mechanism of excitation) [1,2].

(2) Self-excited vibrations which may occur at high velocities of the jet-flow oriented along the vertical surface

of the gate (the jet-flow mechanism of excitation). Such conditions are observed, for example, in the gate shaft with a skimmer wall when the water moves over the partially-opened gate (Fig.1). In this case, the horizontal vibrations of the gate cause time-dependent variations in the discharge of the flow between the gate and the rigid walls of the structure and set up the inertia component of hydrodynamic pressure with a phase-shift by $\pi/2$ relative to the motion of the gate. Though here eddies do form in the wake behind the gate, their effect, having a random character, must contribute to "quenching" of self-excited vibrations [3].

The difference between the two categories of self-excited vibrations consists in the fact that those due to eddies occur in the direction normal to the flow under the gate, while those set up in case of overflow are normal to the flow along the downstream face of the gate, i.e. they occur in the direction of the minimum gate rigidity. The existence of such flow-induced vibrations is confirmed by the full-scale tests carried out in 1967 [4].

The field tests were performed at the design head of 14,7 m on a vertical-lift gate of welded rods intended for a bottom outlet of a 3,5 m internal width and 5.7 m height. From the oscillograms of self-excited vibrations the lowest frequency of the horizontal oscillations of the gate in the air was found to be $N_0 = 58$ cps, and the logarithmic decrement $\lambda_0 \approx 0.1$. In still water the frequency and the damping decrement at the relative opening of the gate $n = a/h = 0.1$ were $N_1 = 25$ cps and $\lambda_1 \approx 0.4$ to 0.6, respectively. It was found that damping due to the external hydrodynamic resistance is almost proportional to the square root of vibration velocity. In addition, the calculations showed that the vibrations of an elastic bar with hinged ends having a uniformly distributed mass and a moment of inertia of the cross-section equal to the total moment of inertia of the actual gate can be adopted as a calculation scheme.

The flow-induced vibrations recorded during the lift of the gate indicated that the character of oscillations alters considerably as the relative opening of the gate increases.

Within the time interval, from the start of the gate lift till the moment when the seal left the skimmer wall, low-frequency random vibrations with a mean amplitude of $A_x \approx 0.027$ mm and a frequency of $1.5 \div 2$ cps were recorded. These vibrations should, perhaps, be classified as forced vibrations induced by the turbulent fluctuations of pressure of the gate.

Henceforth, starting from the moment when the seal left the skimmer wall, intensive horizontal vibrations of the gate were observed. On the average, the amplitude of these vibrations was $A_x \approx 0.47$ mm, which is 17 times the initial amplitude. The characteristic feature of the above vibrations was that in their shape they were very close to biharmonic ones having a frequency equal to the frequency component of the self-excited vibrations of the gate in still water ($21 \div 25$ cps) plus the superimposed low-frequency component ($1.5 \div 2$ cps). It should be noted that the vibrometers mounted on the top and bottom rods of the gate recorded practically the same vibration amplitude, without a marked displacement in phase. This is another evidence for the similarity between the vibrations of the gate and those of an elastic bar supported at the ends. With further lifting of the gate, intensive periodic oscillations again gave place to random ones with an insignificant amplitude, this character of oscillations being unchanged till the gate was fully opened. In Fig. 2, the vibration amplitude of the gate is plotted versus its opening. The cross-hatched area corresponds to intensive biharmonic oscillations.

The fluctuations of the hydrodynamic pressure acting on the top rod in the gap between the downstream skin-plate of the gate and the wall of the shaft were recorded simultaneously with the gate vibrations. The observed data showed that during biharmonic oscillations, pressure fluctuations occurred with an amplitude of the order of 8 m w.g. at a frequency of about 25 cps. The pressure fluctuations coincided in phase with the velocity of the gate travel, which pointed to the fact that the reactive component of the pressure due to the interaction between the oscillating structure and the water flow in the gap was recorded, rather than the external disturbing pressure.

Thus, the vibrations observed may be considered as self-excited vibrations. This concept of the nature of the self-excited vibrations of the gate induced by the overflow is used as a basis for further analysis aimed at determining the conditions for the existence of self-exciting oscillations.

Theoretical Solution of the Problem[1]

In a two-dimensional formulation of the problem, the gate vibrations are reduced to small vibrations of an elastic bar described by the following equation:

$$EI \frac{\partial^4 w}{\partial y^4} + m \frac{d^2 w}{dt^2} = p(y,t) , \tag{1}$$

where $w(y,t)$ is the deflection of the gate; EI is the flexural rigidity; m the mass per unit area of the gate; $p(y,t)$ the nonstationary hydrodynamic pressure on the gate, equal to

$$p(y,t) = \delta p(t) + p'(t) - m_1 \frac{d^2 w}{dt^2} - r_1 \left(\frac{dw}{dt}\right)^2 . \tag{2}$$

Here, δp is the reactive component of hydrodynamic pressure on the a - b portion of the gate within the shaft (Fig.3); p is the random component of hydrodynamic pressure on the b - c portion of the gate within the conduit, the component being induced by the turbulent flow past the gate and uniformly distributed over the gate surface; m_1 is the virtual mass of water, uniformly distributed over the gate surface related to the unit area of the surface; and r_1 is the coefficient of nonlinear hydrodynamic resistance to oscillations. The m_1 and r_1 values can be experimentally found from the analysis of self-induced vibrations of the gate in still water.

[1] The present paper assumes a slightly different procedure of the solution of the problem in comparison with [4].

By writing the solution to Eq. (1) as a series of natural-mode shapes

$$w(y,t) = \sum_{k=1}^{\infty} x_k(t)\cos\frac{k\pi y}{B}, \tag{3}$$

we arrive at a set of conventional differential equations in the generalized coordinates:

$$\ddot{x}_k + 2\beta_k \omega_k \dot{x}_k + \omega_k^2 x_k = Q_k(t) \qquad (k = 1,2,\ldots\infty). \tag{4}$$

The second term of Eq. (4) is introduced upon the transition to the conventional differential equations. Here, β_k is the coefficient of the internal inelastic resistance of the structural material, calculated by Sorokin's theory [5]; ω_k is the circular frequency of the k-th natural mode shape; Q_k is the generalized force determined for a general case by the formula [6]

$$Q_k(t) = \frac{\iint\limits_S p(x,y,z,t)w_k(x,y,z)dS}{\iiint\limits_V \rho w_k^2(x,y,z)dV}. \tag{5}$$

For an approximate analysis, the dissipative coupling between the individual mode shapes is assumed to be negligible. Hence, the set of Eqs. (4) is decomposed into separate equations, which allows to investigate the dynamic stability of the system subjected to small disturbances for each of the generalized coordinates independently. The stability criteria applied to one-degree-of-freedom systems appear to hold true for the system under consideration.

Further, the solution of the problem reduces to the determination of the components of the hydrodynamic pressure δp and p' acting on the gate, the evaluation of the generalized force from Eq. (5), and the subsequent analysis of the stability to small disturbances of the system described by Eq. (4).

For evaluating the hydrodynamic pressure component δp on the a - b portion of the gate, the equation of the overflow discharge may be written as in [7], i.e.

$$q_o = \eta S_2 \sqrt{2g(H - h_1 - h)} = \eta S_2 \sqrt{2gH_d}, \tag{6}$$

where η is the coefficient of the discharge through the gaps between the gate and the shaft walls; S_2 is the cross-sectional area of the gap between the downstream skin-plate of the gate and the skimmer (Fig.1); H is the difference between the upstream water level and the conduit bottom level; h_1 the head losses at the conduit inlet before the gate; h the piezometric head in the constricted cross-section of the flow under the gate; g the gravity acceleration; and H_d the design head on the gate.

With the gate vibrating horizontally, the flow of water over the gate will be unsteady and the equation of discharge (6) will assume the form

$$q(t) = \eta S_2 \sqrt{2g(H_d - h_i)}. \tag{7}$$

Here h_i is the inertia head due to the non-stationary regime of the flow through the gaps between the gate and the shaft walls, which can be obtained from the formula:

$$h_i = \frac{\alpha_o}{g} \frac{\delta q}{\delta t} \int_L \frac{dz}{S(z)} = \frac{\alpha_o l'}{gS_2} \frac{\delta q}{\delta t}, \tag{8}$$

in which α_o is a momentum corrective; l' is the reduced length of the waterway designed for the discharge of water q; and L is the full (actual) length of the waterway.

According to the calculations, the ratio h_i/H_d in Eq. (7) is of the order of $\lambda_1/\pi = 10^{-2}$ (here λ_1 is the logarithmic decrement of flow-induced vibrations of steel structures). Therefore $h_i/H_d \ll 1$, and thus Eq. (7) can be written to a first approximation as

$$q(t) \approx \eta(t) S_2 \sqrt{2gH_d}.$$

Substituting this expression into Eq. (8), we obtain the formula

$$\delta p = \gamma h_i = \alpha_o \rho_o l' \sqrt{2gH_d} \frac{\partial \eta}{\partial t} . \qquad (9)$$

After Naudascher [7] the η-value for the gate located in the shaft is equal to

$$\eta = \frac{\eta_2}{\sqrt{1 + (\eta_2 S_2/\eta_1 S_1)}} ,$$

where η_1 and η_2 are the discharge coefficients for the upstream and downstream gaps between the gate and the shaft walls, respectively.

Generally, for an actual gate with the seal on its downstream face $S_1 \gg S_2$, therefore $\eta \approx \eta_2$, i.e., the discharge capacity of the waterway, through which the q discharge is passed, depends mainly on the local resistances and the cross-sectional area of the downstream gap between the gate and the skimmer wall.

The discharge coefficient η_2 is mostly dependent on the relation between the dimensions of the seal and the skimmer, S_2 and S_3. Moreover, the discharge coefficient η_2 is known to be a nonlinear function of the gap cross-section and, consequently, for an oscillating gate, it is nonlinearily-dependent on the generalized coordinate $x(t)$. Let the function $\eta_2 = \eta(x)$ be continuous and differentiable at $x(t) = 0$. By expanding the function into a series we obtain

$$\eta_2(x) = \eta_2(0) + x \frac{d\eta_2(0)}{dx} + \frac{x^2}{2} \frac{d^2\eta_2(0)}{dx^2} \ldots$$

For small vibrations of the gate let us consider only the linear terms of the series

$$\eta_1(x) = \eta_2(0) + x \frac{d\eta_2(0)}{dx} .$$

Designating $d\eta_2/dx$ by K_η and l'/L by K_l, Eq. (9) can be replaced by

$$\delta p(t) = \alpha_o K_\eta K_l \rho_o L \sqrt{2gH_d}\, \frac{dx}{dt} \; . \tag{10}$$

Earlier studies indicate that the fluctuation of the hydrodynamic pressure $p'(t)$ on the b - c portion of the gate (Fig.1) is a stationary random process whose distribution function nearly obeys the normal law [8,9]. To describe such a process, one needs information on autocorrelation and spectral density functions. Based on generalized experimental data [8] and the approximating expressions for autocorrelation functions of hydrodynamic pressure fluctuations, a correlation is established between the parameters, $\sigma^2_{p'}$, α, and β, of the autocorrelation functions, and the approximate characteristics, p'_{max}, N_n of pressure pulsations, obtained earlier [8]. Thus, for a submerged vertical-lift gate, located in the shaft, this correlation is given by

$$\alpha = 1{,}07\ \pi\lambda N_n \; , \tag{11}$$

$$\beta = 2{,}27\ \pi\lambda N_n \; , \tag{12}$$

$$\sigma^2_{p'} = (0{,}03 \div 0{,}04) \left[\frac{P'_{max} K}{F} \right]^2 , \tag{13}$$

$$\omega^{max} \approx \sqrt{\alpha^2 + \beta^2} \; , \tag{14}$$

where α is the autocorrelation function parameter for the damping of the correlation between individual outshootings of pressure pulsations; β is the autocorrelation function parameter for the "master" frequency of the process; $\sigma^2_{p'}$ the dispersion of hydrodynamic pressure pulsation; N_n, P'_{max} the "dominant" frequency and total amplitude of hydrodynamic load pulsations on the vertical-lift gate, as evaluated from [8]; λ, K the empirical coefficients of the transition from the total load pulsation to the pressure fluctuation at a given point on the gate surface, as determined in [10]; ω^{max} is the circular frequency, corresponding to the maximum of the spectral density function of pressure fluctuation on the gate; and F is the area of the vertical face of the gate.

The spectral density function of the hydrodynamic pressure fluctuation on the gate is approximated by the expression

$$S_{p'}(\omega) = \frac{\alpha\sigma_{p'}^2}{\pi} \frac{(\omega^2 + \alpha^2 + \beta^2)}{(\omega^2 - \alpha^2 - \beta^2)^2 + 4\alpha^2\omega^2} . \qquad (15)$$

Let the generalized coordinate $x_k(t)$ describe the first-mode shape of oscillations. Then, after appropriate transformations, Eq. (4), with Eqs. (5) and (10) taken into account, will give the nonlinear stochastic differential equation:

$$\ddot{x} + 2\delta\dot{x} + \omega_1^2 x + \varepsilon(\dot{x})^2 = \eta p'(t) , \qquad (16)$$

where

$$2\delta = \left[2\beta\omega_o - \frac{\rho_o L^2 K_1 K_\eta B\sqrt{2gH_d}}{m_o}\right]\left[1 + \frac{m_1 h_1 B}{m_o}\right]^{-1} , \qquad (17)$$

$$\omega_1^2 = \left[1 + \frac{m_1 h_1 B}{m_o}\right]^{-1} \cdot \omega_o^2 , \qquad (18)$$

$$\varepsilon = \frac{8r_1 B h_1}{3\pi m_o}\left[1 + \frac{m_1 h_1 B}{m_o}\right]^{-1} = \frac{8r_1 B h_1}{3\pi m_o}\left(\frac{\omega_1}{\omega_o}\right)^2 , \qquad (19)$$

$$\eta = \frac{4(h - a)B}{\pi m_o}\left[1 + \frac{m_1 h_1 B}{m_o}\right]^{-1} = \frac{4(h - a)B}{\pi m_o}\left(\frac{\omega_1}{\omega_o}\right)^2 . \qquad (20)$$

Here, B is the design span of the gate and m_o is the total mass of the gate.

For determining the conditions of inception and quenching of the self-excited oscillations of the system described by Eq. (16), use is made of the statistical linearization method [11]. According to this method the nonlinear Eq. (16) with the random right side is solved as a sum of the periodic and the random components:

$$x(t) = A_x \sin \Omega t + x'(t) \tag{21}$$

in which A_x and Ω are self-induced vibration amplitude and frequency, respectively, and $x'(t)$ is a random low-frequency component. According to [11] the nonlinear term of Eq. (16) is

$$\varepsilon(\dot{x})^2 = K_S \varepsilon \dot{x} , \tag{22}$$

where K_S is the statistical linearization coefficient as given by the condition of the minimum standard deviation of the substituting function from the assumed one, i.e.

$$\langle [\varepsilon(\dot{x})^2 - K_S \varepsilon \dot{x}]^2 \rangle = \min . \tag{23}$$

When the external random disturbance $p'(t)$ acting on the system is a stationary random normal process, with the spectral density function $S_{p'}(\omega)$ having nonzero values only within the frequency range of $0 \leqslant \omega < \Omega$, the linearization coefficient can be found from the relationship [11]:

$$K_S = \frac{q_S + (2\sigma_{x'}^2/A_x^2)K_{1m}}{1 + 2\sigma_{x'}^2/A_x^2} , \tag{24}$$

where q_S and K_{1m} are the sinusoidal and the random component "transfer" coefficients, respectively; $\sigma_{x'}^2$ is the dispersion of the amplitude fluctuations of self-excited oscillations.

Using Eq. (22) we can write Eq. (16) in the linearized form:

$$\ddot{x} + (2\delta + K_S \varepsilon)\dot{x} + \omega_1^2 x = \nu_1 p'(t) , \tag{25}$$

and the corresponding characteristic equation as

$$p^2 + (2\delta + K_S\varepsilon)p + \omega_1^2 = 0 , \tag{26}$$

where $p = d/dt$ is the differentiation operator. Substituting $p = j\Omega$ into Eq. (25) and separating the real and the imaginary parts yields

$$2\delta + K_S\varepsilon = 0 \tag{27}$$

and

$$\Omega = \omega_1 . \tag{28}$$

The condition of the existence of self-excited oscillations is established by Eq. (27) as the condition, when this equation gives a real positive value for the unknown A_x^2:

$$\delta \leqslant - \frac{2\sigma_{x'}\varepsilon\omega_1}{\sqrt{2\pi}} , \tag{29}$$

whereas in the absence of the external random disturbance, self-excited vibrations would take place at any value of

$$\delta \leqslant 0 . \tag{30}$$

The dispersion of the amplitude fluctuations of self-excited oscillations is computed by the formula

$$\sigma_{x'}^2 = \frac{\sigma_{p'}^2 \eta^2}{2(\omega_1^2 - \alpha^2 - \beta^2)^2} . \tag{31}$$

The physical meaning of condition (29) is that the self-excited vibrations of the system can exist only at a certain level of the amplitude fluctuation which does not exceed the "threshold" value

$$\sigma_{x^1}^{thr} = \frac{\delta\sqrt{2\pi}}{2\varepsilon\omega_1} , \tag{32}$$

beyond which, at $\sigma_{x'} > \sigma_{x'}^{thr}$, the self-excited oscillations will cease.

As reported elsewhere [3, 11], the phenomenon of "quenching" of self-excited vibrations by an external random force is, in its physical nature, similar to the phenomenon known in the vibration theory as "trapping", when an extraneous harmonic excitation is introduced into a self-oscillating system. In this case at a certain level of the external disturbance the transition takes place from biharmonic self-excited vibrations to forced vibrations having the frequency of the external disturbing force.

References

1. Naudascher, E.: On the Role of Eddies in Flow-Induced Vibrations. The Xth IAHR Congress, London, 1963, R. 3.9.

2. Rumyantseva, A.N.: Dynamic Action of Eddies on Self-Excited Vibrations of Submerged Tainter Gates. M., VINITI AN SSSR, 1968.

3. Pervozvanski, A.A.: Self-Oscillating Control Systems Subjected to Random Forces. Trudy LPI, Vyp. 210 (1960).

4. Abelev, A.S.; Dolnikov, L.L.: Investigation of Nonstationary Hydrodynamic Forces Induced by a Plate Oscillating in Liquid Flow (Two-Dimensional Problem). The XIV IAHR Congress, Paris, 1971.

5. Sorokin, E.S.: Internal and External Resistance of Oscillating Solid Bodies. TsNIISK, Nautchnye soobshcheniya, Vyp. 3 (1957).

6. Bolotin, V.V.: Statistical Methods in Structural Mechanics. Gosstroyizdat, 1965.

7. Naudascher, E.: Hydrodynamische und hydroelastische Beanspruchung von Tiefschützen. Stahlbau, No. 7, No. 9 (1964).

8. Abelev, A.S.: Determination of Total Hydrodynamic Load Pulsation on Submerged Vertical-Lift Gates. Izvestia VNIIG 68 (1961).

9. Lysenko, P.E. et al.: Dynamic Problems of Mechanical Equipment of Hydraulic Structures and Dynamic Calculation Methods. Trudy koordinatsionnykh soveshchanii po gidrotekhnike, Vyp 54 (1970).

10. Abelev, A.S.: Correlation between Pressure Fluctuation at Individual Points on the Gate and Total Hydrodynamic Load Pulsation Acting on the Gate. Izvestia VNIIG 89 (1962).

11. Pervozvansky, A.A.: Approximate Method of Investigation of Self-Vibrating Systems Subjected to Random Force. Izvestia AN SSSR, OTN, No. 3 (1968).

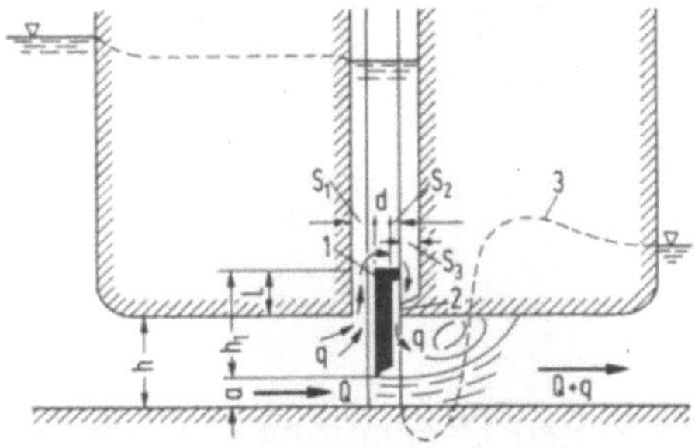

Fig. 1. Layout of the gate in the conduit.
1 Top horizontal seal; 2 Short skimmer; 3 Hydraulic gradient line.

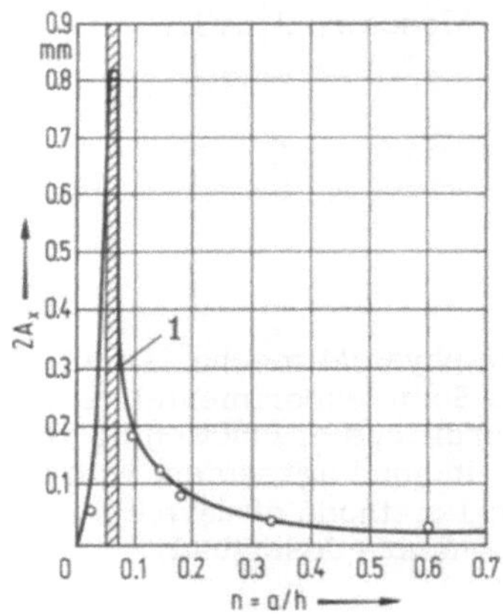

Fig. 2. Relationship between the amplitude of the gate vibrations and the gate opening.
1 Domain of self-excited vibrations.

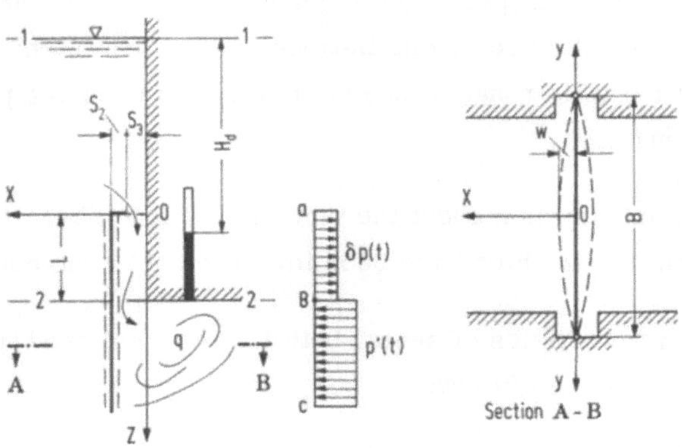

Fig. 3. Calculation scheme of the self-excited vibrations of the gate.

On Self-Excited Oscillations of Gate Seals

By

P.E. Lyssenko, G.A. Chepajkin

Moscow, U.S.S.R.

Summary

The paper deals with the physical mechanism of self-excited oscillations of gate seals. Some experimental results obtained in laboratories and under conditions of functioning structures are presented. The results are in good agreement with conclusions of theoretical analysis. General methods of developing seals not subjected to self-excited oscillations are described.

Introduction

Many cases are known in practice when gate seals develop self-excited oscillations. Especially the seals of gates, dams and navigation works are affected by oscillations as they are intended to operate within a wide range of heads or when the pressing process of seal is controlled.

Table 1 gives information about the vibration of a number of seals of quite different types which have been investigated by the authors.

Vibration characteristics of seals, both for the mentioned types and for other ones, are as follows:

(a) Spontaneous initiation and disappearance of oscillations at exactly defined heads for each type of seal and for every regulation;

(b) periodicity of oscillations with evident dependence of frequency on head;

(c) stability of the oscillation mode.

Besides of unpleasant noise, the seal oscillations causes intense and high-frequency vibration of the gate as a whole. Petrikat and Unny [1] indicated that periodical leakage succeeding the seal vibration results in "hydraulic resonance"; this in its turn causes overloading of the gate and conduit. A similar case was observed by the authors at one of the constructions under operation. Usually, the seal oscillations occur "suddenly" in the course of operation, and the same type of seal will vibrate on one gate but not on an other, which operates under the same conditions as the first one. It is evident that all types of seals of practical use may vibrate under certain conditions. The conditions which are necessary for the vibration to occur seem to involve some factors which are hardly discoverable by external analysis of the gate and seal structure. To discover these factors, special analysis is required.

This paper deals with the theoretical and experimental investigations on "self-excited" vibrations of seals and attempts to clarify the conditions of its occurrence.

Analysis Linear Approximation

All well-known types of gate seals, including those shown on Figs. 1 to 4, are similar in operation: the elastic element is provided with a rubber detail. When tightening the gap between the gate and the embedded parts, the rubber element is pressed to the sealing surface due to difference of elasticity force and water pressure on the seal. The displacement trajectories both of the sealing element and the mass center of the seal moving parts may be either curvilinear (Figs.1,2,4) or rectilinear (Fig.3) or even of more complicated forms. The system of force application to a seal and the strained state of the elastic element are also rather complicated. (For example the seal shown in Fig. 4 makes flat-parallel displacements and has "the bend" in horizontal direction caused by the deformation of the seal element being under head). However, for all cases, allowing for seal oscillation, we may compose an equation of its dynamic balance with displacement in direction "x"; this direction we shall now assume for the trajectory of the seal displacement while closing the gap. For certain cases, this equation (owing to the seal structure) may be regarded as an equation of force balance, and in other cases, as an equation of momentum balance.

We present the seal scheme in generalised form as an elastic system with elasticity coefficients K_e and reduced mass M (accounting for the so-called "apparent mass"). The system oscillates along the coordinate "x" round the mid-position "x_0" and is affected by the hydrodynamic force F, by linear resistance forces $(h_L + \delta_*)\dot{x}$[1] and non-linear (relative to vibration velocity) resistance $(h_d + S)\dot{x}|\dot{x}|$. Assuming that the corresponding coefficients of force factors reduced to the selected equilibrium center are taken into account in estimating the constants of the basic equation of motion (later on we do it), the generalised equation of seal motion has the following form:

$$M\ddot{x} + (h_L + \delta_*)\dot{x} + (h_d + S)\dot{x}|\dot{x}| + (x - \bar{x})K_e = F. \qquad (1)$$

According to the remarks above, the values M, h_L, δ_*, h_d, S, K_e, F should be estimated based on the system, and these values may be assumed to be masses, damping coefficients and others, or the moments of these values relative to selected center of equilibrium. The displacement value x may be either a linear displacement or a rotation angle of the system. This assumption does not change the outcome of the further analysis.

The coefficients h_L and h_d take into account the water medium, and δ_* and S take into account the material properties and seal structure. The neutral position of seal elastic system not being under action of water or other deformation restraints is $\bar{x}$.

---

[1] As usual points mean the time differentiation.

Thus the initial equation in its general form is the nonlinear differential equation of motion of a dynamic system with one degree of freedom.

The hydrodynamic force F which acts on the seal is equal to the integral based on the hydrodynamic pressure p. Integration is performed along the streamlined seal surface S_u. Coefficients K_S reduce the elementary forces pds to selected equilibrium centers.

$$F = \int_{S_u} pK_S dS = \int_{S_u} (p_B - \Delta p)K_S dS , \tag{2}$$

where p_B is the water pressure upstream of the seal; and Δp is the difference between the pressure p_B and the pressure on the seal over the area dS.

Figure 5 gives two possible seal schemes concerning the qualitative picture of pressure distribution over the seal; the example of Δp estimation is given.

Let us represent the pressure differences Δp in the form of dimensionless coefficients $\xi_S = 2\Delta p/\rho u^2$

where $u = \mu(x)\sqrt{2gH_m}$ is the velocity of flow leaking through the seal;
$\mu(x)$ is the coefficient of discharge through the gap with area $\omega(x_0)$;

$H_m = H + H_i$ is the full head acting on the seal accounting for the unstationarity of flow;

$H_i = \frac{1}{g}\int_0^L \frac{\partial u(l)}{\partial t} dl$ is the inertial head between sections "0" and "L" caused by the unstationarity of flow; and

$u(l)$ is the flow velocity in the section with coordinate "l".

Performing the integration in Eq. (2) and remembering that $\int_{S_u} p_B K_S dS = 0$ we obtain:

$$F = - \mu^2(x)\gamma[H + H_i] \int_{S_u} \xi_S K_S dS .$$

Let us introduce the "reduced" length of the gap L between the seal and the sealing surface by the relationships:

$$H_i = \frac{1}{g} \int_0^L \frac{\partial u(l)}{\partial t} dl = \frac{L}{g} \frac{du}{dt} \simeq \frac{L}{g} \sqrt{2gH} \frac{d\mu(x)}{dt} . \qquad (3)$$

Here we use the experimental observation proving that usually $H_i \ll H$; therefore, when estimating du/dt, we can assume that $\sqrt{H + Hi} \simeq \sqrt{H} = \text{const}.$[1]

As a result, we obtain the expression for the generalized force F:

$$F = -\mu^2(x)\gamma \left[H + \frac{L}{g} \sqrt{2gH} \frac{d\mu(x)}{dt} \right] \int_{S_u} \xi_S K_S dS . \qquad (4)$$

To analyse the oscillation conditions, Eq. (1) may be linearised; the seal oscillations are assumed to be small and harmonical:

$$x = x_0 + x_a \sin \omega t . \qquad (5)$$

When the value x_a is "small", the following relationships can be assumed to

$$\mu(x) \simeq \mu(x_0) + K_\mu (x - x_0) , \qquad (6)$$

$$\mu^2(x)\gamma \int_{S_u} \xi_S K_S dS \simeq m(x_0) + K_m (x - x_0) . \qquad (7)$$

Solving these relationships, we estimate the "small oscillations" for the problem under consideration. The harmonic character of "small oscillations" (5) follows from the expected equation of linearization (1) for which Eq. (5) is introduced.

---

[1] It would be wrong to think that it follows from $H_i \ll H$ that we should always neglect H_i. Later we shall show that H_i is of great importance in the vibration process inspite of the fact that inertial head (absolutely) is small.

Substitution of Eqs. (5), (6), (7) into (1) and (4) yields an equation which contains the terms $\dot{x}|\dot{x}|$ and $\dot{x}|x|$. Instead of these terms, let us introduce their linear equivalents with linear damping. Condition for equivalence is the equality of works of true and equivalent resistance forces for each quarter of a harmonic oscillation period. Thus,

$$\int_0^{\pi/2\omega} x_a^3\omega^2 \cdot \cos^2\omega t \cdot \sin\omega t \cdot dt =$$

$$= K_{\varepsilon 1} \int_0^{\pi/2\omega} x_a^2\omega \cos\omega t \cdot \sin\omega t \cdot dt = K_{\varepsilon 1} \cdot I, \tag{8}$$

$$\int_0^{\pi/2\omega} x_a^3\omega \cos\omega t \cdot \sin^2\omega t \cdot dt = K_{\varepsilon 2} \cdot I. \tag{9}$$

Using Eqs. (8) and (9) we obtain $K_{\varepsilon 1} = 2/3\omega x_a$, $K_{\varepsilon 2} = 2/3x_a$; then instead of terms $\dot{x}|\dot{x}|$ and $\dot{x}|x|$ let us substitute $2/3\omega x_a \cdot \dot{x}$ and $2/3x_a \cdot \dot{x}$ into the linearized general equation.

Damping linearization will not change the result of the main equation at least at any half-period of oscillation; and it will be the more exact, the smaller the value x_a.

After substitution of Eq. (4) into (1), taking into account the relations (5), (6), (7), (8), (9) and their consequences, and after individual setting to zero the sums of terms of the obtained relationships depending and not depending upon time, we obtain:

(a) Equilibrium condition of time-averaged forces acting on the seal:

$$K_e(\bar{x} - x_0) - Hm(x_0) = 0, \tag{10}$$

(b) condition of instantaneous equilibrium forces acting on seal:

$$\ddot{x} + 2\beta\dot{x} + \omega_0^2 x = 0, \tag{11}$$

where

$$\beta = \frac{1}{2M}\left\{0{,}66x_a\left[\omega(h_d + S) - K_m K_\mu \frac{L}{g}\sqrt{2gH}\right] + \right.$$

$$\left. + \left[(h_L + \delta_*) - m(x_0)K_\mu \frac{L}{g}\sqrt{2gH}\right]\right\} \tag{12}$$

and

$$\omega_0 = \sqrt{\frac{K_e - K_m H}{M}} . \tag{13}$$

Direct physical interpretation of Eq. (10) does not present a problem as it represents an equation of seal equilibrium due to time-averaged forces.

Equation (11) is an equation of free oscillation of a system with one degree of freedom. General analysis of such an equation is well known. Figure 6 shows all cases of possible motions described by Eq. (11) at different relations ω_0 and β. The expression (12) for β shows that the system can basically undergo any motion given in Fig. 6 - i.e., periodical and aperiodical, stable and unstable, depending on the system parameters.

In particular, any accidental small impact on the seal being under head H at $x_0 > 0$ may cause:

(a) A gradual (aperiodical) closing of the gap and a stopping of water flow through the gap between the seal and the sealing surface at $\beta < 0$ and $\omega_0^2 < 0$ (i.e. $K_e < K_m H$) or $\omega_0^2 < \beta^2$;

(b) a return of the seal into the initial state in which the seal has been prior to impact - at $\beta > 0$ (aperiodical or damped oscillations);

(c) undamped harmonical oscillations with amplitude of original impact - at $\beta = 0$;

(d) diverging periodical oscillations of the seal ($\beta < 0$, $\omega_0^2 > \beta^2$).

However, the mentioned analysis is valid only for "small" oscillations; therefore, the conclusion concerning the undamped oscillations due to any small impact is quite reliable. Other conclusions are valid only so far as the above mentioned relations between ω_0 and β hold when the system deviates from $x = x_0$.

Equation (12) at $\beta = 0$ and $x_a = 0$ makes it possible to define the condition of specified seal stability against undamped oscillation:

$$g(h_L + \delta_*)[m(x_0) \cdot K_\mu L \sqrt{2gH}]^{-1} > 1. \qquad (14)$$

If condition (14) does not apply but relation $\omega_0^2 > \beta^2$ holds, the oscillations due to small initial impact will develop with diverging amplitude x_a at frequency $\omega = \sqrt{\omega_0^2 - \beta^2}$. This results in an immediate reduction of the value β.

With the current value $\beta(x_a, \omega) = 0$, the process of further oscillation becomes stables with final amplitude x_{as}:

$$x_{as} = 1,5 \frac{m(x_0) \cdot K_\mu \cdot \sqrt{2gH}\, L/g - (h_L + \delta_*)}{\omega(h_d + S) - K_m K_\mu \sqrt{2gH}\, L/g}. \qquad (15)$$

Equation (14) shows that "unsuccessful" combinations of seal parameters which disturb the conditions of dynamic stability of the seal are quite real (for example by increasing the head on the seal, or if the gap length between the gate and the embedded parts is too large etc.).

The results obtained describe all features of seal oscillations which were mentioned in the introduction of this paper; and they clearly elucidate their origination. For example, we can readily indicate the cause of the dependence of oscillation frequency, excitation, and separation conditions upon the acting head (see Eqs. (13), (14)); alse the dependence of stable amplitude upon head can be readily shown.

The vibration under consideration belong to the so-called linear self-excited oscillations with all inherent features [2] such as:

(a) Energy source keeping undamped oscillations with final amplitude is the upstream head. This source is not of oscillatory nature.

(b) "Regulator" of periodical supply of energy to the dynamic system is the seal itself; its displacements change the flow through the gap and hence the loads.

(c) Oscillations are in agreement with the natural oscillation of the seal.

(d) Negative feedback between seal oscillations and acting forces is realized by inertial head. Inertial head H_i (see Eq.3) is phase shifted in time by the angle $\pi/2$ in comparison with the displacement. It means that H_i is in phase with the resistance forces acting against the dynamic system oscillation. The work due to inertial head per cycle of oscillation is close to the work of resistance forces at the same time, and sometimes exceeds it. In the latter case, the oscillations are amplified. The work equality of resistance forces and forces due to inertial head per cycle of oscillation corresponds to the stable undamped vibration ($\beta = 0$) - The above mentioned is basic in description of physical mechanism of oscillation.

In accordance with the described mechanism, the seal is an oscillating system with so-called "smooth self-excitation". In fact, when condition (14) does not apply, the self-excited oscillations are caused by initial impact with an infinite small amplitude; oscillation parameters depend only upon head (see Eq.15). This is in good agreement with direct observations (see introduction) and experimental results.

Analysis Approximate Consideration of Non-Linearity

The characteristics of a dynamic system are changed with its position "x". All coefficients in Eq.(II) are also variable. It follows that common methods do not make it possible to solve the main equation of oscillation at significant amplitudes of seal motion. For this case, the use of approximate solutions is recommended. During a small period of time τ, the Eq. (11) is supposed to be linear and with constant coefficients. It may be solved by well-known methods for initial conditions v^0 and x^0 (where v^0 and x^0 are the velocity of motion and the position of seal at the beginning of the time interval τ). The values of coefficients in Eq. (9) during the interval τ are functions of v^0 and x^0. Using the common methods of solution with constant coefficients for given initial conditions we can evaluate x and v by the end of the time interval. The obtained values are assumed to be the initial conditions in estimating the system behavior at the subsequent time interval. Solving the problem successively and going from time interval τ_n to time interval τ_{n+1} $(n = 1,2,3,\ldots)$, we are thus led to the law of motion for a seal exposed to significant oscillations (i.e. if linearization used in Section 2 is not acceptable). The solution will be the more exact, the smaller the time interval τ.

The main formulae have the following forms:

$$x_n^0 = x_{n-1}^0 + v_{n-1}^0\tau , \qquad (16)$$

$$v_n^0 = v_{n-1}^0\left[1 - 2\beta\left(x_{n-1}^0, v_{n-1}^0\right)\tau\right] - $$

$$- x_{n-1}^0\left[\beta^2\left(x_{n-1}^0, v_{n-1}^0\right) + \omega_0^2\left(x_{n-1}^0, v_{n-1}^0\right)\right]\tau , \qquad (17)$$

where n is the number of interval, and τ is the "small" time interval. τ may be looked upon as small if we are able to neglect the terms with powers higher than one in the expansions in τ of the functions $\exp - \beta\tau$, $\cos\omega_0\tau$, and $\sin\omega_0\tau$. The term $\beta(x_{n-1}^0, v_{n-1}^0)$ is estimated from Eq. (12) considering the dependence of K_m, K_e, and K_μ on x:

$$\beta\left(x_{n-1}^0, v_{n-1}^0\right) = \frac{1}{2M}\left\{\left[v_{n-1}^0(h_d + S) - K_m K_\mu\, L/g\sqrt{2gH}\cdot x_{n-1}^0\right] + \right.$$

$$\left. + \left[(h_L + \delta_*) - mK_\mu\, L/g\sqrt{2gH}\right]\right\}. \qquad (18)$$

Using Eq. (17) for a practical solution of the problem at significant values of "n", a computer should be applied; this is done step-by-step until steady flow is achieved upon introducing the given initial disturbance x_1^0 and v_1^0. By the way of example, Fig. 7 gives some results of the described solution for initial conditions $v_1^0 \neq 0$, $x_1^0 = 0$ for the seal shown in Fig. 1.

It is established that sinusoidal oscillations could be expected to change abruptly and to change their periods, if non-linearity is taken into account. Oscillations become "saw-shaped" and this is in agreement with observations at hydraulic structures under operation.

The cases where the oscillating seal impacts against the sealing surface are of great interest. After contact between seal and sealing surface is achieved, the dynamic system is abruptly changed. The equivalent mass and rigidity of the system would be considerably changed, if a support is attached to the sealing surface. Further motion of the seal is accomplished at the expense of additional deformation of the sealing device - in particular, of a rubber sealing element.

This deformation effect is responsible for a continuous change of the water pressure on a seal and for damping the additional inertial head which arises with closing of the gap. Provided that continuity in the displacement of the center of seal mass with two time derivatives exists, and making use of elastic impact theory, we can easily establish that the center of mass is somewhat displaced. Now the displaced center is in accord with a half-period of free oscillations that is peculiar to a new dynamic system with initial conditions which exist at the moment of impact origin. Using Eq. (17), we obtain the same result.

Similarly, we can analyse the case where the seal withdraws out of the gap between the gate and sealing surface at significant amplitudes of oscillations (this is typical for the seal shown in Fig. 1).

In contrast to the previous case, the load is shock-changed when the seal withdraws out of the gap (i.e. values ω_0 and β are changed).

As for the kinematics of an elastic system, it remains unaffected. Figure 8 gives the results concerning the determination of the mode of oscillation for the seal shown in Fig. 1. This is just the case when the seal withdraws alternatively out of the gap and impacts against the sealing surface. The oscillations obtained are non-sinusoidal, and occur with abrupt changing of motion velocity (called "relaxation self-excited oscillations"). In this case, the stable value of the oscillation period is achieved by a greater number of cycles than in the previous case (Fig.7). The value of the stable period is much smaller. This circumstance in combination with the presence of impacts against the sealing surface makes the latter mode of oscillation to be the most dangerous both for the gate and the structure as well as for the seal itself.

Experimental Results

Experiments (Fig.9) were carried out with a seal (Fig.1) under the condition of operating structure. Gate oscillations which were simultaneously recorded show that the mentioned conditions are dangerous for the structure. Initially, oscillations occur at a limited length of seal - just in the middle of the gate span. However, within some cycles the whole seal, extending over a 18 m length, appears to oscillate. The phenomenon of oscillation synchronization over the span has been observed in the laboratory. The method of oscillation elimination applied to the given seal (see Table 1), lead to the development of a "negative natural frequency" of the seal (by reducing K_e); this follows directly from the theory (see Eq.13).

Table 1

Fig. No.	Type of seal	Seal oscillation	Remark
1.	Rubber seal on elastic plate for sluice balanced radial gate. $H_{max} = 2 \div 4$ m	Periodical non-sinusoidal relaxation. Frequency - 10 cps. Also 30 cycl. natural frequency of gate is pronounced.	Oscillations are stable; repeated at each gate opening. Eliminated by reduction of spring rigidity.

Table 1 (Continuation)

Fig. No.	Type of seal	Seal oscillation	Remark
2.	Metal seal provided with spring. Seal is intended for sluice culverts $H_{max} = 23$ m	Sinusoidal. Frequency 5 cps. Vibration of gate and stem at the same frequency is pronounced.	Oscillations occurred at $1.8\,m \leqslant H \leqslant 5.3$ m; spring failed and nuts screw off.
3.	Rubber deformable seal for outlet radial gate. The seal is tightened through displacement Δ of eccentric. $H_{max} = 33$ m	Sinusoidal. Frequency 25-30 cps. Pressure pulsation upstream of seal at the same frequency is pronounced.	Oscillations caused extraction of the seal out of its housing. Eliminated by rounding the upstream edge of the sealing element.
4.	Controlled hydraulically operated rubber seal for outlet radial gate. $H_{max} = 95$ m	Sinusoidal and relaxation. Frequency - 30-130 cycles. Pressure pulsation upstream and downstream of seal at the same frequency is pronounced.	Oscillation occurred at $10\,m \leqslant H \leqslant 60$ m; Eliminated by rounding the upstream edge of the sealing element.

Laboratory experiments were carried out with the sealing devices shown on Figs. 3 and 4. A closed water tunnel was used with a 0,4 m seal in its working section. The seal was designed as a prototype. The seal displacements and pressure pulsations were recorded by non-inertial gages and by means of speed filming.

All tests show the stable mode of self-excited oscillations (Fig. 10) and the "smooth self-excitation". For a given position of seal, the oscillations could be repeatedly reproduced with just the same amplitude and frequency only through changing the head. In all points upstream of, downstream of, and under the seal, an out-of-phase pulsation relative to the seal oscillation was observed. Figure 13 presents Lissajous figures for the scheme "pulsation-sealing element displacement". The stability of the figures and their nonzero area indicate the presence of perceptible inertial head on the seal; thanks to inertial head, the system "is charged" with energy. The experiments help to clarify the existence of a mechanism of linear self-excited oscillations. For the seal shown in Fig. 4, we investigated the region of self-excited oscillations (Fig.12). As follows from the theory the region of instability is enclosed by abc:

(a) Left boundary "ab" is based on $\beta \geqslant 0$. With heads being less than those acted on line "ab", the self-excited oscillations do not occur (see Eqs. 12, 14). The line inclination "ab" is based on the variable coefficients $m(x_0)$ and K_μ as a function of initial gap x_0.

(b) Lower boundary "ac" is based on $\omega_0^2 \leqslant 0$, i.e., on the hermeticity of the seal after any accidental shock which takes place in the state of equilibrium (unstable equilibrium).

(c) Line "bc", judging by the test results and by the film results, is based on the formation of stable flow separation region under the sealing element with slightly changed average pressure being close to the pressure downstream of the seal. In this connection, the value "m" in Eqs. (12) and (14) is abruptly reduced due to considerable decrease of ξ_S under the sealing element (see Eq.4).

The experiments show that the self-excited oscillations and the induced oscillations due to turbulent pressure pulsation develop independently. Fig. 14 gives the normalized energy spectra of the seal element displacements and the pressure pulsations on the sealing surface downstream of the seal. Naturally, pulsation dispersion grows abruptly as the self-excited oscillations set in.

Under certain conditions (close to the instant of oscillation separation) obviously non-harmonical processes similar to relaxation-self-excited oscillations (Fig.11) were observed. Speed filming shows that the vibrating sealing element makes complicated displacements in two directions in a trajectory with a perceptible angle relative to the sealing surface. Under certain conditions, the approaching of the sealing element to the surface is accomplished by the formation of an instantaneous "cavitation torch"; it corresponds to negative peaks of pressure pulsations being cut off by the level of absolute vacuum ($p_{abs} \simeq 0$).

The oscillation excitation and maintenance mechanism does not change as the cavitation occurs. However, the formation of "cavitation torch" under the sealing element causes additional pressure pulsation and "suction" of the element; sometimes "suction" causes the element to be extracted out of the housing. The latter phenomenon may be intensified by "self-synchronization" of oscillation along the whole seal as was mentioned above.

Conclusions

In all cases discussed, the initiation of self-excited oscillations should be regarded as inadmissible. The oscillations are dangerous as the seal frequency and the natural frequency of the gate and its parts can coincide; resonance may also occur in the structure or the conduit. Table 1 shows that the range of natural frequency of a seal ($10 \div 30$ cycles) may coincide with that of a gate being in water (10 to 60 cycles) and its parts (100 to 200).

Universal recommendations for elimination of seal self-excited oscillations are hardly possible; however, some general measures derived from theory and experiment should be proposed:

(a) The seal has not to work at small gap openings (under any operation conditions, the seal is not permitted to be non-hermetic, or, on the contrary, to be opened quickly, if the gate operates for a full-rated gap opening);

(b) create damping by means of special damping devices, by limited leakage from pressure chambers (increasing h_L and h_d);

(c) create inner friction within allowable limits in accordance with seal elasticity (increasing δ_* and S);

(d) use streamlined sealing elements (decreasing m and K_μ);

(e) mount two seals in sequence; the second seal is not tight but it is designed as non-vibrating. Such arrangement reduces the leakage through the main seal (decreasing m and K_μ);

(f) diminish the length and increase the gap height between the gate and structure (decreasing L);

(g) reduce the rigidity of seals (decreasing K_e) in order not to permit the existence of periodical oscillations ($\omega_0^2 < 0$).

The effect of some recommendations is seen from Table 1.

When designing self-excited oscillations in advance (using, for example, Eq. (14) and, if necessary, the non-linear approximation). Necessary data for this purpose may be readily obtained from laboratory experiments with seal models without creating the self-excited oscillation conditions (because special complicated equipment is required for this purpose).

However, sometimes it is useful to carry out the experiments like those mentioned above.

When selecting the dynamic characteristics for seals, special attention should be given to the natural frequency of the seal; this has to be higher than for the case of the gate being in water and for water being in the conduit. In this case, the seal oscillations - occurring due to accidental dehermetization of the seal - appears to be not dangerous for gate or conduit.

References

1. Petrikat, K., Unny, T.: Vibration in Bottom Outlet Structures of Dams Excited by the Elasticity of Gate Seals. XI Congress IAHR, 1965, Leningrad.

2. Kharkevitch, A.A.: Self-excited Oscillations. Gostechizdat, 1954.

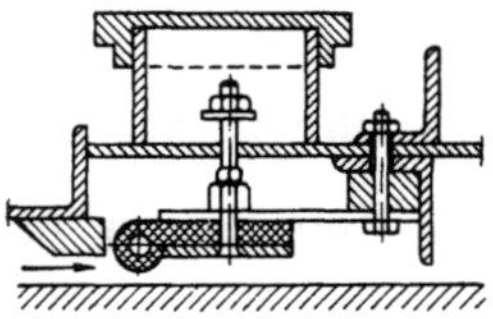
Fig.1

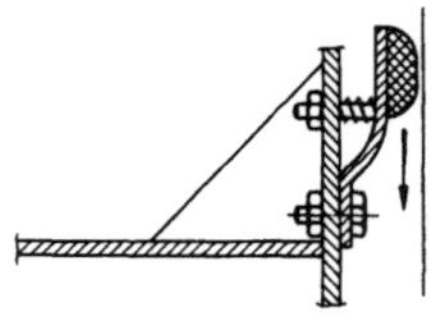
Fig.2

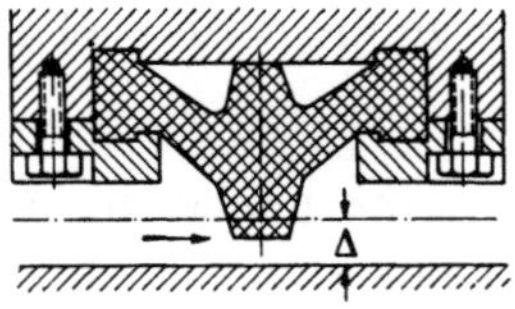
Fig.3

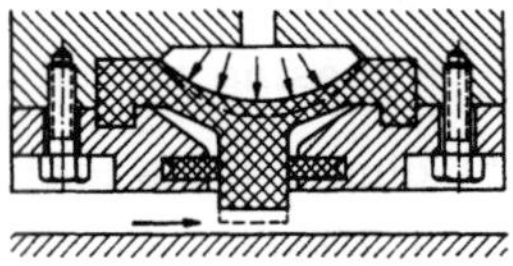
Fig.4

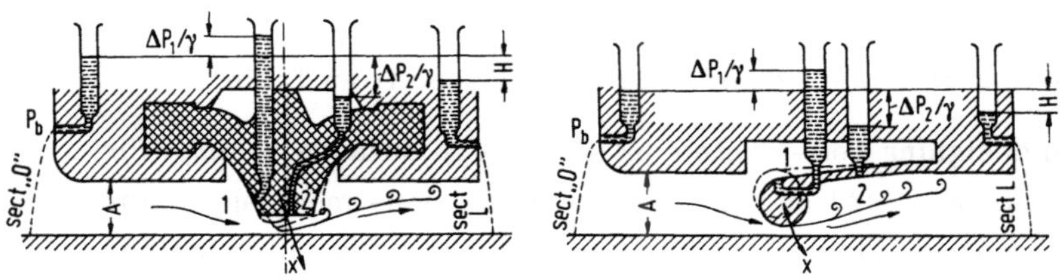

Fig.5

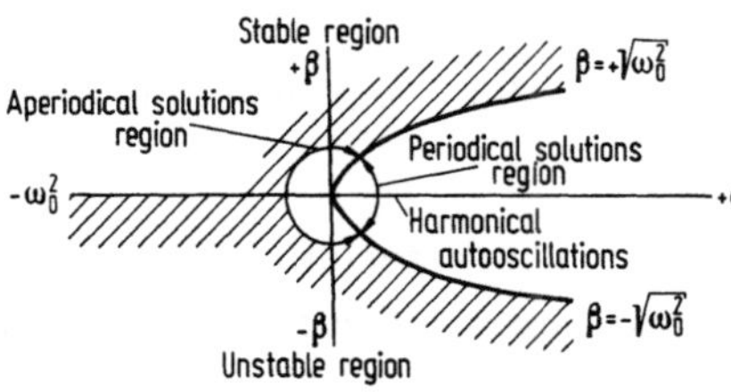

Fig.6

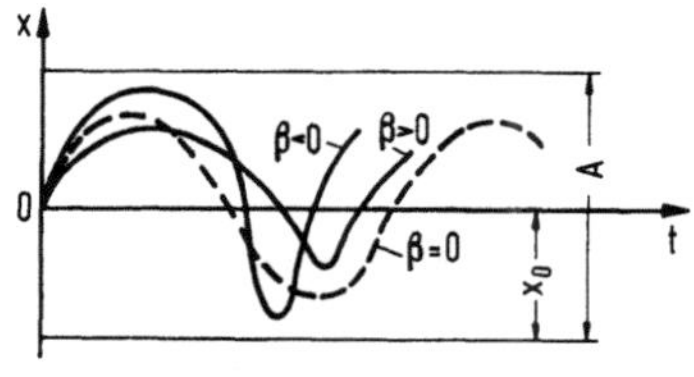

Fig.7

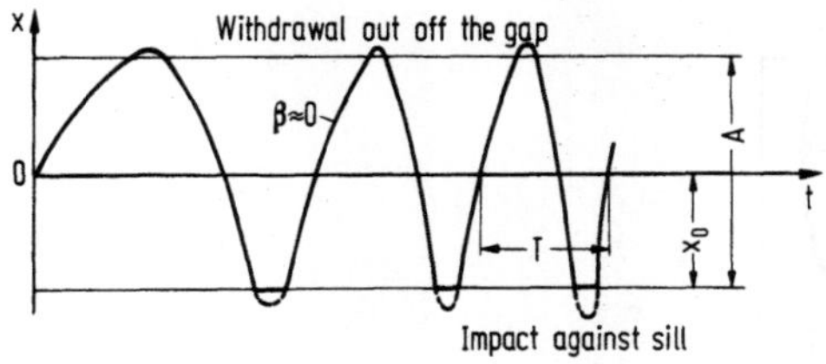

Fig.8

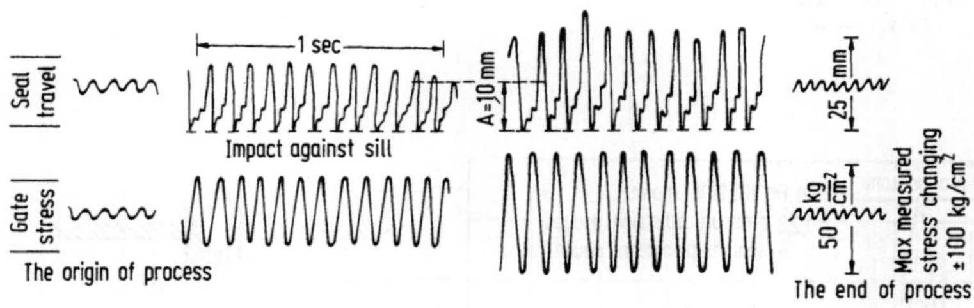

Fig.9

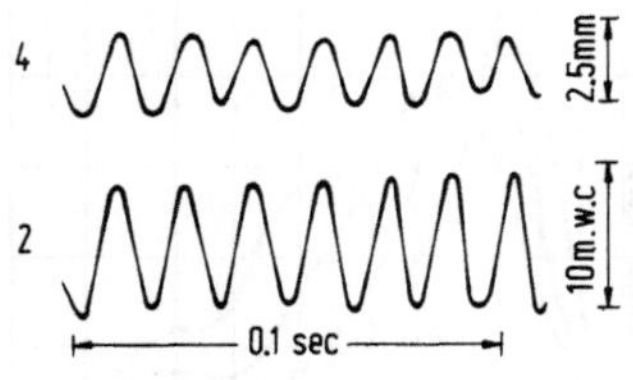

Fig.10

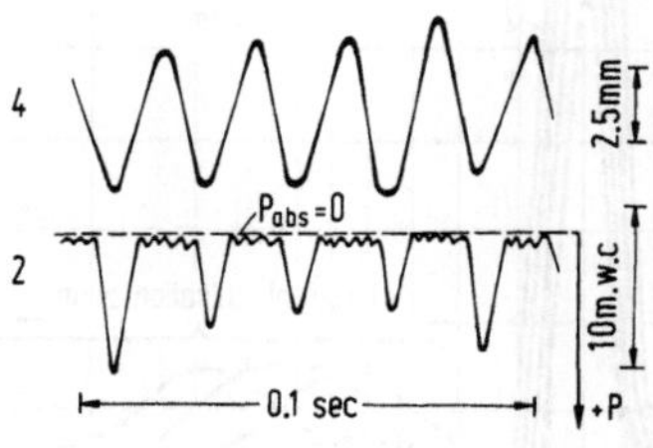

Fig.11

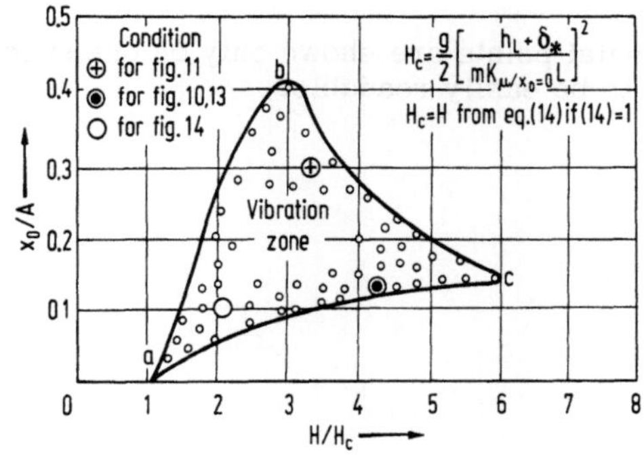

Fig.12. "Critical head" H_c has been obtained by solving Eq. (14) relative to head at $x_0 \to 0$, and after the unequality in Eq. (14) has been substituted by the equality to unity.

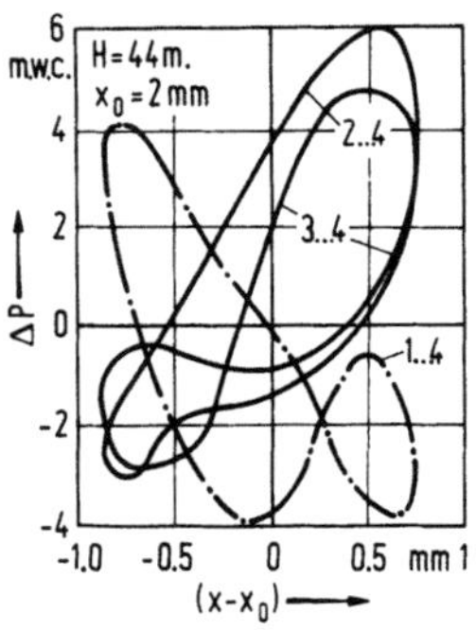

Fig. 13. Lissajous figures were obtained by photography of the screen of an electronic oscillograph.

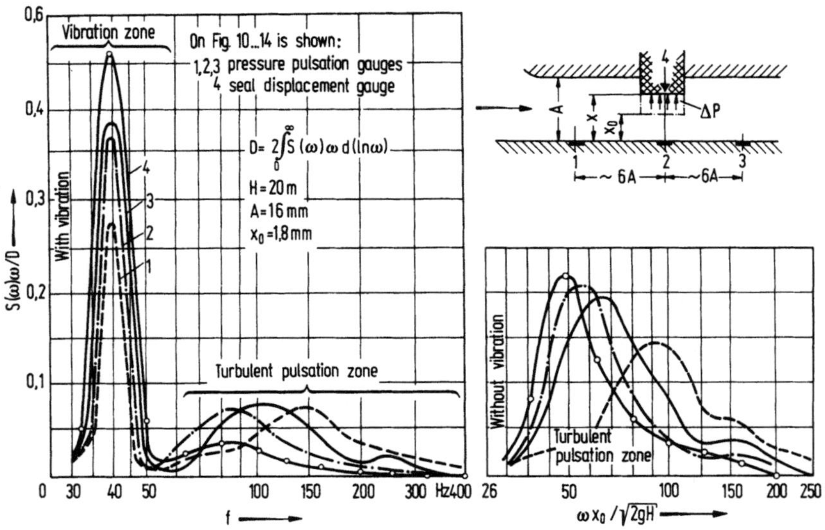

Fig. 14. Experimental points are shown only on one spectrum for the schemes to be easily readable.

Field Investigations of Dam and Gate Vibrations

By

L.A. Goncharov, V.M. Semenkov

Moscow, U.S.S.R.

Summary

The paper deals with some results of experimental field investigations of vibrations of dams and gates. Among the vibration characteristics observed are a great variety of flow effects and reactions of structural members, the high dynamic response capacity of oscillatory systems leading to multiple-resonant and pseudoresonant phenomena and to auto-oscillations, and instability with time of parameters of vibrations of hydraulic structures on soft-soil foundations resulting from changes of soil properties under the effect of vibrations.

Experimental data on the influence of a vibrating gate upon the flow characteristic downstream from the gate are presented; some problems concerning adequate procedure and techniques for vibration tests in hydraulic installations are discussed.

Introduction

Systematic field investigations of vibrations of hydraulic structures, which started in the Soviet Union with the pioneer studies of the Seismological Institute of the U.S.S.R. Academy of Science in 1938 to 1940, received further development and are at present carried out by a number of scientific research organizations (VNIIG, Hydroproject, Mosgydrostal, etc.). These field investigations extend our ideas about actual operation conditions of hydraulic structures; they comprise complex multipurpose dynamic investigations including the study of a number of processes accompanying vibrations, such as dynamic stresses in structural elements and pressure pulsations. Their objective was comparison of design assumptions with real conditions of the operation of hydraulic structures and equipment, development and improvement of recommendations for the optimization of service regimes, accumulation of data for use in the future practice of design studies, etc. Field investigations performed on a number of hydraulic structures permitted to develop a rational procedure of investigation and to choose adequate measuring devices. Some peculiarities of hydraulic structures and their vibrations were revealed by the analysis of the experimental data. The results enhance our concepts about actual conditions of vibrations of existing structures.

Dam Vibrations

The success of laborious field experiments depends mainly on a rational choice of measuring equipment and procedure. Since all measuring systems possess some filtering and phase-shifting properties, any information obtained will reflect the nature of the measuring system.

Poor choice of measuring-apparatus parameters may also lead to erroneous conclusions. For example, vibrations of a dam, 12 m high (specific power of waste-water discharge $W = 2.2$ MW/m), which were measured with the help of an apparatus having a passband of 15 to 100 Hz, were within the range 45 to 85 Hz, and the maximum double amplitude $2A_{max}$ did not exceed 3 μm. Repeated measurements with an apparatus having a passband of 1 to 100 Hz showed that no less than 90 % of the vibration spectrum lies in the range of 2.5 to 4.0 Hz and $2A_{max} \simeq 30$ μm. Here and below the specific power of waste-water discharge is $W = kqH$, where q is the specific discharge in m^3/sec per m and H is the head acting upon the structure.

Rational choice of parameters of measuring apparatus permits to separate such oscillations which interfere with analysis, e.g., forced vibrations of the structure induced by storm waves, and to record the natural vibrations of a massive structure excited in this case. The above-mentioned method gave the possibility to record natural vertical vibrations of a section of the spillway dam with a weight equal to more than 140,000 t [1], with a dam-base area of more than 3,700 m^2, erected on fine sand (F_z = 1.05 ± 0.05 Hz, logarithmic decrement = 0.44).

On the basis of the measurement it was found that the source of vibration induced in spillway dams when the flood was being passed over the spillway lies in the dissipation of kinetic energy on the apron. The location of the vibration source was determined by two methods: by the intersection in the horizontal plane of the main axes of the elliptical trajectories of movements of the spillway-dam sections adjacent to the one in operation; and by the hodograph method usually used in seismic investigations. The results received were confirmed by the following vibration measurements on the apron of the dam. The mechanism of spillway-dam vibrations on soft soils arising under the action of waste-water flow is described in the papers by L. Maksimov [2], V. Ljatkher [3], and others.

Hydrodynamic pressure pulsations of waste-water flow induce mainly vertical vibrations of the apron. In this case, elastic surface waves which involve neighboring sections of the apron and spillway dam are generated, and also secondary "resonance" effects appearing in dam elements are superimposed.

It is interesting to note that hydraulic structures and their elements have high dynamic response capacities:

$$K_{dyn} = \frac{1}{\Pi(0)} \sqrt{\frac{\int_0^\infty |\Pi(\omega)|^2 S(\omega)d\omega}{\int_0^\infty S(\omega)d\omega}} > 1,$$

where $S(\omega)$ is the spectral-load density and $\Pi(\omega)$ and $\Pi(0)$ are the moduli of the transfer functions of the oscillatory system.

The behavior of hydraulic structures is similar to the "bandpass filter" action, which selects a narrow band from a wide-band effect near the natural frequencies of the system, that accounts for about 70 to 95 % of the power spectrum. Therefore, auto-correlation functions of dam vibrations are well approximated by

$$R(\tau) = \sigma^2 \exp(-\alpha|\tau|)\cos\beta\tau .$$

As a whole, the spillway-dam vibrations induced by waste-water discharge have a wide frequency range according to the writers' experiments, i.e., a range from 0.1 to 0.25 Hz up to 60 to 80 Hz (Fig.3). Low-frequency vibrations appear due to the low-frequency component of the pressure pulsations on the apron, the correlation radius of which is of the order of the width of the spillway bay. These vibrations have not been sufficiently studied yet. They are not very intense. The ratio $2A_{max}/W$ of the maximum double amplitude of these vibrations $2A_{max}$ (μm) to the specific power of the spillway flow W does not exceed 1 to 2 (μm per MW/m).

Dam vibrations recorded in a frequency band within the range of 1 to 2 Hz to 4 to 8 Hz are chiefly induced by elastic waves in the foundation generated due to pulsations of the downstream apron. The double amplitude of them is $2A_{max}/W$ = 10 to 40 μm per MW/m. The contribution to this value of induced natural vibrations of the dam sections (the logarithmic decrement of which, according to the writers' experiments, does not exceed 0.35 to 0.50) and of localized vibrations of dam elements (the latter occurring with frequencies of 10 to 100 Hz) is insignificant. The direct effect from the flow on the spillway face is not considerable; it does not exceed 1 to 2 % of the total.

According to the writers' measurements on the spillway of a high dam ($H \simeq 100$ m), spillway-face slope, 1:0.8, $W \simeq 30$ MW/m), the pressure pulsations on the spillway face have a wide frequency range within 1 to 3 Hz and 200 to 300 Hz (Fig.7) and their root-mean-squares do not exceed $0.03\ \rho U^2_{max}/2$. Their correlation radius has a very low value: in the flow direction, it is equal to more than 0.5 to 0.7 of the flow depth on the spillway crest, and perpendicular to the flow it reaches 0.15 to 0.20 of that value (Fig.8). Hence, the averaged dynamic effect of the flow on the spillway is very low.

Vibrations of dams on rock foundation with a ski-jump are considerably less intense than of dams on soft foundation with an apron or a stilling basin. Relative amplitudes of vibration of massive dams on rock amount to $2A_{max}/W = 0.1$ to 0.4 μm per MW/m. Vibrations of the dam increase almost 1.5 times when the flow on the spillway is thrown off by added ski jumps above the concrete surface and falls back on it. These ski jumps were designed to saturate the wall-boundary layer with air in order to prevent cavitation erosion (U.S.S.R. patent No. 300565). In this case, the pressure-pulsation spectrum is somewhat displaced toward low frequencies in comparison with the non-aerated flow. Thus, the main source of vibration disturbances is the zone of kinetic-energy dissipation. The distance from this zone to the dam and the foundation properties are the main factors determining the structural vibrations.

High dynamic-response capacity of hydraulic structures brings about interesting peculiarities. Thus, as already mentioned, it creates possibilities for inducing and recording natural vibrations of massive hydraulic structures on soft soils. By utilizing these possibilities [1,4,5], one of the difficulties of experimental dynamic investigations of massive structures as pointed out by D.E. Hudson [6] was partially overcome (Figs. 9 and 10).

The change with time of vibration parameters of spillway dams recorded during waste-water discharge can be explained by the high dynamic-response capacity (selectivity). The vibration measurements of a spillway dam about 35 m high, on soft foundation ($W = 3.2$ MW/m) carried out under the same conditions but with an interval of two years (Figs. 1 to 5) revealed that:

- the root-mean-squares of the vibration-displacements increased by 15 ÷ 30 %;

- the prevailing frequency of vibration, determined by the first zero crossing of the auto-correlation function increased by almost 20 % [4].

It seems that this change of vibration parameters resulted from a converging of partial frequencies of the system "apron slab/foundation soil/spillway-dam section" (the natural frequencies of the apron and dam sections were equal to almost 2.1 Hz and 1.1 Hz, respectively; the logarithmic decrement amounted to not more than 0.50). This converging in frequencies was due to a change of the mechanical properties of the soft-soil foundation on account of consolidation under the effect of vibrations. A similar effect was noted by V.M. Ljatkher [7], and it was observed in the writers' experiment by recording the velocity of propagation of the elastic waves in the structural foundation. These observations show that there are not only irreversible changes with time of properties of the foundation soil when the velocities of Rayleigh-surface waves V_R increase but also reversible changes; the velocities V_R decrease at reservoir drawdown. These results are in good agreement with F. Gassmann's theory, with D. White's and R. Sengbush's experiments, and with findings of many other authors [8,9].

Field investigations of spillway dams show an interesting peculiarity of vibrations of separate sections of downstream aprons, concrete dams, massive power house, etc., which apparently is also due to high dynamic-response capacity of these structures. Thus, when we have a source of narrow-band excitation (apron-slab) or a harmonic vibration (power-house section), the rest of the massive blocks vibrates with almost the same frequency and with a slight phase shift which is determined by the time of wave propagation from the source of vibration to the "recipient". In case of adding one or more inductors of narrow-band vibrations (it is most distinctly observed when "adding" inductors of harmonic vibrations of the same frequency), the doubling or multiplying of vibration frequencies of not only "inductors" but also "recipients" does not occur, as it would, for instance, in the case of adding a few harmonic vibrations of the same frequency but different phase shifts. There occurs a "readjustment" of oscillatory bodies or a self-synchronization (with time lag) of a mechanical system, of which separate links oscillate with a phase shift that is constant in the probabilistic or deterministic sense. The magnitude of the shift depends on the nature of excitation.

This phenomenon was often observed when measuring dam vibrations induced by the operation of hydropower units. The frequency of vibration of the dam body remained constant and close to the revolution frequency of the hydropower units. The phenomenon is to some extent analogous to the phenomenon of self-synchronization of a pendulum clock observed in the XVIIth century by Ch. Huygens and close to the so-called effect of synchronization of rotors [10]. It is the cause of fictitious decrease of oscillatory-system damping. Figure 6 shows two dimensionless auto-correlation functions of vertical vibrations of one and the same dam section (on soft-soil foundation) corresponding to the case of waste-water discharge through one bay of the spillway (1) and through a number of equally opened spillway bays (2). The self-synchronization effect is most distinctly observed, if one has vertical vibrations, and less distinctly for rotational vibrations.

The self-synchronization phenomenon must be taken into account at dynamic calculations of structures, because mutual readjustment of oscillatory elements increases the vibration intensity.

Gate Vibrations

One of the most dangerous vibrations of hydraulic gates with overflow and underflow are self-excited vibrations. It is known that auto-oscillations appear in nonlinear systems which transform aperiodic effects into periodic vibrations. The nonlinearity, and therefore the absence of the superposition of processes, necessitates a simultaneous study of the whole complex of processes (synchronous multipoints investigations). Because of the nonlinearity of real oscillatory systems, the conditions of auto-oscillations induced in them are not unambiguous. In model tests, this was already noted by O. Müller, observed by R. Hart and I.E. Prins [11], and described by P. Lisenko and L. Kuznetsov.

Similar conditions were also observed in writers' field investigations of gate vibrations. In cases with "rigid" characteristics, the oscillatory system displayed "jumps". The excitation of auto-oscillations is accompanied by radical changes of the nature and the intensity of vibrations: random vibrations are replaced by almost periodical vibrations with constant frequency or amplitude and their intensity is increased several times (Fig.11).

During investigations of a vertical-lift gate in a navigation double-lift lock culvert ($H_{max} = 27$ m, $S = 15$ m^2), four main modes of gate vibrations were recorded (Fig.11):

(a) Auto-oscillations due to a leakage in the gate sealings on the sill. A. Abelev and L. Dolnikov [12] discussed one of the auto-oscillations of this mode for $n > 0$, where n is the relative gate opening ($n = a/a_0$). According to these investigations, auto-oscillations arise at $0 < n < 0.5$ and Strouhal numbers above some critical value.

(b) Frictional auto-oscillations are generated during gate lifting or lowering, when the relationship between the friction force R and the velocity of motion v corresponds to $dR/dv < 0$. For the first time, these vibrations were observed on a prototype by D.A. Kharin [13]. Amplitudes of these vibrations were satisfactorily approximated by the following relations:

$$A = \text{const} \quad \text{for} \quad 0.05 < n < 0.25 \quad \text{and}$$
$$A = A_0 + A_1 n \quad \text{for} \quad n > 0.25 .$$

Frequencies of frictional auto-oscillations observed the relations (Figs.12 and 13):

$$f = a - bn \quad \text{for} \quad 0.05 < n < 0.7 \quad \text{and} \quad H = 10.5 \text{ to } 24 \text{ m};$$
$$a = 3.5 \text{ to } 5.5 \text{ sec}^{-1};$$
$$b = 3.7 \text{ to } 1.8 \text{ sec}^{-1}.$$

(c) Vibrations of random nature induced by the action of turbulent flow.

(d) Self-induced vibrations or auto-oscillations of the gate of the conventional flutter type which occur at $n = 0.73$ to 0.90 and $H = 5$ to 20 m [14].

The last mentioned mode of vibration is the most intensive; it is characterized by an almost constant frequency and represents essentially rotatory motion of a low-deformable gate around a horizontal axis parallel to gate sill (Fig. 14); the height h of this axis is determined by the mass distribution including the "added" mass of water. From the test results it follows that the relative height $h/a_0 = 1.6 \pm 0.15 - 0.4\,n$, where a_0 is the height of the navigation lock culvert and n is the relative gate opening a/a_0.

"Resonance" properties of vertical-lift high-head gates clearly manifested themselves at some heads larger than a critical one, that is $H > 15.5 - 13\,n$. Amplitude relationships of the $A = f(\omega/2\pi)$ type, where ω is the frequency of vibration at $n = \text{const}$, show that by increasing n, converging of partial frequencies occurs, and relationships $A = F(\omega/2\pi)$ gain a distinct resonance nature. A simple mathematical model of this system and approximate analytical conditions of stability were suggested earlier [14].

By one of the experimental curves $A(\omega)$ at $n = 0.081$ and $5\text{ m} < H \leqslant 16\text{ m}$, frequencies of "resonance" gate vibrations were found, and from the width of the resonance curve the corresponding logarithmic decrement ($\delta \simeq 0.25$) was obtained; this low value indicates a high response capacity of the system (Fig. 15).

Averaged frequency-response characteristics of gate vibrations, that is Strouhal numbers $Sh_a = Fa/v_0$ (where F is the average vibration frequency, a is the gate opening, v_0 is the velocity of flow under the gate), determined for regimes $n = \text{const}$, $H \neq \text{const}$ (the change of effective head is due to the filling of the lower lock chambers from the upper one) are larger than the commonly accepted values Sh_a characterizing pressure pulsations of averaged hydrodynamic loads, as determined by A. Abelev in his experiments on rigid (nonvibrating) models [15] (see Fig. 16). The amplitude and frequency characteristics of pressure pulsations measured on a prototype gate as well as in a navigation lock culvert (Fig.16) are substantially different from those determined in the corresponding and in other model tests. As E. Naudascher notices [16], a vibrating gate radically transforms the pressure-pulsation spectrum in the flow downstream of the gate. In some cases, at $n \leqslant 0.2$, the writers noted that frictional auto-oscillations of a gate moving upward induce intensive periodic compression waves with double amplitudes up to 25 % of the head (Fig.20) and with a correlation radius comparable to the wave length[1]. Such phenomena are accompanied by a transformation of a wide-band pulsation spectrum into a narrow-band and even a monochromatic spectrum, as well as with a considerable increase of intensity and radius of correlation; they may therefore easily contribute to fatigue failures of steel linings of penstocks.

Resonance effects can play an important role in amplifying periodic pressure pulsations generated by frictional auto-oscillations. They can take place, if a known relation [17] is observed between the frequency of pulsation (F), the water-conduit length (L), and the sound velocity in the flow of water or water-air mixture (C), that is:

$$\frac{4FL}{C} = 2K + 1, \qquad K = 0,1,2,\ldots$$

In the example investigated, the pulsation frequency was "set" by the gate vibration and was equal to almost 4 to 5 Hz at $n \leqslant 0.2$; the length of the navigation lock culvert to the first culvert branch in the chamber floor was equal to L = 60 to 62 m. If we assume C = 1400 to 1200 m/sec (the latter value corresponds to about 1 % of flow aeration), we obtain a quite satisfactory justification of the above relation for the fundamental mode (K = 0). While gate lifting frequency F is decreased. The relation is not fulfilled. Amplitudes of pressure pulsations are decreased. (See Fig.20.)

---

[1] In these tests, the pressure gauges in the navigation lock culvert were located at a distance of 1 m, 14 m and 22 m from the gate and record almost synchronous pressure pulsations.

In conclusion it is to be noted that the operating conditions of hydraulic structures as observed in the prototype differ from the ones of even the most refined models. For example, the averaged pressures measured on the walls of a navigation-lock culvert downstream from a gate are three and more times smaller than the values which V.M. Ljatkher gives with reference to K. Hajek [19].

Auto-oscillations of gates have been found impossible to eliminate in a number of cases. Therefore, the necessity arises in engineering practice to estimate the danger for gate vibrations. K. Petrikat suggested the use of the first norms for turbine vibrations (Fig.19) which he adopted from T.C. Rathbone [20,21]. However, Rathbone's investigations are based mainly on measurements with mechanical seismographs and, therefore, his recommendations are at present quite imperfect. Besides, the range of application of these norms cannot be artificially extended without taking into consideration structural-design differences, the nature of forces, and the nature of induced vibrations. For example, Rathbone's recommendations normalize a certain vibration parameter of mode $A \cdot F^{\alpha(f)} = \text{const}$, where A is the vibration amplitude, f the frequency, and $\alpha(f)$ some function of the frequency with the range $0.2 < \alpha < 1.2$. In the frequency range 20 to 60 Hz, $\alpha \simeq 1$, that menas the normalized vibration parameter is the vibration velocity. Normalization of machine vibrations by the amplitude of the vibration velocity or its effective value is very common [22]. But it may be used only in the range of natural frequencies of vibrating members of one or another structure. In reality, the amplitude of deformation vibrations of the i-th element which a foundation experiences for harmonic vibrations of mode $A \cdot \sin \omega t$ will be as follows:

$$x_i = A \frac{\omega^2}{\omega_{0i}^2} \frac{\sin \omega(t - \rho)}{\sqrt{\left(1 - \frac{\omega^2}{\omega_{0i}^2}\right)^2 + 4D_i^2 \frac{\omega^2}{\omega_{0i}^2}}},$$

where ω, ω_{0i} are circular frequencies of excitation and natural frequencies of the i-th element and D is the damping coefficient ($D = \ln\nu/\pi$).

If the excitation frequency ω is much less than the natural frequency of the element, $\omega \ll \omega_{0i}$, and $D \ll 1$ (the latter is practically always fulfilled), then the amplitude of deformation vibrations of the i-th element is in proportion to the acting acceleration, that is $x_i \approx A\omega^2/\omega_{0i}^2$. At $\omega \gg \omega_{0i}$, we obtain $x_i \approx A$. In case the excitation frequency ω is close to the natural frequency ω_{0i} (resonance), the deformation amplitude is in proportion to the acting vibration velocity, that is $x_i \approx A\omega/\omega_{0i}$. Therefore, normalization of vibration by the vibration velocity may be used only in the case of sharp-resonance systems ($D \simeq 0$) and only within a band of natural frequencies of their elements. However, the latter are different for gates and turbines.

Assuming the gate as a beam on two supports (the justification of such an assumption is shown by V.A. Palunas and P.E. Lisenko [23]) and separating the deformation vibrations from the transfer vibrations, the following relation should be obtained between the amplitude of deflection A in the middle of the span L and the amplitude of the dynamic stress σ, measured at a distance y from the neutral axis (Fig.17):

$$\frac{\sigma}{A}\frac{l^2}{yE} = K = \text{const} ,$$

where E is the modulus of elasticity. The constant coefficient K must have the following values: at concentrated load in the middle of the span - 12, at uniformly distributed load - 9.6, at load placed nearer to the supports and on the cantilever - 8. However, in model tests by P.E. Lisenko and the writer and in prototype measurements [14, 24], the observed values of K range within 2 and 3 or 6 and 8, varying sharply from experiment to experiment (Fig.18). It is possible that these discrepancies can be explained by incompleteness of the scheme of measurement. At the same time, however, it shows that the evaluation of gate vibrations is a more complicated problem than was considered by K. Petrikat. In practice, usually the parameter which is used for estimation is measured. These measurements are complemented by synchronous measurements of gate vibrations. In this case the latter is supposed to be non-deformable.

The problem discussed above vividly demonstrates that the operating conditions of hydraulic structures can substantially differ from one or the other idealized and simplified model. Reliable operation of hydraulic structures under conditions of high flow velocities, vibration, and cavitation is possible only when exhaustive information about actual conditions of operation is available, so that the prediction or control of one or another dangerous effect of high-velocity flow action on structures is possible [25, 26].

References

1. Maksimov, L.S., Goncharov, L.P.: Gydrotekhnicheskoe stroitelstvo, No. 10, 40-41 (1961).

2. Maksimov, L.S., Tishchenko, V.G., Kharin, D.A.: Trudy hydroproject, Issue No. 7, Moscow, 1962, pp. 99-107.

3. Ljatkher, V.M.: Turbulence in Hydraulic Structures, Moscow: Energija Publishing House 1968.

4. Goncharov, L.A.: Trudy koordinatsionnykh soveshchanii po gidrotekhnike, Issue 54, Moscow-Leningrad: Energija Publishing House 1970, pp. 373-377.

5. Loginov, V.N.: Trudy koordinatsionnykh soveshchanii po gidrotekhnike, Issue 20, Moscow-Leningrad: Energija Publishing House 1965, pp. 5-16.

6. Hudson, D.E.: Engng. Mech. Div. Proc. ASME, No. 3, Part I, 1-19 (1964).

7. Ljatkher, V.M., Didukh, B.I.: Trudy Hydroproject, Issue 20, Moscow 1971.

8. Gassmann, F.: Geophysics, 16, 673-683 (1951).

9. White, J.E., Sengbush, R.L.: Geophysics, 18, 54-69 (1953). pp. 54-69.

10. Blekhman, I.I.: Izvestia, Acad. of Sc. of the USSR, OTN, No. 8, 1954.

11. Hart, R., Prins, I.E.: Proc. VIII Congr. IAHR, Montreal, 1959, Rep. A-27.

12. Abelev, A.S., Dolnikov, L.L.: Proc. XIV Congr. IAHR, Paris 1971, Vol. 2, Rep. 32, pp. 263-270.

13. Tishenko, V.G., Kharin, D.A.: Trudy Seismitheskogo Instituta, Acad. Sc. USSR, No. 106, 85-90 (1941).

14. Goncharov, L.A., Maksimov, L.S.: Trudy koordinatsionnykh soveshchanii po gidrotekkhnike, Issue 7, Moscow-Leningrad: Gosenergoizdat 1963, pp. 85-96.
Bal, B.A., Goncharov, L.A.: Proc. XIVth Congress of IAHR, Paris 1971, Vol. 5, pp. 61-64.

15. Abelev, A.S.: Proc. Xth Congr. IAHR, London 1963, Rep. 3.21.

16. Naudascher, E.: Proc. Xth Congr. IAHR, London 1963, Rep. 3.9.

17. Jaeger, C.: Water Power 15, No. 1-3 (1963).

18. Moshkov, L.V.: Izvestiya VNIIG, 85, 34-40 (1967), 88, 48-55 (1969).

19. Ljatkher, V.M., Smirnov, L.V.: Izvestiya VNIIG 85, 186-200 Engng. (1967).

20. Petrikat, K.: Der Stahlbau, No. 9, 12 (1955).

21. Rathbone, T.C.: Power Plant Engng. 43, 721-724 (1939).

22. Federn, K.: Konstruktion im Maschinen-, Apparate- und Gerätebau 10, No. 10, 289-298 (1957).

23. Palunas, V.A., Lisenko, P.E.: Trudy koordinatsionnykh soveshchanii po gidrotekhnike, Moscow-Leningrad: Gosenergoizdat 1963, pp. 97-111.

24. Freishist, A.R.: Energeticheskoe stroitelstvo, No. 7, 50-53 (1967).

25. Galperin, R., Semenkov, V.: Proc. XIVth Congr. IAHR, Paris 1971, Vol. 5, pp. 45-48.

26. Goncharov, L.: Energeticheskoe stroitelstvo, No. 2, 71-74 (1972).

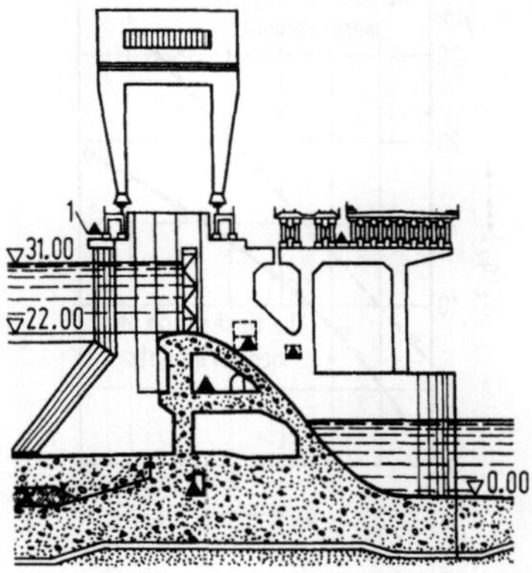

Fig.1. Scheme of vibration measurement of spillway dam section.
Vibration pickups on structure.

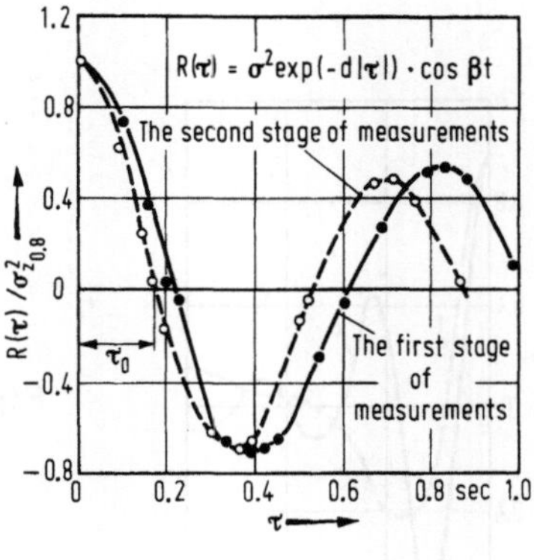

Fig.2. Auto-correlation function of vertical vibrations of spillway dam section.

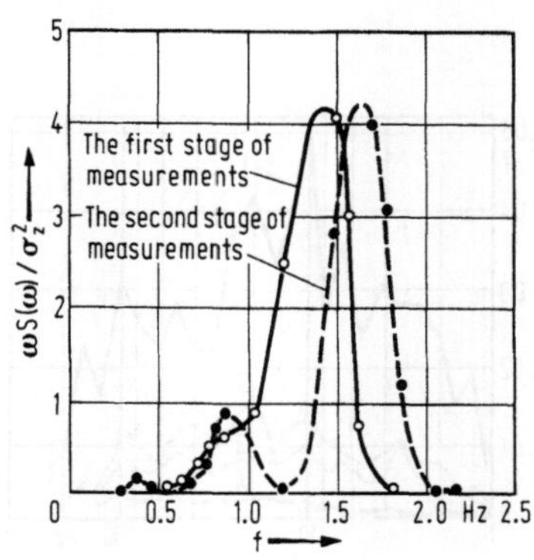

Fig.3. Functions of spectral density of vertical vibrations of spillway dam section.

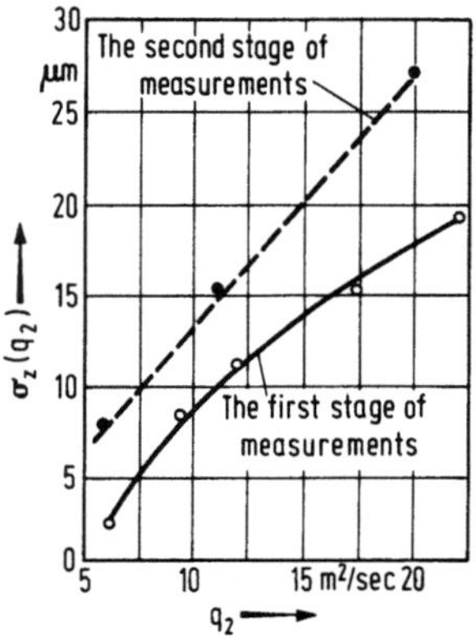

Fig. 4. Root-mean-squares of vertical vibrations of spillway dam section.

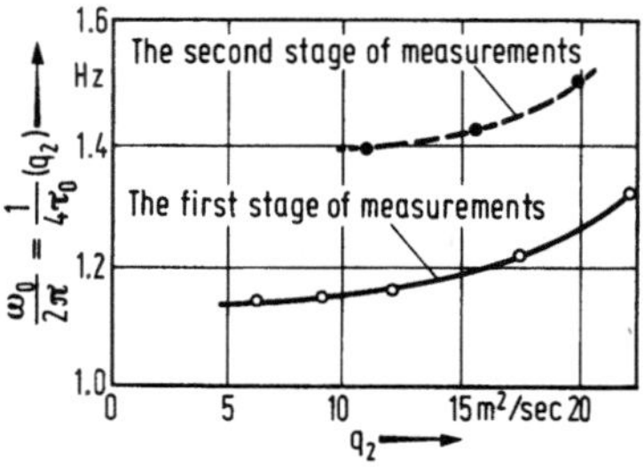

Fig. 5. Prevailing frequencies.

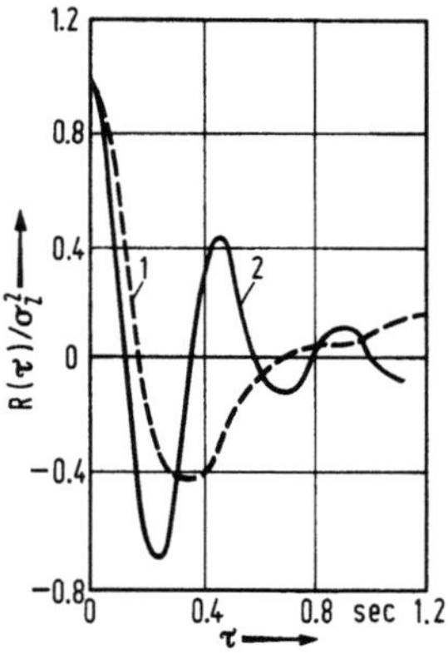

Fig. 6. Auto-correlation functions of vertical vibrations of spillway dam section.
1 Water discharge through one bay;
2 Water discharge through several bays.

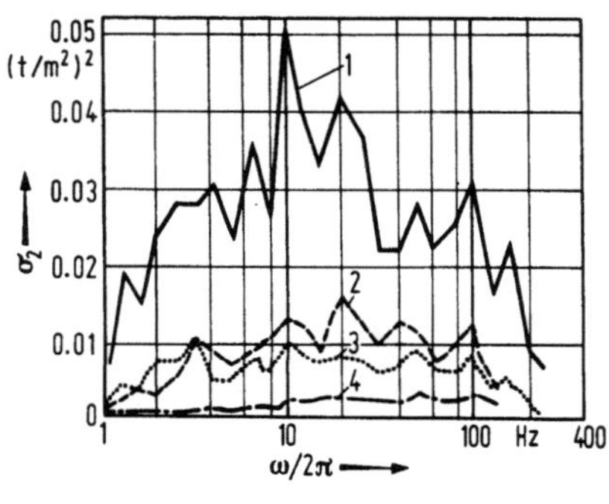

Fig. 7. Spectral densities of pressure pulsation according to measurements on spillway face of a high dam. Measured at a point located 72,0 m (1), 58,8 m (2), 46,0 m (3), 30,5 m (4) below the spillway crest.

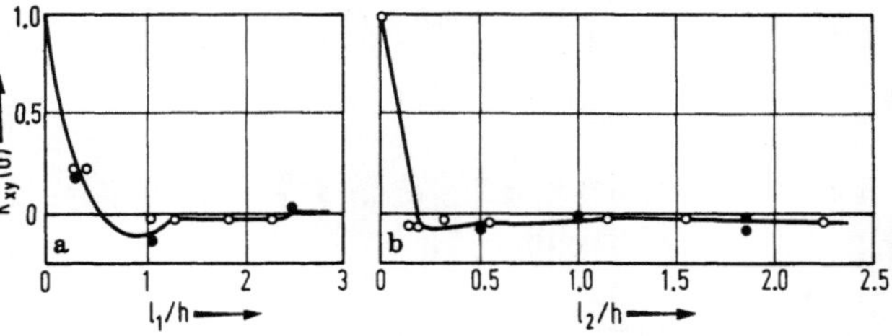

Fig.8. Correlation coefficients of pressure pulsations on a spillway of a high dam.
a) Along flow; b) Across the flow (measured 58,8 m below the spillway crest); l = Distance; h = Flow depth in the measuring point.

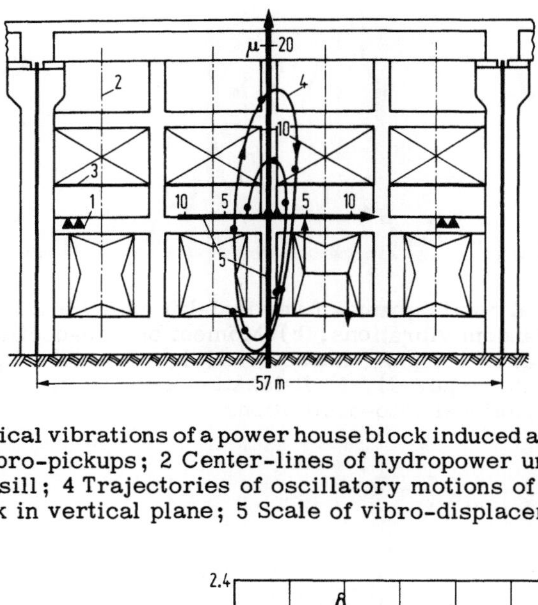

Fig. 9. Vertical vibrations of a power house block induced at load-off.
1 Vibro-pickups; 2 Center-lines of hydropower units; 3 Spillway sill; 4 Trajectories of oscillatory motions of power house block in vertical plane; 5 Scale of vibro-displacements.

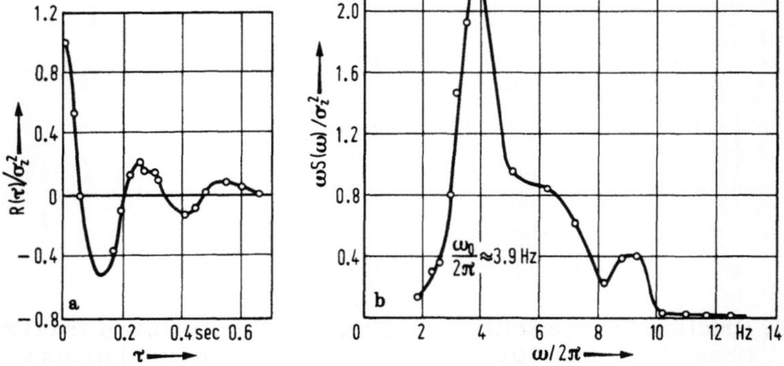

Fig. 10. Auto-correlation function (a) and function of spectral density (b) of vertical vibrations of power house block at load-off.

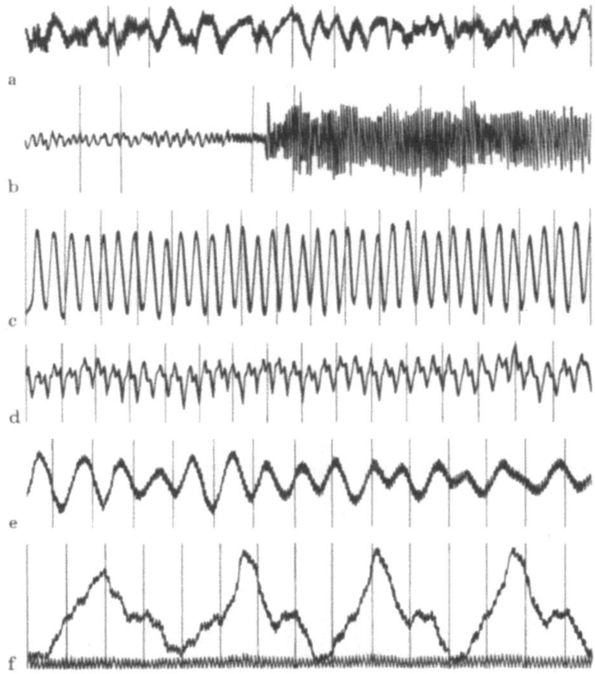

Fig.11. Typical oscillograms of vertical-lift gate vibrations. a) Random vibrations; b) Moment of auto-oscillations excitation ("rigid" characteristic); c) Auto-oscillations of flutter type; d), f) Relaxation auto-oscillations; e) Frictional auto-oscillations.

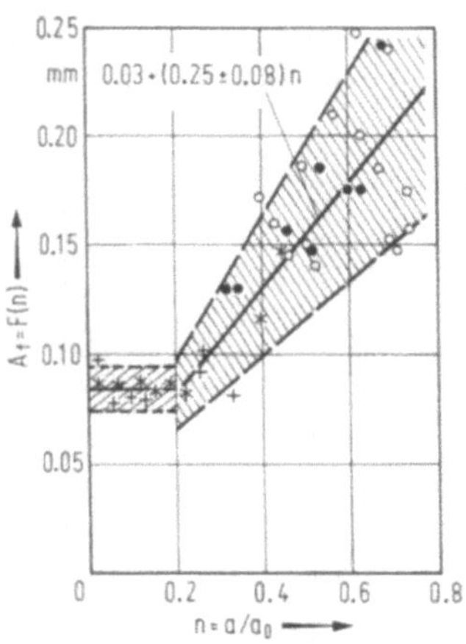

Fig.12. Frictional auto-oscillations of vertical-lift gate at its travel with different velocities.

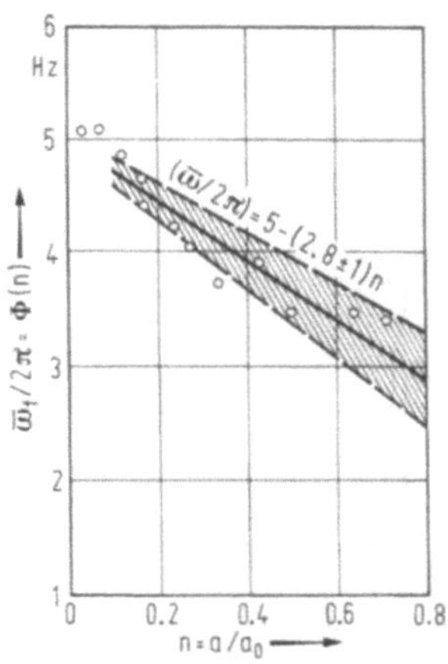

Fig.13. Frequencies of frictional auto-oscillations of vertical-lift gate.

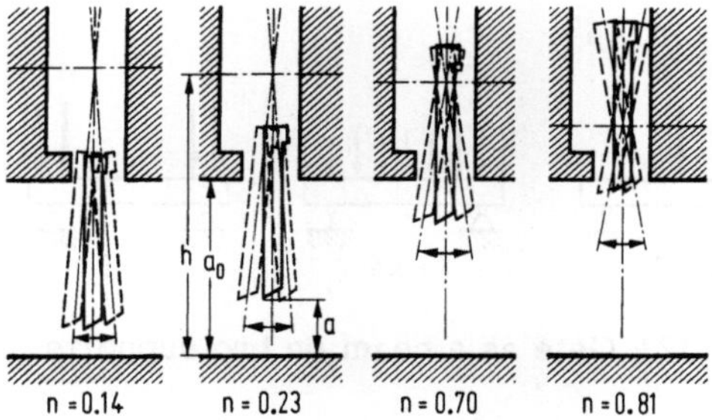

Fig. 14. Types of vibratory motions of vertical-lift gate at $n = a/a_0 =$ = const.

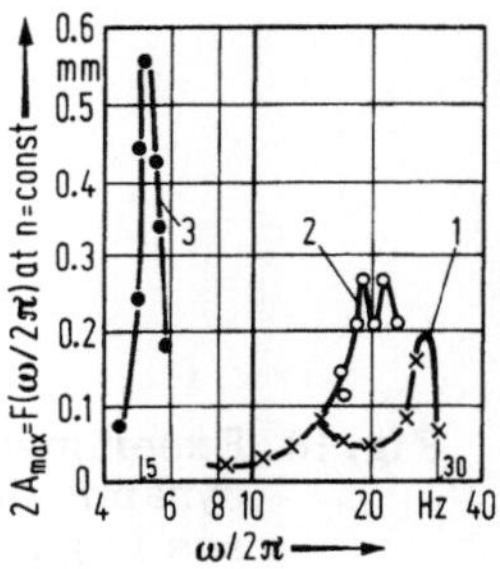

Fig. 15. Experimental "resonance" curves of vertical-lift gate vibration at n = const.
1) n = 0.23 H = 7 to 23 m;
2) n = 0.45 H = 10 to 23 m;
3) n = 0.81 H = 5 to 16.5 m.

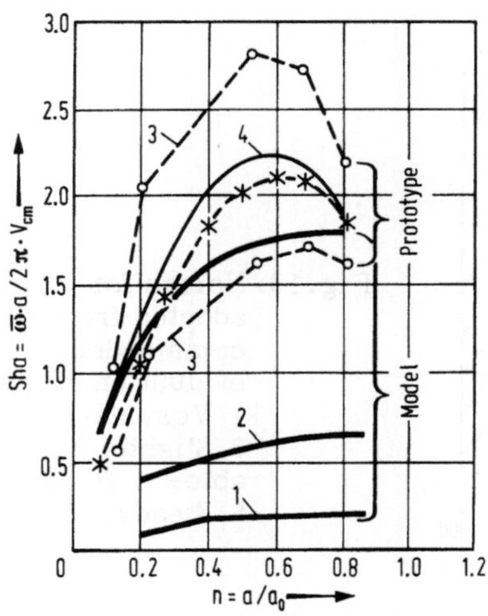

Fig. 16. Strouhal number according to results of measurements on "rigid" models and on prototypes.
1, 3 Boundaries; 2, 4 Average values.

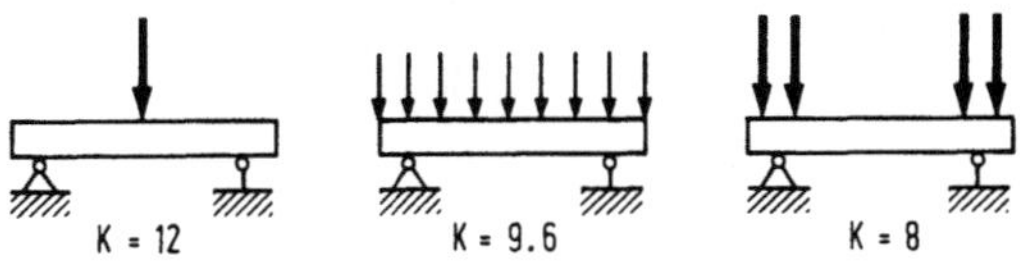

Fig.17. Gate as a beam on two supports.

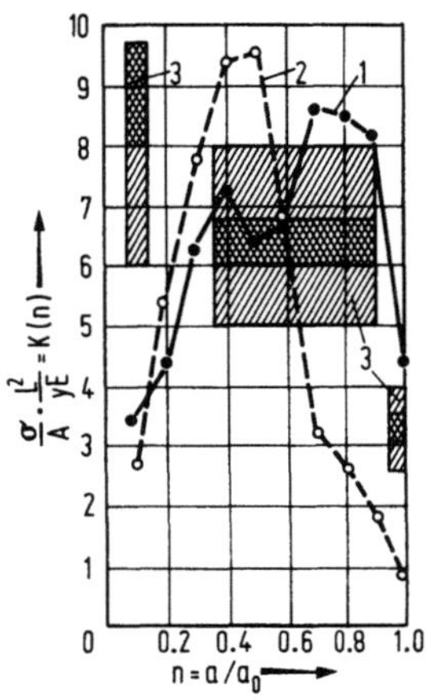

Fig.18. Experimental relations between dynamic deflections and stresses in gate transom.
1,2 Ship lock culvert gate;
3 Spillway gate (measurements by A. Freitshist).

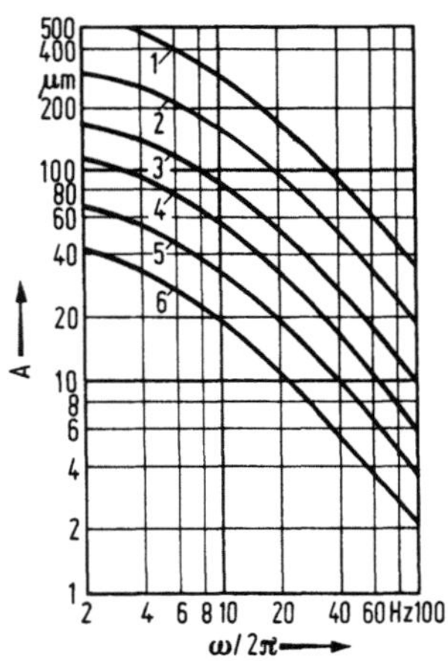

Fig.19. Norms for turbine vibrations adopted from Rathbone and recommended by K. Petrikat for evaluation of gate vibration.
1 Very unsteady; 2 Unsteady; 3 Slightly unsteady; 4 Acceptable; 5 Reasonably steady; 6 Steady.

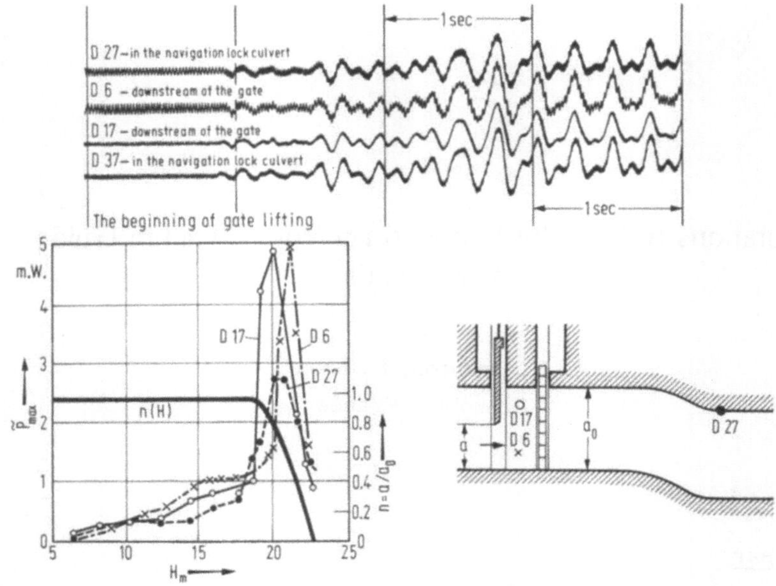

Fig.20. Pressure pulsation in ship lock culvert induced by frictional auto-oscillations of a vertical-lift gate.
H_m Head acting onto the gate;
P_{max} Maximum amplitude of pressure pulsation.

Vibrations Induced by Von Kármán Vortex Trail in Guide Vane Bends

By

R. Brepson, P. Léon

Grenoble, France

Summary

Immersed two-dimensional plate vibration due to a vortex street behind the trailing edge has been investigated experimentally. Research on vaned bend vibration at SOGREAH has shown that the vanes behave as a two-dimensional plate, and especially that the trailing edge configuration is most important and that one must know the natural vane frequencies in order to ensure that they are not the same as the excitation frequencies. A program has been developed for the computation of natural vane frequencies and the associated modes, and the numerical data and results of tests on an actual vaned bend have been compared. The theoretical and experimental results agreed closely and yielded design data enabling the vibration risk for guide vane bends to be reduced to a minimum.

Introduction

The formation of vortex streets behind objects immersed in flowing fluids gives rise to oscillation frequencies which are liable to cause structural vibration. This effect is well-known and vibration of tubes perpendicular to the flow has been extensively studied. Flat plates in parallel flow have also been studied experimentally. However, the conclusions that can be drawn from this work are not of a general nature. This is probably due to the effect of additional variables such as the length and thickness of the plate and the shape of the trailing edge. The case of sharply-curved plates such as the guide-vanes inside elbow-bends with a small radius of curvature does not seem to have been investigated. Nevertheless, it would seem that increase in flow velocity and reduction of structural weight make for more intense vibration and may even cause momentary shut-down of large plants for repairs. It is for this reason that we have been studying ways of avoiding vibration of guide-vanes in elbow-bends. The following study is based on:

(a) Analysis of experimental results obtained by various laboratories on flat plates,

(b) experience gained by ourselves in the field of guide-vane vibration in bends,

(c) use of computer programs to calculate natural frequencies and vibrational modes of guide-vane bends.

Studies on Flat-Plate Vibration in Parallel Flow

In a study on vanes which sing in water, C.A. Gongwer [1] recorded sound given out by a plate moving through a stationary liquid. The maximum velocity of the plate was 30 m/sec. The plate was in fact a double-walled blade, each wall being slightly curved (30 cm chord). The rounded trailing edge was 0.32 cm thick, the thickness being gradually reduced in the course of the work. The maximum thickness of the plate was 1.9 or 0.95 cm for the two blades tested. The wake of the blade was traced and photographed.

Studying vibration in hydraulic turbines, R.M. Donaldson [2] measured vibration of an aluminum plate submerged in a water tunnel. The plate thickness was 0.635 cm. The flow velocity did not exceed 10 m/sec. The aim of the study was to investigate the effect of the shape of the trailing edge on the amplitude and frequency of plate vibration. An application to hydraulic turbines is mentioned.

G. Heskestad and D.R. Olberts [3] have also studied the influence of trailing-edge shape on vibration of a plate in a water tunnel. Their flow velocities varied between 3 and 18 m/sec. Vibration characteristics were measured and the flow was visualized. The vortex strength was calculated. The plate thickness was 0.635 cm.

Finally, G.H. Toebes and P.C. Eagleson [4] demonstrated the hydroelastic nature of submerged-plate vibration. Their tests were based on a rigid plate fixed to a torsion spring, i.e. a system with one degree of freedom. The plate thickness was 0.6 cm, four different trailing-edge shapes were tested. A mathematical analysis of plate vibration was carried out with a view toward explaining the self-excitation mechanism reported previously [5] by members of the same research team (P.S. Eagleson, J.W. Daily and G.K. Noutsopolous).

(1) Source of Excitation. Research workers are unanimous in ascribing the source of excitation to the formation of vortex streets on the trailing edge of the plate. Dominant vortex frequency is given by the following general formula:

$$f = \frac{SV}{D},$$

where V = flow velocity, S = Strouhal number, and D = some crosswise dimension of the object.

(2) Strouhal Number. In the case of cylindrical objects perpendicular to the flow, it is well known that the Strouhal number remains constant and equal to 0.2 within the Reynolds number region from 10^3 to 2.10^5. This is not so for plates. Plate Strouhal numbers seem to depend on the shape of the trailing edge and on plate thickness. GONGWER concluded that, for constant Strouhal number, trailing-edge thickness should be increased by a virtual thickness adapted to that of the boundary layer. This virtual thickness is given by:

$$\delta_V = \frac{0{,}0297\ x}{\left(\frac{Vx}{\nu}\right)^{0,2}},$$

where x = plate length, V = flow velocity, and ν = kinematic viscosity.

Under these conditions, the average Strouhal number for trailing-edge thicknesses between 0.038 and 0.32 cm is given by:

$$S_m = 0.186.$$

Extreme values of S are 0.177 and 0.2. The plate tested by Gongwer had rounded trailing edge. According to Donaldson who tested eleven different trailing-edge shapes, this factor has no effect on vortex frequency.

According to Heskestad, vortex frequency cannot be accurately determined by the Strouhal formula on account of the effect of trailing-edge shape. However, one can deduce from Heskestad's published data that the Strouhal number for ten different shapes varies between 0.155 and 0.34. In these tests, trailing-edge thickness was 0.6 cm and the virtual thickness due to the boundary layer was not taken into account. Comparison with Gongwer's data is impossible since rounded shapes were not studied by Heskestad. Plate thicknesses were also different. Nevertheless, for comparable shapes one finds comparable Strouhal numbers.

Eagleson [5] introduced the concept of a Strouhal number which varies with the type of vibration. This leads to maximum and minimum Strouhal numbers for each shape of trailing edge. The difference between maximum and minimum values is slight, if the maximum amplitude of vibration is low, but it will be large for high amplitudes. The following values may be quoted as an example:

rounded edge: $0,185 < S < 0258$, large vibration,

grooved edge: $0,234 < S < 0,251$, small vibration.

Maximum Strouhal number would seem to occur at the inception of vibration when the velocity gradually increases; the minimum Strouhal value occurs at maximum amplitude. This result may be compared with data obtained for a cylindrical rod [6] whose minimum Strouhal number, corresponding to maximum vibrational amplitude, was 0.166 instead of 0.2 for a stationary rod. It is difficult to compare Eagleson's data with those of other workers. However, it is worth noting that Eagleson's minimum number for a rounded edge is equal to that given by Gongwer.

(3) Vibration Amplitude. The shape of the trailing edge is of the greatest importance as regards amplitude of vibration. Its effect has been specially studied by Donaldson and also Heskestad using fifteen different trailing-edge shapes. In the following table, maximum vibration amplitudes are given as relative values in increasing order of magnitude. Toebes's data [4] on four different shapes confirm these values. The following conclusions may be drawn from the foregoing:

(a) Sharp trailing edges (e.g. 30° apex angle) give practically no vibration amplitude at all.

(b) A favorable effect on amplitude also occurs from using rounded shapes with large radii of curvature in order to reduce plate thickness gradually. This avoids having a sharp transition to the side faces. Only the sharp angle at the downstream end of the trailing edge need be kept.

(c) When the trailing edge is symmetrical with respect to the median plane of the plate, the effect of a sharp angle at the edge is considerable. Amplitude is negligible for an angle of 30° whereas a 60° angle gives the largest amplitudes.

(4) Type of Vibration. Eagleson [5] attempted to discover the nature of plate vibration. To do this he chose a system with one degree of freedom consisting of a rigid plate, the front edge of which was attached to a torsion bar. The existence of three velocity regions was demonstrated:

(a) In the first, increasing-velocity region, vibration is sinusoidal and amplitude is variable. The Strouhal number is constant, i.e., vibration frequency increases linearly with velocity.

(b) In the second region, vibrational amplitude is large and stable. However, analysis of the vibration signal shows that a harmonic frequency is superimposed on the fundamental frequency. The amplitude of the harmonic increases with frequency. Strouhal number varies inversely with velocity.

(c) In the third velocity region, amplitude becomes small and variable and the Strouhal number increases with velocity.

It should be stated, however, that this description of vibration phenomena is only valid for plates with trailing edges which are the source of intense vibration.

It seems then that, in the second velocity region, the frequency no longer increases with velocity but stays more of less in the neighbourhood of the natural frequency, i.e., the exciting frequency seems in fact to depend on vibration. It may be said that the nature of this type of vibration is hydroelastic inasmuch as it is due to an interaction between structural vibration and exciting frequencies in the fluid. These considerations show how important it is to know the natural modes of vibration and the natural frequencies of guide vanes.

Study of Guide-Vane Bend Vibration

Generally speaking, guide-vane vibration gives rise to a musical sound, the frequency of which is equal to one of the natural frequencies of the guide-vane. The structure is said to sing. Vibration does not necessarily cause damage, especially if the flow velocity is low, but quite often considerable damage results and a number of examples of this might be quoted. We had the occasion, for instance, to investigate guide-vane vibration in a plant which comprised many such elbow-bends for which no particular effort had been nade to avoid vibration. After thirty of forty hours of operation, cracks began to appear in the pipe to which the guide-vanes were welded. This meant that the installation had to be shut down to enable elbow-bends of different design to be fitted. For this particular example, a test program was set up to discover what made the first type of bend vibrate and to make sure that the new design would not do so. Strain gauges were used to measure the characteristics of guide-vane vibration for different flow velocities. Noise level was measured simultaneously using a sound meter equipped with a frequency analyser. Guide-vanes of the new design were longer and thinner (2 mm instead of 6 mm) than the initial vanes. Also, the old design had a straight trailing edge whereas the tips of the new vanes were chamfered symmetrically at 30°.

The measurements showed that the old vanes vibrated at frequencies between 265 cps and 315 cps (depending on vane length) whereas the new vanes, within the design range of velocities, vibrated at random frequencies with a much smaller amplitude. Vibration frequencies of the old vanes were precisely equal to the natural frequencies of vanes. Noise measurements on the old vanes varied between 88 and 102 decibels, depending on flow rate, within the 250 cps occurred at a lower velocity than at 315 cps. It may be assumed that these vibrations were excited by the fluctuating vane wake. Strouhal numbers for the longest vane are $S = 0.25$ (vibration inception) and $S = 0.17$ (maximum vibrational amplitude). For the shortest vane, calculation gives $S = 0.26$.

The above Strouhal values may be compared with those of Eagleson [5] for flat plates with the same shape of trailing edge. Eagleson's values were 0.25 and 0.214. Some interesting observations were made on the new vanes at flow rates less than the design value. Forced oscillations were measured for two different-length vanes at 33% and 75% of design flow. Vibration frequencies were 350 and 800 cps. However, vibration only occurred within a narrow velocity range and amplitude (especially in the second case) was low. The Strouhal number was $S = 0.32$ for both these vibrations.

For the same shape of trailing edge, Heskestad [3] reports a Strouhal number of 0.31 for a flat plate. Various musical notes were heard when the velocity was increased up to 75% of the design value, though in each case the velocity range was small. Each note was probably associated with shortlived vibration inception of successive vanes, all of which had different natural frequencies. The small velocity region in which each vibration occurs is probably related to the shape of the trailing edge - as Heskestad concluded for flat plates. However, no periodic vibration or musical sound was observed in the region of design flow. It would therefore seem that it was not possible to excite higher modes of guide-vane vibration, at least with the shape of trailing edge tested. Conclusions from these observations might be as follows:

(a) Guide-vanes in elbow-bends behave like flat plates when subjected to oscillating-wake excitation at the trailing edge.

(b) Strouhal number depends on trailing-edge shape.

(c) Amplitude and noise level also depend on shape of trailing edge. A symmetrical shape with an apex angle of 30° has a favorable effect on vibration.

(d) Guide-vane design must consider guide-vane natural frequencies in order to avoid resonance with the vortices in the vortex streets.

Computer Program for the Determination of Natural Frequencies of Guide Vane Bends.

Guide-vane bends are three-dimensional structures, and, as shown in the previous section, it is necessary to establish their successive natural frequencies and associated natural modes. The vanes are usually cylindrical panels of varying thickness at the leading and trailing edges. In some installations, they are stiffened by bracing members or struts (see Fig.1). For the frequency calculation, the continuous medium of the vanes is simulated by a discrete model, the definition of which is based on the equivalence method, the principles of which are, briefly, that the stress and strain tensors (σ_{ij} and e_{ij}) are related by the elasticity equation:

$$\sigma_{ij} = \lambda \, e \, \delta_{ij} + 2 \mu e_{ij}, \quad (1)$$

where λ and μ are characteristic Lamé constants for the material.

The expression for the energy stored per unit body volume is

$$U_0 = 1/2\lambda\left(\sum e_{ii}\right)^2 + \mu\sum\left(e_{ij}\right)^2. \tag{2}$$

Considering a given two or three-dimensional figure formed by a given bar assembly, an expression can be found for the strain energy for all the bars forming the figure.

Consider a bar whose unit elongation ε in a given direction defined by the cosine of its angle α_i is given by the following relationsphip:

$$\varepsilon = \sum \alpha_i \alpha_j e_{ij}. \tag{3}$$

Then, associating a parameter $\rho = ESL$ with a bar of cross-section S and length L, where E is Young's modulus for the bar material, the expression for the strain energy of the bar is as follows:

$$1/2\,\rho\,\varepsilon^2. \tag{4}$$

Substituting Eq. (3) into Eq. (4), one obtains

$$1/2\,\rho\left(\sum \alpha_i \alpha_j e_{ij}\right)^2. \tag{5}$$

Summing Eq. (5) for all the bars in the equivalent figure, an expression is obtained for the total strain energy W, which corresponds to the bars with hinged nodes.

The equivalence principle is based on the equality of the total strain energy of the continuous body and the equivalent model for the exterior and interior forces. As, however, the displacements at the boundaries of the continuous body and the equivalent figure are equal, it is more convenient to equate the internal strain energy of the continuous medium to that of the equivalent medium formed by the bar system. V being the volume of the body, the equivalence principle is expressed by the following relationship, where U_0 is the energy stored per unit volume:

$$W = VU_0. \tag{6}$$

Compatibility of the models is a basic condition for the equivalence principle. A justaposition of various models about the same running node or one on the contour must satisfy the equilibrium conditions expressed as a function of the strain tensor e_{ij} in the considered node. If these conditions are not met, the models are mutually incompatible.

The final requirement is that the models be stable. The model bar cross-sections and inertia can be calculated by the equivalence principle. Two-dimensional models may be rectangular, square, triangular, or in the shape of a regular trapezium or a lozenge. Available three-dimenional model shapes are the parallelepipedon, the cube and the regular octahedron.

The first step in representing a braced vane structure is to define the optimal mesh complying with the geometrical contour shape. The mesh dimensions depend on criteria applicable to shells and shell elements, especially the characteristic length of the inherent shell parameter.

After plotting the mesh system, the nodes are numbered in an order such that the interval between any two connected points is a minimum. Then, the natural vibration frequencies and modes can be calculated by the three following programs:

The ASSIMIL program, which calculates the cross-section and inertia of each bar of each elementary figure. These cross-sections and inertias can be cumulated with those of discrete elements (bracing members). A list of the cross-sections and inertias is printed. The model nodes are hinged for membrane forces and fixed for transverse flexural forces.

The CADRE program, which calculates the stiffness of each bar with respect to three coordinate axes related to the bar, assembles and classifies the coefficients and establishes the general rigidity band matrix for the whole structure. This matrix relates the exterior forces to the internal forces expressed in terms of the generalised displacements by the following:

$$(F) = [A](\delta).$$

The CADRE program solves the 6N algebraic equations for the three-dimensional structure by elimination, from which the following symbolic expression is obtained:

$$(\delta) = [A]^{-1}(F).$$

The program calculates the forces in the bars from the displacements (δ) and the membrane and flexural stresses in each facet of the discrete model by the equivalence relationships.

The OLSTRI[1] program determines the natural values and modes of (A), in which the lowest natural value is the natural frequency, the next higher the first harmonic, and so on. The usual practice, however, is to try to determine the highest natural value of the resolved matrix $(A)^{-1}$.

The program is of the iterative type, the calculation being stopped when two natural vectors are parallel to within the required degree of accuracy.

Comparison with Experimental Data

These programs were checked by natural-frequency measurements on various structures, among which is a 900 mm diameter guide-vane bend, with average flow velocities of up to 12 m/sec.

The bend was designed for the required temperature and pressure conditions, and as it consisted of four identical quarter-bends, the measurements were confined to the first quarter only.

The model quarter-bend with vanes and vane struts comprised 493 nodes. Vane thickness variations were represented by bar cross-sec-

[1] "Oscillations Libres des Structures Tridimentionnelles" (Free oscillations of three-dimensional structures).

tion and inertia variations. As the vanes were thin, the models were represented by thin facets but were three-dimensional with respect to the degree of freedom.

Local increases in thickness due to welds were neglected as the calculation was carried out before the last bend was made, but they could easily have been allowed for. It will be shown that discrepancies between the theoretical and test data were partly due to this omission.

The basic frequency was calculated to be 670 Hz the corresponding natural mode is shown in Fig. 1 (leading or traling edge of a quarter-bend) and Fig. 2 (longitudinal section parallel to the central supporting plate, half-way between the plate and the elliptical bend section). As Fig. 1 shows, the most highly stressed vane is the longest one, but the other vanes and even the supporting plate also have some effect. Fig. 2 shows a nodal line half-way between the leading and trailing edges of each vane.

As the calculation was carried out while the bend was still at the design stage (i.e., before it was made), some discrepancies arose between the theoretical and experimental data due to manufacturing details, especially a number of particularly bulky welds which added to the stiffness to the extent of increasing the natural freqency by 13 to 14 %. Except for these discrepancies, agreement between the theoretical and experimental data can be considered very satisfactory. The effect of the added liquid mass on the natural frequency was allowed for in the calculation. The calculated natural frequency of the immersed bend worked out at 491 Hz, giving a theoretical natural frequency reduction of 26.7 % due to the liquid, compared to the true figure of 32.2 % found experimentally. The difference of 5.5 % between these figures must have been due to the added mass of the supporting plate, which was neglected in the calculation.

Conclusion and General Recommendations

The investigation described in this report, which was based on a number of experimental investigations on the behavior of flat plates in parallel flow and the observed behavior of guide vane bends, seems to show that vane excitation is governed by the same relationships as flat plate excitation.

In particular, both vane vibration frequency and amplitude appear to depend on the shape of the trailing edge and the relative vane thickness. For instance, a trailing edge with an open angle of 30° should give a lower vibration amplitude and a smaller velocity area in which vibration occurs. There are also a number of other satisfactory configurations (see table annexed). As the excitation frequency depends on both velocity and the flow pattern (frequency of the vortex street forming at the trailing edge) it is important to ensure that one of the natural vane frequencies does not lie close to the excitation frequency. This is why it is necessary to calculate the natural frequencies and to modify them where required (e.g., by altering the vane thickness or unsupported lengths, etc.), though not without the effect of vane thickness allowing for on the excitation frequency. With the programs described, these natural-frequency calculations can be carried out and the natural modes determined with due allowance for all the mechanical characteristics involved (i.e., vane curvature, cross-sectional variations, boundary conditions, etc.) and the virtual mass of the fluid.

These conclusions apply exclusively to the effects of excitation on the trailing edge, and in particular, they are based on the assumption of head-on flow (zero angle of incidence) at the leading edge.

References

1. Gongwer, C.A.: A Study of Vanes Singing in Water. J. Appl. Mech., December 1952.

2. Donaldson, R.M.: Hydraulic Turbine Runner Vibration. Amer. Soc. Mech. Engs., Paper No 55 A, 130.

3. Heskestad, G., Olberts, D.R.: Influence of Trailing-Edge Geometry on Hydraulic Turbine-Blade Vibration Resulting from Vortex Excitation. ASME, Paper Nő 59, Hyd. 7.

4. Toebes, G.H., Eagleson, P.S.: Hydroelastic Vibrations of Flat Plates Related to Trailing Edge Geometry, ASME, Paper No 61, Hyd.16.

5. Eagleson, P.S., Daily, J.W., Noutsopolous, G.K.: Flow Induced Vibration of Flat Plates: the Mechanism of Self-Excitation. Hydrodynamics Laboratory, Report No 58, Massachussetts Institute of Technology, February 1963.

6. Léon, P.: Vibration of Rod in Perpendicular Flow. IAHR XI Congress, Leningrad 1965.

Vibration relative amplitudes

Trailing edge	Author		Trailing edge	Author	
	Donaldson	Heskestad		Donaldson	Heskestad
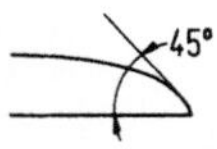 45°		0		(43)	31
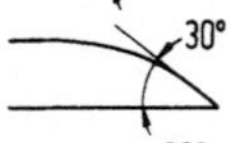 30°		0	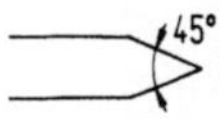 45°		43
 30°		0	 60°	48	
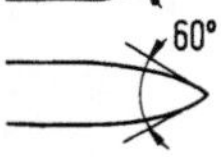 60°	# 0			100	100
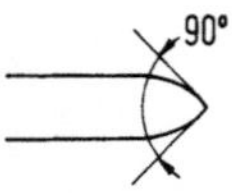 90°	# 0		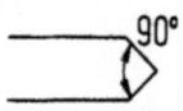 90°	230	190
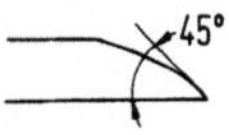 45°	# 0	3		260	
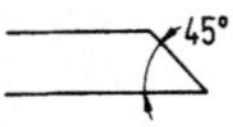 45°	20	38	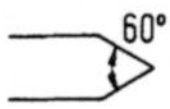 60°	360	380
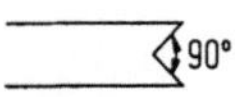 90°	22				

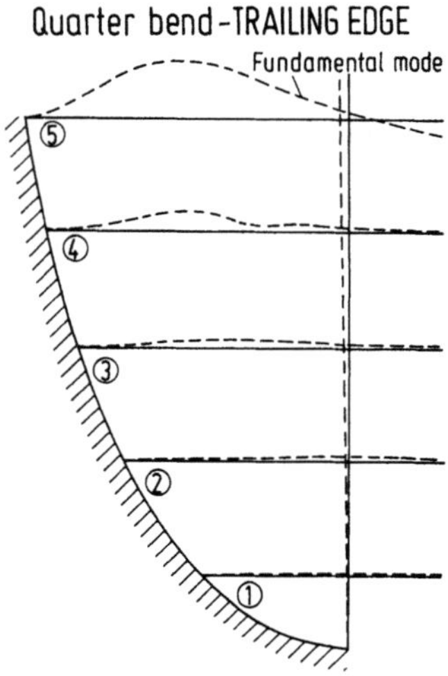
Quarter bend-TRAILING EDGE
Fundamental mode
⑤
④
③
②
①

Fig. 1

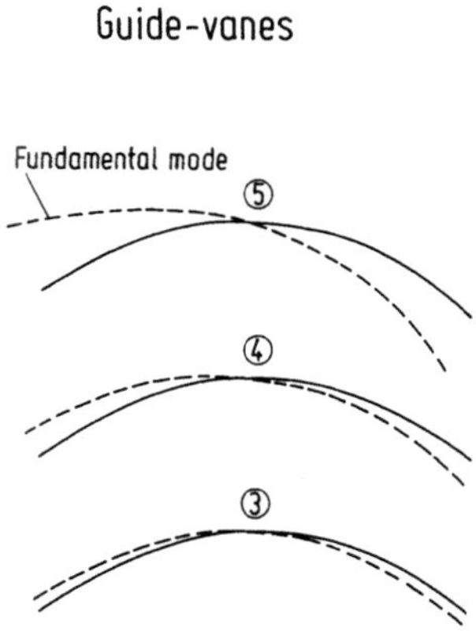
Guide-vanes
Fundamental mode
⑤
④
③

Fig. 2

On the Hydroelastic Vibration of a Swing Check Valve

By

D. S. Weaver
Hamilton, Ontario,
Canada

N. Kouwen
Waterloo, Ontario,
Canada

W. M. Mansour
Waterloo, Ontario,
Canada

Introduction

This paper examines the behavior of a swing check valve which was found to vibrate violently when the pump was rapidly shut down. Slamming of the valve disc on its seat under such circumstances is commonplace with many check valves. However, as will be seen, the problem here is hydroelastic in nature - under certain conditions, the elastic and inertia forces of the valve are coupled with the hydrodynamic forces in such a way that energy is transferred from the flow to perpetuate the motion.

Attempts to alleviate the slamming by the addition of a spring damper assembly, as shown schematically in Fig. 1, only aggravated the problem. With no damping several oscillations at a well defined frequency occurred. As the damping was increased the number of oscillations, as well as their amplitude, increased while their frequency decreased. For sufficiently high damping a stable limit cycle oscillation developed which would continue, if allowed, until failure of the valve occurred.

Experimental Procedure

The apparatus was set up as shown in Fig. 2. The valve was turned around so that the flow was backwards, simulating closure. The disc was held fixed by an arm pinned to an extension of the disc pivot shaft. A strain gauge bridge was arranged on this arm and calibrated so that the hydrodynamic torque of the disc about its pivot could be measured. On the other end of the pivot shaft an indicator for determining angle of closure was arranged.

A large number of tests were carried out determining torque, pressure drop across the valve, ΔH, and discharge for a wide range of upstream pressures and angles of closure.

For practical reasons, the quantitative results reported here are for a six inch valve although the same phenomenon occurs in larger valves of the same design. A section through the valve is shown in Fig.3 to assist in interpretation of the results.

Results and Discussion

The ratio of velocity head $V^2/2g$, to head loss across the valve, ΔH, is plotted against closure angle in Fig.4. The solid line corresponds to velocity calculations based on the area of the pipe whereas the dashed curve is based on the minimum flow area, i.e., the maximum velocity occurs between the disc and its seat for angles less than about 13°.

The curve is flat in zone 2 since the flow is governed by the area between the edge of the disc and the valve casing and this remains constant from about 23° to about 13° as seen in Fig.3. In zone 3, the minimum flow area is diminishing as the valve closes. Hence, for a given head loss ΔH, the discharge drops as does the velocity in the pipe. However, as the dashed curve indicates, the velocity through the slot between the disc and seat increases. At some small angle, between 1 and 3 degrees depending on upstream pressure, the flow is choked and the discharge drops rapidly. It is here that the instability develops.

The ratio of measured torque to torque calculated from the product of pressure drop, disc area and pivot arm is plotted against closure angle in Fig.5. Clearly this calculated torque will not give a good estimate of the actual hydrodynamic torque on the valve. However, its use allows comparison of results for a range of upstream pressures and gives an indication when local flow phenomena significantly alter the hydrodynamic loading.

The curve indicates that these flow phenomena can be discussed in terms of two extremes.

(i) At around 6 degrees, the flow path is relatively tortuous as seen from Fig.3. Hence, the turbulence causes a relatively large head loss across the valve. There is even some indication that the flow is impinging on the downstream face of the disc.

(ii) At a small angle, 1 to 3 degrees, the flow has become quite regular and passes as a high velocity jet through the narrow slot around the disc. It is here that the instability develops, preceded always by a sudden peak in the measured torque.

At relatively low upstream pressures, the instability occurs at about 1/2 to 1 degree. For higher pressures, this angle is larger, up to 3 degrees or more, and the instability is considerably more violent. The data in Figs.4 and 5 for small angles corresponds to low upstream pressures since instability at higher pressures usually led to shearing of the pivot arm pin.

Mechanism of Instability

The above results may now be applied to the quasi-steady case of the valve disc closing slowly, being constrained by a spring-damper. Under conditions of normal closing, the upstream pressure remains constant. The important features of the instability are outlined as follows:

(a) As the disc moves through the range from about 13° to 6° there is a marked increase in the hydrodynamic closing load and an even greater increase in head loss due to turbulence.

(b) For smaller angles the closing load continues to increase as turbulence gives way to a more regular flow through the narrow slot between the disc and its seat.

(c) At some small critical angle, depending on upstream pressure, the flow is choked. The sudden reduction in discharge, however small, causes an instantaneous increase in pressure on the upstream face of the disc followed immediately by a lower pressure as the wave reflects upstream. In addition there is an increase in pressure on the downstream face of the disc due to the breakup of the high velocity parallel flow there.

(d) These combined effects cause an abrupt reduction in hydrodynamic closing load whereupon the disc opens rapidly.

(e) As soon as the disc has opened sufficiently to re-establish the flow, the large hydrodynamic pressure starts to close the valve and the cycle repeats itself.

For little or no damping, the velocity of the disc is high enough that a regular flow at smaller angles is not sufficiently established, random turbulence dominates and the motion is curtailed after a few cycles. As damping and/or spring stiffness is increased, the motion of the disk is slowed so that hydroelastic coupling may develop and, finally, stable limit cycle oscillations occur.

It should be noted that although the pressure waves may be instrumental in starting the opening portion of the cycle, its frequency is at least an order of magnitude higher than the frequency of the valve disc. Also, a sudden drop in the hydrodynamic closing torque rather than a change in sign, is sufficient to cause instability. The latter has been confirmed by preliminary analogue computer studies.

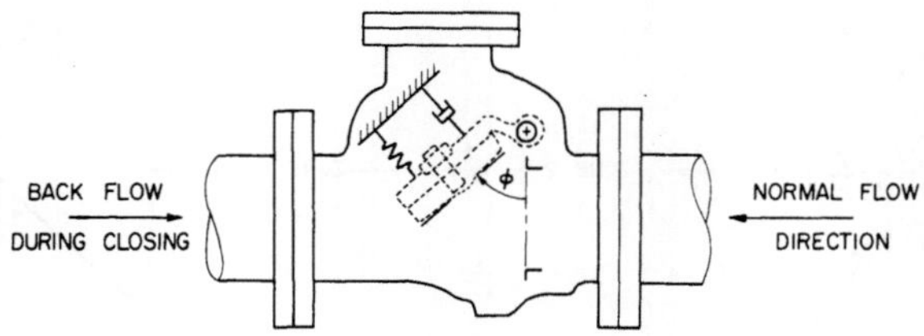

Fig.1

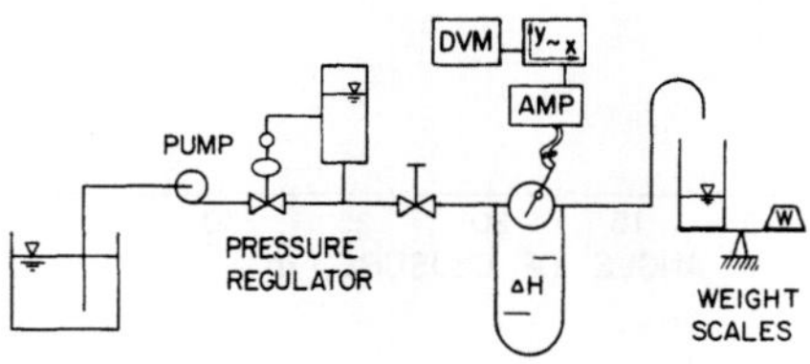

Fig.2

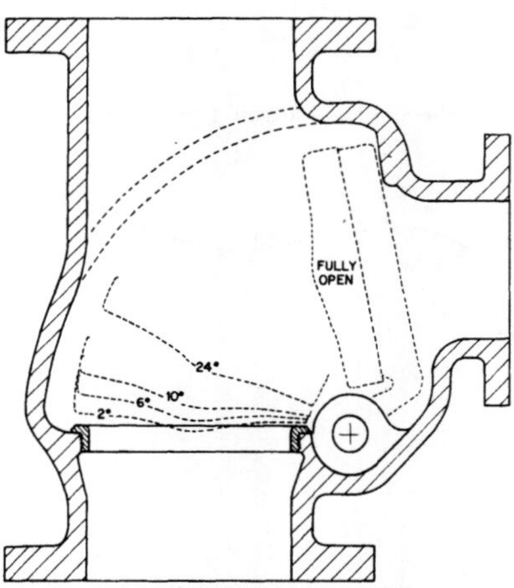

Fig.3

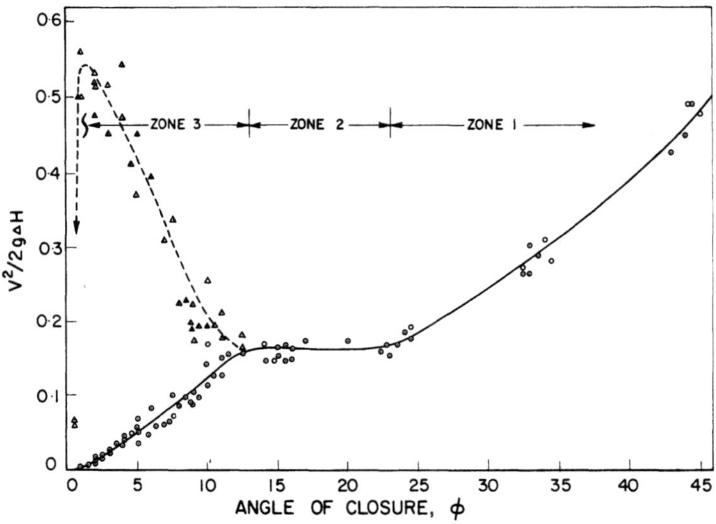
V²/2gΔH
0·6
0·5
0·4
0·3
0·2
0·1
0
ZONE 3
ZONE 2
ZONE 1
0
5
10
15
20
25
30
35
40
45
ANGLE OF CLOSURE, φ

Fig. 4

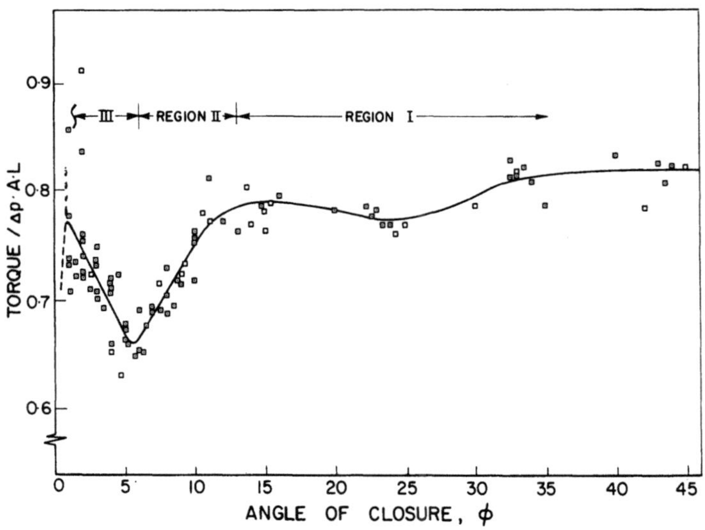
TORQUE / Δp·A·L
0·9
0·8
0·7
0·6
III
REGION II
REGION I
0
5
10
15
20
25
30
35
40
45
ANGLE OF CLOSURE, φ

Fig. 5

TECHNICAL SESSION D

"Flow-Induced Vibrations of Beams and Bridge-Decks"

Chairman: W.D. Baines, Canada
Co-Chairman: W.W. Martin, Canada

CONTRIBUTIONS

Flutter and Aerodynamic Response Considerations for Bluff Objects in a Smooth Flow

By

R.H. Scanlan, K.S. Budlong

Princeton, N.J., U.S.A.

Summary

The paper derives from an experimental study of the flutter stability of suspension bridge decks. Earlier experimental studies have described the smooth-flow flutter stability derivatives applicable to the sinusoidal motion of such decks. The present paper deals with the generalization of this aerodynamic derivative information to the case of general deck motion under variable winds.

Introduction

The "bluff objects" with which the writers have been dealing in the present context are models of suspension bridge deck structures. The primary aim of the aerodynamic tests on such models is, of course, to render the full-scale suspension bridge as safe as possible against undesirable aeroelastic responses or effects under natural wind. Because tests of full suspension bridge models are elaborate and costly, it has become an accepted practice to study spring-mounted section, or two-dimensional, models set between end plates, which are of course much simpler.

An original simple interpretation of the results for such models has been that if they could be rendered aerodynamically stable by adjustment of their geometry then a fair inference of the stability of the full-scale bridge could be made. This remains an important first consideration, and the overall aerodynamic design of important bridges has indeed, in the main, been fixed by such studies. However, some further interesting considerations may ensue.

First, in what sense is the dynamic action of a model truly representative of that of the full bridge? For example, what degrees of freedom and frequencies, and what mechanical damping should be allowed in the model? If model damping is not directly controlled, how should model dynamic response be interpreted relative to full scale? What choice of full bridge modes should a section model represent? Second, the full bridge is enveloped in the natural wind which is characterized by variability and turbulence; how can these important effects be read into the section model situation?

Again, the simple and reasonably reliable answer to the above has been that model stability under laminar flow test conditions with mechanical damping of a few percent (two or three) is probably a conservative assurance of full scale stability in those modes having frequencies corresponding to model frequencies. Insofar as these modes go, one may attempt an explanation of the conservative nature of test results on the basis that turbulence, introducing random forces along the full bridge span as it does, reduces the spatial coherence of induced aerodynamic effects and thus delays to higher mean velocities the occurrence of coherent full-bridge dynamic responses such as flutter. There are quite fair chances that this reasoning is sound if significant bridge frequencies and modes are chosen for modeling and that bridge models, as conventionally tested, can indeed assure reasonable bridge stability under wind. This will be the case particularly if the aerodynamic properties of deck sections can, by design, be rendered so "benign" that only low-level aerodynamic actions occur. Many practical design features and compromises, however, lead to less than ideal deck action under wind, and therefore leave the full and final bridge response question quite open. One is thus led to pursue the matter a bit farther.

One may become interested in attempting the prediction of a variety of full-bridge responses such as sway, buffeting, lateral buckling, flutter, and divergence, all three-dimensional effects which are not necessarily readable from direct tests on section models but which are calculable, given good basic sectional aerodynamic data as starting information. In this way the bridge dynamic response problem parallels quite closely a set of analogous design problems seen in aeronautics.

The present paper starts then from the point of view that the geomtry of the deck section is at least tentatively fixed and that the section model will be employed primarily as a source of purely aerodynamic data, not as a direct predictor of the action of the full bridge per se. The focus of the study will then be the manipulation of the acquired section model aerodynamic data in ways which allow analytic calculation of the responses of interest for the whole bridge.

Some Qualitative Flow Considerations

When a bluff body stands in a lamiar flow, the arriving flow is smooth while the body precipitates the occurrence of a generally turbulent wake. One might term such induced turbulence "signature turbulence" to distinguish it from turbulence due to other sources. Some of the signature turbulence may already appear over the body itself as well. When the same body stands in a turbulent flow the signature turbulence will be more or less affected by the oncoming turbulence. In addition, the cross-wind spatial coherence of the oncoming turbulence will be less than unity from point to point over the body. As a result of these two effects, the body will be subjected to new pressure distributions compared to those developed under laminar flow.

Should the scale of turbulence be large, however, compared to the alongwind dimensions of the body, it may be argued that at least the signature turbulence of the body will not be greatly different from that developed under laminar flow. It may still be reasonable then, under these conditions, to consider that laminar flow section aerodynamic data still hold and that in calculating three-dimensional effects under turbulent wind, all influences of turbulence may be attributed only to local section effects, subject to across-wind spatial coherence changes but not to changes in signature turbulence. This viewpoint will be adopted in the present paper.

When a bluff body offers a large solid obstruction to the wind it is well known that rather coherent vortex shedding can and does occur in its wake. This phenomenon is quite Reynolds-number dependent, and as a result bluff bodies exhibiting this phenomenon are, strictly speaking, not modeled correctly unless Reynolds number can be duplicated, even when other modeling parameters such as Strouhal number are respected. As will be seen in the body of the paper, certain aerodynamic coefficients analogous to airfoil flutter derivatives are experimentally obtained and used for bridge deck sections. Those sections not grossly susceptible to coherent vortex shedding, such as very streamlined sections or open truss-stiffened bridges, are the most favorable for applications of the methods to be developed here, as they permit of reasonable linearization of their responses to wind. The open truss-stiffened deck provides many sharp corners at which the flow may separate from the structure locally. In short, a lot of "vortex shredding" rather than coherent large-scale shedding, takes place. This is thought by the writers to account somewhat more for the possibility of linearization of the response aerodynamics in these cases, as is done in the paper.

It is in fact recognized here at the outset that the detailed motional aerodynamics over the bluff bodies in question cannot be assessed in the strictest sense as linear; however, an overall linear theory appropriate to small motions will be employed nontheless. In this the stability aerodynamics of bridge decks follows in style the classical aerodynamics of the unsteady motion of airfoils. The main difference is that for deck models no theory exists. Experimentally, however, model motional response appears to be quite accurately described by exponentially modified sinusoids; and this central fact permits of a linearization of the equations of motion.

Two methods have then commonly been employed in the field to obtain aerodynamic motional derivatives: one [1] drives the model under wind through controlled sinusoidal displacements and measures the aerodynamic forces which result; the other [2] observes the motion of the elastically supported model under wind and subsequently calculates aerodynamic coefficients appropriate to account for observed motion. There are some subtle points involved in each process which are worthy of detailed discussion, but the latter is beyond the scope of this paper. Results from the method of the freely oscillating model will be used here.

Detailed Aerodynamics

Tomko an Scanlan [2] have employed the free-oscillation test technique to derive aerodynamic coefficients appropriate to bridge section model oscillation. Giving the section model, which is symmetrical in mass and geometry about its roadway centerline, two degrees of freedom: bending (h) and torsion (α) about this centerline, the equations of motion used are

$$m\left[\ddot{h} + 2\zeta_h\omega_h\dot{h} + \omega_h^2 h\right] = L_h,$$
$$I\left[\ddot{\alpha} + 2\zeta_\alpha\omega_\alpha\dot{\alpha} + \omega_\alpha^2\alpha\right] = M_\alpha, \qquad (1)$$

where

m = mass per unit span,
I = mass moment of inertia per unit span,
ζ_h, ζ_α = damping ratios-to-critical,
ω_h, ω_α = natural circular frequencies,

and the fluid forces (of a self-excited nature) are given by

$$L_h = \frac{1}{2}\rho v^2(2b)\left[kH_1^*\frac{\dot{h}}{v} + kH_2^*\frac{b\dot{\alpha}}{v} + k^2H_3^*\alpha\right],$$
$$M_\alpha = \frac{1}{2}\rho v^2(2b^2)\left[kA_1^*\frac{\dot{h}}{v} + kA_2^*\frac{b\dot{\alpha}}{v} + k^2A_3^*\alpha\right], \qquad (2)$$

where $k = b\omega/v$; ω is the oscillation frequency; ρ, v, b, are air density, wind velocity, and deck half-width, respectively; while H_i^*, A_i^* ($i = 1, 2, 3$) are aerodynamic coefficients. The latter are functions of k.

Choosing $B = 2b$ and $N = \omega/2\pi$, H_i^* and A_i^* may be plotted as functions of v/NB as shown in Fig.1, for example. This figure is taken from Scanlan and Tomko [2] for the case of a representative truss-stiffened bridge, the cross section of which is suggested on the figure.

In case the section is being buffeted by random winds, it may be assumed that an additional time-dependent force L(t) and a moment M(t), respectively, contribute to L_h and M_α. In this case it is necessary to identify v as the mean wind velocity $\bar{v}$, so that in this case we may approximate the aerodynamic lift and moment by the expressions

$$L_h = \frac{1}{2}\rho\bar{v}^2(2b)\left[kH_1^*\frac{\dot{h}}{\bar{v}} + kH_2^*\frac{b\dot{\alpha}}{\bar{v}} + k^2H_3^*\alpha\right] + L(t),$$

$$M_\alpha = \frac{1}{2}\rho\bar{v}^2(2b)\left[kA_1^*\frac{\dot{h}}{\bar{v}} + kA_2^*\frac{b\dot{\alpha}}{\bar{v}} + k^2A_3^*\alpha\right] + M(t), \quad (3)$$

where now $k = b\omega/\bar{v}$.

The above formulation assumes that response is in the main exponential sinusoidal, with additional forcing effects caused by the random signals L(t), M(t). This form of the aerodynamic forces was recently suggested by Scanlan [3]. However, it suffers to some extent from the assumption that the motional aerodynamic forces are centrally dependent upon only the mean wind through the parameter k, and it does not take full account of the aerodynamic effects accompanying buffeting motions.

As a result, one seeks a still more adequate theory for depicting bridge response. Before going into the details of this, some remarks will be given on the information expected to be extracted from bridge deck models in the wind tunnel.

Types of Model Test and the Data Expected Therefrom

When a section model between large endplates is available for test in the wind tunnel, it may be expected that at least two types of data may be obtained through its use:

(1) steady lift and moment coefficients versus geometric position angle,

(2) oscillatory aerodynamic coefficients such as H_i^* and A_i^* $(i = 1, 2, 3)$.

In addition, if testing under turbulent conditions is possible it is first necessary to assure that the wind tunnel turbulence spectra and scales correctly simulate their atmospheric counterparts. This is an important consideration dealt with adequately elsewhere [4] but it is indispensable to acquisition of a third necessary type of data, namely:

(3) The terms $L(t)$ and $M(t)$ ascribable to turbulent buffeting as given in Eqs. (3).

For purposes of collecting data on $L(t)$ and $M(t)$ as random force and moment signals, it is necessary that a model be provided which is sensitive to these quantities, and in fact represents only a narrow section of the deck, so that lateral wind coherence problems across the model are not created in the test. A simple concept for accomplishing this is a light narrow model provided with relatively stiff strain gage mounts which are designed to be sensitive to lift $L(t)$ and moment $M(t)$, the narrow model being imbedded within the span of a wider, fixed model of the same deck configuration. This general arrangement serves aerodynamically to ensure the two-dimensionality and "typicalness" of the local section flow over the narrow measuring section. The natural frequencies of the measuring section should be high relative to wind spectral frequencies of interest so as to assure that the model is detuned, i.e. nonresonant. Such a narrow model then serves as the analog converter of random wind velocity components into the random section inputs $L(t)$ and $M(t)$ which must be known for later dynamic analysis.

Altogeter therefore, versions of the section model are counted upon to furnish static, oscillatory, and random force sectional aerodynamic data. At the present time it does not appear possible to dispense with these minimum aerodynamic data acquisitions for the purposes to be described below. The theory to be described here is therefore experimentally based and does not represent derivations from first principles of fluid flow. In what follows, however, considerable guidance and insight is gained by using the theoretical example of the airfoil which itself has been derived from first principles.

Aerodynamic Forces for Arbitrary Motion - General Considerations

In the case of the theoretical airfoil it is well known that the circulatory lift (i.e., the total lift exclusive of quasi-steady and air inertial effects) is calculable for arbitrary airfoil motion through use of a superposition or Duhamel integral. The integrand of this integral is the product of the time derivative of the vertical velocity $w_{3/4}$ of the airfoil 3/4-chord point and the well-known Wagner indicial lift function, the latter being shown in Fig.2. The fact that for the airfoil the entire motion is completely and adequately typified by $w_{3/4}$ is a remarkable fact depending upon circulatory theory for unsteady motion and the satisfaction of the indispensable Kutta condition. Raggett [5] has presented a resonable case for employing $w_{3/4}$ also in the case of those bridge decks which act effectively like plates regarding their aerodynamic stiffness (A_3^*), though not necessarily their aerodynamic damping (A_2^*). Raggett and Scanlan [6] exploited this empirical theory for predicting the bridge torsional flutter coefficient A_2^* in the single (α) degree of freedom. The Raggett-Scanlan theory has the convenience of requiring knowledge of only static aerodynamic derivatives, under the assumption of an empirical Wagner-like function, for the prediction of the aerodynamic coefficients A_2^* and A_3^*. However, it has the drawback of inaccuracy. In the present paper a somewhat different approach will be used - that is, an empirical indicial function will be obtained directly from experimental data.

It is well known in airfoil theory that there is a relation of the nature of Fourier or Laplace transforms between indicial lift (Wagner function) and steady state sinusoidal lift (circulatory) terms. Something of this nature will be considered here, but with an attempt to eliminate too close dependence, for the case of the bridge, upon airfoil-like theoretical considerations. In particular, an attempt will be made to eliminate the postulate that $w_{3/4}$ plays any role, and purely experimental results for aerodynamic coefficients will be retained, as given for example in Fig. 1. Sabzevari [7] has made a study to which the present discussion is closely related, which discusses particularly the transformation of the oscillatory aerodynamic coefficient H_1^*. We shall examine particularly the transformation of A_2^* and A_3^*.

We begin by making some general observations on airfoil theory for unsteady motion. If the angle of attack of an airfoil is $\alpha(t)$ as a function of time and we set $s = vt/b$ as a non-dimensional time, then $\alpha = \alpha(s)$ describes the motion. Let $\Phi_L(s)$ represent the total non-dimensional lift evolution with s after α has been given a step angle of attack. Then for arbitrary $\alpha(s)$ the lift becomes, by the principle of superposition:

$$L = \frac{1}{2}\rho v^2(2b)\frac{dC_L}{d\alpha}\int_0^s \Phi_L(s-\sigma)\frac{d\alpha(s)}{d\sigma}\,d\sigma. \qquad (4)$$

Note that in requiring that $\Phi_L(s)$ be the total non-dimensional indicial lift function (rather than merely the circulatory (Wagner) indicial lift function) $\Phi_L(s)$ will not be identical to the Wagner function. The gain in this formulation is that $w_{3/4}$ is eliminated from consideration. That this important step is actually feasible will be now demonstrated. This will then set the stage for a similar approach to the bridge deck problem.

Aerodynamic Moment for Arbitrary Angle of Attack: Airfoil

For convenience only, attention will now be directed to the aerodynamic moment developed by an airfoil under arbitrary angle of attack change. This is expressed, analogously to Eq. (4), by

$$M_\alpha = \frac{1}{2}\rho v^2(2b^2)\frac{dC_M}{d\alpha}\int_0^s \Phi_M(s-\sigma)\alpha'(\sigma)d\sigma, \qquad (5)$$

where C_M is the static moment coefficient ($dC_M/d\alpha = \pi$ for the airfoil) and $\Phi_M(s)$ is the indicial moment function desired. Now for the case of oscillation $\alpha = \alpha_0 \cos \omega t$; the corresponding moment is [2];

$$M_\alpha = \frac{1}{2}\rho v^2(2b^2)\left[kA_2^*\frac{b\dot{\alpha}}{v} + k^2A_3^*\alpha\right]. \tag{6}$$

Hence, writing $S_M = dC_M/d\alpha$, we have the equality:

$$S_M\int_0^s \Phi_M(s-\sigma)\alpha'(\sigma)d\sigma = kA_2^*\frac{b\dot{\alpha}}{v} + k^2A_3^*\alpha. \tag{7}$$

A means for eveluating Φ_M from physical data is suggested by the closely approximate form earlier employed by Jones [8] for the actual Wagner function, i.e.:

$$\Phi(s) = 1 - 0.165e^{-0.041s} - 0.335e^{-0.320s}. \tag{8}$$

Hence we assume that the desired Φ_M may be expressed in the fairly general form

$$\Phi_M(s) = a + ce^{sd} + fe^{sg}, \tag{9}$$

where a,c,d,f,g are constants to be determined. Setting this form into Eq.(7) and performing the integrations results in the equivalences:

$$\begin{aligned} A_2^* &= -(S_M/k)\left\{\frac{cd}{d^2+k^2} + \frac{fg}{g^2+k^2}\right\}, \\ A_3^* &= (S_M/k^2)\left\{a + \frac{ck^2}{d^2+k^2} + \frac{fk^2}{g^2+k^2}\right\}. \end{aligned} \tag{10}$$

In some applications, analytical or experimental values of A_2^* and A_3^* are plotted as function of v/Nb (see, for example Ref. [2]). Hence, for convenience in using data so presented and letting $1/k = v/(2\pi Nb) = v/(\pi NB)$, we may obtain and use the alternative forms

$$\begin{aligned} A_2^* &= -S_M\varkappa^3\left\{\frac{cd}{1+d^2\varkappa^2} + \frac{fg}{1+g^2\varkappa^2}\right\}, \\ A_3^* &= S_M\varkappa^2\left\{a + \frac{c}{1+d^2\varkappa^2} + \frac{f}{1+g^2\varkappa^2}\right\}. \end{aligned} \tag{11}$$

Since in the zero-frequency limit ($\varkappa \to \infty$), $A_2^*/\varkappa$ should approach the steady result S_M, it may be concluded that here, just as in the Jones approximation, one must have a = 1. This leaves four constants c, d, f, g to be determined.

These are determined by applying the four collocation conditions that three representative points on the curve of A_2^* vs v/Nb and one point on the curve of A_3^* vs v/Nb be matched by the expressions (11). An important incidental fact is that the resulting four conditions constitute four nonlinear equations to be solved for c, d, f, g. In the present case these are conveniently solved by a direct search carried out in a few steps by computer. In the search, one pair of values (d, g) is assumed at a time and the resulting four equations solved in pairs for two pairs of tentative solution values (c, f). For those individual pairs (d, g) giving two pairs (c, f) approaching coincidence, refined search nets of values for (d, g) permit rapid convergence to the final solution set (c, d, f, g). For the airfoil the theoretical value of S_M is π, and the matched indicial moment function, using the process described above, turns out to be

$$\Phi_M(s) = 1 - 0.362\,e^{-0.116\,s} - 0.581\,e^{-1.522\,s}. \qquad (12)$$

Inversely, use of this function Φ_M, placed in Eq. (7), yields A_2^* and A_3^* values which plot as indicated against theoretical values in Fig. 3, constituting a good check. $\Phi_M(s)$ from Eq. (12) is plotted in Fig. 2, on the same graph as the Wagner function. It generally resembles the Wagner function but starts effectively near zero. Physically, one might argue in addition that actually a negatively infinite spike should be postulated conceptually at the start (s = 0) of the indicial lift function, this being consistent with an initially infinite negative moment associated with the snapping of the airfoil instantly to a fixed angle of attack.

Aerodynamic Moment for Arbitrary Angle of Attack: Bridge Deck Section

Having illustrated the feasibility of the development of an indicial moment function for the theoretical airfoil case for use in a superposition integrand which is wholly independent of $w_{3/4}$, we now may apply the same technique to a representative bridge deck. Experimental coefficients A_2^* and A_3^* are given for such a deck in Fig. 1. Again we postulate a functional form for $\Phi_M(s)$ as given in Eq. (9). The resulting expressions (11) are now matched to the A_2^* and A_3^* curves of Fig. 1. Note that in the example chosen A_2^* departs exceptionally strongly (though typically [2]) from the airfoil trend, in fact exhibiting a reversal in sign. Using the curve-matching technique already described above for the case of the airfoil, we now abtain the following indicial moment function for the particular bridge under examination:

$$\Phi_M(s) = 1 + 5.41\,e^{-0.71\,s} - 10.19\,e^{-2.4\,s}, \tag{13}$$

which is the third curve plotted in Fig. 2. It is observed that in this more physical case the indicial function does indeed begin with a fast-rising negative part. It then "overshoots" its steady state value, becoming asymptotic to the latter from above as $s \to \infty$. Such an indicial moment function is sharply different from the Wagner function and in fact is very characteristic of proclivities toward single-degree-of-freedom torsional instability, as was pointed out by Raggett [5]. We further note that in this case the slope of the model moment coefficient for small angles of attack is $S_M = 0.8$. Using these results, leading again to expressions (11), we may, as a check, reevaluate A_2^* and A_3^* as indicated by the broken curves in Fig. 1.

Section Aerodynamic Expressions for General Bridge Motion

Assuming that analogous indicial functions Φ_L can be obtained for lift (see Ref. [7]) as well as moment on the basis described above, (or on any other basis), we now have assembled the necessary information to write the complete section lift and moment expressions for bridge response involving h and α freedoms:

$$\begin{aligned} L_h &= \frac{1}{2}\rho v^2 (2b) S_L \int_0^s \Phi_L(s-\sigma)\left[\frac{h'(\sigma)}{b} + \alpha(\sigma)\right]' d\sigma + L(t), \\ M_\alpha &= \frac{1}{2}\rho v^2 (2b^2) S_M \int_0^s \Phi_M(s-\sigma)\left[\frac{h'(\sigma)}{b} + \alpha(\sigma)\right]' d\sigma + M(t). \end{aligned} \tag{14}$$

These represent strip or section values. To proceed to full bridge response one may proceed by the general modal methods mentioned in Ref. [9] or [3]. Essentially these require, first, the expression of h and α as

$$h = \sum_{i=1}^{\infty} \varphi_i(x) h_i(t),$$
$$\alpha = \sum_{i=1}^{\infty} \psi_i(x) \alpha_i(t), \tag{15}$$

and insertion of these results into the equations of motion. Subsequent multiplication of each equation by φ_i (or ψ_i, as appropriate) and integration over the span L yields a theoretically infinite set of equations of motion $(i = 1, 2, \ldots)$ with generalized forces defined by

$$\int_0^L L_h \varphi_i(x) dx = L_i,$$
$$\int_0^L M_\alpha \psi_i(x) dx = M_i. \tag{16}$$

By delimiting the actual number of modes to a finite set of modes of interest the three-dimensional bridge response problem is formulated in real time. Subsequent studies concerned with input and response spectra, etc. can be started from this point.

We note that the problem as formulated is particularly suited to analog computer treatment, and the buffeting, flutter, and divergence problems are contained in it in a single unified formulation.

In the present context, direct Fourier transformation of the experimental curves A_2^* and A_3^* was not attempted due to lack of precise knowledge as to their limiting behavior as $\varkappa$ approaches infinity. If this shortcoming can be overcome, direct numerical Fourier transformation may be preferable to the method employed here.

Acknowledgement

The writers are grateful to Norman J. Sollenberger of the Department of Civil and Geological Engineering, Princeton University, for assistance in the computer methods associated with the determination of the constants in Eq. (11).

References

1. Tanaka, H., Ito, M.: Trans. Japan Soc. civ. Engs. 1, Part 2, 209 - 226 (1969).

2. Scanlan, R.H., Tomko, J.J.: J. Eng. Mech. Div., Proc. ASCE, 17 - 1737 (December 1971).

3. Scanlan, R.H.: Proc. Third Int. Conf. on Wind Loads on Buildings and Structures, Tokyo, October 6 - 10, 1971, paper lV - 29.

4. Davenport, A.G., Isyumov, N.: Proc. Sympos. on Wind Effects, Ottawa, Canada, September 1967, pp. 201 - 230.

5. Raggett, J.D.: Doctoral Dissertation, Princeton University, October 1970.

6. Raggett, J.D., Scanlan, R.H.: Torsional Flutter Instability of Stalled Surfaces. Preprint 1367, ASCE National Structural Meeting, Baltimore, Md., April 1971.

7. Sabzevari, A.: Proc. Third Int. Conf. on Wind Effects on Buildings and Structures, Tokyo, October 1971, Paper lV - 37.

8. Jones, W.P.: British Aero. Res. Council R and M 2117 (1945).

9. Davenport, A.G.: I. Struct. Div., Proc. ASCE, 233 - 264 (June 1962).

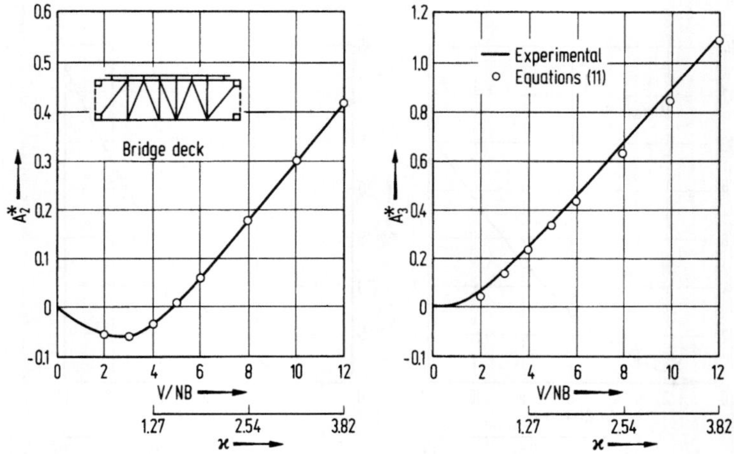
Bridge deck
Experimental
Equations (11)
A_2^*
A_3^*
V/NB
1.27
2.54
3.82
$\varkappa$

Fig.1

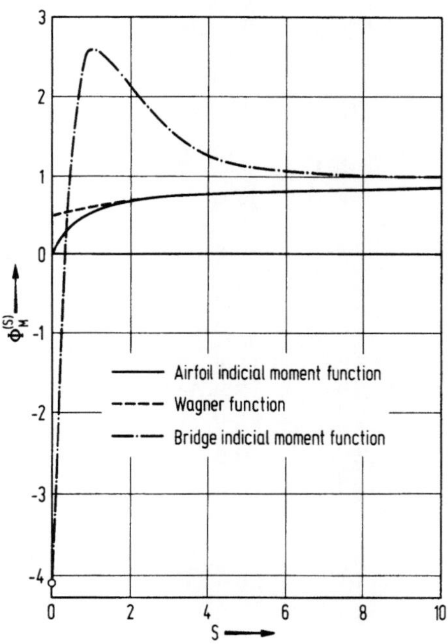
Airfoil indicial moment function
Wagner function
Bridge indicial moment function
$\Phi_M^{(S)}$
S

Fig.2

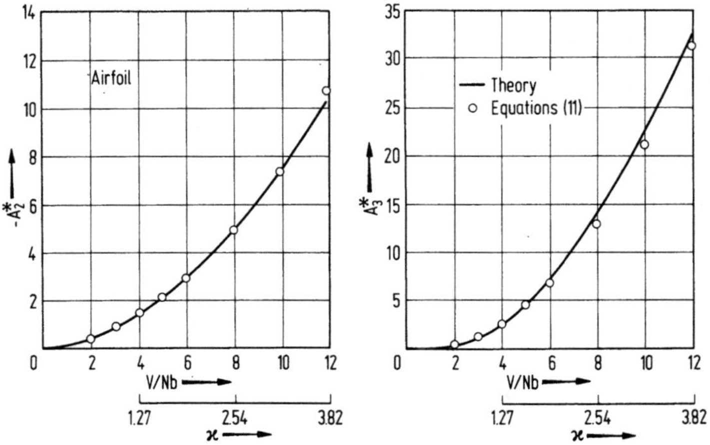

Fig.3

Quasi-Steady Analysis of Torsional Aeroelastic Oscillators

By

V. J. Modi, J. E. Slater

Vancouver, B. C., Canada

Summary

The galloping, aeroelastic instability of bluff cylinders in torsion is analyzed using a quasi-steady approach. The analysis indicates the presence of various singularities which are generally centers or foci. Furthermore, there exist limit cycles, the stability of which can be determined from the Liapunov criterion. The amplitude of the galloping motion increases with wind velocity accompanied by a slight shift in oscillation frequency. The experimental results on a structural angle substantiate the analysis.

Introduction

The oscillation of aerodynamically bluff bodies, when exposed to a fluid stream, has been a subject of considerable study. In general, the oscillation may be induced by vortex resonance or geometric-aerodynamic instability called galloping with the vibration occurring, predominantly, in one of the two degrees of freedom; flexure transverse to the flow direction, or torsion about the longitudinal elastic axis.

The distinct character of vortex excited and galloping oscillations should be emphasized. The former is essentially a resonance phenomenon where the vortex formation frequency, and hence the fundamental frequency of the forcing function, coincides with the natural frequency of the system under consideration. This type of oscillation is referred to as forced vibration since the sustaining alternating force exists independent of the motion. Numerous theoretical and experimental studies have been reported by investigators such as Strouhal, Benard, von Kármán, Roshko, Kovasnay, Rosenhead and others. Marris [1] has presented an excellent review of this literature.

The second form of instability, referred to as galloping, represents a type of self-excited vibration. Geometrically bluff sections may exhibit a galloping type of oscillation because of the nature of the aerodynamic forces or moments. If the aerodynamic loading shows an increase with model attitude and the amount of resulting energy input by the fluid force exceeds that dissipated through various forms of damping, then oscillations will continue to grow until a net energy balance is established. In the ananlysis of such a motion, a quasi-steady approach is often adopted with instantaneous forcing function on the oscillating body replaced by its steady value on the stationary model oriented at the same apparent angle of attack. This assumption implies that no aerodynamic hystersis effects exist in the force or moment characteristics and the vortex shedding frequency is far removed from the cylinder freqency.

From the nature of the problem, the lateral force or twisting moment turn out to be highly nonlinear functions. However, certain simplifications are possible in the aeroelastic problem since:

(1) the ratio of air densiy to model density is small and consequently, the aerodynamic force is small compared with the elastic and inertial forces of the system;

(2) the frequency of oscillation is close to the natural frequency;

(3) the equilibrium motion is nearly sinusoidal.

Under these assuptions the problem becomes quasi-linear which can be solved by various analytical techniques.

The mechanism of galloping excitation of bluff cylinders was probably first described by Den Hartog together with the determination of his plunging stability criterion. Originally, however, it was Lord Rayleigh who indicated the inadequacy of linear theory and proposed a nonlinear equation to explain the "sustained" oscillations. Van der Pol 's development of his classical nonlinear equation in 1920 led to a flurry of research activity on this subject by Appleton, Greaves and others. Excellent reviews of these developments are summarized by van der Pol and Le Corbeiller, with many recent references found in Minorsky [2]. Wardlaw extended Den Hartog 's analysis by developing a generalized stability criterion for coupled torsional-flexural motion. For the plunging degree of freedom, the quasi-steady theory has been established by Scruton [3], Parkinson et al. [4 to 6], and Novak. Sisto and Ii, considering the problem of stall-flutter, incorporated the downwash condition at the three-quarter point to simplify the analysis. The galloping oscillations of existing structures, mainly of transmission conductor lines, were observed and studied by Scruton, Richardson, et al. , Cheers, Dryden and Hill and others. Davenport suggested the possibility of galloping instability of tall thin buildings which may be constructed in the near future. Dynamic amplitude measurements of two-dimensional bluff cylinders of various cross sections have been reported by Brooks. Both Smith and Santosham concentrated their studies to aeroelastic galloping of rectangular cylinders in a plunging degree of freedom; while Chaun and Otsuki have presented results on torsional oscillations of airfoils and prismatic bars, respectively. Parkinson [7] and Slater [8] have discussed the aeroelastic behaviour of bluff bodies and presented useful reviews of the literature.

The lack of a suitable thwory for studying torsional galloping of bluff cylinders is apparent. The paper aims at filling this gap through a modified theory following the plunging quasi-steady approach. However, the torsional anlysis is somewhat complicated through the nonlinear aerodynamic moment which is now a function of the instantaneous angular position as well as the velocity.

Formulation of the Problem

Consider an arbitrarily shaped two-dimensional bluff cylinder with linear spring and viscous damping and restrained to a torsional degree of freedom about the longitudinal elastic axis (Fig. 1). Let the system be subjected to an external fluid moment M_ϑ. The governing equation of motion for the sytem can be written as (for list of symbols see below):

$$I\overset{\circ\circ}{\Theta} + r_\theta\overset{\circ}{\Theta} + k_\theta\Theta = \frac{1}{2}\rho V^2 h^2 l C_{M_\theta}(\Theta, \overset{\circ}{\Theta}). \tag{1}$$

Following the quasi-steady approach, established by Parkinson and associates [4 to 6], for the plunging degree of freedom, the instantaneous excitation moment M_θ can be related to the steady moment M, by

$$C_{M_\theta} = -\left(V_{rel.}/V\right)^2 C_M(\alpha), \tag{2}$$

where

$$C_M = M/\left(\frac{1}{2}\rho V_{rel.}^2 h^2 l\right), \tag{3}$$

$$\alpha = \alpha_0 - \Theta + \gamma_r. \tag{4}$$

It is apparent that V_{rel} is different from V, both in magnitude and direction, because of the angular velocity $\dot{\Theta}$. Furthermore, each surface element on the contour experiences a different relative velocity governed by its position with respect to the center of rotation. It is assumed that an effective relative velocity can be written as

$$V_{rel.} = \overset{\circ}{\Theta} r_r e^{i\eta_r}, \tag{5}$$

where r_r is the effective radius of rotation and η_r is the direction of V_r. For different angles of attack, it is convenient to obtain the η_r from the relation

$$\eta_r = \eta_{r_0} - (\alpha_0 - \Theta), \tag{6}$$

where η_{r_0} is the value η_r at $\alpha_0 = 0^\circ$ for the stationary system. From the velocity vector diagram shown in Fig. 1, the relative velocity can be written as

$$(V_{rel.}/V)^2 = 1 - 2\frac{\overset{\circ}{\Theta}}{V} r_r \cos\eta_r + \left(\frac{\overset{\circ}{\Theta} r_r}{V}\right)^2 \tag{7}$$

and the representative angle γ_r becomes

$$\gamma_r = \tan^{-1}\{(-\mathring{\Theta} r_r \sin \eta_r)/(V - \mathring{\Theta} r_r \cos \eta_r)\}. \tag{8}$$

Application of this approach to a flat airfoil at zero angle of attack gives an effective downwash velocity at the three-quarter chord point which is in agreement with the value adopted in the stall flutter analysis by Sisto and Ii.

The governing equation of motion (1) can be non-dimensionalized and reduced to the form

$$\ddot{\Theta} + \Theta = \mu_\theta f_\theta(\Theta, \dot{\Theta}), \tag{9}$$

where

$$\left.\begin{aligned} \mu_\theta &= n_\theta a_1 \\ f_\theta(\Theta, \dot{\Theta}) &= \mu_0 \dot{\Theta} + \frac{U^2}{a_1} C_{M_\theta}(\Theta, \dot{\Theta}). \end{aligned}\right\} \tag{10}$$

Correspondingly, the auxiliary expressions transform to

$$\left.\begin{aligned} C_{M_\theta} &= -(U_{rel}/U)^2 C_M(\alpha), \\ (U_{rel}/U)^2 &= 1 - 2(\dot{\Theta} R_r/U)\cos\eta_r + (\dot{\Theta} R_r/U)^2, \\ \gamma_r &= \tan^{-1}\left\{\left(-\frac{\dot{\Theta} R_r}{U}\sin\eta_r\right)\Big/\left(1 - \frac{\dot{\Theta} R_r}{U}\cos\eta_r\right)\right\}. \end{aligned}\right\} \tag{11}$$

Based on the aforementioned assumption for a system in air, $\mu_\theta \ll 1$; therefore Eq. (9) is a quasi-linear differential equation of the autonomous type.

Singularities and Stability in the Small

Analytically, the condition of stability of the motion in the small can be determined [9] by investigating the singularities in the phase plane. For the torsional system, the governing equation (9) becomes

$$\left.\begin{aligned} \dot{\Theta} &= z, \\ \dot{z} &= -\Theta + \mu_\theta f_\theta(\Theta, z), \end{aligned}\right\} \tag{12}$$

giving

$$\frac{dz}{d\Theta} = (-\Theta + \mu_\theta f_\theta(\Theta, z))/z. \tag{13}$$

The singularities of Eq. (13) are located on the Θ axis at the points given by the zero of the function $f_\theta(\Theta, 0)$. For aeroelastic galloping, the origin (i.e., $\Theta = \dot{\Theta} = 0$) generally represents one of the singularities. The remaining singular points, being relatively large in Θ, do not affect the condition of stability unless the torsional disturbances are severe.

The nature of the singular point and the phase plane trajectories in its vicinity can be studied through a linear analysis of the system equation. This would involve the reduction of Eq. (12) to linear form

$$\dot{\Theta} = z, \quad \dot{z} = -(1 - \mu_\theta a)\Theta + \mu_\theta bz, \tag{14}$$

where a and b are coefficients of the linear terms of the function $f_\theta(\Theta, \dot{\Theta})$. Note that the Θ term does not appear in the expression for $\dot{\Theta}$. The characteristic roots are, therefore, given by

$$\lambda_1, \lambda_2 = \frac{\mu_\theta b}{2} \mp \left\{ \left(\frac{\mu_\theta b}{2} \right)^2 - (1 - \mu_\theta a) \right\}^{1/2}. \tag{15}$$

Since $\mu_\theta \ll 1$, the roots are complex conjugate indicating the singularity at the origin to be either a centre or a focus depending on the magnitude of the real part. For the case of $b = 0$, the singular point is a centre. This value represents the critical condition of the instability. All other values of b give rise to a focus, the stability of which is determined by the criterion

$$\left.\begin{aligned} b &< 0 \quad \text{stable}, \\ b &> 0 \quad \text{unstable}. \end{aligned}\right\} \tag{16}$$

Thus, since the magnitude of the coefficient b is given by the nature of the function f_θ in Eq. (10), pertinent information about the critical oscillatory conditions can be obtained by examining the moment distribution.

Limit Cycles and Build-Up Time

For this system, since the oscillation amplitude varies slowly with time, it is convenient to use the method of variation of parameter for studying the galloping motion. The generating solution, when $\mu_\theta = 0$, is of the form

$$\Theta = \bar{\Theta}\sin(\tau + \Phi). \tag{17}$$

Considering

$$z = \bar{\Theta}\cos(\tau + \Phi), \quad \Psi = \tau + \Phi, \tag{18}$$

and assuming $\bar{y}$ and Φ to be functions of time τ, the first order solution when $\mu_\theta \ll 1$ can be determined by reducing Eq. (12) to

$$\left.\begin{aligned} \dot{\bar{\Theta}} &= \mu_\theta f_\theta[\bar{\Theta}\sin\Psi, \bar{\Theta}\cos\Psi]\cos\Psi, \\ \dot{\Phi} &= (-\mu_\theta/\bar{\Theta}) f_\theta[\bar{\Theta}\sin\Psi, \bar{\Theta}\cos\Psi]\sin\Psi. \end{aligned}\right\} \tag{19}$$

Since $\dot{\bar{\Theta}}$ and $\dot{\Phi}$ are proportional to μ_θ, $\bar{\Theta}$ and Φ represent slowly varying functions of time. Hence, they can be considered, approximately, constant during one cycle. Eqs. (19) can then be replaced by their average values and written in the form

$$\dot{\bar{\Theta}} = -\bar{\Theta}\,\delta_\theta(\bar{\Theta}), \tag{20a}$$

$$\dot{\Phi} = -K_\theta(\bar{\Theta}), \tag{20b}$$

where

$$\delta_\theta = -\frac{\mu_\theta}{2\pi\bar{\Theta}} \int_0^{2\pi} f_\theta[\bar{\Theta}\sin\Psi, \bar{\Theta}\cos\Psi]\cos\Psi d\Psi, \tag{21a}$$

$$K_\theta = \frac{\mu_\theta}{2\pi\bar{\Theta}} \int_0^{2\pi} f_\theta[\bar{\Theta}\sin\Psi, \bar{\Theta}\cos\Psi]\sin\Psi d\Psi. \tag{21b}$$

The amplitude of the limit cycle can be obtained from Eq.(20a) using the condition $\dot{\bar{\Theta}} = 0$ and evaluating for the real positive roots, $\bar{\Theta}_j$, of the algebraic equation

$$\delta_\theta(\bar{\Theta}) = 0, \tag{22}$$

The stability of the sustained motion can be determined by Lyapunov's stability criterion

$$\left.\frac{\partial\dot{\bar{\Theta}}}{\partial\bar{\Theta}}\right|_{\bar{\Theta} = \bar{\Theta}_j} \begin{cases} < 0 & \text{stable,} \\ > 0 & \text{unstable,} \end{cases} \tag{23}$$

or in the light of Eq.(20a)

$$\left.\frac{\partial\delta_\theta}{\partial\bar{\Theta}}\right|_{\bar{\Theta} = \bar{\Theta}_j} \begin{cases} > 0 & \text{stable,} \\ < 0 & \text{unstable.} \end{cases} \tag{24}$$

On determining the limit cycle amplitude, Φ can be evaluated using Eq.(20b) as

$$\Phi = -K_\theta\tau + \Phi_0, \tag{25}$$

where Φ_0 is the constant of integration. Thus, the steady-state oscillation is given by

$$\Theta = \bar{\Theta}\sin\Psi,$$

where

$$\Psi = \tau + \Phi = (1 - K_\theta)\tau + \Phi_0. \tag{26}$$

From Eq.(26) it appears that the frequency of oscillation is reduced from the natural frequency by the amount K_θ. Therefore, $(1 - K_\theta)$ can be denoted as the reduced frequency parameter.

The time of amplitude build-up is readily obtained from Eq.(20a) by evaluating the integral

$$\tau = \int_{\bar{\Theta}_0}^{\hat{\Theta}_j} - \frac{d\bar{\Theta}}{\bar{\Theta}\delta_\theta(\bar{\Theta})}, \tag{27}$$

where Θ_0 represents a small initial displacement and $\hat{\Theta}_j$ is some prescribed fraction (say 95%) of the limit cycle amplitude, $\bar{\Theta}_j$.

Illustrative Example

For various orientations of a 1 in. and 3 in. angle section models with a variety of damping and mounting configurations, no galloping instability in the torsional degree of freedom could be induced either from rest or with large initial amplitude at any wind speed below approximately 70 ft/sec. The absence of the galloping oscillations can be explained using the quasi-steady analysis. For example, the relevant portion of the steady aerodynamic moment distribution as obtained from stationary model balance test is plotted in Fig. 2 for $\alpha_0 = 45°, \eta_{r_0} = 5\pi/4$. Evaluation of the moment coefficient for the oscillating system using Eq. (11) gives variation of $C_{M_\theta}(\Theta, \dot{\Theta})$ as shown in Fig. 3. The representative curves are lines of constant C_{M_θ} which appear nearly linear and parallel for this range of Θ. As a result, for simplicity, a new Cartesian coordinate system, ξ and ζ, is established with the ζ-axis along the contour line $C_{M_\theta} = 0$ which is at an angle λ to the $\frac{\dot{\Theta}}{U}$ axis. Therefore, to a first approximation, C_{M_θ} can be expressed as a function of ξ which is related to Θ and $\dot{\Theta}$ by the coordinate transformation

$$\xi = \left[\frac{\dot{\Theta}}{U}s - \Theta c\right], \tag{28}$$

where $s = \sin\lambda$,

and $c = \cos\lambda$.

Let the aerodynamic moment coefficient be expressed as a polynomial of the form

$$C_{M_\theta}(\xi) = a_1\xi + a_2\xi^2 + a_3\xi^3 + \dots + a_N\xi^N. \qquad (29)$$

The degree of the polynomial used is based on a least square error criterion. It should be emphasized that the simplification introduced by taking C_{M_θ} as a function of ξ is not imperative for the analysis. For the angle section at $\alpha_0 = 45°$, the resulting distribution of $C_{M_\theta}(\xi)$ is plotted in Fig. 4 and approximated by a 17th degree polynomial. Only the positive values of ξ are shown since the moment is symmetric about the origin for this angle of attack. Therefore, substituting Eq. (29) into Eq. (10), $f_\theta(\Theta,\dot{\Theta})$ becomes

$$f_\theta(\Theta,\dot{\Theta}) = U^2\left\{\left(\frac{U_s - U_0}{U^2}\right)\dot{\Theta} - c\dot{\Theta} + \frac{a_2}{a_1}\left(\frac{\dot{\Theta}}{U}s - \Theta c\right)^2 \right.$$
$$\left. + \frac{a_3}{a_1}\left(\frac{\dot{\Theta}s}{U} - \Theta c\right)^3 + \dots + \frac{a_N}{a_1}\left(\frac{\dot{\Theta}s}{U} - \Theta c\right)^N\right\}. \qquad (30)$$

From the singularity analysis, the origin of the phase plane is a singular point with the coefficient

$$b = U_s - U_0. \qquad (31)$$

In particular, when the singular point is a center, $U = U_0/s$ represents the critical wind velocity for onset of the galloping instability. For all other values of U, stability is determined from Eq. (16).

For the evaluation of the sustained motion, δ_θ and K_θ can be determined by substituting Eq. (30) into Eq. (21) and evaluating the integrals. The expressions become

$$\delta_\theta(\bar{\Theta}) = -\frac{\mu_\theta}{2} U^2 \left[\frac{U_s - U_0}{U^2} \right] + \frac{a_3}{a_1} \left\{ \left(\frac{s}{\bar{U}}\right)^3 b_{3,1} + \left(\frac{s}{\bar{U}}\right) c^2 b_{3,3} \right\} \bar{\Theta}^2$$

$$+ \ldots + \frac{a_i}{a_1} \left\{ \left(\frac{s}{\bar{U}}\right)^i b_{i,1} + \left(\frac{s}{\bar{U}}\right)^{i-2} c^2 b_{i,3} + \ldots + \left(\frac{s}{\bar{U}}\right)^{i-(j-1)} c^{j-1} b_{i,j} \qquad (32)$$

$$+ \ldots + \left(\frac{s}{\bar{U}}\right) c^{i-1} b_{i,i} \right\} \bar{\Theta}^{i-1} + \ldots \Big],$$

$$K_\theta(\bar{\Theta}) = -\frac{\mu_\theta}{2} U^2 \left[c + \frac{a_3}{a_1} \left\{ \left(\frac{s}{\bar{U}}\right)^2 c t_{3,2} + c^3 t_{3,4} \right\} \bar{\Theta}^2 \right.$$

$$+ \ldots + \frac{a_i}{a_1} \left\{ \left(\frac{s}{\bar{U}}\right)^{i-1} c t_{i,2} + \left(\frac{s}{\bar{U}}\right)^{i-3} t_{i,4} + \ldots \left(\frac{s}{\bar{U}}\right)^{i-j} c^j t_{i,j+1} \qquad (33)$$

$$+ \ldots + c^i t_{i,i+1} \right\} \bar{\Theta}^{i-1} + \ldots \Big],$$

where $\quad i = 3, 5, 7, \ldots\ldots, N; \quad j = 1, 3, 5, \ldots\ldots, i.$

Note, the existence of a nonzero K_θ expression is in contrast to the plunging case analysed by Parkinson et al. [9 to 11]. The coefficients $b_{i,j}$ and $t_{i,j+1}$ are given by

$$b_{i,j} = c_{i,j} d_{i,j}, \quad t_{i,j+1} s_{i,j+1}. \qquad (34)$$

Here the $c_{i,j}$ are the binomial coefficients of the terms in Eq. (30), and $d_{i,j+1}$ and $s_{i,j+1}$ are constants obtained from the integration of Eq. (21). These coefficients can be determined from the following expressions:

$$c_{i,1} = c_{i,i+1} = d_{i,1} = g_{i,1} = e_{i,i} = 1,$$

$$c_{i,k+1} = [(i-(k-1))/k]c_{i,k},$$

$$g_{i,j+2} = [j/((i+1)-(j-1))]g_{i,j},$$

$$e_{i,i-j-1} = [(j+2)/(j+3)]e_{i,i-j+1},$$

$$d_{i,j+2} = g_{i,j+2}\, e_{i,j},$$

$$s_{i,j+1} = d_{i,i-j+1},$$

where

$$i = 1,3,5,..,N;\ j = 1,3,5,...,i;\ k = 1,2,3,..,i-1. \qquad (35)$$

Using the coefficients of the $C_{M_\theta}(\xi)$ polynomial in the expressions for δ_θ and K_θ, together with Eqs. (22) and (24), the displacement and reduced frequency values for a range of U can be determined. Figure 5 summarizes the results for a 3 inch angle model at $\alpha_0 = -45°$ with typical values of the damping and inertia parameters. The amplitude curve consists of an initial unstable limit cycle region (indicated by the dotted portion) followed by a stable limit cycle motion. The prediction of a positive starting velocity, U_0/s, by the galloping theory indicates that the angle section at this orientation possess a soft oscillator characteristic, since when $U > U_0/s$, the origin is an unstable focus and, therefore, the motion will build up from rest in the presence of any small disturbance. Furthermore, as can be seen from the plot as well as from the expression for U_0, the galloping instability shifts to higher velocities with increased damping or reduced inertia parameter. However, as suggested by the curves or by direct examination of the equation for δ_θ, the torsional galloping system does not collapse to a reduced or universal form which is independent of the system parameters, η_θ and β_θ, as in the plunging case.

From Fig. 5, further information concerning the nature of the galloping motion in torsion can be obtained. The reduced frequency parameter $(1-K_\theta)$ illustrated that the frequency of oscillation is slightly higher than the natural frequency. This frequency shift increases with wind-velocity and is directly proportional to the inertial parameter n_θ as indicated by the expression for K_θ. For case 1 or 3, the relatively large value of starting velocity, $U_0/s = 7.76$, would correspond to a wind velocity of approximately 200 ft/sec in the practical situation. Therefore, the quasi-steady analysis indicates that galloping in the torsional degree of freedom will occur only at very high wind speeds or for systems of very low damping, which are outside the range of practical interest.

Concluding Remarks

The quasi-steady approach can be used to study torsional galloping of bluff cylinders. The problem is complicated by the dependence of the nonlinear aerodynamic moment function of the instantaneous angular position as well as the angular velocity, however, linear and parallel moment contour lines introduce useful simplification. The analysis shows the presence of singularities and various stable and unstable limit cycle motion. In general, for the aeroelastic problem, there is one predominant singular point at the origin of the phase plane with the existence of a main stable limit cycle, the amplitude of which increases with wind velocity. The galloping motion in the torsional degree of freedom is accompanied by a slight shift in oscillation frequency.

List of Symbols

C_M — measured steady moment coefficient, $M/(\rho V_{rel}^2 h^2 l/2)$

C_{M_θ} — moment coefficient for torsional system, $M_\theta/(\rho V^2 h^2 l/2)$

I — mass moment of inertia about elastic axis

K_θ — reduced frequency parameter

M — measured steady moment

M_θ — moment for torsional system

N	degree of polynomial curve fit
R_r	dimensionless representative radius parameter, r_r/h
U	dimensionless fluid velocity, $V/\omega_{n_\theta} h$
U_0	dimensionless critical velocity, $2\beta_\theta/n_\theta a_1$
V	fluid velocity relative to stationary system
V_{rel}	fluid velocity relative to oscillating system
a_i	$(i = 0,1,2,3,\ldots,N)$ coefficient of polynomial curve fit
c	$\cos\lambda$
h	characteristic width of the cylinder
k_θ	torsional stiffness of the system
l	length of the cylinder
n_θ	dimensionless mass moment of inertia parameter, $\rho h^4 l/(2I)$
r	radial distance from elastic axis to surface element
r_θ	torsional viscous damping coefficient
s	$\sin\lambda$
t	real time
v	surface element velocity for torsional oscillation
Θ	torsional displacement of oscillating system
$\bar{\Theta}$	amplitude of torsional displacement
$\bar{\Theta}_0$	initial amplitude of torsional displacement
α	instantaneous angle of attack of oscillating system or attitude of stationary cylinder
α_0	mean angle of attack of oscillating system
β_θ	dimensionless damping parameter, $r_\theta/2I\omega_{n_\theta}$
γ	angle between relative and approaching fluid velocities for oscillating system
ε	increment in angle of attack of oscillating system, $\alpha - \alpha_0$
ζ	coordinate along contourline of C_{M_θ}, passing through origin

η angle between surface element velocity and approaching fluid velocity for oscillating system

η_{r_0} value of η_r at $\alpha_0 = 0°$

λ angle between ζ and $\mathring{\Theta}/U$ axis

μ modified inertia parameter, $n_\theta a_1$

ξ coordinate perpendicular to ζ axis

ρ density of the fluid

τ dimensionless time, $\omega_{n_\theta} t$

Φ phase angle

ω_{n_θ} natural circular frequency in torsion, $(k_\theta/I)^{1/2}$

Subscript

r average value of the parameter for torsional oscillations

Superscripts

(.) derivative with respect to dimensionless time τ

(°) derivative with respect to real time t

References

1. Marris, A.W.: A Review of Vortex Streets, Periodic Wakes, and Induced Vibration Phenomena. J. Basic Engng., ASME 86, 185 - - 196 (1964).

2. Minorsky, N.: Introduction to Non-Linear Mechanics, Ann Arbor, Michigan: Edwards Brothers 1947.

3. Scruton, C.: On the Wind-Excited Oscillations of Stacks, Towers, and Masts. Proc. First Int. Conf. on Wind Effects on Bldgs. and Structures, NPL, England, 1965, Vol. II, pp. 797 - 832.

4. Parkinson, G.V., Brooks, N.P.H.: On the Aeroelastic Instability of Bluff Cylinders. Trans. ASME, J. Appl. Mech. 83, 252 - 258 (June 1961).

5. Parkinson, G.V., Smith, J.D.: The Square Prism as an Aeroelastic Non-Linear Oscillator. Quart. J. Mech. and Appl. Math. 17, Part 2, 225 - 239 (May 1964).

6. Parkinson, G.V., Santosham, T.V.: Cylinders of Rectangular Section as Aeroelastic Nonlinear Oscillators. ASME Vib. Conf., Paper No. 67-VIBR-50 (March 1967).

7. Parkinson, G.V.: Aeroelastic Galloping in One Degree of Freedom. Proc. First Int. Conf. on Wind Effects on Buildings and Structures, NPL, England, 1965, Vol. II, pp. 581-609.

8. Slater, J.E.: Aeroelastic Instability of a Structural Angle Section. Ph.D. Thesis, Univ. of British Columbia, March 1969.

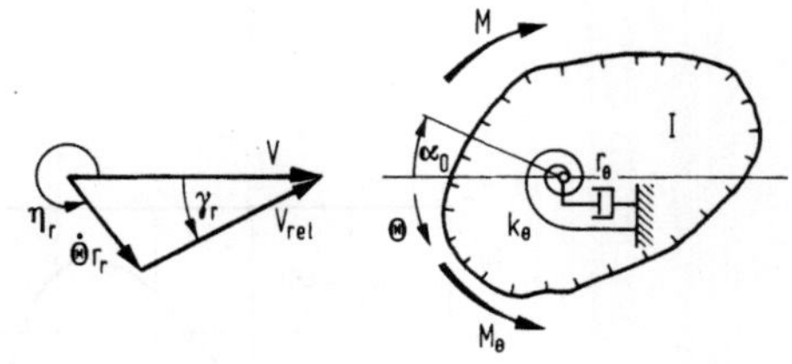

Fig.1 . Aerodynamically excited motion of a bluff body in torsional degree of freedom.

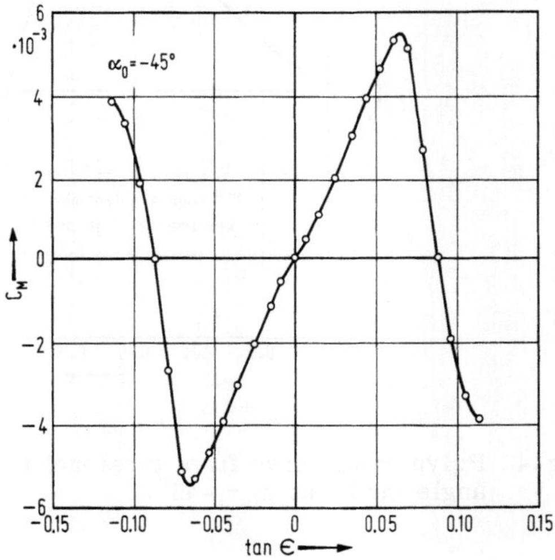

Fig.2. Moment coefficient distribution for stationary model at $\alpha_0 = -45°$.

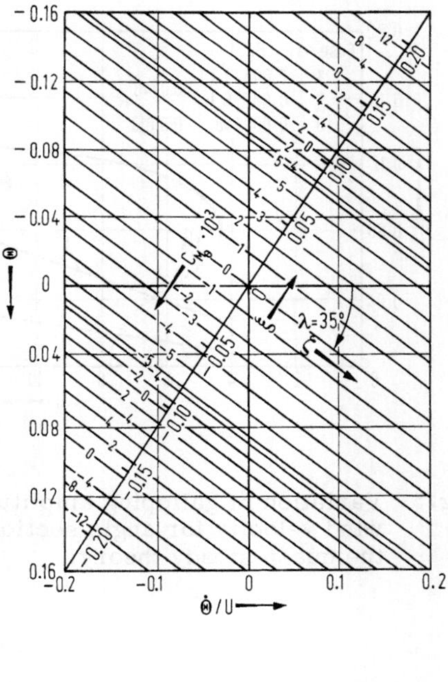

Fig.3. Contour plot of torsional moment coefficient as a function of Θ and $\dot{\Theta}$ for $\alpha_0 = 45°$.

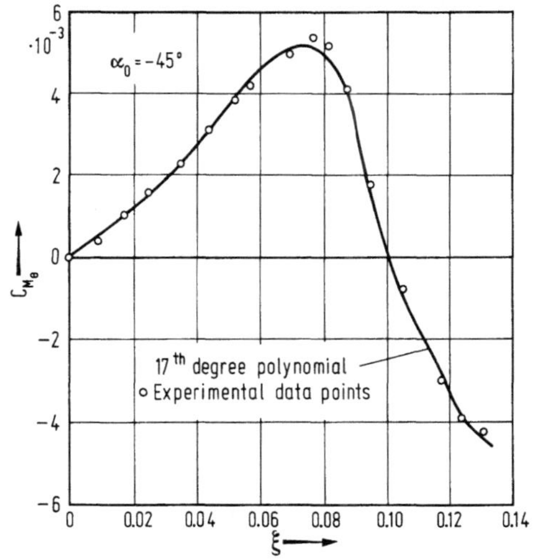

Fig.4. Polynomial curve fit of torsional moment coefficient data for angle model at $\alpha_o = -45^\circ$.

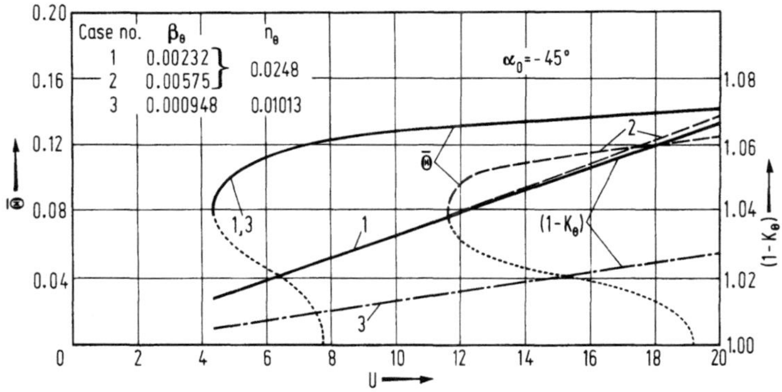

Fig.5. Variation of galloping amplitude and reduced frequency with wind velocity for angle section at $\alpha_o = -45^\circ$ as predicted by the quasi-steady theory.

The Use of Taut Strip Models in the Prediction of the Response of Long Span Bridges to Turbulent Wind

By

A. G. Davenport

London, Canada

Introduction

The purpose of this paper is to describe a new experimental approach for the prediction of the response of long span bridges to turbulent wind. The approach makes use of so-called "taut strip models", a term used to describe simplified aeroelastic models of a bridge deck system [1].

As illustrated in Fig. 1, these models are constructed using two or more taut wires (or strips) stretched span wise between anchor blocks and suitably spaced apart and tensioned to reflect the relative torsional, vertical and (if necessary) horizontal stiffness of the bridge system. To these wires is attached a light flexible model of the deck system which is ballasted appropriately to provide inertial mass scaling. The span length is variable.

As shown in Fig. 2, vibrations of such a model are characterized by torsional (θ), and horizontal (ξ) and vertical (η) translations of the deck section. As with any taut string the span wise mode shape consists of an integral number of half waves.

Such models are intended for test in a turbulent boundary layer wind tunnel simulating the scale and intensity of the tubulence of the natural wind. The response of the model for different wind speeds is measured; the objective is to deduce from these response measurements of this simplified model the response of another bridge of more complicated structural form such as a suspension bridge.

The testing of taut strip models appears to offer certain advantages over the testing of conventional section models now used routinely in the assessment of performance of suspension and other long span bridge systems.

The principal advantage is that the taut strip model permits the direct measurement of the response in a three-dimensional turbulent wind which does not seem possible with the essentially two-dimensional section model. A further advantage is that it can be tested in winds oblique to the bridge axis which the section model cannot. A possible disadvantage is that normally it must be constructed at a smaller scale than a section model. The taut strip model can be assembled and tested with no greater difficulty than the section model and notably, greater facility than a replica of the complete prototype structure in which simulation of the entire tower-cable-hanger deck system is attempted.

In summary, the taut strip model is a simplified aeroelastic instrument for measuring sinusoidal mode response. It avoids in its construction several of the costs and difficulties attendant upon the construction of a replica three-dimensional models of a complete bridge while retaining the essential features of its aeroelastic behaviour in turbulent wind. In so doing, some of the difficulties and limitations of section models are avoided.

The Response of a Long Span Bridge to Turbulent Wind

Consider a long span bridge which has mode shapes defined at station x by functions $\mu_j(\eta, \xi, \theta; x)$. Since in almost all bridge structures the vertical, horizontal and torsional modes are relatively uncoupled mechanically (but not necessary aerodynamically) the modes can be written as sets of modes $\mu_{\eta j}(x)$, $\mu_{\xi j}(x)$ and $\mu_{\theta j}(x)$ with corresponding frequencies $\omega_{\eta j}$, $\omega_{\xi j}$ and $\omega_{\theta j}$. The equations of motion in each of the modes (written in the frequency domain) can be written

Vertical/Lift

$$q_{\eta j}(\omega)\, I_{\eta j}(\omega) = P_{\eta j}(\omega), \tag{1a}$$

Horizontal/Drag

$$q_{\xi j}(\omega)\, I_{\xi j}(\omega) = P_{\xi j}(\omega), \tag{1b}$$

Twist/Pitching moment

$$q_{\theta j}(\omega)\, I_{\theta j}(\omega) = P_{\theta j}(\omega) \qquad j = 1,2\dots \tag{1c}$$

In this $q_{rj}(\omega)$, $(r = \eta, \xi, \theta)$ are the dimensionless mode amplitudes at frequency ω. $I_{rj}(\omega)$ are the complex impedances of the system and given by

$$I_{rj}(\omega) = \left\{\left(1 - \omega^2/\omega_{rj}^2\right) + 2i\alpha_{rj}\omega/\omega_{rj}\right\}, \tag{2}$$

in which α_{rj} is the damping coefficient.

$P_{rj}(\omega)$ is the generalized force acting, given by (x) dx

$$P_{rj}(\omega) = \left[\int_S p_r(x,\omega)\,\mu_{rj}(x)dx\right] \Bigg/ \left(\omega_{rj}^2 b \int_S m(x)\mu_{rj}^2(x)dx\right). \tag{3}$$

In Eq. (3) $p(x,\omega)$ is the aerodynamic force component acting at station x.

In general terms we can distinguish three contributions to the aerodynamic force thus:

Total	=	Self	+	Wake	+	Turbulent,
force		excitation		forces		forces

$$p_r(x,\omega) = p_r^*(x,\omega) + p_r^W(x,\omega) + p_r^T(x,\omega). \tag{4}$$

The self excitation forces are functions of the motion itself and can be expressed

$$p_\eta^*(x,\omega) = L_\eta(\omega)\eta + L_\xi(\omega)\cdot\xi + L_\theta(\omega)\theta + \varepsilon_\eta, \tag{5a}$$

$$p_\xi^*(x,\omega) = D_\eta(\omega)\eta + D_\xi(\omega)\cdot\xi + D_\theta(\omega)\theta + \varepsilon_\xi, \tag{5b}$$

$$p_\theta^*(x,\omega) = G_\eta(\omega)\eta + G_\xi(\omega)\xi + G_\theta(\omega)\theta + \varepsilon_\theta, \tag{5c}$$

in which L, D, and G are complex functions of the frequency and ε are the non linear residual terms. For small amplitude motions, $\varepsilon_{\eta,\xi,\theta} \to 0$.

At this point it is possible to introduce several practical factors which simplify the problem, namely:

(1) The mode shapes of the same order in the η, ξ and θ directions are generally sufficiently similar that we may write

$$\mu_{\eta j} \approx \mu_{\xi j} \approx \mu_{\theta j}. \tag{6}$$

(2) The cross coupling of the vertical and torsional movements with drag movements is generally small, and consequently

$$L_\xi(\omega) \to 0; \quad G_\xi(\omega) \to 0; \quad D_\theta(\omega) \to 0 \text{ and } D_\eta(\omega) \to 0. \tag{7}$$

(3) The mass per unit length is usually constant and, consequently,

$$m(x) = m = \text{const.} \tag{8}$$

Substituting Eqs. (5a) to (5c) into Eq. (3), using the or thogonality of the modes and results (6), (7) and (8), and recalling that

$$\eta(x,\omega) = \sum_{j=1}^{\infty} q_{\eta j}(\omega)\mu_j(\omega), \tag{9}$$

we can write the equations of motion (1) in the form

$$q_{\eta j}(\omega)\left\{A_{\eta j}(\omega) + L_{\eta j}^*(\omega)\right\} + q_{\theta j}(\omega)L_{\theta j}^*(\omega) = F_{\eta j}(\omega), \tag{10a}$$

$$q_{\eta j}(\omega)G^*_{\eta j}(\omega) + q_{\theta j}(\omega)\left\{A_{\theta j}(\omega) + G^*_{\theta}(\omega)\right\} = F_{\theta j}(\omega), \quad (10b)$$

$$q_{\xi j}(\omega)D^*_{\xi j}(\omega) = F_{\xi j}(\omega). \quad (10c)$$

In these equations the starred quantities represent complex generalized aerodynamic force derivatives and are typically given by

$$L^*_{\eta j}(\omega) = L(\omega)/\left(\omega^2_{\eta j}mb\right). \quad (11)$$

The forces $F_{\eta j}(\omega)$, $F_{\theta j}(\omega)$ are generalized expressions for the wake and turbulence-induced forces and given typically by

$$F_{\eta j}(\omega) = \left\{\int_S p_\eta^{WT}(x,\omega)\mu_{\eta j}(x)dx\right\} \Big/ \left\{\omega^2_{\eta j}bm \int_S \mu^2_{\eta j}(x)dx\right\}, \quad (12)$$

where

$$p_\eta^{WT}(x,\omega) = p_\eta^{W}(x,\omega) + p_\eta^{T}(x,\omega). \quad (13)$$

Equations (10a) and (10b) represent the coupled vertical and torsional equations; Eq.(10c), the drag equation, is, it seems, generally uncoupled. The aerodynamic coefficients on the left-hand side are generally dependent on wind speed and the reduced frequency $\omega b/\overline{U}$: that is $L(\omega) = L(\overline{U}, \omega b/\overline{U})$. The vanishing of the determinant of the coefficients gives the critical speed for instability. Even below this speed, however, the aerodynamic terms may have sufficiently modified the effective stiffness and damping of the structure so as to amplify the response to the wake and turbulent forces.

It is noted that the nonlinear self-exciting terms (denoted ε) have been dropped. These are likely to be significant only at large amplitudes which are likely to be in excess of permissible limits.

The solution to the coupled vertical torsional equations (10a) and (10b) are in matrix form

$$\begin{vmatrix} q_{\eta j}(\omega) \\ \\ q_{\vartheta j}(\omega) \end{vmatrix} = \begin{vmatrix} A_{\eta j}(\omega) + L^*_{\eta j}(\omega) & L^*_{\theta j}(\omega) \\ \\ G_{\eta j}(\omega) & A_{\theta j}(\omega) + G^*_{\theta j}(\omega) \end{vmatrix}^{-1} \times \begin{vmatrix} F_{\eta j}(\omega) \\ \\ F_{\theta j}(\omega) \end{vmatrix}. \quad (14)$$

As noted, the instability of the system is determined by the vanishing of the determinant of the coefficients. The mean-square response amplitude is given by

$$\overline{\eta^2}(x,\omega) = \sum_j \sum_k q_{\eta j}(\omega)\, q_{\eta k}(\omega)\, \mu_j(x)\, \mu_k(x). \quad (15)$$

Inspection of Eqs. (14) and (15) indicates that the mean-square response involves terms

$$F^2_{\eta j}(\omega),\; F_{\theta j}(\omega)\, F_{\eta j}(\omega),\; F^2_{\theta j}(\omega),\; F_{\eta j}(\omega)\, F_{\eta k}(\omega),\; F_{\eta j}(\omega)\, F_{\theta k}(\omega)$$

and $F_{\theta k}(\omega)\, F_{\theta j}(\omega)$. From Eq. (12) it is seen that these terms are typically of the form

$$F^2_{\eta j}(\omega) = \iint_S \frac{p_\eta(x,\omega)p_\eta(x',\omega)\mu_{\eta j}(x)\mu_{\eta j}(x')\,dx dx'}{\left\{\omega^2_{\eta j} bm \int_S \mu^2_{\eta j}(x)dx\right\}}, \quad (16a)$$

$$F_{\eta j}(\omega)\, F_{\theta j}(\omega) = \iint_S \frac{p_\eta(x,\omega)p_\theta(x',\omega)\mu_{\eta j}(x)\mu_{\theta k}(x')dx dx'}{\left\{\omega^2_{\eta j}\omega^2_{\theta j} b^2 m_\eta m_\theta \int \mu^2_{\eta j}(x)dx \int \mu^2_{\theta j}(x)dx\right\}}. \quad (16b)$$

These expressions involve the convolution of the mode-shape functions with the cross product of the forces at the two stations x and x'. This implies a knowledge of the cross correlation of the aerodynamic forces - both lift and pitching moment – at different span-wise stations. It turns out that the contributions to the response from cross-mode terms of Eq. (15) are in general significantly weaker than the squared terms in Eq. (15) for two reasons:

(1) Unless the frequencies of the two modes coincide, the admittance of the aeroelastic system associated with resonances is significantly lower;

(2) orthogonality of dissimilar mode shapes tends to reduce, but not eliminate, their contribution.

Thus, the contribution of the cross-product terms is considerably weaker, if the order of the mode is different, that is $j \neq k$.

Unlike the self-excited aerodynamic forces, the generalized wake and turbulence induced forces are dependent on mode shape. This suggests the difficulty in assessing quantitatively the response of a complete bridge to turbulence and wake induced forces from the response of a section model. Nevertheless, the section model will successfully indicate critical speeds for incipient instability and vortex shedding - provided the aerodynamic effects are not modified by turbulence.

Application of the Taut Strip Model

As remarked, the taut strip model is a simplified aeroelastic device for measuring the response of an elongated body having a half wave span-wise mode shape. Not infrequently, the fundamental mode of the full-scale structure is itself a half wave; in this case, the response of the taut strip model exactly corresponds. In general, however, this is not the case as suggested by the typical suspension-bridge mode shapes given in Fig.3. Here the mode shapes consist of a series of roughly sinusoidal half waves. The prediction of the response in this mode shape from information obtained from the half wave response of a taut strip model is now explained.

Consider a long span structure, having a mode shape consisting of a series of "waves" as in Fig.4, approximated by sinusoidal half waves of spans $s_1, s_2 \ldots . s_n$ and amplitudes $a_1, a_2 \ldots . a_n$, the sign corresponding to the direction of the wave. It is assumed that this mode shape is similar for both the lift and pitch movements - and, possibly, the drag also. It is also assumed that the coupling between modes of different shape is small as we have suggested above. Let us suppose that a taut strip model is constructed such as to resemble the bridge in question in all respects except the similarity in the mode shape: that is to say that geometrically the bridges are similar, the mass distributions and damping characteristics correspond, and the frequencies are in the same rate. Further, let us suppose that this taut strip model is tested for several span length so that estimates of the response of waves having spans $s_1, s_2 \ldots . s_n$ can be made as functions of reduced velocity $U/b\omega$. The mean-square responses of the taut strip models so determined can be written $\Phi_\eta^2(s_1/b, \overline{U}/b\omega)$,

$$\Phi_\eta(s_1/b, \overline{U}/b\omega)\Phi_\theta(s_1/b, \overline{U}/b\omega) \text{ and } \Phi_\theta^2(s_1/b, \overline{U}/b\omega).$$

Let us turn now to the complete bridge. Examining the double integrals in Eq. 16a for F_η^2, we see that the integral consists of the products of two aerodynamic force terms at x and x' which may, or may not, be contained in the same wave of the mode shape. Let us suppose that both x and x' lie within the ith wave: Then the contribution to the modal response q_η^2 from this wave is given by

$$\left(q_\eta^2\right)_{ii} = a_i^2 s_i^2 \Phi_\eta^2(s_i/b, \overline{U}/\omega b)/\left\{\sum a_i^2 s_i\right\}^2. \tag{17}$$

If x and x' are not in the same wave, the contribution to q_η^2 cannot be measured directly from the taut strip model. However, its contrbution will be smaller and for this separation of the points it seems reasonable to assume that the correlation of forces is similar to that for the oncoming turbulence. It has been found that the wind speed correlation at two points Δx apart at a frequency f is of the form

$$R(x, x'; f) = e^{-C\frac{|x-x'|f}{\overline{U}}}$$

where C is a constant found to be generally $\simeq 8$. With this assumption a reasonable estimate of the total response including the "cross-wave" contribution is

$$q_\eta^2 = \frac{\sum_i \sum_j a_i a_j s_i s_j \Phi_\eta^2\left(\frac{s_i+s_j}{2b}, \frac{\overline{U}}{\omega b}\right) e^{-C\frac{|\overline{x}_i-\overline{x}_j|f}{\overline{U}}}}{\left\{\sum a_i^2 s_i \sum a_j^2 s_j\right\}^2}, \tag{18}$$

where $|\bar{x}_1 - \bar{x}_2|$ represents the distance between the crests of the waves.

For the cross term $q_\eta a_\theta$, the covariance of the lift and pitch, $\Phi_\eta \Phi_\theta$ measured on the taut strip model is used in place of Φ_η^2. This expression permits the response of taut strip models to be applied to more complicated bridge deck systems.

Reference

1. Davenport, A.G., Isyumov, N., Miyata, T.: The Experimental Determination of the Response of Suspension Bridges to Turbulent Wind. Proc. 3rd. Int. Conference on Wind Effects on Buildings and Structures, Tokyo, September 1971.

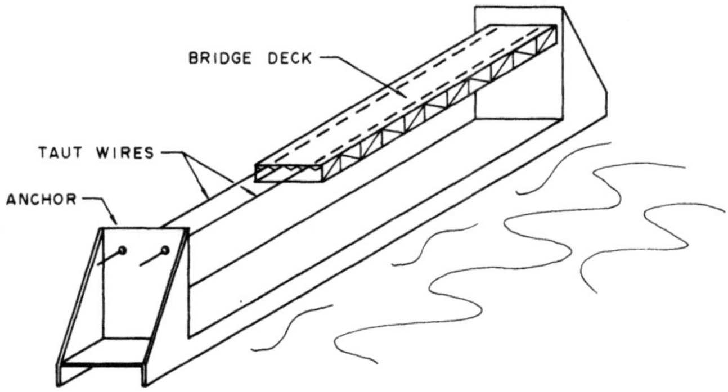

Fig.1. Taut strip model of long span bridge.

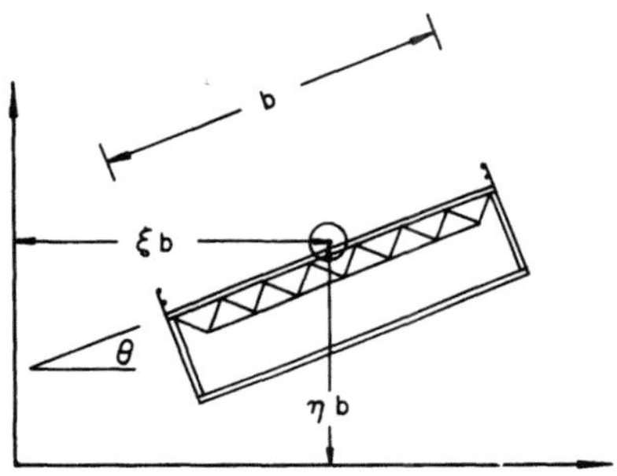

Fig.2. Deck movements: Notation.

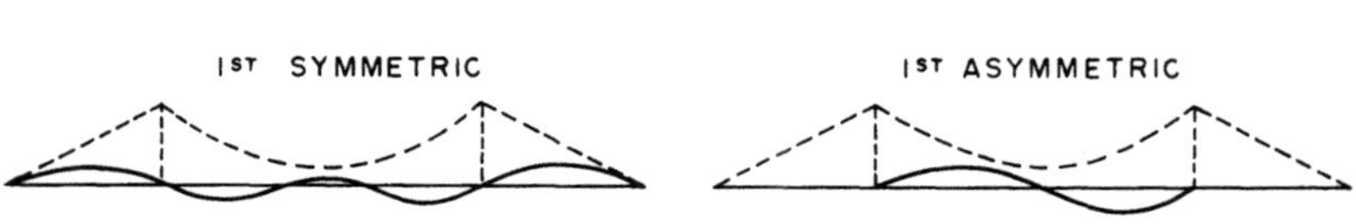

Fig.3. Suspension bridge mode shapes.

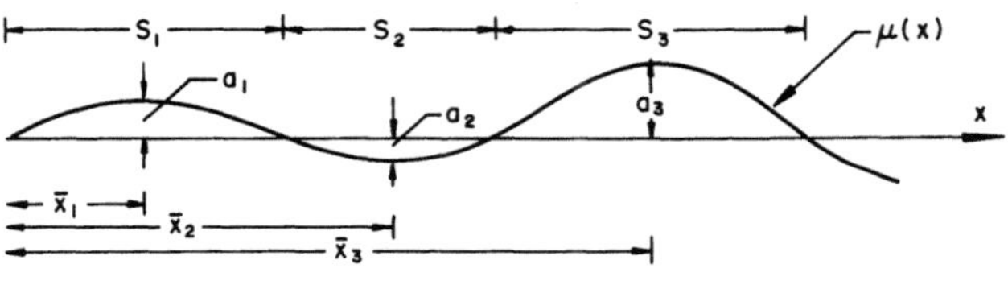

Fig.4. Generalized mode shape.

The Effect of Twist Motion on the Dynamic Multimode Response of a Building

By

F.H. Durgin, P. Tong

Cambridge, Mass., U.S.A.

Summary

Most areroelastic wind-tunnel tests of buildings have only considered the first two bending modes of a building. When the twist mode is added, it is found experimentally in uniform flow that it is possible to have what is apparently a superharmonic instability due to vortex S shedding that will occur at $V/ND \simeq 3.2$ which corresponds to a wind speed of 60 to 70 mph for a typical 40-story building. It is also shown, both theoretically and experimentally, that when the damping in the appropriate bending mode is about 0.1 % of critical, a classic flutter instability can occur at $V/ND \simeq 1.7$.

The current tests not only have demonstrated both instabilities, but also have shown that the superharmonic instability is very sensitive to both small misalignments (less than 1 b of the building length) of the building's center of gravity and elastic axis, and to the ratio of the twist and bending frequencies. Further tests must be run to determine the effect of the superharmonic instability on rms response and the ratio of peak-to-rms response of buildings in typical rough flow.

Introduction

About a year ago the Wright Brothers Wind Tunnel became involved in a study of the dynamic response of a 40-story rectangular building. On-site measurements showed that the twist motion of the building was important, so the dynamic model built for simulating the building

in the wind tunnel had provision for simulating the first twist as well as the first two orthogonal bending modes of the building (usually only the two bending modes are considered [1,2]). The study was privately sponsored and is proprietary, but certain general results were obtained which led to the current study.
They are:

(1) If the twist and short axis bending periods are nearly the same the twist motion can be important.

(2) While all motion provides audible cues, twist motion also provides visual cues from windows and therefore can be very noticeable to the occupants of upper stories.

(3) Slight asymmetries in the model involving the mass distribution and effective shear center (elastic axis) drastically alter the response of the building for winds from opposite directions along the long axis of the rectangle.

It was the third observation which led to the current investigation. This was to examine the effect of the ratio of the short axis bending to twist frequencies, of small mass unbalance along the long axis and of changes in damping in each mode on the responses of a rectangular building in uniform flow.

The Experimental Equipment

The Wind Tunnel

The experiments were performed at the Massachusetts Institute of Technology in the Wright Brothers Wind Tunnel. It is a continuous-flow closed-return wind tunnel with a 7.5 × 10 ft elliptical test section about 15 ft long. The tunnel is capable of speeds from 0 to 140 mph. For this test a ground board was installed to obtain an 8 ft wide floor. The ground board extended from one foot into the upstream contraction to three feet into the diffusor. The model was mounted four feet from the ground board leading edge. Velocity surveys showed that the flow at the model was uniform except for the boundary layer on the floor which was less than an inch thick.

The Model

The model was a $3\times5\times15.5$ inch tall rectangular prism. It approximated the volume and height of the 1:400 scale model of the 40-story building used in the original test. The mass, spring and damping characteristics of this model were designed to match those of the original aeroelastic model. Thus the following aeroelastic parameters were approximately matched as between the original real building and model for each of the three modes considered.

(1) $K = V/ND$ reduced frequency,

(2) θ; $\beta = \Delta y/h$ ratio of deflection due to static aerodynamic loading to a building dimension[1] [3],

(3) η the percent of critical damping.

These parameters define a scaling for velocity and frequency once the dimensional scale (1-400) and model flexibility are chosen, and these were:

$$V_{model}/V_{building} = 1/6,$$

$$N_{model}/N_{building} = 400/6 = 66.7.$$

Matching β is straightforward for the two bending modes where the building and model mode shapes are similar. For the twist mode, the model and building have very different mode shapes-the building's twist-angle increases nearly linearly with height whereas the model's is constant. The model and building twist deflection angles due to a typical aerodynamic load were matched at 0.6 of the building and model height.

Figure 1 is a schematic of the aeroelastic model. Note that the three types of motion (two bending and one twist) are measured by the strain gages attached to the bottom flexure. The necessary additional twist flexibility was obtained using the bearings and twist spring above.

---

[1] This is equivalent to having the same density ratio between building and air but is more realistic when only the first three modes are being matched.

Damping is provided by the paddles in the pots of viscous fluid. Since the paddles are very flexibly mounted for lateral motion, twist damping is obtained through rocking motion of the paddles. Thus in a rough way the damping for the bending modes is determined by the area of the paddles, whereas for twist it is the shape-i.e. width to length. Using this technique, the damping for each mode could be separately controlled. Overall damping was also controlled by varying the damping fluid and height of paddles in the pots.

In the experiment the damping, the mass unbalance, and the torsional stiffness were varied. The torsional stiffness was varied by changing the upper twist spring. The unbalance was varied by drilling the hole in the "added mass" off center (see Fig.1). In the experiment the unbalance was always placed along the long axis of the building in the positive x-direction.

While the damping in each of the vibration modes, the twist frequency and the unbalance varied during the program, changes are referred to the initial model characteristics given below:

Mode	N-Natural frequency Hz	Stiffness ft lb/rad	I-Inertia slug ft^2	η-percent critical damping
Long axis bending	11.4	120	0.0234	1.3
Short axis bending	12.6	145	0.0234	0.8
Twist	12.8	294	0.00045	1.6

Fig.2 shows the building axis system and defines many symbols.

The Experiment

All the tests were conducted in uniform flow in the direction of the long axis of the prism.

Basic Prism Tests

The first wind tunnel tests were run to check the overall symmetry of the basic model. The results are presented in Fig.3 and show essentially the same rms amplitude versus V/ND for both the β(i.e., y) and θ responses for the wind from either +x or -x. Only the y and

θ results are shown, but the rms x amplitude was also measured and found to be small. Note that the first sharp rise in amplitude for both β and θ occurs at $V/ND \simeq 2.5$ and has a peak rms y response of $\beta \simeq 0.007$ in bending and $\theta \simeq 0.05$ in twist. The second rise in amplitude starts at $V/ND \simeq 4.7$ and at $V/ND \simeq 5.9$ both β and θ have about the same rms value of 0.02. Because the y amplitude was so great at this point, no attempt was made to find the peak rms amplitude for fear of damaging the strain gage balance.

We shall refer to the sharp rises in amplitude as instabilities. For both instabilities the β and θ frequencies became the same with a fixed phase between the two of approximately 0° or 180°. In that case we shall refer to the two motions as being locked. For $V/ND \simeq 3$ the model appeared to rotate about the windward end of the model, whereas at $V/ND \simeq 6.5$ it was about the lee end.

When the two motions (β and θ) locked, the common frequency was nearly the same as the β frequency at $V/ND \simeq 3$. However, at $V/ND \simeq 6.5$. the common frequency was generally less than either the original β or θ frequencies.

The Unbalanced Model Tests

The model was then tested with the "added mass" weight displaced 0.125 inches from the model center, i.e., $\varepsilon/D = 0.007$. This is a displacement of the center of gravity of less than 1 % of the length of the model. None of the other parameters was changed significantly. The effect of having the wind first from +x and then from -x is compared with the results for the balanced model in Fig.4. With positive unbalance and the wind from +x both the β and θ responses are relatively unchanged compared to the balanced model including the locking of the β and θ motion. However with wind from -x and positive unbalance, the instability at $V/ND \simeq 3$ is eliminated, and the locked β and θ responses are increased at the second peak.

Similar results (not shown) were also obtained for $\varepsilon/D = 0.014$.

The Twist Frequency Tests

As noted in the figures the ratio of the twist to y frequencies was 1.02 up to this point. The effect of changing the twist frequency so that N_θ/N_β = 1.02, 1.04 and 3.27 is shown in Fig.5. For the β motion even the small change from N_θ/N_β = 1.02 to 1.04 not only reduces the rms β response at V/ND $\simeq$ 3 by about a factor of 2, but also the range of V/ND over which the instability occurs. Further, the peak moves from V/ND $\simeq$ 3.5 to 4.0. On the other hand the twist amplitude increases slightly, and its range of instability in V/ND is unchanged.

Finally increasing the N_θ/N_β to 3.27 again eliminated the instability in β and θ at V/ND $\simeq$ 3.5, reduced the stable rms twist response by a factor of 10, and eliminated any locking of the bending and twist motion at the instability at V/ND $\simeq$ 6.5.

Fig.6 shows the effect of adding a small amount of unbalance to the model when the ratio of the θ to β frequencies is 1.04. Again when the unbalance was along +x and the wind from -x the instability at V/ND $\simeq$ 3 was eliminated. With the wind from +x, the lower instability returns but the V/ND of its peak is reduced from 4.0 to 3.2 and both the β and θ amplitudes are reduced. The V/ND of the second instability is essentially unchanged with the wind from +x but decreased with wind from -x.

Damping Tests

The last series of tests was made to determine the effect of small and large changes in damping. The results for small increases at N_θ/N_β = 1.04 are shown in Fig.7.

Increasing the β damping from 0.8 to 1.4 % of critical caused a drop in the rms peak amplitude of the β motion at V/ND $\simeq$ 3.5 roughly inversely proportional to the change in damping. The θ amplitude was also decreased, but not so much. When the β and θ damping were further increased to 1.7 and 2.6 % of critical respectively, the instability at V/ND $\simeq$ 3.2 was eliminated, but the one at V/ND $\simeq$ 6.5 remained.

Some tests were also run on the original model ($N_\theta/N_\beta = 1.02$) with the smallest possible damping--$\eta = 0.1$ and $0.4\ \%$ of critical in β and and θ respectively. The results are shown in Fig.8. The first instability now occurs at $V/ND \simeq 1.8$ and there is a reduction in the rms β amplitude at $V/ND \simeq 3.2$ although none appears in θ.

Similar data was also obtained for $N_\theta/N_\beta \simeq 3.27$ and as can be seen in Fig.9 the first β instability again occurs at $V/ND \simeq 1.8$. No data was taken above $V/ND = 2.5$ because the amplitude of the β motion was becoming large enough to damage the strain gages.

From the test data, it seems the sharp rise in response occurs at a slightly higher V/ND for the case $N_\theta/N_\beta = 3.27$ than for $N_\theta/N_\beta = 1.02$. The rms response (in β) of the former case is also larger.

Aeroelastic Analysis

By standard procedure of aeroelastic analysis, we approximate the motion of the building by two degrees of freedom, namely bending in the direction ($\beta = \Delta y/h$) and twist about the elastic axis, θ. The equations of motion are:

$$\begin{aligned} I_\beta \frac{d^2\beta}{dt^2} - I_{\beta\theta} \frac{d^2\theta}{dt^2} + 2I_\beta \omega_\beta \eta_\beta \frac{d\beta}{dt} + I_\beta \omega_\beta^2 \beta &= M_\beta, \\ I_\theta \frac{d^2\theta}{dt^2} - I_{\beta\theta} \frac{d^2\beta}{dt^2} + 2I_\theta \omega_\theta \eta_\theta \frac{d\theta}{dt} + I_\theta \omega_\theta^2 \theta &= M_\theta, \end{aligned} \tag{1}$$

where ω_β and ω_θ are the bending and the twisting frequencies respectively, I_β, $I_{\beta\theta}$, and I_θ are moments of inertia, and M_β and M_θ are the aerodynamic moments. If quasi-static aerodynamic theory is used for small angle of attack, M_β and M_θ can be approximated as

$$\begin{aligned} M_\beta &= -\frac{\xi}{2}\rho_a V^2 Dh\, z_{ac}\left(\frac{dC_L}{d\alpha} + C_D\right)\left[\theta + \frac{\delta_{11}}{V}\frac{d\beta}{dt} + \frac{\delta_{12}}{V}\frac{d\theta}{dt}\right], \\ M_\theta &= +\frac{\xi}{2}\rho_a V^2 Dh\, x_{ac}\left(\frac{dC_L}{d\alpha} + C_D\right)\left[\theta + \frac{\delta_{21}}{V}\frac{d\beta}{dt} + \frac{\delta_{22}}{V}\frac{d\theta}{dt}\right], \end{aligned} \tag{2}$$

where ρ_a is the air density, V the air velocity, D the prism's length and z_{ac}, x_{ac}, and $\left(\frac{dC_L}{d\alpha}+C_D\right)$ determined from static aerodynamic measurements. ξ is the lift deficiency factor to account for the unsteady effect. The quantities δ_{ij}, are not known for the prism. In the thin airfoil theory ξ is the Theodorson's function, $\delta_{12} = D/4$, $\delta_{22} = D/4$ and h/2 is usually used for δ_{11}. (Note that $z_{ac}\,\delta_{12} = x_{ac}\,\delta_{21}$.)

Introducing the nondimensional parameters

$$\mu = \frac{1}{2}\rho_a D^4 h/I_\beta, \qquad K = V/ND = 2\pi V/\omega_\beta D,$$
$$I = I_\theta/I_\beta, \quad I_{12} = I_{\beta\theta}/I_\beta, \tag{3}$$

and substituting (2) and (3) into (1), we obtain

$$\ddot\beta + g_{11}\dot\beta + \beta - I_{12}\ddot\theta + g_{12}\dot\theta + \Omega_{12}\theta = 0,$$
$$I_{12}\ddot\beta + g_{21}\dot\beta + I\ddot\theta + g_{22}\dot\theta + I\Omega_{22}\theta = 0, \tag{4}$$

in which

$$(\dot{\ }) = \frac{d}{d\tau} = \frac{1}{\omega\beta}\frac{d}{dt},$$

$$g_{11} = 2\eta_\beta + \mu\frac{K}{2\pi}\xi\left(\frac{dC_L}{d\alpha} + C_D\right)\frac{z_{ac}}{D}\frac{\delta_{11}}{D},$$

$$g_{12} = \mu\frac{K}{2\pi}\xi\left(\frac{dC_L}{d\alpha} + C_D\right)\frac{z_{ac}}{D}\frac{\delta_{12}}{D} = -g_{21},$$

$$g_{22} = 2I\frac{\omega_\theta}{\omega_\beta}\eta_\theta - \mu\frac{K}{2\pi}\xi\left(\frac{dC_L}{d\alpha} + C_D\right)\frac{x_{ac}}{D}\frac{\delta_{22}}{D}, \tag{5}$$

$$\Omega_{22} = \left(\frac{\omega_\theta}{\omega_\beta}\right)^2 - \frac{\mu}{I}\left(\frac{K}{2\pi}\right)^2\xi\left(\frac{dC_L}{d\alpha} + C_D\right)\frac{x_{ac}}{D},$$

$$\Omega_{12} = \mu\left(\frac{K}{2\pi}\right)^2\xi\left(\frac{dC_L}{d\alpha} + C_D\right)\frac{z_{ac}}{D}.$$

Equation (4) can be solved in the form

$$\beta, \theta \sim e^{\lambda\tau}, \tag{6}$$

where λ is determined by the characteristic equation

$$\left(\lambda^2 + g_{11}\lambda + 1\right)\left(I\lambda^2 + g_{22}\lambda + I\Omega_{22}\right) = \\ = \left(-I_{12}\lambda^2 + g_{12}\lambda + \Omega_{12}\right)\left(-I_{12}\lambda^2 + g_{21}\lambda\right). \tag{7}$$

If the real part of the root λ in Eq. (7) is positive, the solution is unstable. Thus the critical condition is

$$\mathrm{Re}(\lambda) = 0,$$

i.e., the real part of λ going to zero defines the onset of an instability which is usually called flutter.

Calculations have been performed for the tested model using $\xi = 0.85$, $\delta_{12} = D/4$, $\delta_{22} = D/4$ and $\delta_{11} = h/2$. The quantities x_{ac}, z_{ac} and $(dC_L/dx + C_D)$ are based on the measured static data, namely

$$x_{ac}/D = 0.195, \; z_{ac}/D = 1.43, \; (dC_L/d\alpha + C_D) = -1.94 .$$

For the very small damping, e.g., $\eta_\beta = 0.1$ % and $\eta_\theta = 0.4$ %, the critical condition occurs at $K = V/ND = 1.7$ for $N_\theta/N_\beta = 1.014$ and at $K = 1.6$ for $N_\theta/N_\beta = 3.27$. These numbers seem to agree (within 10 %) with the experiment for the onset of the instability (fig. 8 and 9). When the two frequencies are very close, the coupling of the bending and the torsion delays the occurrence of aeroelastic instability slightly. For higher damping, say $\eta_\beta = 0.8$ %, $\eta_\theta = 1.6$ % and for $N_\theta/N_\beta = 1.014$, the aeroelastic analysis predicts the onset of flutter at $K = 13.5$ which is much higher than that observed in any of the tests.

Discussion of Results

Two types of instability are believed to have been observed in these experiments-vortex shedding and flutter. The vortex shedding instability occurs when the natural aerodynamic vortex shedding frequency or a multiple of it coincides with one of the building's natural frequencies (bending perpendicular to the wind or twist). Flutter on the other hand occurs when the negative aerodynamic damping becomes equal to or greater than the structural damping as noted in the analysis above.

At very low structural damping (e.g. $\eta_\beta = 0.1\ \%$), a flutter instability occurs at very low V/ND (~ 1.8) where N is referred to the bending frequency in y-direction. The onset of this instability is predicted by an aeroelastic analysis based on the quasi-static aerodynamic theory using a lift deficiency factor of 0.85. The addition of the torsional degree of freedom slightly delays the onset of instability and decreases the rms response in the y-direction (Figs.8 and 9). This effect is contrary to that found for thin airfoils and is due to the fact that the lift slope is negative for the blunt prism.

At higher structural damping (e.g., $\eta_\beta = 0.7\ \%$ and $\eta_\theta = 1.6\ \%$) the present theory no longer predicts the instability. A vortex shedding instability occurs before the flutter type of instability considered in the analysis. Vortex shedding has not been included in the analysis.

The tests of the balanced prism (Fig.3), with $N_\theta/N_\beta = 1.02$ and moderate damping, show instabilities at V/ND of about 3.2 and 6.5. For a building, the instability at V/ND = 3.2 is of great interest because it occurs at a wind speed of 60 to 70 mph. Both instabilities are believed to be due to vortex shedding. The first appears to be a superharmonic resonance where the building frequency is twice the shedding frequency. When based on the prism's dimension perpendicular to the wind, the V/ND of the second instability is about 0.1, which is in agreement with previous investigators [2,5]. Fung cites a subharmonic response [4].

The tests on the unbalanced model (Figs.4 and 6) show the extreme sensitivity of the superharmonic instability to small diplacements of the prisms center of gravity with respect to its elastic axis. When the unbalance is toward the lee end of the building the superharmonic response is eliminated, but not so when it is toward the windward end.

The data show that for unbalances equivalent to a displacement of center of gravity between 0.007 and 0.014 of the building's length, the superharmonic instability was eliminated for twist to bending frequency ratios between 1.02 and 1.04 when the wind was from - x. Unfortunately it was not possible in this preliminary study to investigate smaller and larger unbalances or somewhat larger frequency ratios. Only a limited range of damping for the two modes was investigated.

The locking of the bending and twist frequencies for both the superharmonic and harmonic instabilities is significant and needs further investigation. Frequently when the locking of frequencies became steady there was a step rise in amplitude. Step changes in amplitude occurred both near the lowest and highest V/ND of an instability.

The superharmonic-harmonic vortex shedding ideas given above seem to offer a reasonable explantion of much of the data, but some features of the prism's response are anomalous: for instance, the change in the V/ND of the superharmonic instability from about 3.2 to 4.0 when the twist frequency was increased by about 2 %, and its drop back to V/ND = 3.2 when unbalance was added to the model. Another consistent bit of data that is hard to explain is the apparent decrease in the V/ND of the harmonic vortex shedding instability due to unbalance when the superharmonic response was eliminated.

Conclusions

Most tests on buildings have only considered the two heaving modes of the buildings. When the twist mode is added it is found experimentally that it is possible to have what is apparently a superharmonic instability due to vortex shedding that will occur at V/ND = 3.2 which corresponds to a wind speed of 60 to 70 mph for a typical 40-story building.

The current tests not only have demonstrated the instability, but have also shown that this superharmonic instability is very sensitive to both small misalignments (less than 1 % of the building length) of the building center of gravity and elastic axis and the ratio of the twist and bending frequencies.

The current tests define a range of short axis bending to twist frequencies, of unbalance, and of damping over which this superharmonic instability can occur for a 1.67, to 1, to 5 building. Additional information is needed to define the complete rages of building shape (rectangular to square), of ratio of bending to twist frequencies, damping, and unbalance for which the superharmonic instability can occur.

One may question the usefulness of the tested results in a uniform flow for a building. As has been shown in Ref. [2] a square prism in uniform flow has a peak rms response in bending at V/ND = 10, but the turbulence eliminates the resonance peak completely. In other words, what is found in uniform flow may not be found in the real world where the flow is always turbulent. It is true that there may not be sharp peak rms response in turbulence flow, when the rms value is taken over a long period of time, because of lack of coherence in the turbulence. However, sharp peaks will exist if rms value of the response is taken of a much shorter period. A large gust closely resembles uniform flow. This gust may well start to excite the instability. It is this type of sharp peak resonance which annoys people in the building and can cause structural damage. Therefore, it is important to have information in a uniform flow. Further tests must be run to determine the effect of the superharmonic instability on rms response and the ratio of peak to rms response of prisms in typical rough flow.

References

1. Davenport, A.G., Isyumov, N.: The Application of the Boundary Layer Wind Tunnel to the Prediction of Wind Loading. Paper No.7, Proceedings of the International Seminar on Wind Effects on Buildings and Structures, National Research Laboratory, Ottawa, Canada, September 1967.

2. Rosati, P.A.: Wind Action on Finite Rectangular Cylinders in Turbulent Boundary Layer Flow. Engineering Sci. Res. Report, BLWT-8-70, The University of Western Ontario, July 1970.

3. Whitbread, R.E.: Model Simulation of Wind Effects on Structures. Paper No. 2, National Physical Laboratory, International Conference on Wind Effects on Buildings and Structures, London, England, 1963.

4. Fung, Y.C.: An Introduction to the Theory of Elasticity, New York, Dover Publications 1969.

5. Vickery, B.J.: Fluctuating Lift and Drag on a Long Cylinder of Square Cross-Section in a Smooth and in a Turbulent Stream. J. Fluid Mech., 25, Part 3 (1966).

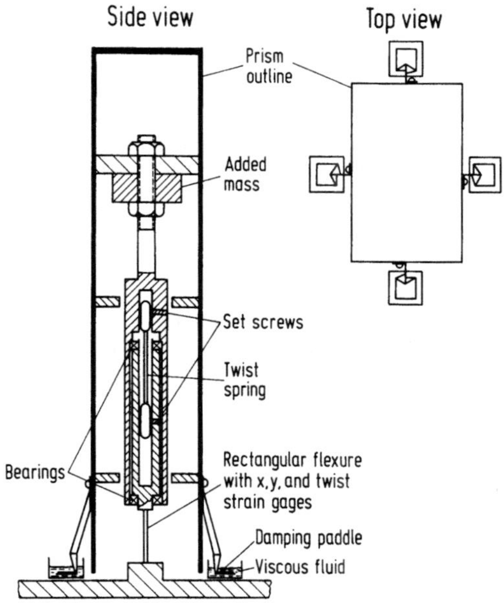

Fig.1. Schematic of aeroelastic prism.

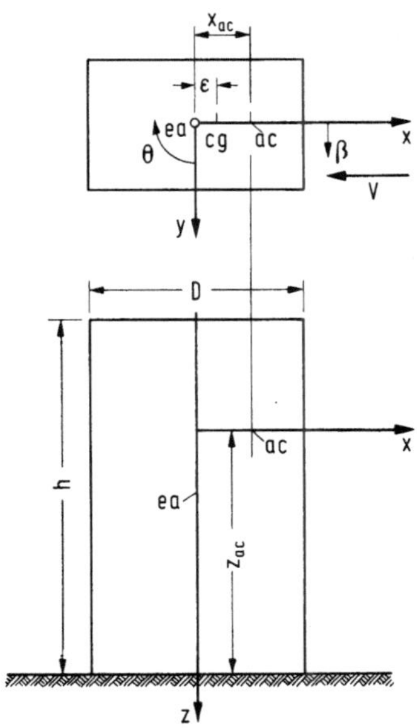

Fig.2. Axis system and symbol definition.

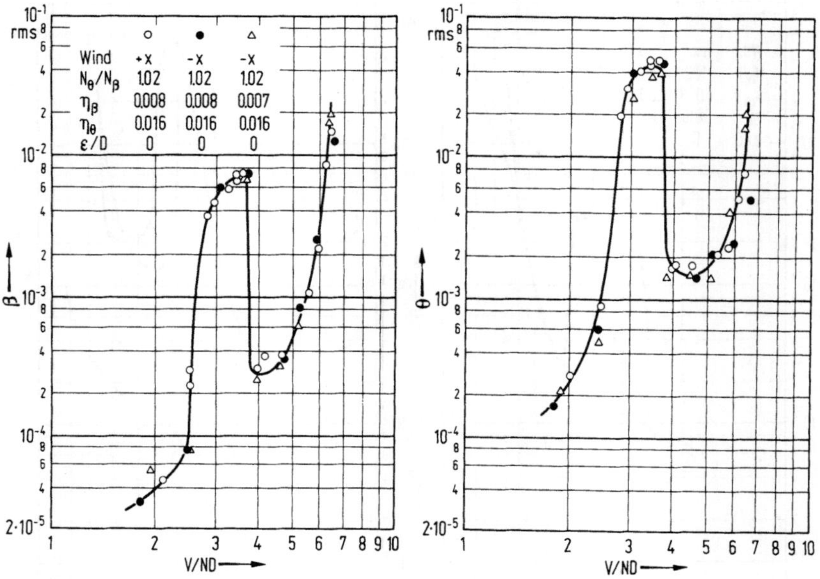

Fig.3. Response of balanced prism; N_θ/N_β = 1.02.

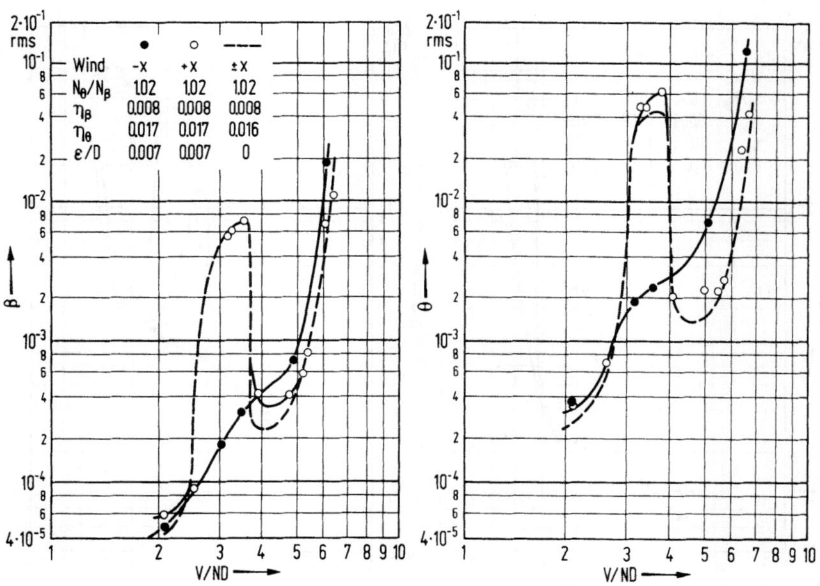

Fig.4. Response of unbalanced prism; N_θ/N_β = 1.02.

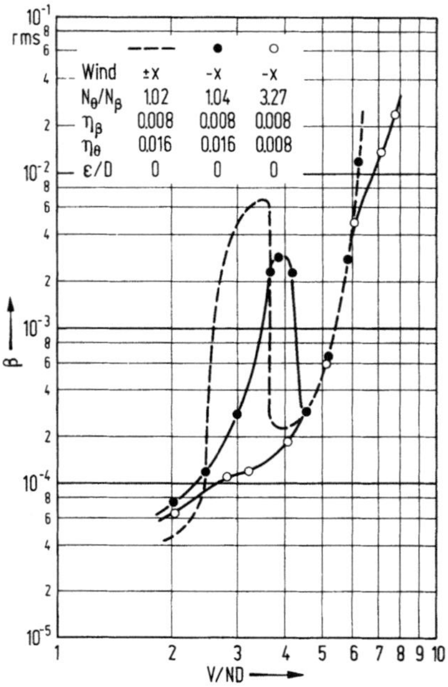

Fig.5. Response of balanced prism at three twist frequencies.

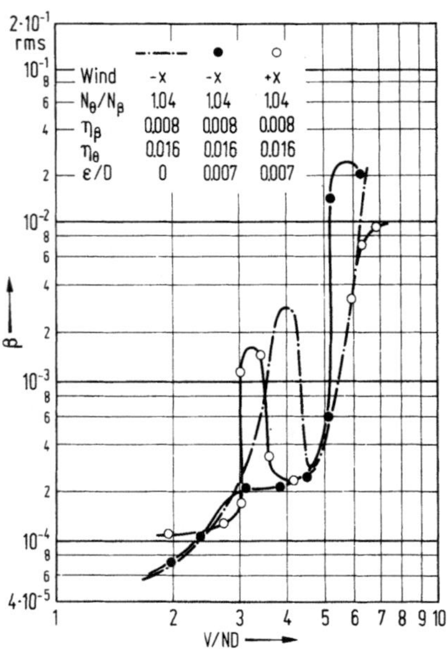

Fig.6. Response of unbalanced prism; $N_\theta/N_\beta = 1.02$.

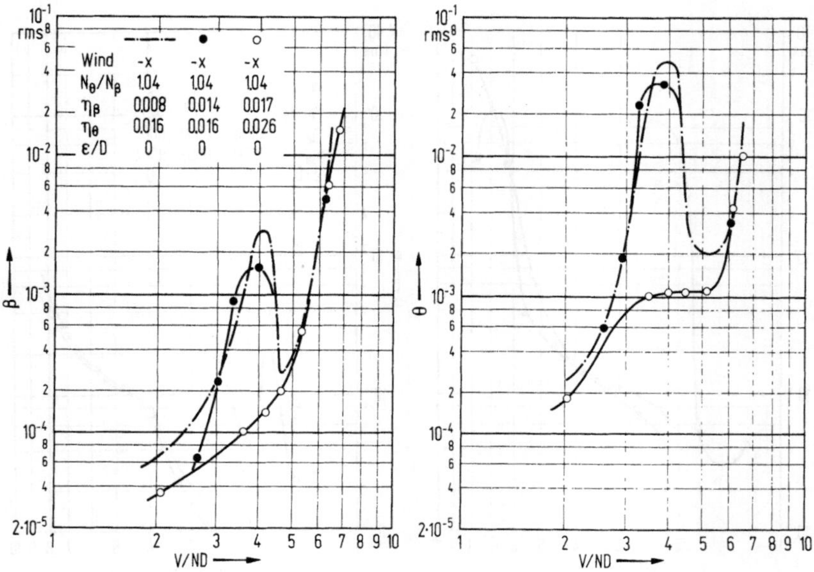

Fig.7. Effect of increased damping on balanced prism; N_θ/N_β = 1.04.

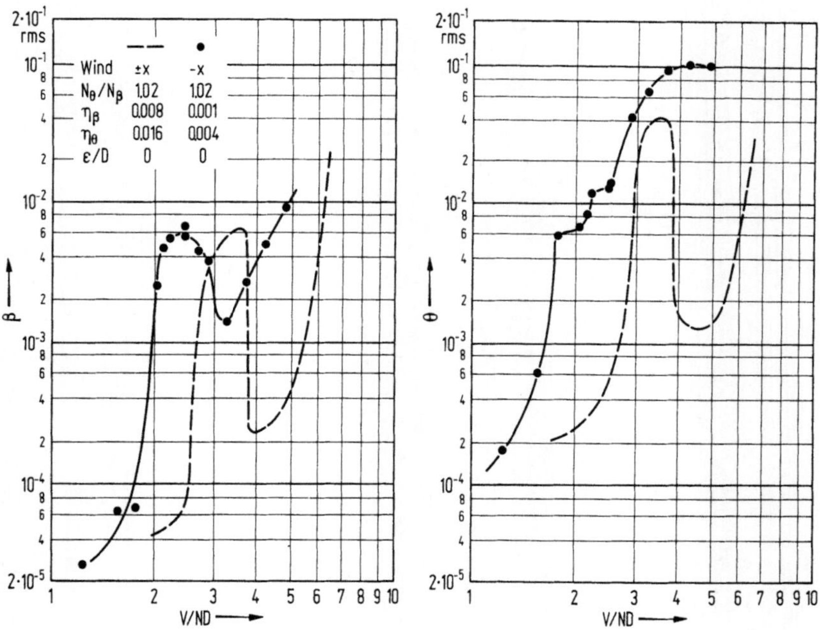

Fig.8. Effect of decreased damping on balanced prism; N_θ/N_β = 1.02.

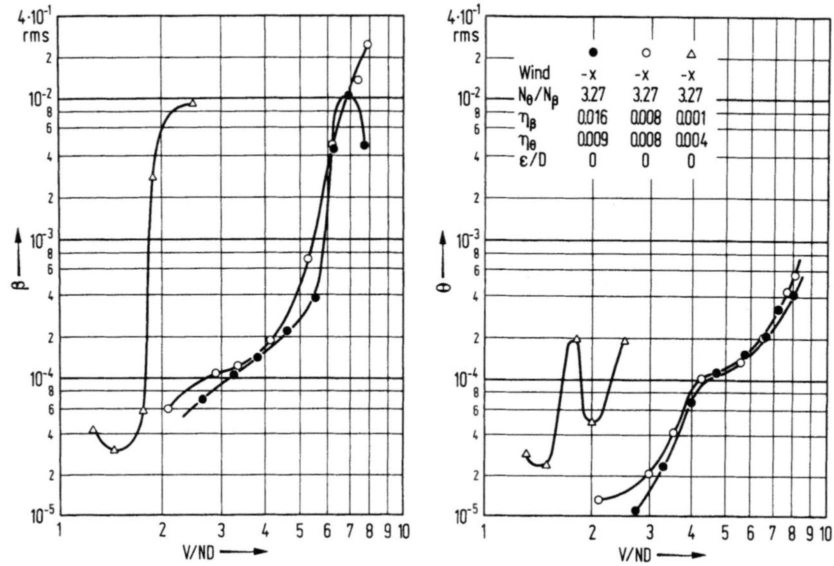

Fig.9. Effect of decreased damping on balanced prism; $N_\theta/N_\beta = 3.27$.

On Aerodynamic Responses of Truss-Stiffened Bridge Sections in Fluctuating Wind Flows

By

N. Shiraishi

Kyoto, Japan

This investigation is concerned with aerodynamic responses of truss-stiffened bridge structures in fluctuating gusts. In similarity with aerodynamic responses of a flat plate in turbulent flow, a fundamental investigation is performed in connection with one of the proposed bridge sections of the Honshu-Shikoku Connecting Suspension Bridge in Japan.

In this study, aerodynamic responses of structures in turbulent flows are assumed to comply with the mean velocity and fluctuating velocity independently. Aerodynamic coefficients or unsteady aerodynamic forces are obtained by the free vibration method as shown in Fig. 1 in which the full lines indicate those of a flat plate evaluated from the Theodorsen's Circulation Function. The results disclose clearly the differences between an actual bridge section and ideal flat plate by which one can note that most aerodynamic coefficients associated with bridge sections are much smaller than those associated with plates. Attention should be placed on the fact that A^*_2, the coefficient proportional to uncoupled torsional deformation rate, is remarkably small and H^*_2, the coefficient proportional to coupled torsional deformation rate, remains negative, which means opposite sign to the case of flat plate.

As an example of aerodynamic responses due to transversely fluctuating wind velocity, the normalized power spectra of deflectional and torsional deformations of a plate model are analyzed as shown in Fig. 2 by which deflectional responses are known to have two peaks on account of aerodynamic coupling effects. Experimentally obtained power spectra of deflectional and torsional deformations of bridge section models, denoted as $S_{\eta\eta}$ and $S_{\varphi\varphi}$, respectively, are shown in Fig.3, in which the power spectra of longitudinally and laterally fluctuating wind velocity are also shown up to 100 cps. It is interesting to note that at low mean-velocity, power spectra of responses have dominant peaks at natural frequencies though the second peak of $S_{\eta\eta}$ does not coincide with the one of $S_{\varphi\varphi}$. As the mean velocity increases, the second peak of $S_{\eta\eta}$ coincides with the peak of $S_{\varphi\varphi}$ and above a certain critical velocity there appears significant fluttering at which state dominant frequencies in both modes accord with each other and decreasies slightly. Thus, the frequency characteristics of power spectra of responses are considered to play an important role in analyses of initiation of aerodynamic instability of bridge sections in fluctuating gusts.

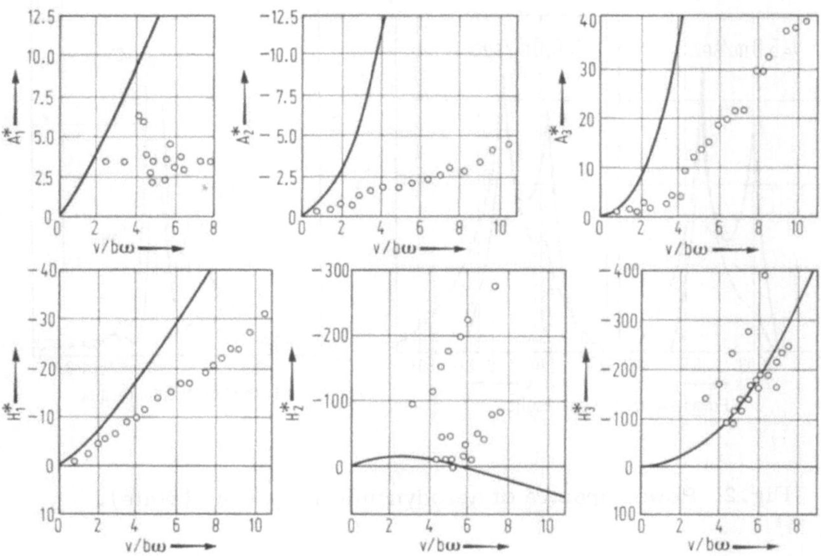

Fig. 1. Aerodynamic coefficients of plate.

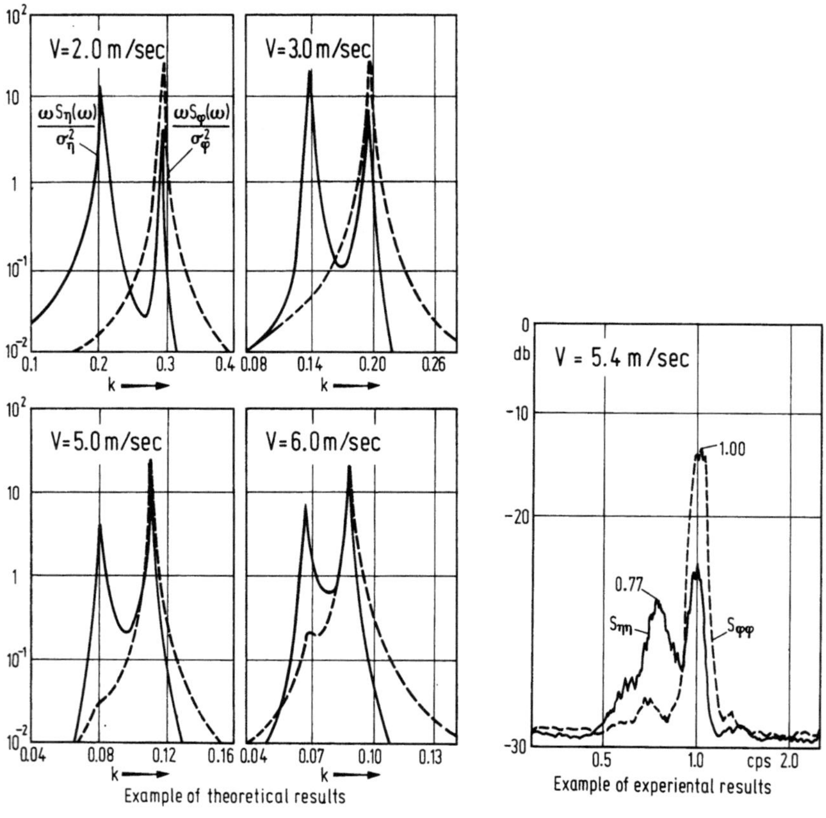

Fig. 2. Power spectra of aerodynamic responses (plate).

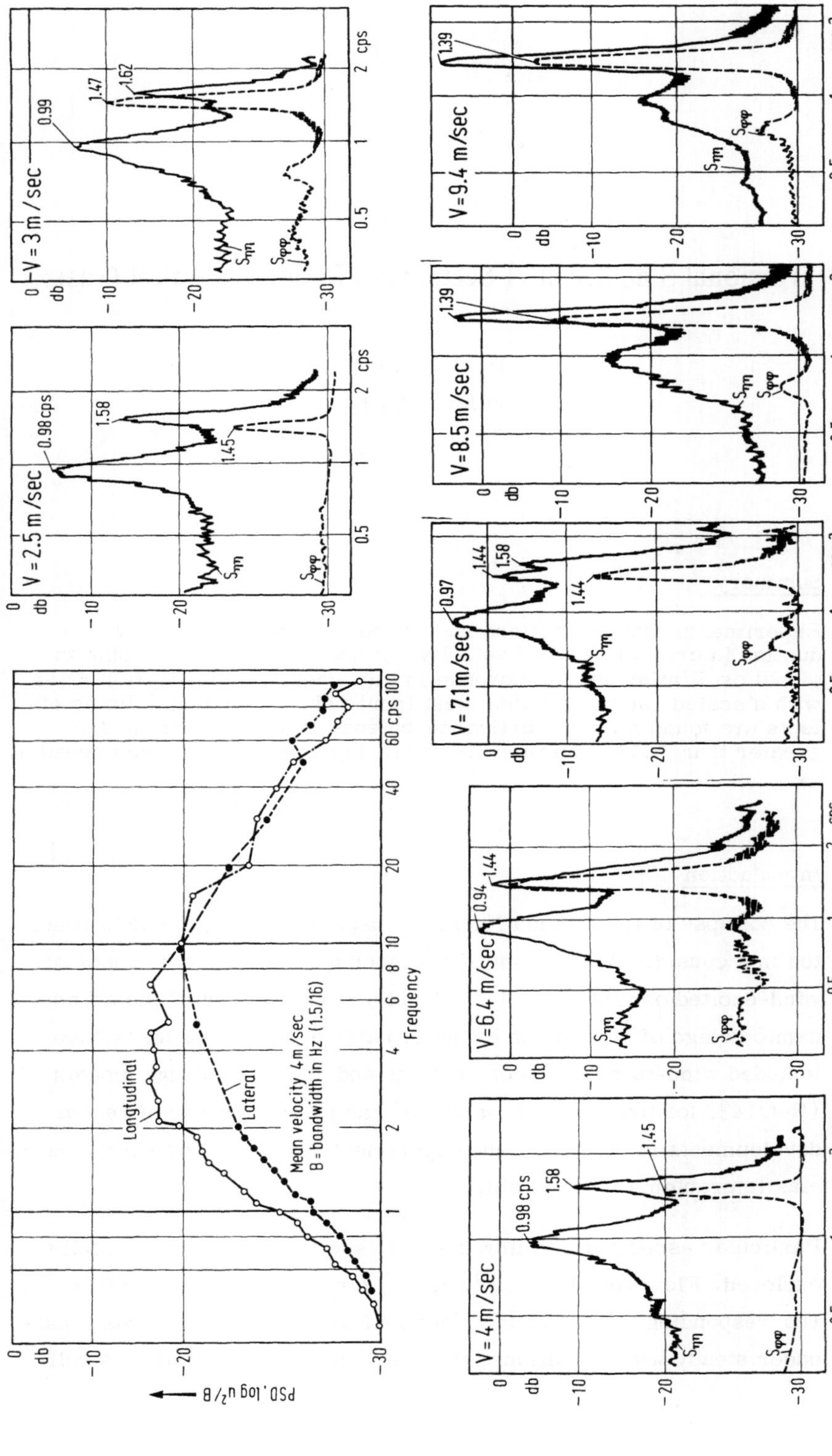

Fig.3. Power spectra of aerodynamic responses (truss-stiffened bridge section).

Torsional Stability of H-Sections in Random Vertical Gusts

By

R. A. Sawyer

Salford, U. K.

Summary

Experiments with H sections suspended in a gust tunnel show a reduction in critical reduced velocity for the onset of catastrophic torsional oscillation and an increase in incremental amplification rates with a scaled random vertical gust input. Maximum destabilising effects are found for 5 % vertical turbulence level and section d/b greater than 0.15, where d is girder depth and b is section breadth.

Introduction

The collapse of the Tacoma Narrows suspension bridge in 1940 resulted in a considerable amount of research activity into the causes of wind-excited oscillations [1,2,3] and a reassessment of the aerodynamic design of suspension bridges. Later design procedures have included wind-tunnel tests on sections and complete bridge models (Ref. [4], for example). Scanlan [5] has given a review of recent developments in analytical and experimental methods for solving aeroelastic problems in this field.

Particular aspects of the flow about H sections have not been fully explored. Flow visualisation tests [6] for a section of d/b = 0.2 (corresponding to the original Tacoma Narrows section) reveal that under steady wind conditions, the main torsional oscillation is self-excited by a stall flutter mechanism at small amplitudes. This involves an out-of-phase growth and decay of the recirculation regions of flow behind the front girder, with cyclic reattachment of the separated flow to the top of the rear girder. At larger amplitudes the reattachment location varies between the bridge deck and the rear girder, until finally an eddy shedding mechanism sets in whereby the air which feeds into the recirculation region behind the front girder is shed into the main stream. This final phase of the motion produces a violent increase in the rate of growth of torsional oscillations. However, the initial instability does not originate from a vortex-shedding mechanism.

Since the origins of the torsional instability lie in cyclic variations in the geometry of separated flow regions, it is to be expected that H sections will be sensitive to vertical components of the wind. The experiments described in this paper concern the response of such sections suspended in the air stream of a blower tunnel (Fig. 1).

Experimental Method

The gust tunnel [1] produces at the model an airstream of steady longitudinal velocity but variable vertical velocity which may be controlled by a function generator or magnetic tape input to an electro hydraulic actuator which determines the incidence of a linked array of aerofoils placed across the tunnel exit. The random gust spectrum used for these experiments is shown in Fig. 2 and has been chosen to represent a typical scaled atmospheric turbulence spectrum. In the experiments the tunnel velocity U was varied to obtain the required range of reduced velocities U/nb, while the model natural frequency n and the gust frequency spectrum remained substantially invarient.

The H sections have constant deck dimensions (span l, 50.8 cm; breadth b, 32.4 cm; thickness, 1.1 cm) with alternative sets of girders of 1.1 cm thickness. End plates are used to give approximately two-dimensional flows and the models are suspended on vertical wires and springs at each of the four corners. The static moment coefficients C_m of the sections shown in Fig. 3 were obtained by integrating pressure distributions.

Steinman's Theory

The basis of Steinman's theory [2] is the Kármán and Sears thin aerofoil theory [8] for aerodynamic forces in unsteady motion. Most of the difficulties in applying attached-flow theory to bridge sections with separated flows were dealt with by Steinman who used pressure distributions on static straight and curved models to determine the unsteady moment functions. Steinman's expression for logarithmic increment δ is

$$D = \frac{I\delta}{\rho b^4 l} = -\frac{1}{8}\left(\frac{G_2 S_2}{\pi}\frac{U}{nb} + F_4 S_4\right)\frac{U}{nb},$$

where $G_2 = \int_0^1 y_1 \sin\Phi\left(\frac{1}{2} - \frac{x}{b}\right) d\left(\frac{x}{b}\right)$, $y_1 = \frac{\partial}{\partial x}(Cp_L - Cp_U)$, straight model,

$S_2 = \int_0^1 y_1 d\left(\frac{y}{b}\right)$, $y_3 = \frac{\partial}{\partial \beta}(Cp_L - Cp_U)$, curved model,

$F_4 = \int_0^1 y_3 \cos\Phi\left(\frac{1}{2} - \frac{x}{b}\right) d\left(\frac{x}{b}\right)$, $\Phi = \frac{x}{b}\frac{2\pi nb}{U}$, lag function,

$S_4 = \int_0^1 y_3 d\left(\frac{x}{b}\right)$, x is the distance from front girder.

α is the straight model incidence, β is the curved model deck incidence at front girder relative to mid-deck.

The expression for D therefore contains an out-of-phase incidence and an in-phase angular velocity effect.

Results and Discussion

Figures 4 and 5 show that the critical reduced velocities for onset of the main torsional oscillation. The wind-off residual damping of the models is given by $D_R = 1.26$ and includes both mechanical and static aerodynamic damping. For the model $I/\rho b^4 l = 30.1$, where I is the movement of inertia and ρ is the air density. It is found that the vertcal gusts are destabilising, particularly for sections with d/b greater than 0.15. A vertical turbulence level of 5 % gives maximum instability; for higher levels the sections are thrown out of oscillation by the larger gusts before substantial amplitudes can build up.

Figure 6 shows critical double amplitudes a_{crit} below which oscillations are damped; and incremental amplification rates D. The destabilising effects of the gusts give increments in D comparable with the residual damping D_R and lead to considerable uncertainty as to the precise valves of δ predicted by Steinman. The Tacoma Narrows section is particularly sensitive in this respect. It appears that the transverse gusts cause excursions into regions of instability above the critical amplitudes for a significant proportion of the time. The ratio of α_{rms} to a_{crit} is 0.56 for instability corresponding to a probability of exceedence of critical amplitude of 0.38 for a Gaussian input.

Concerning Steinman's theory, a scrutiny of flow-visualisation photographs reveals both incidence and angular velocity effects to be present, although the complete changes in flow pattern which occur with increasing amplitude make it difficult to justify the sweeping assumptions of the linear analysis.

References

1. Farquharson, F.B. et al.: Aerodynamic Stability of Suspension Bridges. Bulletin No. 116 (1950 - 1954) University of Washington Experimental Station.

2. Steinman, D.B.: Aerodynamic Theory of Bridge Oscillations. Trans. ASCE 115 1180 (1950).

3. Steinman, D.B.: Rigidity and Aerodynamic Stability of Suspension Bridges. Trans. ASCE 110, 439 (1945).

4. Scruton, C.: An Experimental Investigation of the Aerodynamic Stability of Suspension Bridges with Special Reference to the Proposed Severn Bridge. Pro. Inst. Civil. Engrs. 1, 189 (1952).

5. Scanlan, R.H.: The Suspension Bridge: its Aeroelastic Problems. ASME Paper 71-Vibr-38, Vibrations Conference, Toronto, 1971.

6. Sawyer, R.A.: The Aerodynamic Stability of H Sections. University of Salford, Mech. Engng. Report (1972).

7. Sawyer, R.A.: 0.7 m x 1 m Low Speed Gust Tunnel. University of Salford, Mech. Engng. Report (1970).

8. Karman, Th., Sears, W.R.: Airfoil Theory for Non-Uniform Motion. J. Aero. Sci. 5, 379 (1938).

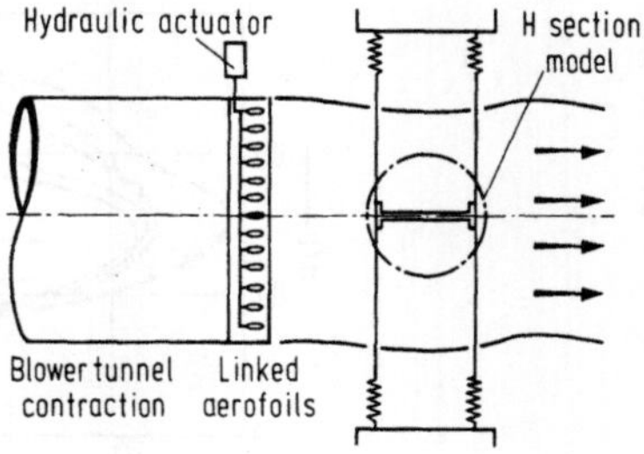

Fig.1. Schematic arrangement of gust tunnel and H section suspension.

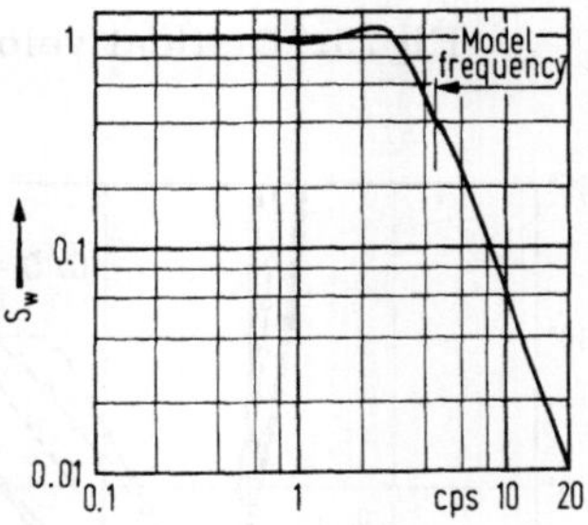

Fig.2. Gust power spectrum.

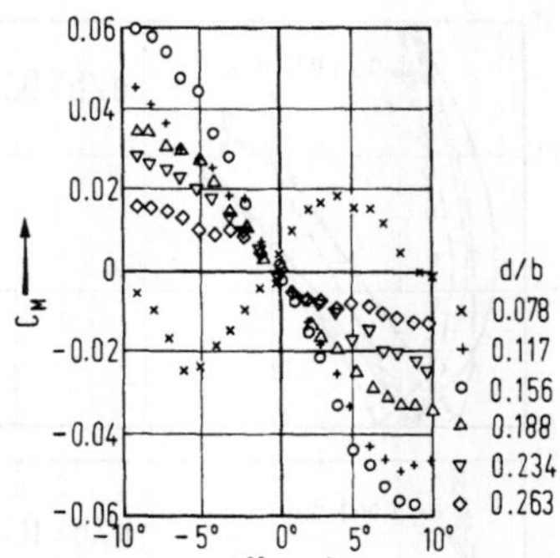

Fig.3. Section static moment coefficients.

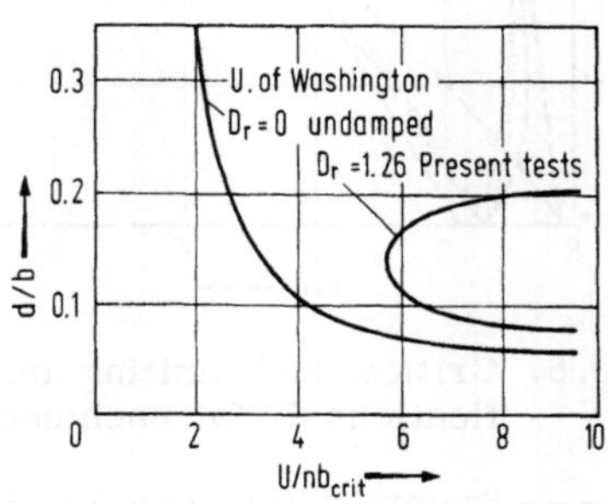

Fig.4. Critical velocities for flow onset of torsional oscillations in steady flow.

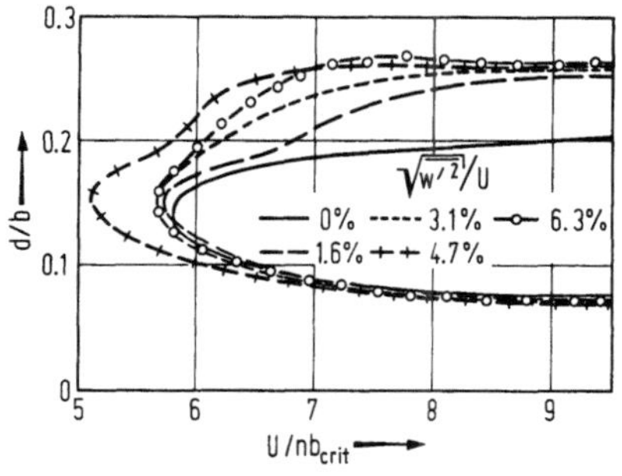

Fig. 5. Critical velocities for flow with vertical gusts.

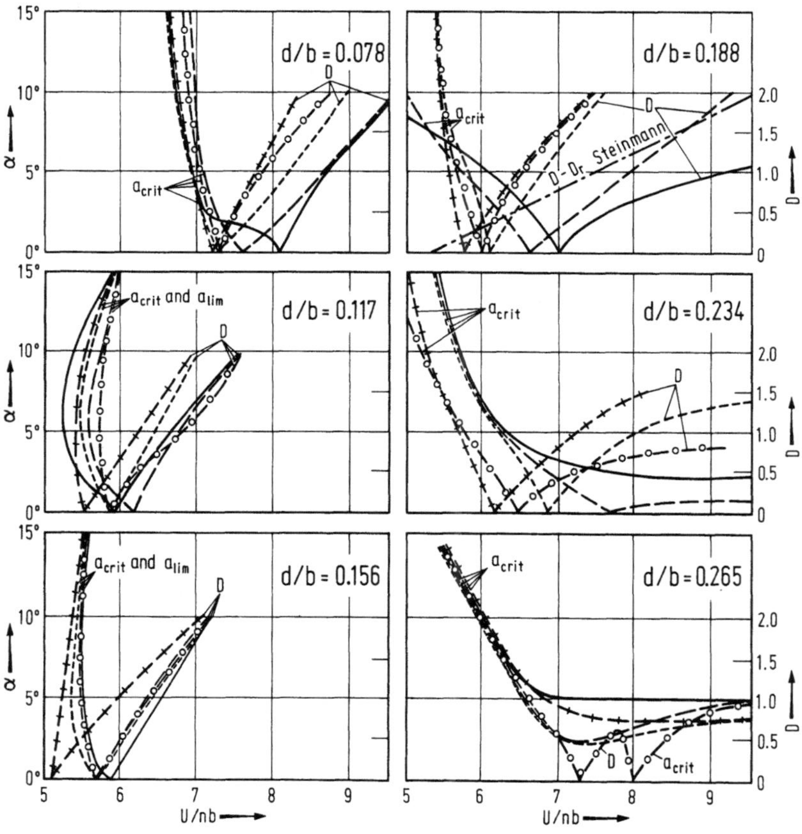

Fig. 6. Critical and limiting amplitudes a, with incremental amplifications D for each section.

----- Steinman's theory for d/b = 0.2

TECHNICAL SESSION E

"Flow-Induced Vibrations of Bluff Bodies"

Chairman: A. Roshko, U.S.A.

Co-Chairman: J. Renner, Germany

GENERAL LECTURE

Unsteady Aerodynamics of Wings and Blades

By

Roland Dat

Châtillon, France

Summary

The method used in flutter analysis for evaluating the unsteady aerodynamic forces on a wing subjected to harmonic vibrations of small amplitude in a uniform flow is briefly described. Then it is shown that the theory can be extended to a lifting surface with an arbitrary motion in still air: the linearisation is based on the assumption that the velocity component normal to the lifting surface is small compared with the tangent velocity, but both components can vary arbitrarily. The generalized formulation so obtained can be used for solving aeroelastic problems more complex than the wing problem. Results obtained for a helicopter rotor with an advancing ratio of 0.3 are used to illustrate the agreement between theory and experiment.

Introduction

In most cases, the source of flow-induced vibrations of wings and blades is a variation of the lift, i.e., of the local or overall force acting in a direction normal to the velocity of the main flow. These variations are produced by the motion of the structure, by the flow turbulence, or by a flow separation.

The structures considered in this general lecture, e.g., the airplane's lifting surfaces, wings, tailplanes and fins, are those on which the lift is generated in the best conditions for the theoretician, because the angle of attack remains everywhere small and there is no flow separation. A linearized theory can be formulated in view of that particular application. The lifting surfaces are represented by thin plates or surfaces of pressure discontinuity producing a perturbation velocity field of small amplitude. The fluid is assumed inviscid, and a boundary value problem can be defined, in which the boundary conditions are obtained by stating that there is no flow separation.

The theory is currently used in connection with numerical procedures in the calculations performed by the designers to predict the flutter boundaries of airplanes, or the response to the atmospheric turbulence. Wind-tunnel tests performed in several countries show that the theoretical results agree remarkably well with experiment when the Mach number is not too close to unity and when the angle of attack remains effectively small everywhere. This paper shows that a lifting surface theory based on the same fundamental assumptions of inviscid fluid, small perturbations and no flow separation can be formulated for configurations or motion more complex than those of the wing, in view of the application to compressor blades and propeller and helicopter blades.

The conditions in which the lift is generated on the blades are less satisfactory for the theoretician than they are for the wing, because the angle of attack is generally higher, but it appears that the flow separation occurs only exceptionally. Then the linearized theory can be expected to be valid for the prediction of a range of flutter instabilities occurring in compressor-blade grids, such as bending-torsion flutter or instability due to blade interference. A numerical application to helicopter blades shows that the periodic aerodynamic excitation acting on the rotor in advancing flight can be predicted as well. Other instabilities, such as panel flutter, which is not considered in the paper, have also been treated successfully with a theory based on the same assumptions.

Apart from the aeronautical structures, the approach of the lifting surface theory can be used for the prediction of the vibrations occurring in various structures such as suspension bridges, hydrofoils, or ship propellers. Obviously, the vibrations of blunt bodies, buildings or smoke-stacks, which are considered in the symposium, are due mainly to flow separation and their prediction cannot be based on the same simplifying assumptions. But even if those structures are excepted, the number of remaining problems which can be treated with the lifting-surface theory is large enough to justify a general lecture in the symposium.

Unsteady Aerodynamics of Wings

For many years, the flutter instability of airplane wings, tailplanes, and fins has been the subject of worldwide research, leading to considerable developments of the lifting-surface theory.

The integral equation given by H.G. Küssner [1] and the formulation of the kernel function of C.E. Watkins [2] are generally used for the determination of the pressure distribution over wings of arbitrary planform vibrating with small amplitude in a uniform subsonic or supersonic flow. The theory is based on the assumptions of inviscid fluid and isotropic transformations, and the boundary conditions are provided by stating that the velocity component normal to the wing surface is the same for a point P belonging to the wing surface and for the fluid particle in contact with it. The non-linear effects of thickness and mean angle of attack are neglected and the wing is assumed to be an infinitely thin plate lying in the xy plane of a Cartesian coordinate system moving with the wing at a uniform velocity V (Fig. 1).

Integral Formulation of the Lifting Surface Theory

For the determination of the flutter boundaries, the analysis can be restricted to harmonic motions. The deflection of the lifting surface is defined by:

$$h(x,y,t) = H(n,y)e^{i\omega t} .$$

Similarly the perturbation velocity potential can be written as:

$$\varphi(x,y,z,t) = \Phi(x,y,z)e^{i\omega t}$$

and the pressure difference over the wing:

$$p(x,y,0^+,t) - p(x,y,0^-,t) = \Delta p(x,y)e^{i\omega t} .$$

The downwash or normal velocity φ_z in contact with the lifting surface is directly related to the deflexion law:

$$\varphi_z(x,y,z,t) = \frac{dh}{dt} \quad \text{for} \quad z = 0^{\pm} \text{ on } (A) ,$$

where d/dt is the substantial derivative $\left(\frac{d}{dt} = V\frac{\partial}{\partial x} + \frac{\partial}{\partial t}\right)$.

We get:

$$\Phi_z(x,y,0^{\pm}) = V\frac{\partial H}{\partial x} + i\omega H = w(x,y)$$

where w is used for the given downwash.

The potential Φ can be related to the lift distribution by an integral equation:

$$\Phi(x,y,z) = \iint_{(A)} \mathcal{K}(x-\xi,\, y-\eta,\, z)\, \Delta p(\xi,\eta)\, d\xi d\eta , \qquad (1)$$

where the kernel function $\mathcal{K}$ depends on the Mach number and on the frequency parameter. For a subsonic flow, it is given by:

$$\mathcal{K}(x,y,z) = \frac{z\,e^{-ikx}}{4\pi p_\infty V}\left[\frac{M\,e^{ikX}}{R\sqrt{r^2+x^2}} + \frac{1}{r^2}\int_{-\infty}^{X/r} \frac{e^{ikru}}{(1+u^2)^{3/2}}\,du\right]$$

with

$$M = \text{Mach number} , \qquad R = \sqrt{x^2+\beta^2 r^2} ,$$

$$k = \omega/V , \qquad \beta = \sqrt{1-M^2} ,$$

$$r = \sqrt{y^2+z^2} , \qquad X = \frac{x-MR}{\beta^2} .$$

The proof of this equation will not be given here, since a more gen-

The downwash can be derived from Φ:

$$w(x,y) = \lim_{z=0} \Phi_z(x,y,z)$$

$$= \lim_{z=0} \frac{\partial}{\partial z} \iint_{(A)} \mathcal{K}(x-\xi, y-\eta, z)\, \Delta p(\xi,\eta)\, d\xi d\eta .$$

Usually, the limit value is obtained by derivation of the kernel function $\mathcal{K}$, and a singular integral equation is obtained:

$$w(x,y) = \iint_{(A)} K(x-\xi, y-\eta)\, \Delta p(\xi,\eta)\, d\xi d\eta . \qquad (2)$$

The kernel function K is given by:

$$K(x-\xi, y-\eta) = \lim_{z=0} \frac{\partial}{\partial z} (\mathcal{K}(x-\xi, y-\eta, z)) .$$

The symbol $⨍$ is used to show that the spanwise integral is singular and that the finite part must be taken [3].

The Collocation Method

To solve actual problems, the integral equation (2) is generally used in connection with a collocation method.

The unknown function Δp is approximated by linear combinations of given functions Δp_i:

$$\Delta p(\xi,\eta) = \sum_{i=1}^{N} a_i\, \Delta p_i(\xi,\eta) . \qquad (3)$$

For the sake of convergence, the functions Δp_i must have the same behavior as Δp on the edges of the lifting surface: they must be infinite at the leading edge, zero at the side edges, and zero at the trailing-edge (Kutta-Joukowsky condition).

For the determination of the N unknown coefficients a_i, a set of N points, named collocation points, is defined on the lifting surface.

After substitution of (3) into (2), the coordinates of the collocation points (x_j, y_j) are considered, and a set of N algebraic equations relating the downwash $w(x_j, y_j)$ to the unknown coefficients a_i is obtained:

$$w(x_j, y_j) = \sum_{i=1}^{N} \left[\iint_{(A)} K(x_j - \xi, y_j - \eta)\, \Delta p_i(\xi, \eta)\, d\xi d\eta \right] a_i .$$

Comparison of Theoretical and Experimental Results

The agreement between computed results and wind-tunnel test results measured in subsonic or supersonic flow is generally good, provided that the Mach number is not too close to unity and that the angle of attack remains small enough to avoid non-linear effects. Two typical examples are given in Figs. 2 and 3.

Fig. 2 shows the pressure distribution on a trapezoidal low aspect ratio wing model oscillating in pitch and Fig. 3 gives the curves of frequency and damping ratio against stagnation pressure at constant Mach number for the same model mounted with two degrees of freedom.

The same method can be used with configurations including two or more lifting surfaces such as wing and tailplane or tailplane and fin. The results shown in Fig. 4 caracterize the interference between two rectangular models lying in parallel planes, such as a wing and a horizontal tail. The coefficient K_{jA}^{B} gives the lift induced on the downstream model by a pitching oscillation of the upstream model. The figure shows that the agreement between theory and experiment is satisfactory provided that the distance H between the two parallel planes is not too small (Fig. 4a). When model B lies in the wake of model A (H = 0), larger discrepancies are observed (Fig. 4b).

The Potential Kernel Function Method

In order to avoid the difficulties in evaluating the singular integral (2), a finite difference procedure can be used for the calculation of the downwash w in connection with the collocation method.

Since Φ varies linearly with z, in the vicinity of the lifting surface, as is shown in Fig.5, the limit value of Φ_z for $z = 0$ can be derived from the numerical values of Φ computed for two small but finite values of z. Equation (1) is used instead of (2) and w is given by the two equations:

$$\Phi(x,y,z) = \iint\limits_{(A)} \mathcal{K}(x-\xi, y-\eta, z)\, \Delta p(\xi,\eta)\, d\xi d\eta \,, \tag{4a}$$

$$w(x,y) \simeq \frac{1}{\varepsilon}\,[\Phi(x,y,\varepsilon) - \Phi(x,y,0^+)]\,. \tag{4b}$$

Calculations performed at O.N.E.R.A. [4] show that the error due to the finite difference procedure remains negligible for values of ε up to 2 or 3% of the chord length. The integrations giving $\Phi(x,y,\varepsilon)$ and $\Phi(x,y,0^+)$ are easily performed by the Gaussian numerical method.

Extensions of the Lifting Surface Theory

Unsteady Aerodynamics of Compressor Blades

The formulation of the integral equation is not restricted to an airplane's lifting surfaces. For instance, an equivalent formulation can be used to predict the aerodynamic unsteady forces due to the vibrations of a compressor-blade grid. In an analysis given by J. Leclerc [5], the actual grid is replaced by an infinite two-dimensional rectilinear grid (Fig.6). The angle of attack and blade thickness are assumed small and the problem is linearized.

The integral equation derived by the author relates the downwash on a blade, $w(x)$, to the lift $\Delta p(x)$ when the grid is vibrating harmonically with an equal phase difference between any two adjacent blades:

$$w(x) = \frac{1}{2\pi} \oint_{-1}^{1} \Delta p(\xi) K(x-\xi)\, d\xi\,.$$

The symbol $\oint$ is used to show that the integral is singular and that the Cauchy principal value must be taken.

The integral is an extension of Eq. (2) to account for an infinite number of equally spaced lifting surfaces for the particular case of the two-dimensional flow and of an equal phase difference.

The numerical solutions can be obtained by the collocation method presented above but the accuracy is improved with a more sophisticated method in which both the pressure distribution and the downwash are represented by a linear combination of orthogonal polynomials, as described in Ref. [6].

The more general case of a harmonic vibration with an arbitrary phase difference between adjacent blades can be treated easily, since it can be considered as a linear combination of vibrations with equal phase difference.

Lifting Surface Moving with an Arbitrary Motion

The motion of the lifting surfaces considered in the applications to wings and to the rectilinear grid is a combination of a vibration of small amplitude and of a translation at uniform velocity. A more complex motion must be considered for the prediction of the aeroelastic vibrations occurring on helicopter rotor blades.

A numerical procedure based on a discretization of the blade wakes with a network of trailing and shed vortices was developed by R.M. Miller [7] and by R.A. Piziali [8]. The velocity induced by the wakes is related to the vortex strength by the Biot-Savart law, and the vortex strength is determined by the time history and spanwise distribution of the blade lift. The method was formulated for an incompressible fluid and the compressibility effects are approximated by either the Prandtl-Glauert correction or by empirical methods. The rigorous formulation for a compressible fluid would be very tedious.

The approach used at O.N.E.R.A. is based on an extension of the procedure defined by Eqs. (4a) and (4b) to a lifting surface with an arbitrary motion, the only restriction being that of a small angle of attack. The derivation of the integral equation will be summarized.

Since the motion of the lifting surface is arbitrary, the velocity field will be given with reference to a fixed origin 0. Then a velocity potential φ can be defined, as well as an acceleration potential Ψ. The substantial derivative is equal to the partial derivative, and the wave equation, verified by the velocity potential, can be written as:

$$\nabla^2\varphi - \frac{1}{a^2}\frac{\partial^2\varphi}{\partial t^2} = 0,$$

where a is the sound velocity.

The velocity of the fluid at a point P is given by:

$$\vec{V}(P,t) = \overrightarrow{\mathrm{grad}}(\varphi)$$

and the acceleration:

$$\frac{d\vec{V}}{dt} = \overrightarrow{\mathrm{grad}}\,\frac{\partial\varphi}{\partial t}.$$

The acceleration potential is defined by:

$$\Psi = \frac{\partial\varphi}{\partial t}$$

and the reciprocal relation can be written:

$$\varphi(P,t) = \int_{-\infty}^{t} \Psi(P,t_0)\,dt_0. \qquad (5)$$

Ψ verifies the wave equation as well as φ:

$$\nabla^2\Psi - \frac{1}{a^2}\frac{\partial^2\Psi}{\partial t^2} = 0 \qquad (6)$$

and the pressure is given by:

$$p - p_\infty = \rho_\infty \Psi. \qquad (7)$$

Among the fundamental singular solutions of the wave equation (6), which can be used to build an integral solution, the acceleration doublet is particularly interesting because it is equivalent to a lifting element. It can be derived from the acceleration source.

Let us consider a source of strength $q(t)$ moving along a path defined by a vector equation $\overrightarrow{0P}_0(t)$ or, more briefly, $\vec{P}_0(t)$. If the fluid is incompressible, the acceleration potential induced at a point P is given by the equation:

$$\Psi_s(P,t) = \frac{q(t)}{4\pi|\vec{P} - \vec{P}_0(t)|} .$$

The acceleration doublet can be built with two sources of opposite strength located at an infinitely small distance along the axis of the doublet defined by a unit vector $\vec{n}_0(t)$.

The acceleration potential at P is derived from Ψ_s:

$$\Psi_D(P,t) = \overrightarrow{\mathrm{grad}}\,(\Psi_s) \cdot \vec{n}_0 = \frac{q(t)\left[\vec{P} - \vec{P}_0(t)\right] \vec{n}_0(t)}{4\pi|\vec{P} - \vec{P}_0(t)|^3} . \tag{8}$$

The pressure field is given by substitution of Ψ_D into Eq. (7):

$$p(P,t) - p_\infty = -\rho_\infty \frac{q(t)\left[\vec{P} - \vec{P}_0(t)\right] \vec{n}_0(t)}{4\pi|\vec{P} - \vec{P}_0(t)|^3} \tag{9}$$

and the velocity potential is obtained by substitution of Ψ_D into Eq. (5):

$$\varphi_D(P,t) = \frac{1}{4\pi}\int_\infty^t \frac{q(t)\left[\vec{P} - P_0(t_0)\right] \vec{n}_0(t_0)}{|\vec{P} - \vec{P}_0(t_0)|^3}\, dt_0 \tag{10}$$

so that φ_D is determined by the time-history of the position, $\vec{P}_0(t_0)$, orientation, $\vec{n}_0(t_0)$, and strength, $q(t_0)$, of the doublet. The integration path $\vec{P}_0(t_0)$ is the wake of the doublet.

It can be shown from Eq. (9) that a lift $F(t) = \rho_\infty q(t)$ is acting on the doublet along the axis $\vec{n}_0$. Consequently, an element of lifting surface with area $d\sigma$, on which a lift $\Delta p d\sigma$ is acting along the normal direction $\vec{n}_0$, is equivalent to a doublet of strength

$$q(t) = \frac{\Delta p(t) d\sigma}{\rho_\infty} \tag{11}$$

and with axis $\vec{n}_0$.

The velocity potential due to a lifting surface is obtained after substitution of Eq. (11) into Eq. (10) and after integration over the surface (Fig.7):

$$\varphi(P,t) = \frac{1}{4\pi\rho_\infty} \iint\limits_{(A)} \int\limits_{-\infty}^{t} \frac{\Delta p(P_0,t_0)\left[\vec{P}(t) - \vec{P}_0(t_0)\right]\vec{n}_0(P_0(t_0))}{|\vec{P}(t) - \vec{P}_0(t_0)|^3} dt_0 d\sigma_0 . \tag{12}$$

When the compressibility effects cannot be neglected, account must be taken of the time delay necessary for the transmission of perturbation from P_0 to P.

For a non-moving source, the acceleration potential would be given by:

$$\Psi_s(P,t) = \frac{q(\tau)}{4\pi|\vec{P} - \vec{P}_0|}$$

with

$$t - \tau = \frac{|\vec{P} - \vec{P}_0|}{a} .$$

When the source is moving, Ψ_s depends on the velocity of the source at time τ (Cf. Feshbach and Morse [9]).

$$\Psi_s(t) = \frac{q(\tau)}{4\pi|\vec{P} - P_0(\tau)| \left| 1 - \dfrac{\vec{V}_0(\tau)\left[\vec{P} - \vec{P}_0(\tau)\right]}{a|\vec{P} - \vec{P}_0(\tau)|} \right|} \tag{13}$$

with:

$$t - \tau = \frac{|\vec{P} - \vec{P}_0(\tau)|}{a},$$

where $\vec{V}_0$ is the velocity of the source at time τ $\left(\vec{V}_0 = \frac{d\vec{P}_0}{d\tau}\right)$.

The acceleration potential due to a doublet moving along a path defined by a vector $\vec{P}_0(\tau)$ with an orientation $\vec{n}_0(\tau)$ and a strength $q(\tau)$ can be derived from Eq. (13) as it is shown in Ref. [9]. Then the velocity potential and the integral equation are obtained in the same way as for the incompressible fluid. Assuming that the velocity remains everywhere smaller than the sound velocity, the equation can be written as:

$$\varphi(P,t) = \frac{1}{4\pi\rho_\infty} \iint_{(A)} \int_{-\infty}^{\tau} \frac{d(\Delta p(P_0(\tau_0)))}{d\tau_0} \mathcal{K}_I(\vec{D},\tau_0)\, d\tau_0\, d\sigma_0 + $$

$$+ \frac{1}{4\pi\rho_\infty} \iint_{(A)} \int_{-\infty}^{\tau} \Delta p(P_0,\tau_0)\, \mathcal{K}_{II}(\vec{D},\tau_0)\, d\tau_0\, d\sigma_0 \qquad (14)$$

with

$$\vec{D} = \vec{P}(t) - \vec{P}_0(\tau_0),$$

$$t - \tau = \frac{\vec{D}(\tau)}{a},$$

$$\mathcal{K}_I = \frac{\vec{D}\vec{n}_0(\tau_0)}{4\pi|\vec{D}|\left[a|\vec{D}| - \vec{V}_0\vec{D}\right]},$$

$$\mathcal{K}_{II} = \frac{\left[a^2 - |\vec{V}_0|^2\right]\vec{D}\vec{n}_0 + \vec{D}\vec{\gamma}_0\vec{D}\vec{n}_0}{4\pi|\vec{D}|\left[a|\vec{D}| - \vec{V}_0\vec{D}\right]^2} + \frac{\vec{D}\,\dfrac{d\vec{n}_0}{d\tau_0}}{4\pi|\vec{D}|\left[a|\vec{D}| - \vec{V}_0\vec{D}\right]},$$

$$\vec{V}_0 = \frac{d\vec{P}_0}{d\tau_0}; \quad \vec{\gamma}_0 = \frac{d\vec{V}_0}{d\tau_0}.$$

Equation (14) is valid, if the angle of attack is small everywhere, i.e., if $\vec{V}_0\vec{n}_0$ remains small compared to $|\vec{V}_0|$. It is a generalized form of Eq. (1) which can be applied to any lifting surface moving arbitrarily. In the same way, Eq. (1) is obtained from Eq. (14), if the lifting surface is moving at constant speed along the z-axis of a Cartesian coordinates system, with the normal vector $\vec{n}_0$ directed along the z-axis and with a harmonic pressure difference $\Delta p(\xi,\eta)e^{i\omega\tau_0}$.

Application to a Helicopter

Equation (14) can be used with a collocation method or an iterative procedure, with a formulation similar to Eqs. (4a) and (4b), for the calculation of the unsteady-lift distribution on helicopter blades. A particular application to a three-blade experimental rotor, built by the S.N.I.A.S.[1] and tested in the S1 wind tunnel of Modane, is described by J.J. Costes in Ref. [10].

The lift and normal velocity on the blades are periodic functions of time (or azimuth angle). The lift on a blade may be represented by a linear combination of given functions:

$$\Delta p(r,\theta) = \sqrt{1-\eta^2}\,\sqrt{\frac{1-\xi}{1+\xi}}\,\sum_{i=1}^{m}\sum_{j=1}^{n} L_i(r)\,F_j(\theta)\,X^{ij}, \qquad (15)$$

where r is the radial (or spanwise) coordinate (see Fig.8), θ is the azimuth ($\theta = \Omega t$), $\eta(r)$ a radial coordinate defined in such a way as to take the value ± 1 at the side edges of the blade, ξ is a chordwise coordinate taking the value -1 at the leading edge and $+1$ at the trailing edge, $L_i(r)$ are given functions of r, and $F_j(\theta)$ are periodic functions of θ.

The function $\sqrt{1-\eta^2}\,\sqrt{(1-\xi)/(1+\xi)}$ has the appropriate singularities on the edges: it is infinite at the leading edge, $\xi = -1$, zero at the trailing edge, $\xi = 1$, and zero at the side edges, $\eta = \pm 1$.

---

[1] Société Nationale Industrielle Aérospatiale.

After substitution of Eq. (15) into (14), the potential at a point P is given in terms of the X^{ij}:

$$\varphi(\vec{P},t) = \sum_{ij} \Phi_{ij}(\vec{P},t)\, X^{ij} , \tag{16}$$

where Φ_{ij} is obtained by substitution of $\sqrt{1-\eta^2}\,\sqrt{(1-\xi)/(1+\xi)}$ $L_i(r)F_j(\theta)$ into Eq. (14).

The normal velocity at a particular point on the blade may be derived from Eq. (16) by the finite difference procedure:

$$w(P,t) \simeq \frac{1}{\varepsilon}\,[q(\vec{P}+2\varepsilon\vec{n},t) - \varphi(\vec{P}+\varepsilon\vec{n},t)] , \tag{17}$$

$\vec{P}$ being the point considered on the blade, $\vec{n}$ the normal vector, and ε a distance of the order of 1 or 2% of the chord length.

m collocation points are selected on each blade and n blade positions or values of the azimuth are considered, in order to define $m \cdot n$ collocation stations on the rotor disk. Then the values of the velocity normal to the blade at the $m \cdot n$ collocation stations can be related to the $m \cdot n$ unknown coefficients X^{ij} by algebraic equations, through Eqs. (16) and (17).

The integrations giving the Φ_{ij} are performed numerically with the Gaussian method. Only one integration point was taken for the chordwise integration (variable ξ), located at the forward quarter-chord ($\xi = -0.5$ according to the rules of the generalized Gaussian integration procedure). It is as if the lift were concentrated on the forward quarter-chord line of the blades.

The calculations were performed for a forward flight with an advancing ratio of 0.3. No flow separation was observed during the test on the blade sections where the pressure distribution was measured. As the motion of the blades was measured, the normal velocity at the collocation points was known and could provide the input of the calculation. The output was the lift.

The variations of the blade lift with azimuth is shown in Fig. 9 for several radial stations. The comparison with experiment is rather satisfactory, and we can infer that the theory is able to predict approximately the periodic excitation of a helicopter rotor in forward flight at moderate values of the advancing ratio.

The lift was not measured for values of the radial coordinate smaller than $r/R = 0.52$. If it had been, larger discrepancies would have been probably observed on the side of the retreating blade, in the region of the reverse circle where the angle of attack takes high values and where flow separation can occur. For the value of the advancing ratio which was considered, the lift remained small inside this circle and it was assumed that it had only a small influence on the generalized forces acting on the rotor.

But if the advancing ratio increases, the size of the reverse circle increases at the same time and the flow separation associated with the high angle of attack can provide an excitation or a negative damping which cannot be predicted by the linearized theory.

Another limitation would occur for flight configurations in which the blade wakes remain in the plane of the rotor disk. Then each blade lies in the wake of the preceding one, a situation similar to that illustrated by Fig. 4b, for which large discrepancies are observed between experiment and theory. But, fortunately, this situation remains exceptional.

Concluding Remarks

The comparison of the theoretical and experimental results, for the helicopter rotor, shows that the lifting surface theory can provide a useful tool for the prediction of highly complex aeroelastic phenomena.

The limitations come from the non-linear effects such as shock-waves at transonic speeds or high angle of attack, and from viscosity effects. In many practical helicopter flight configurations, the blade angle of attack remains small on the major part of the rotor disk, but it may take high values in a region of greater or lesser extent on the side of the retreating blade. In such a situation, the approach using the linearized theory cannot be expected to be as accurate as it is for the wing. Nevertheless, the comparison with experiment is rather satisfactory.

If the aeronautical structures are excepted, this fundamental theoretical tool seems to be used only rarely to solve aeroelastic or hydroelastic problems of actual structures. In their calculations, the designers generally use the two-dimensional unsteady flow theory (which is a simplified form of the lifting-surface theory) in connection with the strip method. But the calculations performed by the airplane designers have proved that the three-dimensional effects are large, except for wings of high aspect ratio, and that the two-dimensional theory is insufficient for most actual structures.

This paper shows that a more general theory can be worked out with modern computers. Its actual possibilities are not yet well established. Considering the fundamental assumption, we might infer that the theory is not valid as soon as non-linear or viscous effects are not negligible. But a few examples show that there are situations differing from the fundamental assumptions, in which it can nevertheless be used in connection with simplified models. For example, in Ref. [11] a simplified boundary-layer model made with an inviscid linearized flow is used to predict the qualitative effect of the boundary layer on panel flutter, and in Ref. [12] the same model is used with limited experimental data to provide a quantitative prediction of the flutter boundaries.

References

1. Küssner, H.G.: General Airfoil Theory. NACA TM 979 (1941).

2. Watkins, C.E., Runyan, H.L., Woolston, D.S.: On the Kernel Function of the Integral Equation Relating the Lift and Downwash Distributions of Oscillating Finite Wings in Subsonic Flow. NACA Report 1234 (1955).

3. William, D.E.: Three-Dimensional Subsonic Theory. AGARD Manual of Aeroelasticity, Vol. II, Chapter 3.

4. Akamatsu, Dat, R.: Calcul par la méthode du potentiel des forces instationnaires agissant sur un ensemble de surfaces portantes. Rech. Aérosp., No. 1971-5.

5. Leclerc, J.: Théorie linéarisée de l'écoulement subsonique instationnaire dans une grille droite bidimensionnelle. Rech. Aérosp., No. 1971-3.

6. Dat, R., Darovsky, L., Darras, B.: Considérations sur la solution matricielle du problème portant instationnaire en subsonique et application aux gouvernes. Note technique ONERA., No. 135 (1968).

7. Miller, R.H.: Rotor Blade Harmonic Air Loading. AIAA J. 2, No. 7 (July 1964).

8. Piziali, R.M.: Method for Solution of the Aeroelastic Response Problem for Rotating Wings. Symposium on the Noise and Loading Actions on Helicopter V/STOL Aircraft (August, September 1965), Institute of Sound and Vibration Research, University of Southampton, England.

9. Morse, Feshbach: Methods of Theoretical Physics, Part 1, p. 841, International Student Edition, McGraw-Hill.

10. Costes, J.J.: Calcul des forces aérodynamiques instationnaires sur les pales d'un rotor d'hélicoptère. Rech. Aérosp. 1972-2.

11. Fung, Y.C.: Some Recent Contributions to Panel Flutter Research. AIAA J. 1, No. 4 (1963).

12. Dat, R.: Influence de la couche limite sur le flottement d'un panneau plan en hypersonique faible. Note technique ONERA, No. 116 (1967).

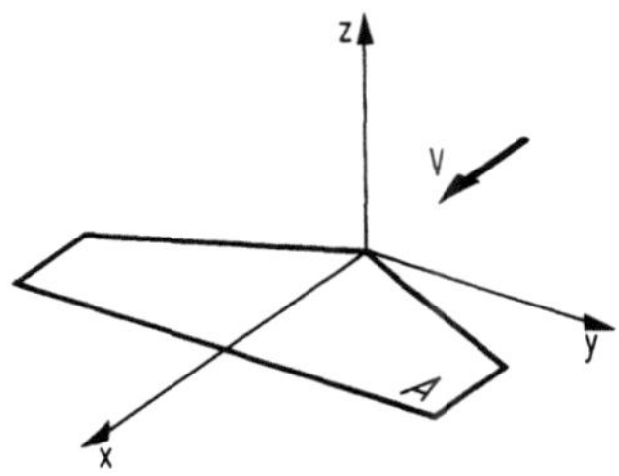

Fig. 1. Definition sketch.

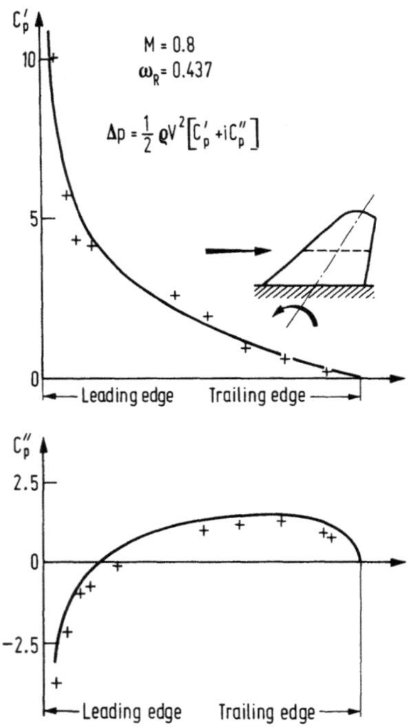

Fig. 2. Semispan chordwise pressure distribution on a model oscillating sinusoidally.

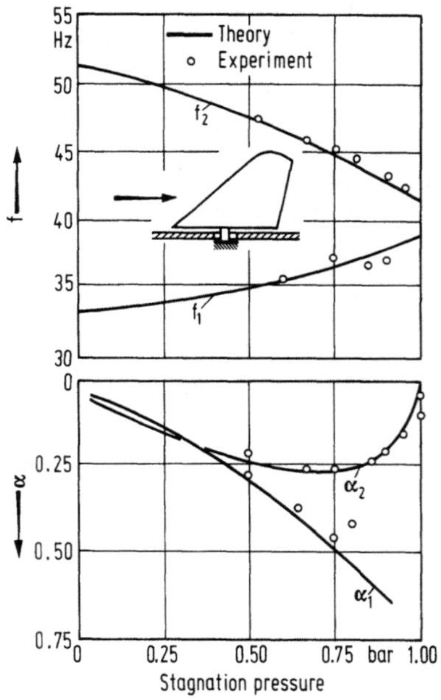

Fig. 3. Frequencies and damping ratio on a two degree of freedom flutter model at Mach 0,8.

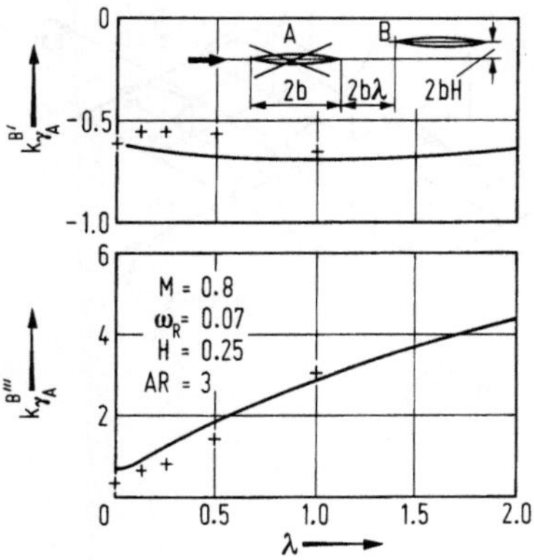

Fig. 4a. Interference between two rectangular models: lift on B due to a pitch-oscillation of A.

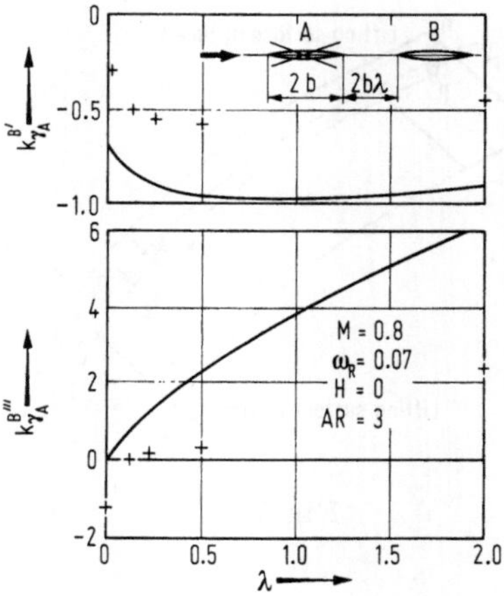

Fig. 4b. Interference between two rectangular models: lift on B due to a pitching oscillation of A.

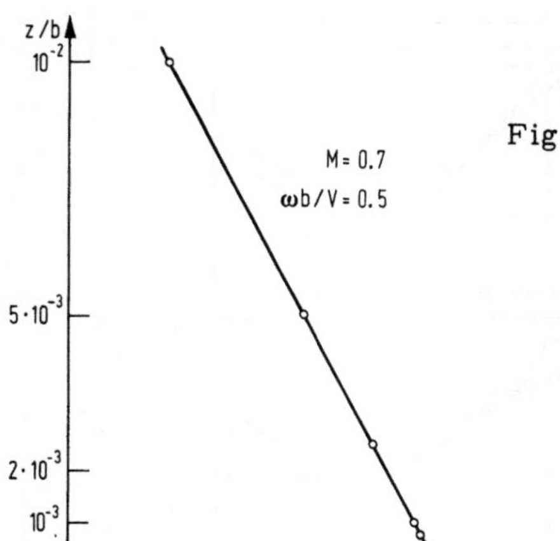

Fig. 5. Velocity potential in the vicinity of a lifting-surface.

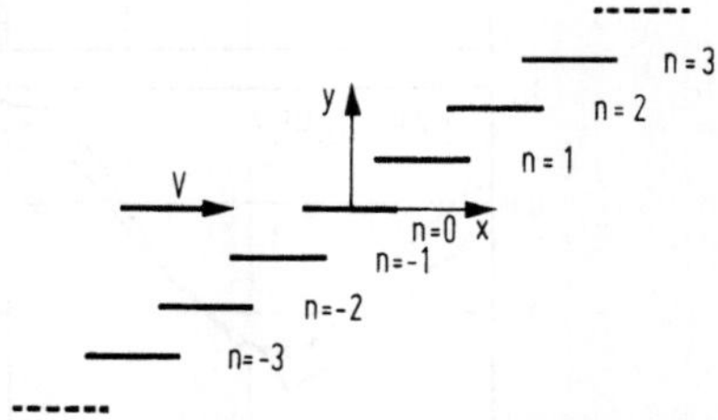

Fig. 6. Two-dimensional rectilinear blade grid.

$h(xt) = H(x)e^{i(\omega t + n\vartheta)}$

$p^+ - p^- = \Delta p(x)e^{i(\omega t + n\vartheta)}$

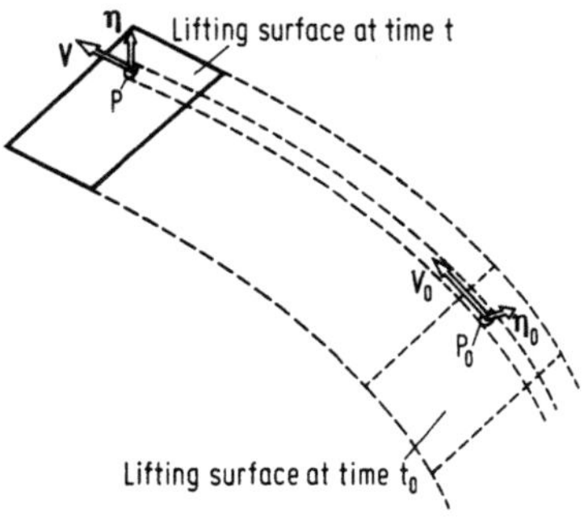

Fig.7

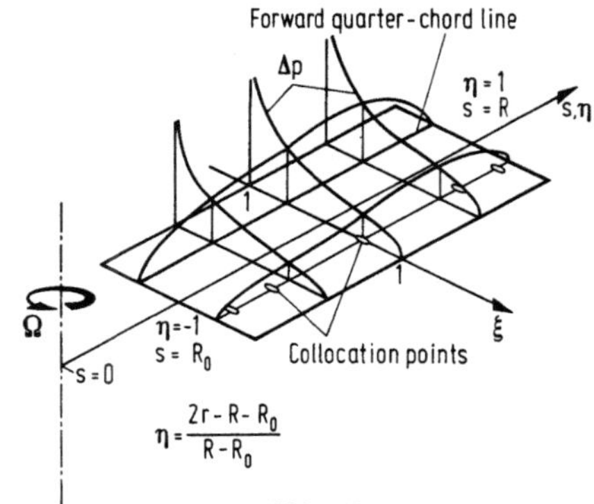

Fig.8

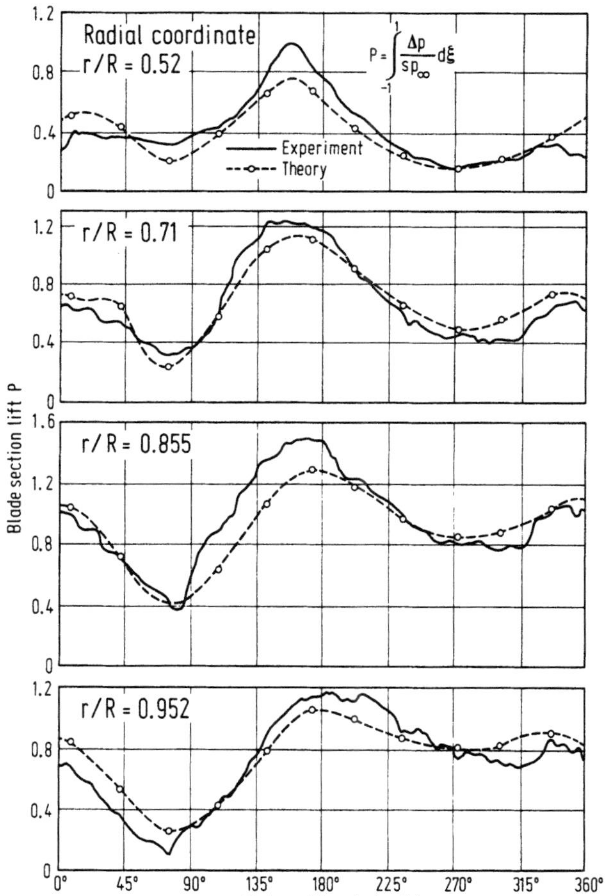

Fig. 9

CONTRIBUTIONS

Unsteady Pressure Distributions Due to Vortex-Induced Vibration of a Triangular Cylinder

By
C. F. M. Twigge-Molecey, W. D. Baines
Toronto, Canada

Summary

A wind-tunnel investigation was conducted on a long cylinder of equilateral triangle cross-section the point being directed upstream. Instantaneous pressure was recorded at twenty points around the section and from this data root-mean-square pressure fluctuation and phase relative to the lift force were determined. Measurements were made for the cylinder fixed rigidly and for the cylinder mounted on springs. The pressure distributions were subsequently integrated to give the rms lift coefficient. It was found that the frequency of shedding of the vortices was independent of the cylinder motion. The lift coefficient was virtually constant except near resonance where a minimum occurred for wake frequency slightly less than the natural frequency of the cylinder and support system and a maximum occurred for the wake frequency slightly greater than the natural frequency.

Introduction

The vibration of a circular cylinder mounted normal to a wind stream at speeds where a von Kármán vortex trail is produced has been the subject of many experimental studies. It has been observed that the exciting force produced by the flowing fluid is not a simple sinusoidal function. Furthermore, an interaction occurs in which the motion of the cylinder modifies the lifting force. Very little progress has been made in defining this integrated fluid response because of the complexity of the fluid flow. For the circular cylinder the separation lines of the boundary layer at the rear vary during the production cycle of the vortices and change as the cylinder moves. As a consequence, the development of the boundary layer on the front part of the cylinder is always unsteady. For this study a simpler shape was chosen, an equilateral triangle with a point directed upstream, so that the boundary layer separation would always occur at the downstream corners. It was hoped thereby that a simpler flow field would result and that the interaction of this with the elastically supported cylinder would be simple enough to allow analysis.

The experimental study was made to determine the magnitudes of the unsteady forces and lateral displacement as the velocity of flow, and elastic characteristics of the cylinder support were varied. For comparison purposes, the static and fluctuating forces on a rigidly mounted cylinder were also measured. Observation of the motion and pressures readily disclosed that small lift and drag were produced by the turbulence in the wake but the dominant components were due to the vortex shedding. The lift force fluctuated at the frequency of shedding or a vortex pair, the Strouhal frequency, while the drag fluctuated twice as rapidly at the frequency shedding of a single vortex. The amplitude of this single component fluctuation did vary apparently because the size and strength of the shed vortices varied. All fluctuations were therefore measured with a root-mean-square voltmeter. Pressure was measured at twenty points around the periphery at the cylinder and integrated to produce the lift and drag coefficients. The phase relationship between the pressure fluctuations at all of the points was also determined and included in the integration as required.

The Strouhal frequency f_s was measured with a hot-wire anemometer mounted close to the vortex train. It was found that the Strouhal number $f_s D/U$ decreased slightly as the Reynolds number increased but was independent of the cylinder motion. This confirms the results of Toebes and Eagleson [2] that there is not a locking of the wake frequency to the cylinder vibration frequency. Because of this independence the results can be treated in the manner standard for problems in mechanical vibrations. Frequencies are expressed as a ratio relative to the natural frequency of the mechanical system and are used in place of a Reynolds number to describe the wind speed.

Description of Experiments

The cylinder was constructed of balsa wood with equilateral triangle cross-section each side of which had a length of four inches. It was 46 inches long and mounted in the center of a wind tunnel of cross-section four feet by eight feet. The velocity range was from zero to twenty-five feet per second. The mass of the cylinder was controlled by the addition of steel weights. Figure 1, a photograph of the model mounted in the wind tunnel, shows the end plates installed to reduce end effects. These reduced the clear length of the cylinder to 36 inches.

Twenty pressure openings were located on one forward face as noted on Table 1 and the downstream face all at the same longitudinal section. The tubes from each tap could be connected to a differential pressure transducer which had a linear response for pressures up to one inch of water and frequencies from 0 to 3000 cps. The d.c. electrical signal output from the transducer was connected to a Bruel and Kjaer RMS voltmeter or to a high-speed chart recorder. Because all of the pressure signals were sinusoidal it was possible to determine the phase relationship between the signals at two points by measuring the RMS of the differential pressure, which together with the RMS value for each pressure yielded the phase from the cosine rule.

Table 1. Pressure-Tap Locations

Pressure tap No.	x_1/D
9	0.06
8	0.20
7	0.30
6	0.40
5	0.50
4	0.57
3	0.68
2	0.78
1	0.96
J	1.00

x_1 is the distance of the pressure tap from the leading edge.

Results for the Rigid Cylinder

With both ends on the cylinder clamped, the rms pressure fluctuations plotted in Fig.2 were obtained. These show the steady decrease of pressure as the Reynolds number is increased for ten points located on the forward face. The amplitude and also the scatter of the points increase steadily in the direction of flow. The phase of the sinusoidal pressure fluctuation relative to tap number 9, which is closest to the leading edge, is presented in Fig.3 for the same range of Reynolds number. The small variation which appears in the plotted points for different Reynolds number cannot be regarded as significant. There is some doubt about the accuracy of the cosine rule for the signals in which there is a turbulence component. For a Reynolds number of 1.9×10^4 the phase shift relative to tap number 9 is given in Fig.4. On this figure is also plotted a solid line which is a phase shift which would result from a pressure oscillation convecting downstream at a velocity of 0.85U, the velocity of the vortices in the wake. It is seen that this simple concept is in agreement with all of the data except for the point at the trailing edge. The pressure in this region is more likely to be controlled by the growing vortex next to the back face and its velocity as it is shed. This is consistent with measurements made of the vortex speed as reported by Twigge-Molecey [3]. The slower speed of the vortex during the shedding process would produce a larger phase shift when viewed as a travelling pulse. The data on Figs.2 and 3 can be combined to produce instantaneous pressure distributions along the front faces of the cylinder. Integration of these curves yield the instantaneous lift force there being no contribution from the back face which is oriented normal to the flow direction. The root mean square of the lift coefficient C_L' is presented in Fig.5 for five values of the Reynolds number. These results produce a trend which agrees closely with the value quoted by Protos et al [1]. The latter was direct measurement of lift and the agreement of integrated value lends confidence to the technique reported herein.

Results for the Cylinder Mounted on Springs

The spring system was arranged to allow a vibration of the cylinder in the direction normal to the uniform flow, that is, in the lift plane.

Results are presented here for one value of the cylinder mass, namely,

$$\frac{M}{\rho D^2 L/2} = 163.$$

The measured maximum amplitude of motion in the lift plane was an rms displacement of 38×10^{-3} in. at the wind speed where the Strouhal frequency was equal to the natural frequency of the support system, f_n.

The measured rms pressure coefficients are plotted in Fig.6 as a function of the frequency ratio together with a dashed line which presents the mean value for the rigid cylinder. The uppermost set of points refer to the pressure tap at the trailing edge and also present the pressure on the back face at the edge. In each case the coefficients decrease as the frequency is increased and reach a minimum at a frequency ratio of around 0.98. A very abrupt rise occurs as resonance is passed and a maximum pressure occurs when the Strouhal frequency is about 1% more than the natural frequency. For further increases of the Strouhal frequency the coefficient approaches the value for the rigid cylinder. Similar behavior is noted in the plots of variation of phase presented on Fig.7. The lower four curves refer to phase relations for the upstream face and the upper curve presents the phase difference between the two edges of the rear face. Significant changes in phase occur for the frequency ratio between 0.98 and resonance. Indicated on the graph are the phase relations for the rigid cylinder. It is evident that over most of the front face the phase is different at all wind speeds compared to that measured on the rigid cylinder.

The combination of Figs.6 and 7 produce instantaneous pressure distributions which give the rms values plotted in Fig.8. The coefficient is divided by the coefficient for a rigid cylinder at the same wind speed. It is interesting to note that the phase variations just below and just above resonance shown in Fig.7 combine in a positive way with the pressure coefficient varations shown in Fig.6 to produce a lower minimum of lift coefficient just below resonance and a higher maximum just above. It was not possible to determine the frequency range in which the lift coefficient increases at resonance and it would appear that a discontinuous jump is a possible explanation of the result. Figure 8 also demonstrates the fact that the lift coefficient for the spring-mounted cylinder is less than that for the rigid cylinder for frequencies larger than $f_s/f_n = 1.05$.

It would appear that if the cylinder is free to move the flow pattern is different from the rigid cylinder and that the fluctuating lift forces are less even though the amplitude of vibration is too small to record. A similar effect was noted on pressures measured on the rear face.

Discussion

The most striking observation from the measurements of lift force is the almost doubling of lift in the very small frequency range next to resonance. One would immediately question whether a very large change occurs in the flow pattern around the cylinder at the same time. Detailed measurements by Twigge-Molecey [3] show that this change is accompanied by very small changes in the strength and lateral spacing of the shed vortices. These changes along with the increase in correlation length for the vortices along the cylinder would produce an increase in lift of less than 10%. The reason for the large change in C_L' must therefore be associated with the fluid reaction to the cylinder motion. This cannot be determined directly without measurements of the fluid velocity and pressure over a large region surrounding the body. Certain observations do, however, give an indication of the properties of this reaction. There is a 180° change of phase in the motion relative to the lift force in the same small range. This, combined with a measured increase in system damping at resonance leads to the conclusion that the fluid reaction is equivalent to an oscillator present in the wake which has small added mass, damping and a large effective spring constant.

References

1. Protos, A.; Goldschmidt, V.W., Toebes, G.H.: Hydroelastic Forces on Bluff Cylinders. ASME Symp. on Unsteady Flows, 1968, Paper 68 FE-12.

2. Toebes, G.H., Eagleson, P.S.: Hydroelastic Vibration of Flat Plates Related to Trailing Edge Geometry. ASME, J. Basic Engng. 83, 671 (1961).

3. Twigge-Molecey, C.F.M.: Ph.D. Dissertation, Department of Mechanical Engineering, University of Toronto, 1972.

Fig. 1. Downstream view of model mounted in wind tunnel showing end plates.

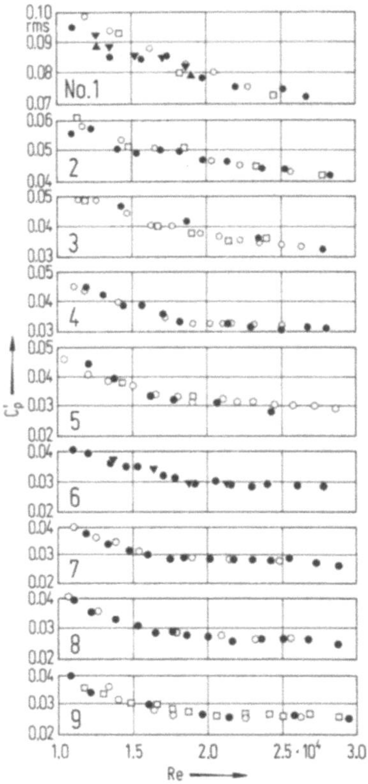

Fig. 2. Variation of fluctuating pressure coefficient on front face vs. Reynolds number (Cylinder fixed).

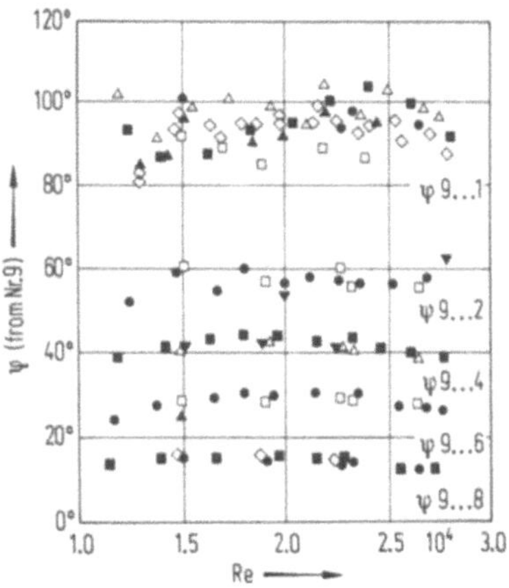

Fig. 3. Upstream-face pressure fluctuation phase separation ψ vs. Reynolds number.

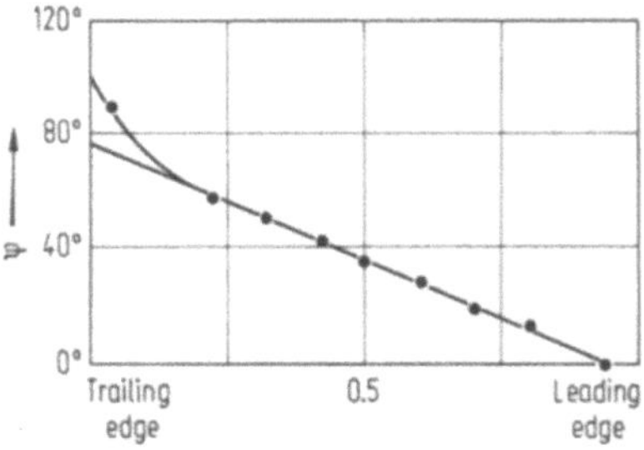

Fig. 4. Phase shift with respect to pressure tap No. 9 on the wedge forward face at Re = 1.9×10^4.

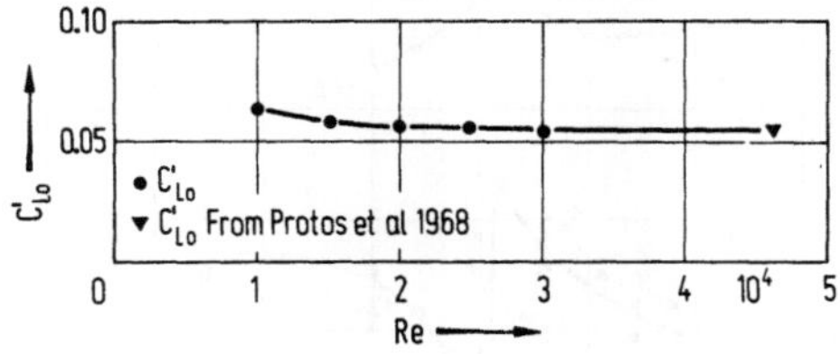

Fig.5. Variation of C'_{Lo}, fluctuating lift coefficient, with Reynolds Number for the fixed wedge.

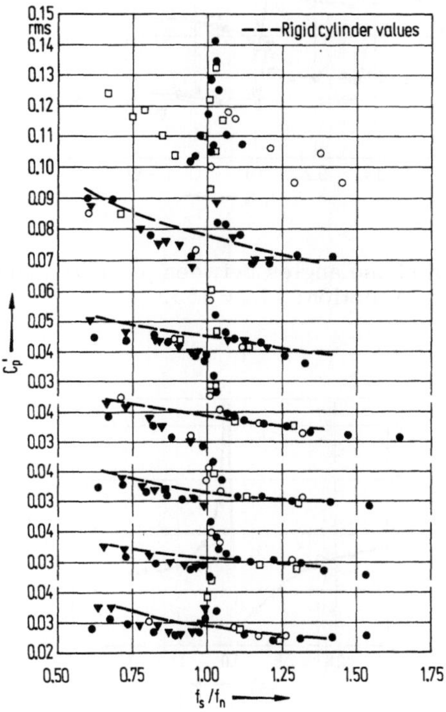

Fig.6. Front face rms fluctuating pressure coefficient vs. f_s/f_n, the non-dimensional wind speed.

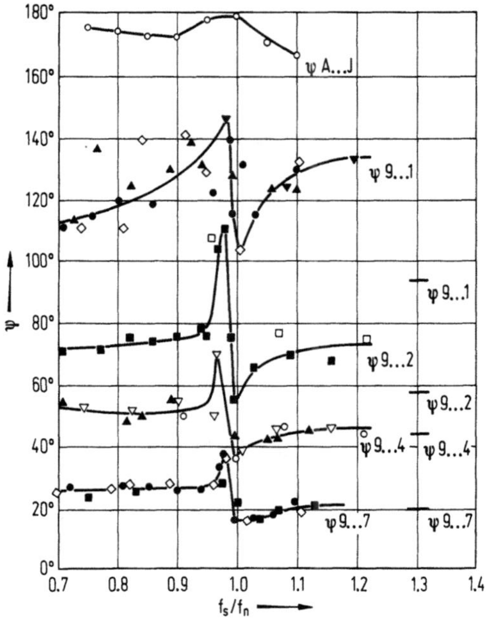

Fig.7. Variation of phase angles between pressure taps vs. f_s/f_n for the cylinder in motion: M_{ro} = 163.

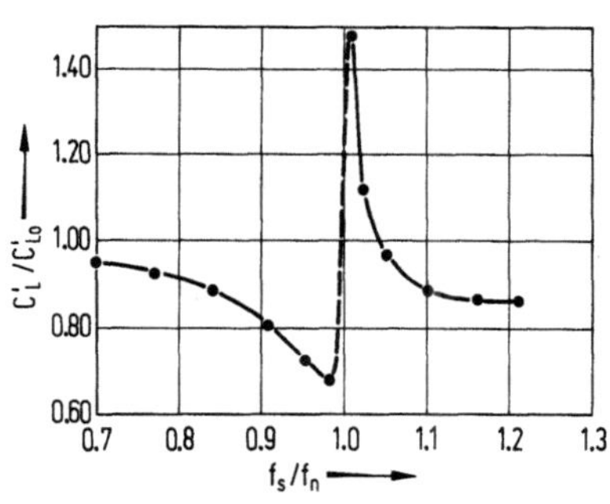

Fig.8. Variation of normalized fluctuating lift coefficient, C_L'/C_{Lo}', vs. f_s/f_n, the nondimensional wind speed, for M_{ro} = 163.

Unsteady Pressures and Forces Acting on an Oscillating Circular Cylinder in Transverse Flow

By

P. Bublitz

Göttingen, Germany

Summary

Extensive experimental studies were performed of the flow phenomena arising with the oscillating circular cylinder in two-dimensional transverse flow. Correlation functions, power spectra, and RMS-values of the unsteady pressure fluctuations for both the stationary and the oscillating cylinder were measured and analyzed at various frequencies of oscillation and amplitudes. The test Reynolds numbers reached from the high subcritical to the low supercritical range. The test set-up is shortly described and the essential results are presented and discussed.

The study of the flow around a circular cylinder is the subject of fluid mechanical research since many years. A great number of publications have been devoted to this subject, and most authors attacked this problem experimentally since the rather complex flow phenomena are so much dependent upon the Reynolds number that a theoretical treatment exhibits great difficulties. Dependent on the Reynolds number, a great variety of particular flow phenomena appear. Correspondingly, experimental and theoretical research work on this problem has always been devoted to special tasks. Concerning the different investigations, one can discern the following special problems: cylinder boundary layer, separation layer and vortex formation, wake and stability of the vortex street, and the pressures and forces acting on the cylinder. The present work has been executed with special regard to practical aeroelastic problems arising with circular cylindrical structures. Therefore, the investigations are mainly concerned with the unsteady pressures and forces acting on the cylinder and with the question how these quantities are affected by forced oscillations of the cylinder.

The experiments have been performed in the 3 by 3 m low-speed windtunnel of the DFVLR, AVA Göttingen [1] within a velocity range of $5\,m/sec < V < 40\,m/sec$. The cylinder had a constant diameter $D = 0{,}25\,m$ and a length $L = 2\,m$. Based on the cylinder diameter, the regime of Reynolds number was between about $1 \times 10^5 < Re < 6{,}7 \times 10^5$. The frequencies of the forced oscillations of the cylinder could be varied from about 5 to 10 Hz; the relative displacement amplitudes attained in the tests reached from $\zeta \simeq 0.01$ to $\zeta \simeq 0.1$, where $\zeta = A/D$ and A is the absolute displacement amplitude.

The test set-up is schematically illustrated in Fig. 1. The cylinder is elastically supported by springs of variable length. Thereby, the resonance frequency of the mass-spring-system can be changed continuously within a wide range. An electro-hydraulic system serves to excite the elastically suspended cylinder to forced oscillations, whereby the frequencies and the displacement amplitudes could absolutely be held constant against the windtunnel flow, see Ref. [2].

A block diagram of the electronic test equipment is shown in Fig. 2. With regard to the regime of Reynolds numbers investigated, statistical methods of data reduction were used [3]. Thereby, the calculations of the power spectra, auto- and crosscorrelation functions were performed on-line and the following quantities were evaluated:

(a) The power spectra and autocorrelation functions of the unsteady pressure fluctuations acting at a typical cross section of the cylinder and the resulting unsteady global lift forces,

(b) the cross-correlation functions of the pressure fluctuations between a reference point and a point at an arbitrary angular position around a typical cross section,

(c) the cross-correlation functions of the pressure fluctuations along a generator of the cylinder at certain axial spacings and an arbitrary angular position.

In addition the test arrangement allowed measurements of the steady values, i.e., the drag and the stationary pressure distribution. Furthermore, the complex lift coefficients could yet be measured [2].

In order to check the accuracy of the test equipment, the mean values of the drag coefficient C_D were measured as a function of the Reynolds number for both the stationary and the oscillating cylinder. The results are plotted in Fig. 3. For comparison, also experimental results of other workers are included. The agreement is satisfactory. The results indicate that oscillations of the cylinder shift the laminar-to-turbulent transition of the boundary layer towards lower Reynolds numbers.

In Fig. 4 the mean values of the pressure coefficients c_p are plotted versus the position angle ψ. They are in good agreement with the results of other authors. Comparing the curves of the oscillating cylinder with those of the cylinder at rest, one finds that the separation point is shifted forward in the case of oscillations.

Fig. 5 shows some calculated power spectra of the unsteady pressure fluctuations at the angle $\psi = 90^\circ$. The variations of the power spectra as a function of growing Reynolds number are obvious, and the same holds for the influence of the cylinder oscillations. In Fig. 5, from left to right, the cylinder is first at rest, then it oscillates at a frequency $f = 5{,}2\,\mathrm{Hz}$ with relative amplitudes $\zeta = 0{,}06$ and $\zeta = 0{,}11$, respectively. Comparing the spectra, one has to mind the different scales of the ordinates. The results of Fig. 5 - representative for many others measured in this research work - lead to the following conclusions:

The wind-excited oscillations of elastic cylindrical structures are obviously *forced* by alternating periodic pressure fluctuations, which are already present with the stationary cylinder. This can clearly be seen from the left column in Fig. 5, where the cylinder at rest is already exposed to unsteady pressures. In the subcritical regime of Reynolds numbers, $Re < 3,3 \cdot 10^5$, the power spectra exhibit a sharp peak indicating the presence of periodic pressure fluctuations. In the supercritical Reynolds number range, $Re > 3,3 \cdot 10^5$, the spectra are random and no predominant frequency of vortex shedding is recognizable. But oscillations can still occur in this regime according to the input-output relation if the mechanical impedance and the damping of the elastic structures are low.

The periodic pressure fluctuations due to the alternating vortex shedding at subcritical Reynolds numbers remain almost unaffected by the forced oscillatory motions up to relative displacement amplitudes $\zeta = 0,11$. This can obviously be concluded from the existence of two clearly discernible peaks in the spectra; one of them is caused by the forced oscillations, the other by alternating vortex shedding. The first one is fixed to the oscillation frequency, the other shifts with growing Reynolds numbers in agreement with Strouhal's relation. An exception of this statement represents the so-called "locking-in" or "locked-in" region, where the oscillation frequency and the vortex shedding frequency are coinciding. This can be seen from the spectra in the second row in the right column of Fig.5. A comparison with the spectrum situated in the second row of the middle column of Fig.5 affirms the findings of G.H. Koopmann [4] that the bandwidth of the locking-in range is dependent on the displacement amplitude of the oscillation. Although Koopmann's regime of test Reynolds numbers was much lower than that of the present work, his result is obviously valid also in the high subcritical Reynolds number range. The relating spectrum shows further that in the case of locking-in the value of the single peak is increased. Cross-correlation measurements in axial direction, which have also been performed but cannot be presented in this short paper (see therefore Refs. [5] and [6]), give reason to suppose that this increment may be due to the enlargement of the correlation length in axial direction on the oscillating cylinder.

For Reynolds numbers higher than about $3,3 \times 10^5$, it is not possible to define Strouhal numbers on the base of frequencies of periodic alternating vortex shedding, see Fig.6. This is evident from the last two rows in Fig. 5. There, the power spectra exhibit a random character, and only the spectra of the oscillating cylinder show peaks, but those are induced by the forced oscillations and not by alternating vortex shedding. The randomness of the spectra for the cylinder at rest demonstrates clearly, that the mechanism of alternating vortex shedding has broken down. From many other power spectra of the unsteady pressure and lift-force calculated during this work it is apparent that for the oscillating cylinder this break-down occurs at lower Reynolds numbers than for the stationary cylinder. This observation is directly related to the discussed results on the steady drag coefficient C_D, see Fig. 4.

Cross-correlation measurements in circumferential direction of a cylinder cross-section at $Re = 1,53 \times 10^5$ yield the time delays τ which are plotted versus the angle ψ in Fig. 7 for the cylinder at rest. The reference point is located at $\psi = 30^\circ$. For comparison, time delays are drawn which are calculated from the theoretical potential velocity distribution and the measured steady pressure distribution at $Re = 1,66 \times 10^5$, given in Fig. 4. The values of the cross-correlation measurements are differing markedly from both the full and the dotted curve. The explanation for that can be given qualitatively by the following consideration (see an interesting series of photographs in connection with that in Ref. [7]): the cylinder boundary layer separates from the cylinder contour at $\psi \approx 110^\circ$. Because of the instability of the separated layer, it starts to roll up forming a logarithmic spiral. During the rolling-up process the pressure disturbances radiated by the increasing spiral reach greater and greater azimut angles ψ on the cylinder until the spiral is wound up to a vortex, which reattaches from the rear of the cylinder and is carried downstream.

Quantitative estimates based on the geometric dimensions of the spiral in the above mentioned reference and on the mean velocity are in fair agreement with the measured values in the range of angles $120^\circ < \psi \leqslant \leqslant 180^\circ$, see Ref. [6]. The pressure fluctuations generated in the region of vortex formation propagate obviously with sound velocity along the cylinder contour in upstream direction. Maximum pressures occur at angles $\psi \simeq 90^\circ$; in upstream direction up to the forward stagnation point, the propagation of the pressure fluctuations becomes progressively damped, as can be seen from Refs. [5] and [6]. This last result may be important with respect to the "Ovalling"-vibration problem.

Conclusion

In this short lecture some results of an extensive experimental study have been presented concerning the unsteady pressures and forces developed on the oscillating circular cylinder in two-dimensional transverse flow in the high subcritical and low supercritical Reynolds number range.

The most important results can be summarized as follows:

(1) The oscillations of elastic cylindrical structures are of a resonance type excited by unsteady aerodynamic forces, which are already present with the stationary cylinder. A self-excitation with aerodynamic feedback coupling as in the case of classical flutter is not existing.

(2) Oscillations of the cylinder cause a laminar to turbulent transition at lower Reynolds numbers than those for the cylinder at rest.

(3) Up to relative amplitudes $\zeta = 0{,}11$, the vortex formation remains almost unaffected by forced oscillatory motions with the exception of the narrow locking-in range, which is dependent on the displacement amplitudes. There the frequency of vortex shedding is influenced by the oscillation.

(4) The increment of the single peak in the power spectra which can be observed in this range seems to be due to the enlargement of the axial correlation length on the oscillating cylinder.

(5) Cross-correlation measurements in circumferential direction on a cylinder crossection revealed that the pressure fluctuations are generated in the vortex formation region at the rear of the cylinder and propagate in the upstream direction along the cylinder contour with sound velocity.

References

1. Riegels, F.W., Wuest, W.: Der 3m-Windkanal der Aerodynamischen Versuchsanstalt Göttingen. Z. Flugwiss. 9, 222-228 (1961).

2. Bublitz, P., Beuermann, R.: Konstruktion einer Meßeinrichtung zur Messung der auf einen ebenen, querangeströmten, schwingenden Kreiszylinder ausgeübten instationären Drücke und Kräfte. AVA-Bericht 69 J 03 (1969).

3. Bendat, J.S., Piersol, A.G.: Measurement and Analysis of Random Data, New York: J. Wiley 1966.

4. Koopmann, G.H.: The Vortex Wakes of Vibrating Cylinder at Low Reynolds-Numbers. J. Fluid Mech. 28, 501-512 (1967).

5. Bublitz, P.: Messung der Drücke und Kräfte am ebenen, querangeströmten Kreiszylinder, Teil I: Untersuchungen am ruhenden Kreiszylinder. AVA-Bericht 71 J 11 (1971).

6. Bublitz, P.: Messung der Drücke und Kräfte am ebenen, querangeströmten Kreiszylinder, Teil II: Untersuchungen am schwingenden Kreiszylinder. AVA-Bericht 71 J 20 (1971).

7. Tietjens, O.: Strömungslehre, Bd.II, Berlin-Heidelberg-New York: Springer 1970, S. 109.

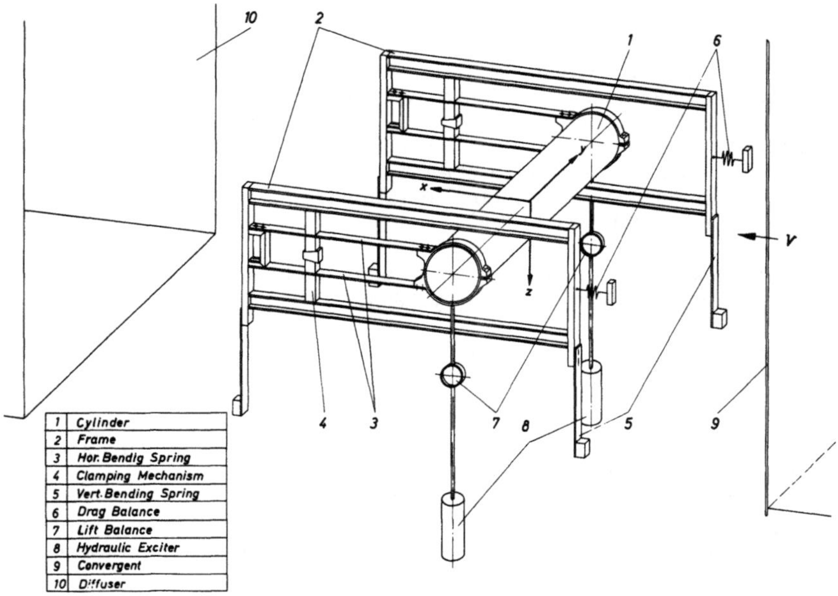

Fig.1. Schematic test set-up.

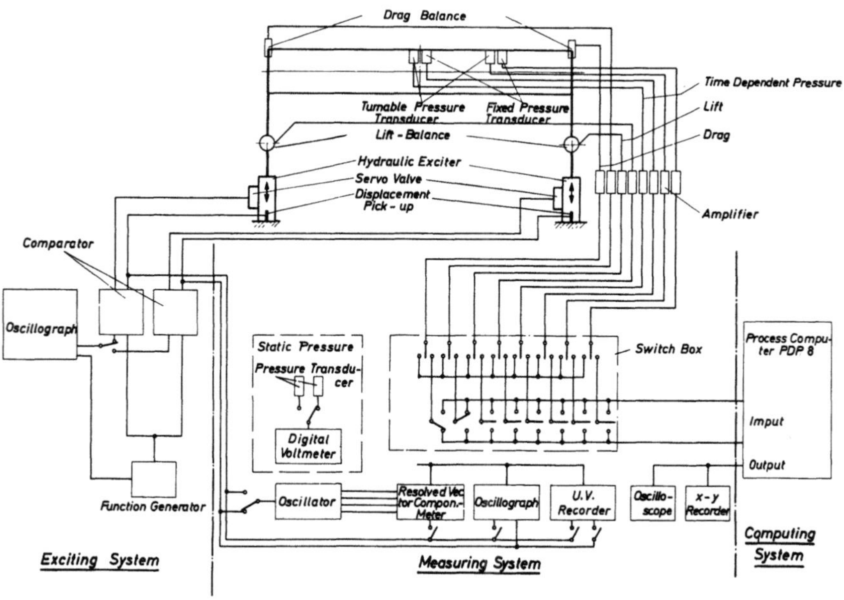

Fig.2. Functional block diagram.

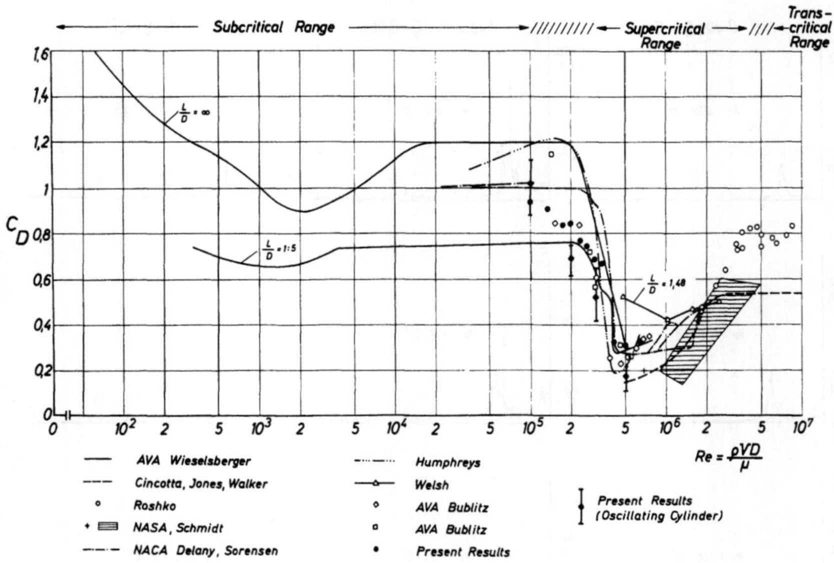

Fig. 3. Drag coefficient C_D versus Reynolds number Re.

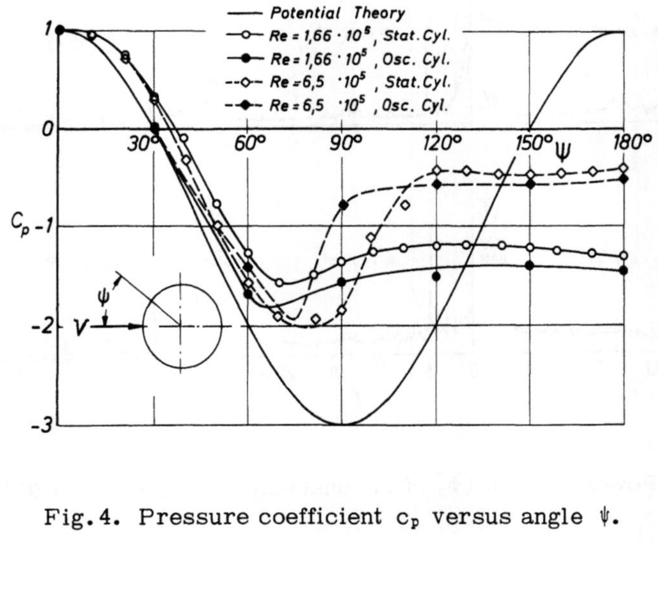

Fig. 4. Pressure coefficient c_p versus angle ψ.

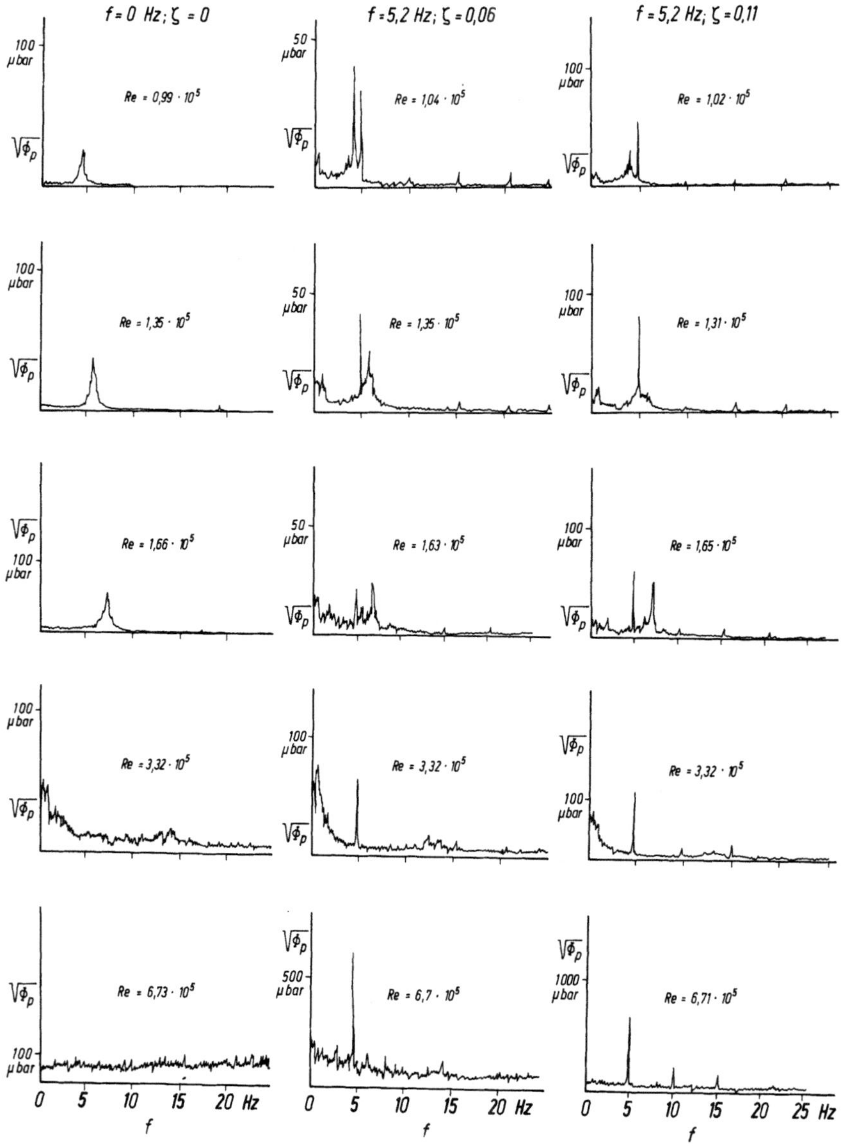

Fig. 5. Power spectra $\sqrt{\Phi_p}$ of the unsteady pressure at $\psi = 90°$.

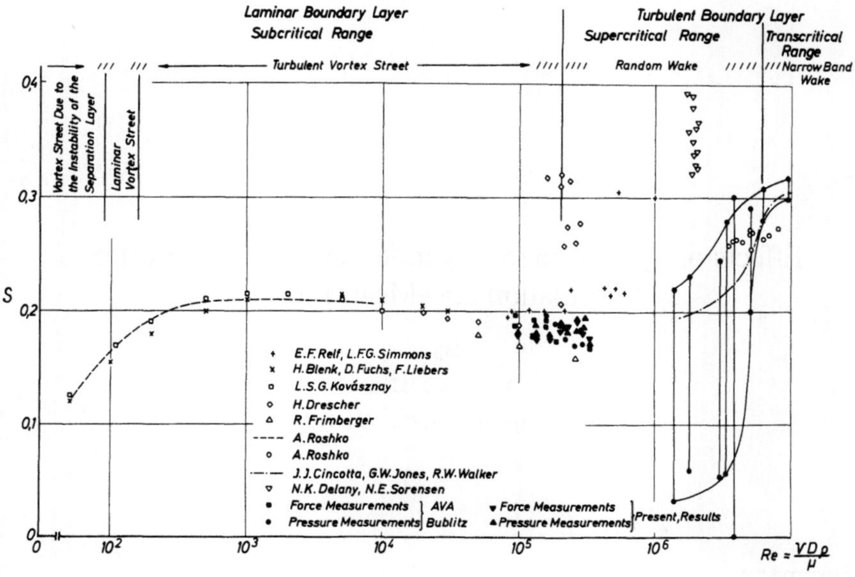

Fig.6. Strouhal number S versus Reynolds number Re.

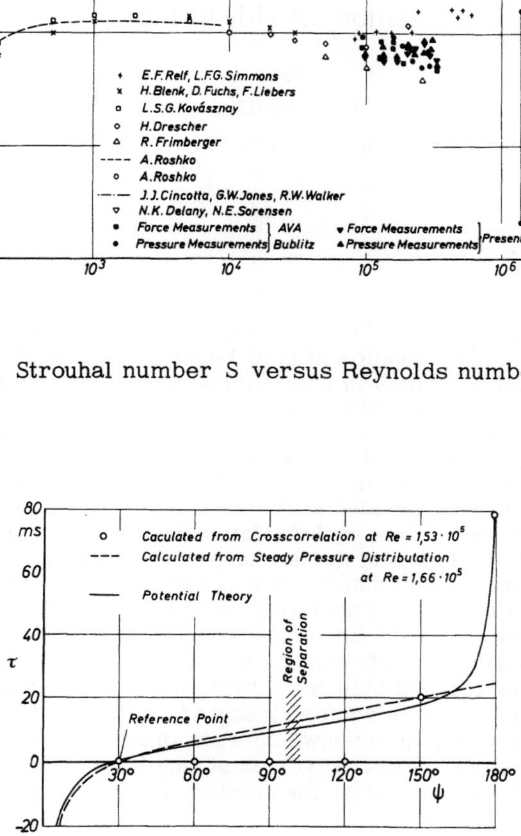

Fig.7. Time delay τ versus angle ψ for stationary cylinder.

Effects of Synchronized Cylinder Vibration on Vortex Formation and Mean Flow

By

O.M. Griffin

Washington, D.C. U.S.A.

Summary

Flow-induced and forced vibrations that synchronize the vortex shedding and vibration frequencies of bluff bodies in a crossflow are related to practical problems in many fields of engineering and related sciences. Measurements have been made in a cylinder wake with a hot wire anemometer for Reynolds numbers between 120 and 350 where vibrations at and near the Strouhal frequency of shedding both synchronize the flow and delay the initiation of turbulence. Results are presented for the changes induced in the vortex formation by different conditions of forced excitation, and the formation length is shown to be a suitable scale for displaying the downstream distribution of fluctuating velocity in the wake. The changes that occur in the amplitude and distribution of both mean and fluctuating velocities are discussed and with the vortex formation length are related to a shedding parameter defined by the amplitude and frequency of the vibrations. It is demonstrated that synchronized cylinder vibrations of half a diameter result in a 65 percent increase in vortex strength from the stationary cylinder value. A correspondence is shown between the decreased formation region length and the increased vortex strength for several conditions for forced excitation at a Reynolds number of 144.

Introduction

It is a well known phenomenon of fluid mechanics that the periodic shedding of vortices due to a cross-flow past a cable or other bluff obstacle can excite the body into resonant, transverse vibrations when the vortex shedding and body frequencies are sufficiently near to one another. There is also a range of frequencies near the Strouhal frequency of vortex shedding where forced transverse vibrations synchronize or "lock-in" with the vortex frequency to control the shedding process. Underwater towing and instrument systems and tube bundle heat exchangers for nuclear power generation are but three of the areas where the knowledge and prevention of cable strumming and other flow-induced, resonant vibration phenomena are important to the improvement of design criteria and to the extension of the present state-of-the-art. The report of a recent Euromech Colloquium calls attention to the need for further knowledge of the synchronization, or locking-in, phenomenon [1].

It has long been surmised that these vibrations are accompanied by increased strength of the vortices formed in the wake of the body, and consequently that a relation exists between changes in the periodic wake flow and the increased lift forces that are known to accompany flow-induced vibrations. Some recent experiments [2,3] have shown and explained many of the changes that are produced in the wake of a body by different forced vibratory conditions. Further, other experiments [4] have shown that flow-excited vibration conditions can be modeled in the laboratory with appropriately designed forced vibration experiments. These experiments, together with earlier work on synchronized vibrations at the Naval Research Laboratory (NRL) [5] and elsewhere [6,7], have led to our recent attempts to determine the changes produced in the strength and spacing of vortices by forced vibrations. The purpose of this paper is to report the results of experiments to determine the changes in the vortex formation region, and the associated changes in the mean and fluctuating velocities at and beyond formation, for Reynolds numbers between 120 and 350. Results that confirm a correspondence between a decreased formation region length and increased vortex strength at Reynolds number 144 are also presented and discussed in terms of the changes that are known to occur in the fluid forces acting on the cylinder.

Experimental and Computational Methods

The experiments reported in this paper were performed in an open jet wind tunnel equipped with a 75×75 mm exit and 20:1 contraction section. DISA anemometers (Model 55D01) and low interference hot-wire probes (Model 55F01) were used for the flow measurements. The hot wire probe outputs were carefully linearized in the speed range 0.25 to 2.5 m/sec so that measurements of mean and fluctuating velocity could be interpreted with confidence. Smooth circular cylinders 2.4 and 3.2 mm in diameter were used for the experiments, and were mounted in a vibration isolated shaker apparatus at the exit jet of the tunnel. The wind tunnel and all its supporting equipment for both hot-wire experiments and flow visualization studies using an aerosol generator have been described previously in detail [2,3,4]. The digital computations for matching the experimental velocity profiles with a flow model for the viscous, laminar vortex street were carried out on the Naval Research Laboratory CDC 3800 computer.

Results and Discussion

The Vortex Formation Region

The importance of the near wake where vortices are formed has long been recognized, and several physical criteria have been defined for determining the initial position of the fully formed vortex [8,9]. Among these are:

(1) the minimum mean pressure on the wake axis,

(2) the maximum of velocity fluctuation at the second harmonic of the shedding frequency, on the wake axis,

(3) the minimum transverse spacing, close to the body base, of the regions of maximum vortex velocity fluctuation.

The measured (criterion 2) length l_F of the formation region, or the initial position of a fully formed vortex, is plotted in Fig.1 as a function of the shedding parameter

$$St^* = \left(\frac{f}{f_n}\right)\left(1 + \frac{a}{d_c}\right)St_n ,$$

where the amplitude and frequency of the vibrations are denoted by a and f respectively, and the naturally occurring Strouhal number is St_n. The Strouhal number $St_n = f_n d_c/U$ where f_n (Hz) is the frequency of vortex shedding caused by the uniform flow speed U (m/sec) past a stationary cylinder of diameter d_c(m). The corresponding Reynolds number $Re = Ud_c/\nu$ where ν is the fluid kinematic viscosity (m^2/sec). The experimental points plotted in Fig. 1 were measured under flow conditions at which the vibration and vortex shedding frequencies were locked together. Frequencies ranged between ±15 percent of the Strouhal frequency and the cylinder amplitude reached values as great as 50 percent of a diameter. A dependence between l_F (measured in terms of the cylinder diameter d_c) and St^* is demonstrated in the figure just mentioned for Reynolds numbers between 120 and 350. The results of these wake formation experiments have shown that the formation length decreases systematically with increased amplitude of cylinder vibration, while the effects of frequency changes are twofold [2]. When the cylinder vibration frequency is decreased to a value less than the Strouhal frequency the distance to formation is extended, and for frequency greater than the Strouhal frequency the distance to formation is decreased. The flow in the near wake is related to the physical model recently postulated by Gerrard [10] for the formation region. The growing vortex is fed by circulation from the shear layer until it becomes strong enough to roll up and draw the opposing shear layer across the wake. This vorticity of opposite sign then cuts off further circulation to the growing vortex, which is subsequently shed and begins to move downstream. The growing vortex behind the vibrating cylinder rolls up more quickly and is shed at a smaller downstream distance, and this process takes place periodically on either side of the wake to produce laminar vortices at Reynolds numbers up to 350.

When the frequency is varied about the Strouhal frequency f_n the formation region is either expanded or contracted, depending on whether the frequency ratio is less than or greater than unity. These results correspond to the observed effects of splitter plates, where interference in the wake formation process with a horizontal splitter plate extends the formation region and results in a decrease in the velocity fluctuation on the wake center line near $x = l_F$. This effect appears in Fig.2b for $f/f_n = 0.9$. The decreased formation length for $f > f_n$ was also observed by Gerrard [10] when a vertical splitter plate was placed in the formation region, and the increased velocity fluctuation on the wake axis corresponding to $f/f_n = 1.1$ appears in the velocity profiles plotted in Fig.2a for $x \simeq l_F$. Recent work by Chen [11] has demonstrated the relation between a decreased scale of the formation region and increased base underpressure and drag for flow past a circular cylinder.

Wake Velocity Fluctuations

Synchronized vibrations appreciably influence the wake velocity fluctuations as well as the distance to formation. Some indication of these effects is given in Table 1 for experiments at a Reynolds number of

Table 1. Maximum Velocity Fluctuations in the Vibrating Cylinder Wake.

(a) Reynolds Number, Re = 200 Relative Frequency, $f/f_n = 0.93$

$\frac{x}{d_c}$	$\frac{x}{l_F}$	$\frac{a}{d_c}$	$\frac{u}{U} \times 100$
2.7	1.0	0.12	34
2.1	1.0	0.30	37
1.7	1.0	0.48	38
5.6	2.0	0.12	19
3.9	1.9	0.30	25
3.8	2.3	0.48	25

(b) Reynolds Number, Re = 200 Stationary Cylinder

$\frac{x}{d_c}$	$\frac{x}{l_F}$	$\frac{f}{f_n}$	$\frac{u}{U} \times 100$
2.8	1.0	1.0	28
3.7	1.3	1.0	21
5.4	1.9	1.0	17

200. When the cylinder is vibrated at 93 percent of the Strouhal frequency and at thirty percent of a diameter, there is a thirty-two per-

cent increase in the amplitude of fluctuations at $x = l_F$ and a forty-five percent increase near $x = 2l_F$, relative to the stationary cylinder wake at the same Reynolds number. The formation length is a suitable scale for normalizing the downstream wake after formation, and velocity fluctuations for a Reynolds number of 144 and two values of vibration frequency are plotted in Fig. 2. The distance from the cylinder is scaled by l_F and the effects of frequency on the fluctuating wake appear in the measured profiles. The maximum fluctuation at each downstream displacement is less for frequency ratio $f/f_n > 1$ than for the condition $f/f_n < 1$, at the same x/l_F. Measurements made on the wake center line, $y = 0$, indicate a different behavior. The fluctuating velocity on the wake center line is greater for $f/f_n > 1$ than for $f/f_n < 1$, and remains greater as the displacement from the end of the formation region is increased. These results are in contrast to the effects of increased cylinder amplitude alone, where a systematic increase in the eddy velocity fluctuation takes place both in the maximum region and on the wake center line. The changes in the velocity field are thus separable into frequency and amplitude effects at each Reynolds number when the wake is normalized by the formation length. The results in the figure just mentioned suggest the use of the formation length as a scale for the distribution of velocity fluctuations in the downstream wake. The regions of maximum fluctuation are plotted in Fig. 3 for $Re = 200$, and the points essentially fall on a single curve marked by three distinct regions. The first section of the curve is the formation region, $x \leqslant l_F$, where the plotted points denote the average spacing of the shear layers from the wake center line. The regions of maximum fluctuation then move apart for $x > l_F$ as the transverse spacing of the vortices begins to increase. This is the stable flow regime. The third regime is the unstable region where the fluctuation peaks begin to converge toward the wake center line. The stable region is characterized by a rapid drop in the velocity fluctuations, and the results in Fig. 2 and Table 1 are indicative of such behavior. The early work of Berger [6] showed that synchronized cylinder vibrations delay the initiation of turbulence in the wake for Reynolds numbers up to 350. The flow in the formation and stable regions as presented in Fig.3 is essentially two dimensional, with virtually no appearance of the low frequency irregularities that are characteristic of the wake behind a stationary cylinder at these Reynolds numbers, and the unstable region is characterized by the appearance of the irregular behavior that usually precedes the turbulent breakdown on the wake [2]. The measurements of vortex formation reported in the previous section were made with a hot wire anemometer to determine the maximum velocity fluctuation on the wake center line as an indicator for the end of the formation region. The results plotted in Fig.3 for Re = 200 demonstrate the correspondence between the criteria 2 and 3 for determining the end of the formation region, since the maximum fluctuation on the wake axis coincides with the downstream distance at which the peaks of maximum eddy velocity fluctuation begin to move apart. Similar results have been obtained at other Reynolds numbers up to 350 for many synchronized vibration conditions.

Mean Flow in the Stable Region

The effects of the vibrations are appreciable not only on vortex formation and velocity fluctuations, but also on the mean flow in the early wake. The velocity profiles plotted in Fig.4 show the early mean flow development for two conditions. The results in Fig.4a were obtained in the stationary cylinder wake and those in Fig.4b were recorded with the cylinder vibrating at thirty percent of a diameter, at the natural shedding frequency. There is a substantial velocity defect in the wake behind the stationary cylinder, with the mean velocity on the wake center line reaching forty five percent of the free stream value at the end of the formation region 3.2 diameters downstream. The end of the formation region is moved upstream by the vibrations to 2.0 diameters, and the mean velocity at this point is increased to sixty-six percent of the free stream value. The effects of the vibrations are presented systematically in Fig.5, where the mean velocity at the end of the formation region is plotted for several values of the parameter St* at a Reynolds number of 144. The increase in St* is accompanied by a decrease in the length of the formation region from 3.2 to 1.7 diameters, and the mean velocity on the wake center line increases from forty-five percent of free stream at $St^* = St_n = 0.178$ to seventy-two percent at $St^* = 0.263$. These measurements demonstrate the systematic decrease in the wake defect as the formation region is reduced in size by the vibration of the cylinder.

Strength of the Vortices

The flow chosen to describe the wake is that of a laminar Kármán vortex street that is infinite in extent and corrected by the insertion of Hamel-Oseen viscous vortices near the point at which the model is matched with experiment. Such a model was used with some success by Berger [12] in his study of the stationary cylinder wake. A position near $x = 2.6\, l_F$ in the stable region of the flow for a Reynolds number of 144 was chosen, and the experimental profiles were matched with the model at nine points between 0.0 and 2.0 cylinder diameters from the wake center line. The solution was defined to be that set of parameters, i.e., vortex strength $K(m^2/sec)$, viscous vortex core radius $r_*(m)$, and transverse spacing of the vortex centers $h(m)$, that produced a minimum mean square error between theory and experiment at the nine chosen points. Bloor and Gerrard [9] employed a related error criterion with success in their measurements of the strength of turbulent vortices behind a stationary cylinder. The present calculations are based on a fit between the model and experiment, with the solution specified by the minimum of the error function

$$E = \sum_{j=1}^{9} \left(\frac{u_{m,j} - u_{c,j}}{u_{m,j}}\right)^2 + \left(\frac{U_{m,j} - U_{c,j}}{U_{m,j}}\right)^2 , \qquad (1)$$

where the u_m and U_m are the measured root-mean-square and mean velocities at the chosen points. The calculated velocities, based on the total velocity at the points, are correspondingly denoted by u_c and U_c.

The potential vortices of the Kármán street are replaced by viscous vortices at those positions nearest the region of matching between theory and experiment, and it is reasonable to suppose that there is no interaction between the viscous cores. An interaction criterion was imposed in the computations, such that

$$\left(\frac{r_*}{l_v}\right)^2 - \left(\frac{h}{2l_v}\right)^2 - 0.0625 \leqslant 0 \qquad (2)$$

to ensure that there is no interaction between the cores and that they do not overlap. A similar criterion was defined by Durgin and Karlsson [13] in a recent study of vortex street breakdown, and both Schaefer and Eskinazi [14] and Goldstein [15] have implied that the stable region of the wake satisfies this condition.

The angular velocity distribution about each Hamel-Oseen vortex is given by the equation

$$\omega = \frac{K}{2\pi r^2}\left[1 - e^{-1.26\left(\frac{r}{r_*}\right)^2}\right], \qquad (3)$$

where r is the radial displacement from the center of a vortex, and the vortex core radius r_* is related to the viscous "age" t of the vortex by the equation

$$r_*^2 = 5.04\,\nu t. \qquad (4)$$

The wake model used in the computations implies that the vortices are circular and have not begun to deform. This assumption is justified so long as the downstream displacement does not increase beyond the stable region of the wake as shown in Fig.3. Flow visualization photographs of the synchronized wake at Reynolds numbers near 200 also confirm the validity of this assumption [3,5]. An example is shown in Fig.6, where the vortex wake formation behind a six-stranded cable is visualized using the apparatus described in a previous section.

Some of the results that were obtained (by simple but time consuming digital computations) for several vibratory conditions are listed in Table 2. There is a 65 percent increase in vortex strength - relative to that which is measured when the cylinder is stationary - for a cylinder vibrating at 0.48 diameter. There is no change in vortex longitudinal spacing with amplitude, and so the increased velocity

Table 2. The Strength of Vortices behind a Vibrating Cylinder

Reynolds Number, $Re = 144$ Strouhal Number, $St_n = 0.178$

Cylinder amplitude	Relative frequency	Wake parameter	Vortex spacing[1]	Vortex core radius	Vortex spacing ratio	Vortex strength	Formation length
$\frac{a}{d_c}$	$\frac{f}{f_n}$	$St_n\left(1+\frac{a}{d_c}\right)\left(\frac{f}{f_n}\right)$	$\frac{l_v}{d_c}$	$\frac{r_*}{l_v}$	$\frac{h}{2l_v}$	$\frac{K}{\pi U d_c}$	$\frac{l_F}{d_c}$
0.0	1.0	0.178	5.4	0.178	-	0.81[2]	3.2
0.12	1.0	0.199	5.4	0.187	0.079	1.10	2.6
0.30	0.9	0.208	5.9	0.195	0.063	1.26	2.3
0.30	1.0	0.231	5.4	0.194	0.068	1.22	2.0
0.30	1.1	0.254	4.9	0.210	0.074	1.32	1.8
0.48	1.0	0.263	5.4	0.205	0.056	1.34	1.7

[1] The length l_v denotes the longitudinal spacing between consecutive vortices of like sign.

[2] Berger [12], stationary cylinder.

fluctuations are predominantly the result of positive changes in vortex strength with increasing amplitude. When the amplitude of vibration is held constant and the frequency varied, the results are more complicated. There are apparently simultaneous changes in both the longitudinal spacing and the strength of the vortices that result in the observed variation in the vortex velocity fluctuations both on and off the wake axis as the vibration frequency changes within the locked-in regime. It is interesting to note that the amplitude and frequency of the vibrations act on the street spacing ratio in different ways. There is a decrease in the apparent spacing ratio as the amplitude is increased, but the spacing ratio increases with increasing frequency of vibration. Earlier measurements [16] in the wake of a cylinder undergoing flow-induced vibrations have likewise shown that the longitudinal spacing of the vortices is increased at frequencies less than the Strouhal frequency. The results of previous flow visualization experiments at NRL [3,5] have indicated many of the effects of synchronized vibrations on the vortex spacing, and an example is given in Fig.6 for vibrations at the Strouhal frequency. If the frequency vibration is held constant and the amplitude is increased, then the transverse spacing h between the vortices on opposite sides of the wake appears to decrease. This behavior is observed in the results listed in Table 2 and Fig.6 for $f/f_n = 1.0$. When the relative frequency is held constant, the vortex spacing $h/2l_v$ decreases from 0.079 to 0.056 as the amplitude is increased from 0.12 to 0.48 diameter. Further, the visualization experiments indicate that the spacing ratio $h/2l_v$ changes as the relative frequency is increased at constant amplitude. This effect is also predicted by the tabulated results.

There is a direct relation between the fluid forces on a bluff body and the length downstream at which vortices are shed, and as the vortex formation region decreases in length there is a corresponding increase in the base underpressure and fluid drag force. Toebes [7] has measured the increased underpressure that accompanies synchronized vibrations. The measured values for the formation region length are listed together with their corresponding vortex strengths in Table 2, and there is an inverse relation between the formation length l_F and the vortex strength K associated with the vibrations. It is thus possible to relate the increased vortex strength to the higher drag that is known to accompany synchronization, and likewise to the increases in the fluctuating lift and pressure that are known from cylinder force measurements to act periodically with the same frequencies as those of the vibrations [16,17].

Summary and Conclusions

The importance of the formation region flow for determining the frequency of eddy shedding has been demonstrated previously by the work of Gerrard [10] and Roshko [18] with splitter plates. The present results demonstrate conversely that the formation and subsequent development of the wake can be controlled and altered by the amplitude and frequency of synchronized cylinder vibrations. More importantly, it is possible to systematically study the effects of the vibration amplitude and frequency and to gain valuable insights into the mechanisms of both forced and flow-induced vibrations, the suppression of turbulence initiation, and wake resonance.

The length of the formation region is decreased at Reynolds number 120 by as much as fifty percent from its value behind a stationary cylinder. There is an inverse relation between the strength of the wake vortices and the formation length and the results of matching the measured velocity profiles with a model for the wake indicate a substantial increase in vortex strength for all vibratory conditions investigated. There is a sixty-five percent increase in vortex strength for cylinder vibrations of one half diameter at a Reynolds number of 144. The formation region length is a suitable scaling parameter for the downstream distribution of fluctuating velocity for Reynolds numbers to at least 350, and the wake is divided into three distinct flow regimes: the formation, stable, and unstable regions.

The effects of frequency on the distribution of fluctuating velocity become apparent when the downstream distance is measured in multiples of the formation length. Frequencies less than the natural shedding frequency decrease the amplitude of velocity fluctuation on the wake axis below that found for vibration frequencies greater than the natural frequency. The maximum velocity fluctuation at each scaled downstream station is greater for the frequency ratio less than unity than for conditions of frequency ratio greater than unity. There is a systematic increase in velocity fluctuation with the amplitude of vibration at each downstream station both on and off the wake center line.

The mean flow distribution in the wake is affected by the synchronized vibrations. The velocity defect on the wake center line at formation is decreased from fifty-five percent behind the stationary cylinder to twenty-eight percent as the amplitude is increased to one half diameter at a Reynolds number of 144.

There is a relation between the fluid forces on a bluff body and the distance downstream at which vortices are shed, there being an increase in base underpressure and drag as the formation region decreases in length. It is thus possible to relate the increased vortex strength to the greater drag that accompanies resonant vibrations such as cable strumming in underwater towing and instrument packages, and to the increased pressure and lift forces that act periodically with the same frequencies as those of the vibrations.

Acknowledgement

The support of the Naval Research Laboratory for the experiments and computations reported here is greatly appreciated. The author wishes to acknowledge his colleagues C.W. Votaw, with whom he collaborated on the flow visualization experiments, and R.A. Skop, who provided the stranded cable test models. A particular note of thanks is due J. Otto for his special processing of the original photographic negatives from the flow visualization experiments.

References

1. Mair, W.A., Maull, D.J.: J. Fluid Mech. 45, Part 2, 209-224 (1971).

2. Griffin, O.M.: Trans. ASME, J. Appl. Mech. 38, Part 4, 729-738 (1971).

3. Griffin, O.M., Votaw, C.W.: J. Fluid Mech. 55, Part 1, 31-50 (1972).

4. Griffin, O.M.: Trans. ASME, J. Engng. for Indus. 94, Part 2, 539-547 (1972).

5. Koopmann, G.H.: J. Fluid Mech. 28, Part 3, 501-512 (1967).

6. Berger, E.W.: Jahr. Wiss. Ges. L & R, Berlin, 1964.

7. Toebes, G.H.: Trans. ASME, J. Basic Engng. 90, 493-505 (1968).

8. Bearman, P.W.: J. Fluid Mech. 21, Part 2, 241-255 (1965).

9. Bloor, M.S., Gerrard, J.H.: Proc. Royal Soc. London, A, 294, 319-342 (1966).

10. Gerrard, J.H.: J. Fluid Mech. 25, Part 2, 401-413 (1966).

11. Chen, Y.N.: Flow-Induced Vibration in Heat Exchangers. ASME, New York, N.Y., 1970, pp. 83-86.

12. Berger, E.W.: Z. Flugwiss. 12, 41-59 (1964).

13. Durgin, W.W., Karlsson, S.K.F.: J. Fluid Mech. 48, Part 3, 507-527 (1971).

14. Schaefer, J.W., Eskinazi, S.W.: J. Fluid Mech. 6, 241-260 (1959).

15. Goldstein, S.: Modern Developments in Fluid Mechanics, Vol. 2, Oxford Press 1943, pp. 563-565.

16. Ferguson, N., Parkinson, G.V.: Trans. ASME, J. Engng. for Indus. 89, 831-838 (1967).

17. Toebes, G.H., Ramamurthy, A.S.: Trans. ASCE, J. Engng. Mech. 93, Part 6, 1-21 (1967).

18. Roshko, A.: NACA Tech. Note 3169, Washington, D.C., 1954.

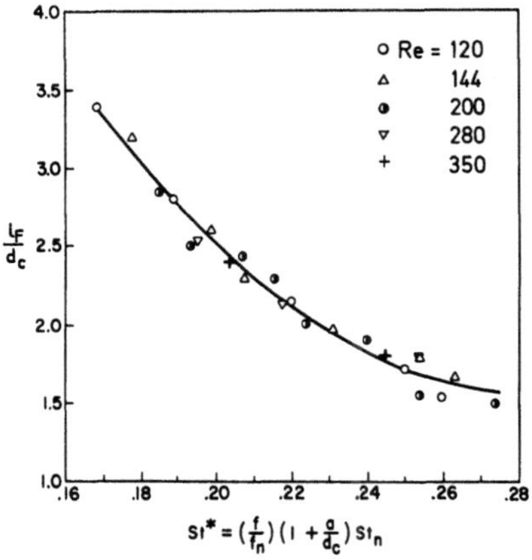

Fig.1. The vortex formulation length l_F, measured in multiples of the cylinder diameter d_c, as a function of vortex shedding parameter St* for Reynolds numbers between 120 and 350.

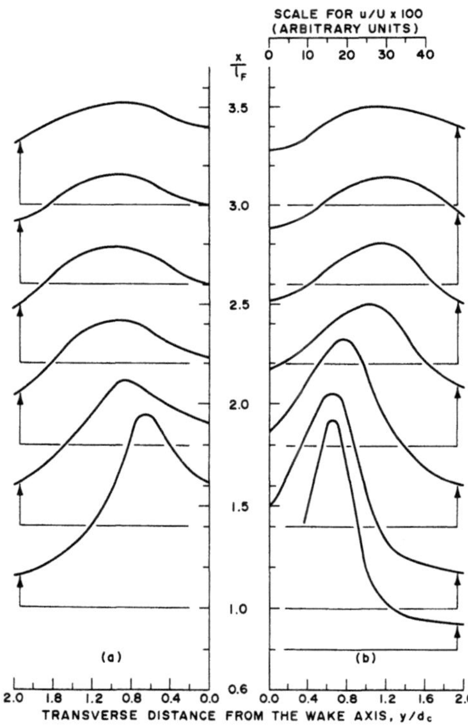

Fig.2. Vortex velocity fluctuations in the vibrating cylinder wake, at Reynolds number 144, as a function of downstream distance measured in multiples of the formation region length l_F.
(a) $f/f_n = 1.1$, $a/d_c = 0.3$; (b) $f/f_n = 0.9$, $a/d_c = 0.3$.

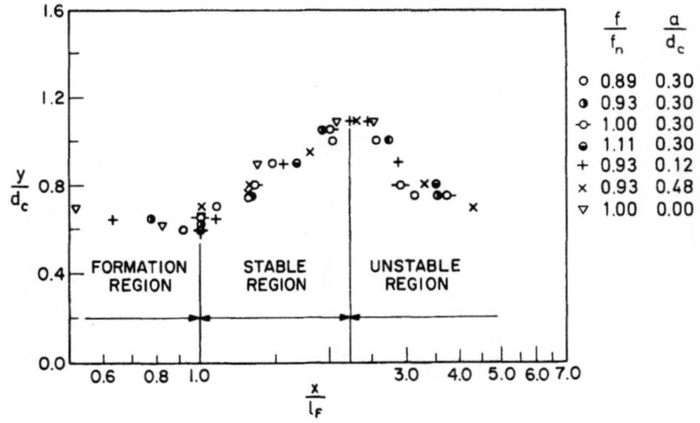

Fig. 3. Distribution of the regions of maximum fluctuation with downstream distance, x/l_F, at Re = 200.

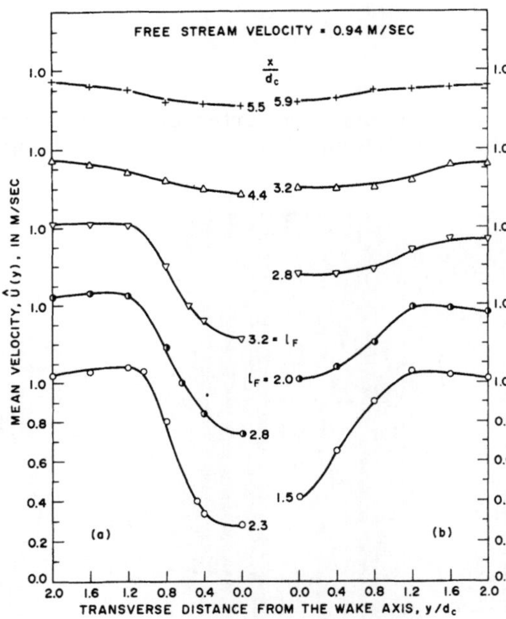

Fig. 4. Mean velocity in the cylinder wake, at Reynolds number 144, as a function of downstream distance x from the cylinder. (a) Stationary cylinder, $f = f_n$; (b) Vibrating cylinder, $f = f_n$, $a/d_c = 0.3$.

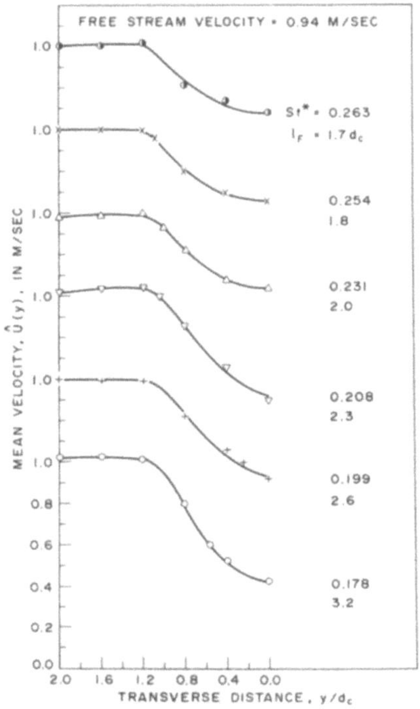

Fig. 5. Mean velocity at the end of the vortex formation region, at a Reynolds number of 144, for six values of vortex shedding parameter St*.

Fig. 6. Vortex shedding from a vibrating six-stranded cable at a Reynolds number of 220. The vibration amplitude was thirty percent of a diameter and the frequency was equal to the Strouhal frequency of 41Hz.

On the Correlation of Dynamic Pressures on the Surface of a Prismatic Bluff Body

By

R.H. Wilkinson
Bristol, U.K.

J.R. Chaplin
London, U.K.

T.L. Shaw
Bristol, U.K.

Summary

The authors give attention to instantaneous pressure distributions over the side faces of prismatic square cylinders. Correlation measurements suggest that the line of transition to turbulence in the shear layer may have a wave-like form, so that cells of coherent shedding may be detected close to the cylinder. Transverse cylinder oscillations are also studied.

Introduction

Recent research has shown that the correlation of fluctuating pressures between different points along a generator of a prismatic bluff body in a fluid flow is an important parameter closely associated with the characteristics of the flow, and that for comparisons between different experiments and for applications of the results, measurements of fluctuating lift and drag lack meaning unless referred to this spanwise correlation of pressure fluctuations.

Between generators there also exists a phase difference in the dynamic pressures associated with vortex shedding. Chordwise correlation is also related to spanwise correlation, i.e. flow characteristics and body shape together determine the overall instantaneous distribution of pressure, and hence the instantaneously imposed force.

But dynamic forces are usually only of consequence if structural response also occurs. Much less is known about fluid/structure interaction for coupled dynamic vibrations, yet this is the physical situation of design importance.

This paper presents some new information on chordwise and spanwise correlations and on the effects of structural movement. Comparisons are made with the results of other researchers, and conclusions are drawn as to the three-dimensional form of the fluid flow accompanying flow-induced forces on structures. All experiments were performed with square section cylinders, face normal to the flow; experimental details are given in Table 1.

Table 1

	Chordwise	Spanwise	F-S Interaction
Reynolds Number	10^4 to 10^5	10^4 to 10^5	10^3 to 10^4
Operating Fluid	Water	Air	Water (Free Surface)
Tunnel Details	15.4cm × 45.7cm rectangular section, open circuit water tunnel	70cm × 70cm rectangular sect. with 13cm corner fillets. Open circuit.	Towing flume, 45cm wide × 30cm deep
Model Details L/D	3 : 1	14 : 1	12 : 1
Blockage Ratio	9 : 1	14 : 1	5 : 1 - 25 : 1
Pressure Tapping	Around mid-section		None
Positions	Arranged to give 15 available on a particular face	Situated at eighth points on span, at sixth points on chord	-
End Conditions	Built-in both ends	Built-in both ends	Full depth, vertical
Transducer System	External Capacitive Type	External Capacitive Type	-
Flow Visualisation	Hydrogen Bubbles	-	Lycopodium Powder

Chordwise Correlation

Experimental determinations of the fluctuating lift force at the vortex shedding frequency on bluff cylinders have in some cases been made by integrating the distribution of unsteady pressure [1,2]. To do this it is necessary either to measure the chordwise correlation of fluctuating pressures, or to assume that the pressures at all points on each side of the cylinder with respect to the wake axis are in phase and perfectly correlated and that pressures on opposite sides are in anti-phase. This leads to direct summation of contributions to the lift force and to cancellation of the greater part of the distribution of unsteady forces in the drag direction.

Measurements of chordwise correlation of pressures on the faces of a square section cylinder at $Re = 1.3 \times 10^4$ [3] suggested that the mean phase lag between two points is such that the corresponding error in calculating the total lift force by adding pressure measurements linearly was less than 2%. This result suggests that the determination of the chordwise correlation is not important for the purpose of summing pressure measurements at the vortex shedding frequency. However, it can provide some information about the characteristics of the flow.

Previous measurements of chordwise pressure correlations were made by Vickery [4], for a square section cylinder at $Re = 10^5$ in smooth and turbulent flows, and by Gerrard [2] for a circular cylinder at $Re = 8.54 \times 10^4$ and 1.14×10^5. Comparison between results suggests that with increasing Reynolds number over the range 10^4 to 10^5, the chordwise correlation for square and circular cross-section cylinders is maintained or improved. Experimental measurements over this range also indicate an increase in fluctuating lift coefficient.

With increasing Reynolds number the transition to turbulence in the separated shear layers moves upstream, and it would be reasonable to expect that this would lead to an increase in the random content of the fluctuating lift force and to a deterioration in chordwise correlation of fluctuating pressure. That the opposite has been observed is a result of the accompanying reduction in the length of the formation region due to increased rate of entrainment by the shear layers, [5]. This has been measured in experiments using hydrogen bubbles in water to visualise the flow around a square section cylinder [6].

A sheet of hydrogen bubbles perpendicular to the axis of the cylinder was generated in the approach flow, and the end of the formation region was found by locating visually the point at which bubbles in the smooth flow outside the shear layers first cross the axis of the wake. This is one of the definitions of the end of the formation region given by Bloor and Gerrard [7]. The results are presented in Fig. 1 together with those by Bloor [8] from velocity measurements for a circular cylinder. Considerable scatter was observed in both cases but to avoid confusion individual points from Bloor's experiments have been omitted.

As successive vortices grow in the formation region, the shear layers after separation oscillate laterally, and this is accompanied by movements of the separation points when they are not fixed by sharp edges on the cylinder's profile. As the rate of entrainment increases and the formation region becomes shorter with increasing Reynolds number, the shear layers approach the cylinder more closely during the oscillation and their amplitude of oscillation at any section increases. This is illustrated by measurements presented in Fig. 2 which were made from observation of the paths of the shear layers from a square section cylinder, visualised by hydrogen bubbles generated from wires parallel to the axis of the cylinder close to the separation points. The scatter in these results (as in Fig. 1) reflects the imprecise method used, but they are sufficient to indicate that the amplitudes of lateral oscillation of the shear layers increase and that the shear layers approach more closely the sides of the cylinder. With a cylinder of square cross-section the regions between the shear layers and the lateral faces are more narrow and more isolated from the wake than in the case of a circular cylinder. Fluid entrained by the shear layers cannot be replenished so readily, and it is therefore reasonable to expect that the flow around a square section cylinder and the chordwise correlation are more sensitive than for a circular cylinder to changes in the rate of entrainment by the shear layers. A rectangular section cylinder would experience periodic re-attachment on the lateral faces if its streamwise dimension exceeded a critical value dependent on Reynolds number.

Associated with the increase in amplitude of oscillation of the shear layers there is an increase in variation of the separation velocity and consequently in the amplitude of the fluctuating lift force. The location of the point of transition to turbulence in the shear layers and the rate of entrainment also depend on the turbulence level in the approach flow. The shortening of the formation region and the accompanying changes in the flow are therefore less marked when the background turbulence is high. This is reflected in measurements of fluctuating lift force.

The instantaneous pressure at a point on the cylinder in the separated region is determined partly by the pressure outside the adjacent shear layer, and partly by local pressure fluctuations in the wake. As a result of the changes in the flow pattern described above with increasing Reynolds number over the range 10^4 to 10^5, the influence of the separated shear layers in determining the pressure on the cylinder increases, the amplitude of the pressure fluctuations increase, and random variations of pressure in the separated region and wake become less significant, despite increased turbulence. Consequently, the chordwise correlation of pressure fluctuations improves.

The correlation between two cross-sections of a bluff cylinder is clearly a function of the chordwise correlation and the sensitivity of the flow to disturbances with components in the spanwise direction. It is therefore reasonable to expect that the improvement in chordwise correlation described above would be accompanied by an improvement in spanwise correlation.

Spanwise Correlation

There have been few quantitative investigations in the past into the three-dimensionality of the flow around a prismatic bluff body. Low

Re flow visualisation studies have revealed that the vortex shedding process does not occur in phase up and down the span of the body [9] and it is now accepted that these flow instabilities will be present at higher Re. The previously obtained values for correlation length are shown together with those obtained in this series of experiments in Fig. 3.

Humphreys [13] has demonstrated that the end conditions of the cylinder are critical to the cross flows induced on the surface of the cylinder, and for this reason, the ends of the cylinder in the present investigation were built in to the wind tunnel walls. Despite this, the values obtained for the pressure correlation length appeared to be dependent upon spanwise location, which suggested that the pressure field on the surface of the cylinder was not homogeneous in a spanwise direction. Tunnel boundary layer effects may account for this. Different correlation lengths were also obtained on different generators on the side face of the cylinder. The results plotted in Fig. 3a are the average values obtained.

Attenuation of the pressure signal in the piping made determination of the absolute value of rms. pressure fluctuation difficult, but as the lengths of all pipes were the same, comparison can be made of dynamic pressure distributions at a constant Re. A complete survey of these was made at Re = 6.3×10^4, which revealed a non-homogeneity of the fluctuating pressure field (Fig. 4). The two typical shapes apparent here are similar to the typical shapes obtained by Vickery [12], although he obtained also a difference in magnitude between the smooth and turbulent flow conditions which was not apparent in the present results. Spanwise rms. pressure traverses at other Re values sometimes revealed also a wavelike structure in the dynamic pressure field.

It is noticeable that the results of el Baroudi [10] in Fig. 3b fail to agree even approximately with the other results; this could be because they were of velocity correlations in the shear layers. Although these perhaps should be comparable with pressure correlations under the same experimental conditions, el Baroudi used an open jet wind tunnel. The comparable curves show an increase in correlation length with Re up to a peak at around Re = 6×10^4, and then the spanwise correlation decreases with further increase of Re. The two curves are in direct opposition to the notion that a square cylinder, with fixed separation points, will display better spanwise correlation than a circular cylinder. Prendergast [11] also measured the pressure correlation length at the rear of the cylinder, and obtained values of about 1d, dropping off to zero as Re increased, and thus an average value taken round one side of the cylinder may reveal correlation lengths more comparable with those now presented.

The values of spanwise correlation length obtained from pressure measurement are as yet not directly comparable with those obtained from velocity measurement in the shear layers downstream of the body, as the two methods have yet to be used under the same experimental conditions. The latter can be expected to give a more precise measure of effective vortex shedding length, whereas the former will be applicable to integration of dynamic pressure to give fluctuating lift and drag coefficients. Both will be decreased by an increase in free stream turbulence, whereas the effective shedding length will be less affected by the conditions of turbulence in the shear layers.

Humphreys [13] obtained evidence by photographic and hot wire measurements of a cell-type structure present in the shear layers at Re = 10^5 stabilized by the presence of threads used to mark the flow. The regular patches of early and late separation were caused respectively by patches of early and late transition to turbulence in the shear layers having reached the surface of the cylinder. This could be the end result of a wavy transition line in the shear layer finally reaching the cylinder surface. If such a wavy transition line were present in the shear layer emanating from a square section, the end result of increase of Re would not be the same, as the separation line is fixed by the front corners of the square. It is suggested that the initial increase in pressure correlation length is due to the mechanism described in the section of this paper concerned with chordwise correlation. As the Re then increases, the transition line in the shear layer moves towards the separation point, and at this stage, the condition of the shear layer above a point on the surface of the cylinder on a particular generator will depend on its spanwise position. Thus the pressure correlation length will decrease as Re increases. The logical conclusion of this is that the transition line reaches the surface of the cylinder, and in the case of the square cylinder, there will be no further drop in correlation length due to this effect (excepting under the effect of free stream turbulence). The separation point on the surface of a circular cylinder, however, is not fixed, and the correlation length will continue to decrease with increase in Reynolds number.

The spanwise correlation of dynamic pressures is thus influenced by two main factors, the condition of the line of transition to turbulence in the shear layer above the surface of the body and the spanwise stability of the shed vortex. It is proposed that the changes in pressure correlation length that occur from $Re = 10^4$ to $Re = 10^5$ are due mainly to the former as the line of transition moves upstream to the separation point. It is also expected that a correlation length computed from velocity measurements may be influenced more strongly by the latter, and that comparison of correlation lengths obtained from velocity measurements with ones obtained from pressure measurements at various generators should provide an interesting line of research.

Transverse Structural Oscillations

The two previous sections of this paper considered static structures. The influence of structural response is an additional, yet fundamental, parameter known to exert sufficient control over the whole motion to render it not only of primary design importance but also of particular scientific interest. Evidence is notably sparse on this topic, but in the last few years has received more prominence. Work at Bristol is now directed towards study of the fluid mechanics of flow-induced structural dynamics; this third part of the present paper reviews progress.

It is by no means uncommon in the field of dynamics for one system oscillating at a particular frequency to come under the influence of a second, stronger system, so that the two become coupled at the frequency of the stronger system. The present subject is not an exception to this phenomenon, although the number of confusing but relevant side issues that occur in this case adds a uniqueness and a challenge to its study. Thus, vortex shedding and its associated circulation and lift force mechanisms [14] impose one frequency (f_v) upon a structure that possesses its own, natural frequency (f_n). Similarity between these frequencies, or between integral multiples of f_n/f_v (or simple fraction [15]), results in resonance at the structural frequency. In particular, the range of frequency ratios ($f_v/f_n = f_r$) within which synchronisation (or "locking-on") may be expected to occur is known to depend upon amplitude ratio (a/D) [9], which in turn is dependent upon damping coefficient [16].

A full appreciation of the sequence of events accompanying locking-on requires that attention be given to the fluid-mechanics of the whole motion, since this is where the driving force has its origin. However a rigorous study remains of questionable value in the absence of a sufficiently clear picture of the constituent parts of the whole motion. In such a situation it is necessary to resort to experimental techniques known to reproduce sufficiently any particular aspect requiring study. Thus, towing-studies remove complications of spanwise correlations and turbulence, allowing the basic interaction between the frequencies to be considered in isolation (Table 1). In this way it is possible to look specifically at the inter-relationship between amplitude, frequency ratio, structural cross-sectional shape and blockage ratio; further, it is a simple matter to observe dynamic wake patterns and vortex arrays.

A simple slow oscillation of an object in a static fluid creates pairs of vortices that pass from the object following reversal of its direction of motion. When fluid flows normal to the plane of body motion it is less easy to discern those vortices alternately generated from the downstream faces of the body, whereas those from the upstream faces become pronounced. They form what amounts to a succession of "starting" and "stopping" vortices, one pair for each cycle of an oscillation, and resembling the pattern associated with aerofoil movement [14]. This succession of alternating "forced" vortices passes into the wake at the structural frequency, unites with the interacting shear layers, and produces a frequency that may or may not resemble that of the structure. If it does then locking-on is present; if it does not then the Strouhal frequency for that amplitude is probably appropriate (see below). For vortex-induced bending oscillations of a circular cylinder (i.e., occurring at natural frequency, f_n), Parkinson [17] quotes the velocity ranges (V_r) corresponding to these frequencies as being $1.0 < V_r < 1.44$ and $0.8 < V_r < 1.0$, $1.44 < < V_r < 1.70$ respectively, where V_r is the ratio between the actual velocity and the velocity required for $S_n = f_n D/V$. For Parkinson's results a/D reached a maximum value of about 0.5 at $V_r = 1.27$.

The second point that must be mentioned, and hopefully resolved, is the definition of Strouhal number (S). For a rigid body $S (= S_v)$ is normally taken to refer to the approach velocity, body width and wake vortex frequency. The "bluffness" of a prismatic member is reflected in its value of S_v, a more bluff body having a wider wake, a reduced frequency of shear layer interaction, and hence lower S_v. It is therefore to be expected that the value of S_v for any structure in transverse (or longitudinal) oscillation will decrease for increasing amplitude as long as the linear dimension used in the calculation of S_v is the body width. The concept of a Universal Strouhal number propounded by Roshko [18] suggests that the dependence of S_v upon a/D could be avoided if some means could be found to account for the influence of amplitude on the wake motion.

To do this it is first necessary to recognise that, for a freely oscillating cylinder, a/D is far from being constant in the locking-on region. Various experimenters, e.g. Parkinson [17] have observed a rise in a/D as fr increases, followed by a peak and a lesser decrease before the wake motion passes out of the locked-on region. To quote Bishop and Hassan [19], ".... each time the velocity of flow is changed, a new wake with a different 'natural frequency' is obtained". To talk of resonance in the locking-on region is therefore inadequate. In fact the peak amplitude condition must correspond to true resonance between the two periodic mechanisms (flow and structure); results for this value of S_v (based on D) have been given for a circular cylinder by Umemura et al [20] for varying a/D that demonstrate the expected fall in S_v resulting from omission of reference to amplitude in S_v.

A curve may be fitted through the results of Umemura et al, apparently being a better fit than that indicated by these authors, that suggests that S_v was taken as $(D + a)$. This conclusion applies to the limit of these results, for which $a/D = 0.3$. Koopman [9] and Toebes [15] also refer to the relevance of amplitude in determining S_v but only in qualitative terms; both authors hold the view that a less than proportional decrease in f_v accompanies an increase in amplitude. Although operating conditions could be the cause of this difference, it is worth mentioning that Koopman's work was done on a very small scale while that of Umemura et al and Toebes was on a 1m to 2m scale.

Taking the finding from Umemura's results allows the "peak" resonance condition to be superimposed upon towing results for a square section cylinder, externally forced, and is shown in Fig. 5. It may be seen that this "peak" resonance line approaches the line marking the lower frequency end of the locking-on range; that this should be so seems a little surprising, but results at large amplitudes against which this could be checked have not been found.

The band-width of locking-on presented in Fig. 5 suggests that for amplitudes up to 0.15D the upper frequency limit rises sharply but then falls away to a value at 0.3D resembling that at 0.05D, and then remains constant at this value. These results are for blockage ratio of 5.5%, and refer to wake frequencies observed 10 body diameters (and greater) behind the cylinder. Closer to the body it is not easy to distinguish a predominant frequency when lateral oscillations occur. Various experimenters claim to have detected frequencies close to the body, but the present authors are in some doubt that the use of instruments in this way satisfactorily distinguishes between two inputs of vorticity to the wake when these are not necessarily exactly in phase (a close phase relationship must exist from the manner in which each component is generated and discharged).

Also marked on Fig. 5 are some results for a circular cylinder, again externally forced, at the same blockage ratio. These results comply with the more detailed observations for the square section. From these results, the pattern for any prismatic body may be inferred with reasonable certainty, following the similarity of wake dynamics for various bodies reported in [21].

The effect of blockage ratio is demonstrated in Fig. 6 for square section cylinders. A marked widening of the locking-on range is evident; in fact, almost all of this increase applies to the upper end of the range whereas the lower end is virtually unchanged. Observations were made consistently at 10 D downstream of the cylinder, and also at 20 D for the higher blockage ratios since it was noted, almost by chance, that in the range 12D to 16D a change of frequency occurs from that relevant to locking-on to the natural value of f_v. This point again suggests that appreciably more attention needs to be given to wake characteristics of dynamic structures and to the associated flow-induced loading.

Conclusions

This paper has dealt mainly with pressure correlation measurements on static cylinders. The work forms a particular part of the general subject of structural response to natural flows, whether of air or water. A complex interaction has been identified, over which there remains considerable speculation. Despite the contributions of a large number of authors, there exists little detailed factual evidence on the nature of fluid-structure interaction.

"Whole face" correlation diagrams giving patterns of instantaneous pressure loading distributions, and the manner of change of these patterns with flow conditions, together with velocity correlation measurements taken in the shear layers and near wake region, are in course of preparation from results resembling those in the earlier part of this paper.

In the third part, some early basic work on dynamic structures is reported. The influence of amplitude ratio on wake form is seen to be substantial, and blockage ratio appears to be just as important here as in other aspects of fluid-structure interaction [21]. Although the results presented apply only to two-dimensional flow, it is probable that this comes close to applying to immersed body conditions, for which the onset of oscillations is known to cause much improved span-wise correlations [15].

References

1. McGregor, D.M.: An Experimental Investigation of the Oscillating Pressures on a Circular Cylinder in a Fluid Stream. IUTAM Techn. Note 14 (1957).

2. Gerrard, J.H.: An Experimental Investigation of the Oscillating Lift and Drag of a Circular Cylinder Shedding Turbulent Vortices. J. Fluid Mech. 11(2), 244 (1961).

3. Chaplin, J.R., Shaw, T.L.: Flow-Induced Dynamic Pressures on Square Section Cylinders. Proc. IAHR Congress, Paris, 1971.

4. Vickery, B.J.: Fluctuating Lift and Drag on a Long Cylinder of Square Cross-Section in a Smooth and in a Turbulent Stream. J. Fluid Mech. 25(3), 481 (1966).

5. Gerrard, J.H.: The Mechanics of the Formation Region of Vortices behind Bluff Bodies. J. Fluid Mech. 25(2), 407 (1966).

6. Chaplin, J.R.: Flow-Induced Forces and Wake Dynamics of Cylindrical Bodies. Ph. D. Thesis. Bristol University, 1970.

7. Bloor, M.S., Gerrard, J.H.: Measurements on Turbulent Vortices in a Cylinder Wake. Proc. Roy. Soc. London, A294, 319 (1966).

8. Bloor, M.S.: The Transition to Turbulence in the Wake of a Circular Cylinder. J. Fluid Mech. 19(2), 290 (1964).

9. Koopman, G.H.: The Vortex Wakes of Vibrating Cylinders at Low Reynolds Number. J. Fluid Mech. 28(3), 501 (1967).

10. el Baroudi, M.Y.: Measurement of Two Point-Correlations of the Velocity near a Circular Cylinder Shedding a Kármán Vortex Street. UTIA Report T.N. 31 (1960).

11. Prendergast, V.: Measurements of Two Point Correlation of the Surface Pressure on a Circular Cylinder. UTIA Rep T.N. 23 (1958).

12. Vickery, B.J.: Fluctuating Lift and Drag on a Long Cylinder of Square Cross-Section in a Smooth and Turbulent Stream. J. Fluid Mech. 25(3), 481 (1966).

13. Humphreys, J.S.: On a Circular Cylinder in a Steady Wind at Transition Reynolds Numbers. J. Fluid Mech. 9(4), 603 (1960).

14. Shaw, T.L.: On the Nature of Fluctuating Circulation, Lift and Flow-Induced Structural Vibrations. J. Sound Vib. 14(2), 251 (1971).

15. Toebes, G.H.: The Unsteady Flow and Wake near an Oscillating Cylinder. Trans. ASME. 91(B), 493 (1969).

16. Wootton, L.R., Scruton, C.: Aerodynamic Stability. ICE. Seminar, The Modern Design of Wind Sensitive Structures 65 (1971).

17. Parkinson, G.V.: Wind-Induced Instability of Structures. Phil. Trans. Roy. Soc. Lond. A. 269, 395 (1971).

18. Roshko, A.: On the Drag and Shedding Frequency of Two-Dimensional Bluff Bodies. NACA, T.N. 3169 (1954).

19. Bishop, R.E.D., Hassan, A.Y.: The Lift and Drag on a Circular Cylinder Oscillating in a Flowing Fluid. Proc. Roy. Soc. Lond. A. 277, 51 (1964).

20. Umemura, S., Yamaguchi, T., Shiraki, K.: On the Vibration of Cylinders Caused by Kármán Vortex. Bull. JSME 14(75), 929 (1971).

21. Shaw, T.L.: Wake Dynamics in Confined Flows. 14th Cong. IAHR, Paris, Paper B6 (1971).

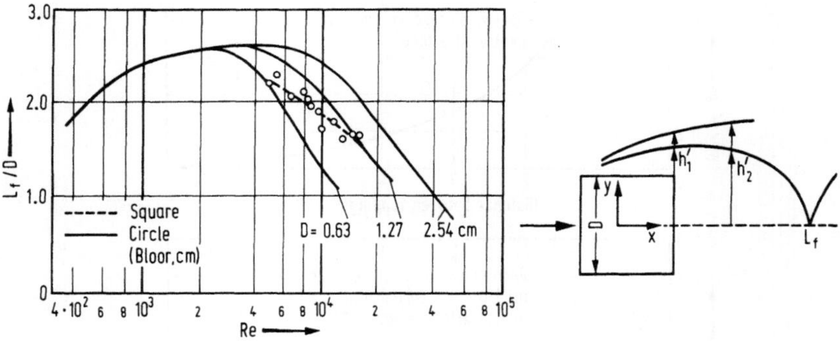

Fig. 1. Length of formation region (L_f) as a function of Reynolds number.

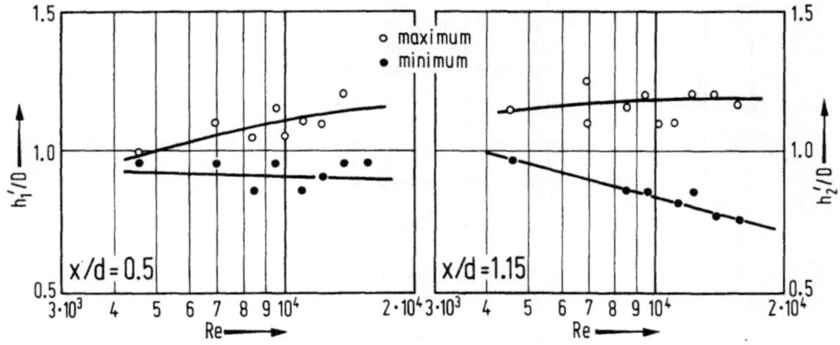

Fig. 2. Positions of the shear layers as functions of Reynolds number.

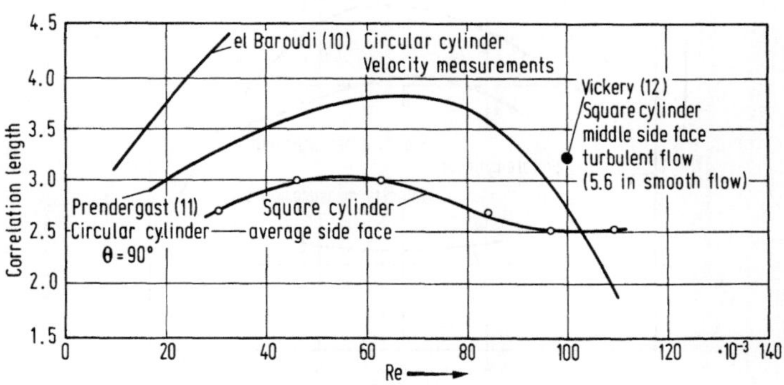

Fig. 3. Correlation length (λ/D) as a function of Reynolds number.

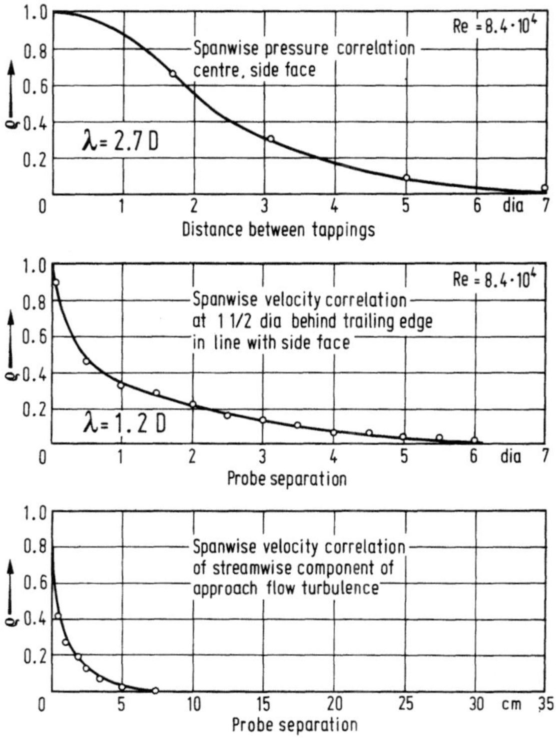

Re = 8.4·10⁴
Spanwise pressure correlation centre, side face
λ = 2.7 D
Distance between tappings
dia
Re = 8.4·10⁴
Spanwise velocity correlation at 1 1/2 dia behind trailing edge in line with side face
λ = 1.2 D
Probe separation
dia
Spanwise velocity correlation of streamwise component of approach flow turbulence
Probe separation
cm

Fig. 3a

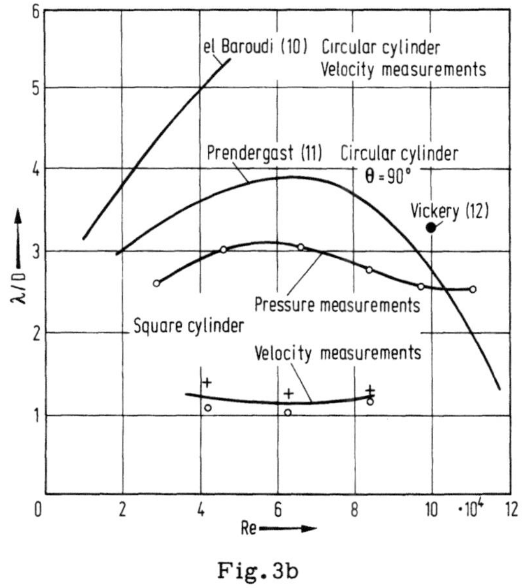

el Baroudi (10) Circular cylinder Velocity measurements
Prendergast (11) Circular cylinder θ = 90°
Vickery (12)
Pressure measurements
Square cylinder
Velocity measurements
λ/D
Re
·10⁴

Fig. 3b

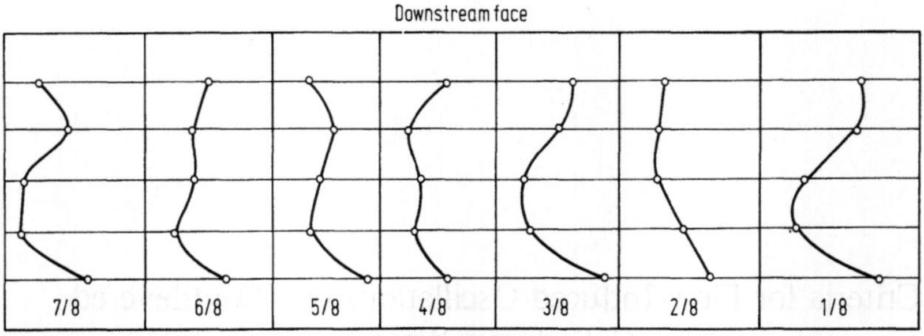

Fig. 4. Shape of RMS pressure distribution on side face, at 1/8 span intervals.

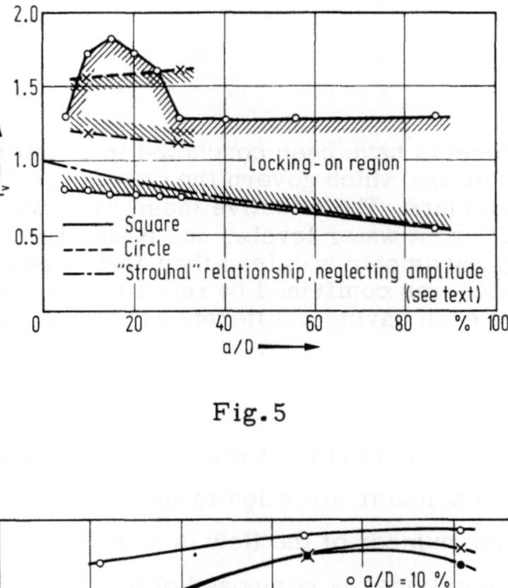

Fig. 5

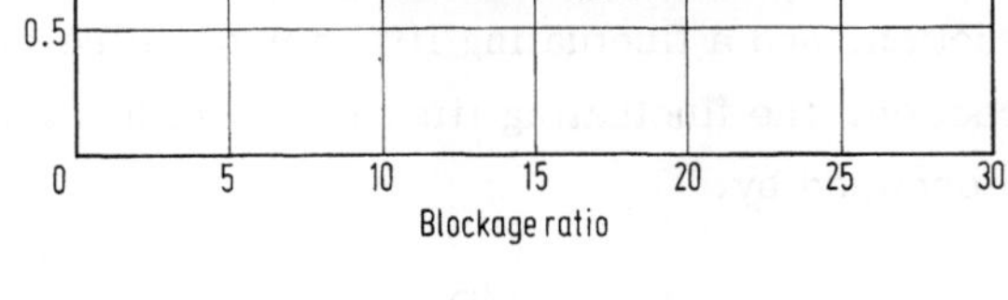

Fig. 6

Criteria for Flow-Induced Oscillations of a Cantilevered Cylinder in Water

By

R. King, M. J. Prosser

Cranfield, Bedford, U. K.

Summary

A series of experiments have been conducted in a water channel to determine the conditions which govern the onset of in-line oscillations of cantilevered cylinders. The effective mass ratio was varied by repeating tests at different water levels, and it was found that when the combined stability parameter was less than 0.8, in-line oscillations occurred. This value was confirmed by repeating the tests using three different cylinders each having a different structural damping.

Introduction

Consider a rigidly mounted cylinder placed in a stream of fluid which is flowing in a perpendicular direction to the axis of the cylinder. As long as the Reynolds number of the flow is above a certain lower limit, the wake behind the cylinder is composed of a series of vortices which have been shed from alternate edges of the cylinder. This unsteady wake produces a periodic force which may be resolved as a steady drag component in the flow direction, a fluctuating drag component in the flow direction, and a fluctuating lift component perpendicular to the flow direction. The fluctuating lift component has a frequency, f_L which is determined by,

$$f_L = S\ V/D, \tag{1}$$

where

$$S = 0.2$$

and corresponds to a pair of vortices being shed from the cylinder for each cycle. The fluctuating drag component has a frequency $f_D = 2f_L$ which corresponds to one vortex being shed from the cylinder for each cycle.

If the cylinder is flexibly mounted, or the cylinder itself is flexible, it may oscillate under the influence of the fluctuating lift and drag components. The frequency of vortex shedding is then no longer described simply by a constant value of S in Eq. (1) and it is found that S may have a maximum of about 0.4. A cylinder having a given natural frequency, N, and diameter D, in a stream whose velocity, V, is gradually increased from zero, may therfore be excited in the cross-flow direction (lift direction) when $V/ND = 2.5$ and in the in-line direction (drag direction) when $V/ND = 1.25$.

However, although one might have a cylinder in a fluid stream where the value of V/ND is above the critical value, there might not be sufficient energy in the stream to cause oscillations of the cylinder. Therefore, there must be a further requirement before conditions are suitable for flow-induced oscillations to occur and this paper describes some experiments which study the onset of in-line oscillations of a flexible cantilevered cylinder in a stream of water.

Wind Excitation

Current practice in industrial aerodynamic problems relating to the oscillation of chimney stacks and other slender structures, is to define stability boundaries in terms of the reduced velocity (V/ND) and a combined stability parameter (k_s) where

$$k_s = 2M\delta s/\rho D^2. \tag{2}$$

In broad terms (e.g., Ref. [1]) if $V/ND > 2.5$ and $k_s < 16$ structural oscillations in the cross-flow direction will occur. Theoretical justification of this procedure arises from dimensional analysis combined with energy considerations at the resonant condition. Ignoring Reynolds number and Froude number effects, assume that the cylinder deflection, y, is a function of N, D, m, δs, V and ρ,

so
$$y = f_1(N, D, m, \delta_s, V, \rho), \tag{3}$$

then we get

$$y/D = f_2\left(V/ND, m/\rho D^2, \delta_s \right). \tag{4}$$

Equating the energy dissipated in the structure by its internal damping, to the energy input from the fluid, it can be shown, (e.g., Ref. [2]) that the last two terms in Eq. (4) can be combined into the parameter $2M\delta/\rho D^2$ where M is the equivalent mass per unit length, defined by

$$M = \int my^2dx/\int y^2dx. \tag{5}$$

Water Excitation

The analagous structure in water has not been studied to the same extent, but similar basic criteria should hold. However, several difficulties arise which make the problem more complex than the air excited case:

(a) The "added mass" effect of the water cannot be neglected and it appreciably lowers the natural frequency from the in-air value.

(b) There is an added damping effect due to the water.

(c) Since the cantilever need not be fully submerged in the water, the effects of (a) and (b) apply to the submerged section only, with corresponding modification to the deflected mode shape.

(d) In practice it is found that the 'in-line' mode of oscillation is quite easily excited in models (Ref. [3]) and on site (Ref. [4]) whereas most of the aerodynamic work has been related to the cross-flow excitation.

If, for a given structure, the flow velocity is progressively increased, in-line oscillations occur when V/ND = 1.2 and indications are that a maximum will be reached at V/ND = 2.5. However, at this value and above, conditions are then suitable for cross-flow oscillation and the cantilever will start to execute combined, or figure-of-eight type motion.

There is very little experimental evidence on the critical values of the combined damping parameters, $2M\delta_s/\rho D^2$, so following a suggestion by Sainsbury (Ref. [5]) some simple flume experiments were undertaken.

Flume Experiments

A series of tests have been made using slender hollow cylinders mounted vertically as cantilevers on the floor of a water channel. Fig. 1 gives a diagrammatic view of the test arrangement. The depth of water and the water velocity could be varied independently. For each cylinder, the water depth, h, was held at a fixed value and the flow velocity was increased from zero whilst observing the cylinder deflection, indicated by strain gauges mounted at the base of the cylinder. For all cases, sustained in-line oscillations were observed for velocities above a certain critical value, V_{crit}, as long as the water depth was comparable with the cylinder length, L. The tests were repeated for different water depths so that corresponding readings of V_{crit} and h were obtained. When h was reduced sufficiently no sustained oscillations were observed even when V was above the calculated critical value of V/ND, so that a critical value of h/L was determined for each cylinder configuration. A summary of the test conditions is shown in the table below:

Table 1

Test No	Length [inches]	Material	m_s [slugs/ft]	δ_s	D [inches]
1	24.5	Brass	3.28×10^{-3}	0.016	0.5
1a	24.5	Brass	5.56×10^{-3}	0.016	0.5
2	41.0	PVC	2.68×10^{-3}	0.080	1.0
2a	41.0	PVC	10.8×10^{-3}	0.080	1.0
3	41.5	Al.	3.91×10^{-3}	0.026	1.0

These five tests represent an attempt to vary the parameters m_s, δ_s and D which are used to evaluate the combined damping parameter k_s. In the cases 1a and 2a the mass per unit length, m_s, was changed by filling the cylinders with water and putting a light cork in the open end to prevent any sloshing effects.

Critical Velocity for In-Line Oscillations

The critical value of V/ND was evaluated for each test and is shown plotted against h/L in Fig. 2. The velocity measurements were not very precise so one should not attempt to analyse these results in detail, but there is the indication that V_{crit}/ND is not a constant value but is a function of h/L. Otherwise, the main conclusion is that V_{crit}/ND lies between 1.2 and 1.9.

The Combined Damping Parameter

The value of M was calculated using Wooton's suggestion (Ref. [5]) that

$$M = \int_0^L my^2dx \Big/ \int_0^h y^2dx$$

for various values of h/L. Note that the numerator is integrated over the cylinder length and the denominator over the water depth. The structural damping was determined experimentally and values of the combined damping parameter $2M\delta/\rho D^2$ were plotted against h/L in Fig. 3 for two cases. For a uniform cylinder Eq. (6) can be re-arranged as follows,

$$M = \frac{\int_0^L (m_a + m_s)y^2dx}{\int_0^h y^2dx} = \frac{\int_0^L m_a y^2dx + m_s\int_0^L y^2dx}{\int_0^h y^2dx},$$

$$= \frac{m_a\int_0^h y^2dx}{\int_0^h y^2dx} + \frac{m_a\int_h^L y^2dx}{\int_0^h y^2dx} + \frac{m_s\int_0^L y^2dx}{\int_0^L y^2dx}.$$

The second term is zero since m_a is zero between h and L; hence

$$M = m_a + m_s \int_0^L y^2 dx \Big/ \int_0^h y^2 dx. \tag{7}$$

The integrals were evaluated numerically by a finite element technique using a digital computer for values of h between 0 and L. They were then used to calculate curves of k_s versus h/L for each cylinder and are shown plotted in Figs. 3 and 4. Fig. 3 shows k_s evaluated using δ_s the structural damping and Fig. 4 shows k_s evaluated using $\delta_a + \delta_s$, the damping in still water. In each case, the measured stability boundary is marked and it is seen that the critical value lies between 0.6 and 1.1 in Fig. 3, and 1.0 and 1.6 in Fig. 4. Probably, Fig. 4 gives a slightly more uniform value of the critical k_s suggesting that the damping in still water is a better criterion to use.

The added damping in still water can be estimated from the theory given in the Appendix and added to an estimated damping in air if it is not practicable to measure them directly.

Damping in Flowing Water

Two interesting effects are shown in Figs. 5 and 6. Fig. 6 shows the critical case (h/L = 0.61) for cylinder No. 2 (see Table 1), where the water depth and so k_s, is just sufficient for sustained oscillation at V_{crit}. From an initial large amplitude deflection the damping is high but as the amplitude reduces in successive cycles, the damping reduces to zero, until a small amplitude sustained oscillation occurs.

For the same cylinder, at a value of h/L which was too small for sustained oscillations, the total damping $\delta = \delta_s + \delta_a$ was measured for various flow velocities. The results in Fig. 5 show that the minimum damping occurs at a V/ND = 1.8, this result having been plotted in Fig. 2, as a stable condition.

Conclusions

The results indicate that the combined damping parameter must be less than about 0.8 for in-line oscillations to be possible if k_s is defined as $2M\delta_s/\rho D^2$. If it is defined as $2M(\delta_s + \delta_a)/\rho D^2$ the critical value is about 1.2. These values have been derived by varying the mass ratio $M/\rho D^2$, where M is defined in Eq.(6), for each cylinder.

Since three different cylinders have been used, three different values of structural damping were used in evaluating k_s, and with the exception of the hollow brass cylinder a reasonably constant value of $(k_s)_{crit}$ was obtained.

The latest results from the experimental study are described in Ref.8.

Appendix

Hydrodynamic Damping in Still Water

A study was made of the hydrodynamic damping experienced by partly immersed cantilevered cylinders. The results of this theoretical and experimental study have led to the formulation of a general equation for the logarithmic decrement of a cantilever vibrating freely in still water.

The theoretical analysis was based on Stokes' mathematically derived equations of the fluid resistance experienced by an element of an oscillating rigid pendulum (Ref.[7]). Stokes showed that the resistance equation consisted of two terms, the first of which reduces the frequency whilst the second produces a dimunition in amplitude.

$$F = \lambda_1 d^2y/dt^2 + \lambda_2 dy/dt. \qquad (8)$$

For transient amplitude decay, $\lambda_2 dy/dt$ is the characteristic force and Stokes derived the following analytical expression for λ_2.

$$\lambda_2 = m_a w\ (2\sqrt{2/G} + 2/G)dx, \qquad (9)$$

where

$$G = wr^2/\nu.$$

By applying Stokes' rigid element analysis to the elastic elements of a vibrating cantilever and integrating $\lambda_2 dy/dt$ with respect to time and distance along the cantilever, the work done per cycle in overcoming viscous damping was calculated, i.e.

$$W = \int_0^h \int_0^T \lambda_2 w^2 X_x^{\,2} \cos^2 wt \, dx \, dt. \tag{10}$$

From this, an equivalent viscous damper was defined and an expression developed for the hydrodynamic contribution (δ_a) to the total logarithmic decrement δ. The total logarithmic decrement is the sum of the structural and hydrodynamic logarithmic decrements, i.e.,

$$\delta = \delta_a + \delta_s.$$

Thus the total logarithmic decrement may be written

$$\delta = \frac{m_a w^2 \,(2\sqrt{2/G})}{\beta EIA(h/L)} \int_0^h \left(X_x/X_e\right)^2 \frac{dx}{L} + \delta_s.$$

X_x = deflection at height,

X_e = maximum deflection,

βEI = section stiffness of cylinder,

$$A(h/L) = \frac{\left(1 + r_f^{\,2}\right)}{r_f^{\,2}} \left\{ \frac{M_e L + 0.243(m_s L + m_a h)}{M_e L + 0.243\left(m_s L + m_a h^4/L^3\right)} \right\},$$

r_f = frequency ratio $\dfrac{\text{second normal mode}}{\text{fundamental}}$,

M_e = tip load.

A computer program was written to calculate the cantilever frequencies and mode shapes using transfer matrix mathematics. A procedure withwithin the program performed the numerical integration over the immersed lengths.

Experimentally, the damping was inferred from measurements of transient amplitude decay when the cantilever tip was released from an initial displacement of up to one diameter.

Frequencies and mode shapes were varied by altering the cylinder mass per unit length, by adjusting the water level and also by attaching masses to the cantilever free end. The tests demonstrated close agreement between theoretical and experimental results. Figures 7 and 8 are graphs of total logarithmic decrement plotted to a base of variable water depth. The experimental results are shown superimposed on the theoretically derived curves.

Conclusions

(a) Hydrodynamic damping of the natural vibrations of a cantilever in still water is purely viscous, and may be calculated by integrating Stokes' element damping equation along the immersed length. This was verified over the Size Number* range 619-21, 600, using free-end deflections of up to one diameter.

(b) Viscous damping increased rapidly with water dempth for $h/L > 0.5$. Below 0.5 the structural damping was the main contributor.

All graphs of total logarithmic decrement versus water depth exhibited approximately similar shapes.

(c) No amplitude effects were detected in the transient decay traces. This verified the predominance of viscous damping effects.

As a further exercise, Stokes' element damping theory was applied to a rigid vertical pendulum (2 in. diameter, 37 in. long) vibrating in varying depths of still water. Conclusions similar to (a), (b), (c) above were reached, confirming the validity of Stokes Theoretical equations at relatively high Size Numbers (up to 16,600).

The results of the pendulum tests are shown in Fig. 9, it will be observed that this graph has a shape virtually identical to the curves of Figs. 7 and 8 of the cantilever tests.

Table 1 Characteristics of the Cylinders Tested

Cylinder Number	Material	Diameter (ft)	EI ($lbf\ ft^2$)	Length (ft)	Frequency Range (Hz)	Size Number Range
1	Fibreglass	0.042	6.84	4.00	4.10- 6.10	619- 834
2	P.V.C.	0.084	95.50	3.42	1.84- 6.31	1080- 3500
3	Fibreglass	0.084	204.50	4.00	5.79- 9.80	3200- 5400
4	Fibreglass	0.151	1072.00	4.00	9.28-10.88	18500-21600
5	Brass Pendulum	0.168		3.10	3.64- 7.52	8050-16600

Addendum

Variation of N, m and δ_s, Using the PVC Cylinder

(a) Filling the cylinder with lead shot had the threefold effect of reducing the natural frequency (N), increasing the structural mass per unit length (m) and decreasing the structural logarithmic decrement (δ_s).

(b) To compensate for the reduction in N, coil springs were attached to the cylinder free end. These springs restored the value of N to the frequency recorded when the cylinder was filled with water; they also produced a reduction in δ_s additional to the one noted in (a).

(c) Consequently the Stability Parameter - Water level graphs of the lead filled cylinder are of considerable interest; the results being shown in Figs.3a and 4a.

(i) The earlier tests suggested a structural stability boundary of between 0.6 to 1.0 for in-line vibrations. Applying this reasoning to Fig.3a would suggest that the lead-filled cylinder should be unstable in the fundamental in-line mode, whether free ended or fitted with tip springs.

(ii) Again, referring to the earlier tests, the stability boundary based on the still-water logarithmic decrement appeared to be between 1.0 and 1.4. When this criterion was applied to Fig.4a it would seem that the lead-filled cylinder would be stable when free-ended and unstable when fitted with tip springs.

(d) When tested in the two conditions described (with and without tip springs) the following results were obtained:

(i) The free ended cylinder was not excited in the fundamental mode in the in-line direction. This confirmed c(ii). However, second normal mode vibrations were readily excited at the higher water levels and fundamental frequencies were also initiated in the cross-stream (lift) direction.

(ii) The stability boundary for the cylinder fitted with tip springs confirmed the results observed in the main series of tests. The stability boundaries are marked on Figs. 3a and 4a.

(e) The conclusion drawn from these tests showed that the Stability Parameter based on the still-water logarithmic decrement could be used for defining the stability boundary for fundamental in-line vibrations.

A further conclusion was that the Stability Parameter for fundamental in-line vibration does not apply to higher normal mode vibrations, neither does it necessarily govern the stability in the cross-stream direction.

Acknowledgement

The work described was undertaken as part of the general research programme of BHRA Fluid Engineering and the authors wish to thank the Director and Council for permission to publish this paper.

Nomenclature

V	fluid velocity
N	cylinder structural frequency
D	cylinder diameter
k_s	combined stability parameter
M	equivalent mass per unit length
ρ	fluid density
δ_s	structural logarithmic decrement
δ_a	additional damping due to water
δ	$\delta_a + \delta_s$

m_s	mass per unit length of cylinder plus any water inside
m_a	added mass
m	$m_a + m_s$
h	water depth
L	cylinder length

References

1. Scruton, C.: On the Wind-Excited Oscillations of Stacks, Towers and Masts. Paper 16, Proceedings of the Conferences on Wind Effects on Buildings and Structures, Teddington, June 1963.

2. Vickery, B.J., Watkins, R.D.: Flow-Induced Vibrations of Cylindrical Structures. Proceedings of the First Australasian Conference. University of Western Australia, December 1962.

3. King, R.: Flow-Induced Vibrations. BHRA Report RR.1093, January 1971.

4. Sainsbury, R.N., King, D.: The Flow-Induced Oscillations of Marine Structures. Proc. of the Institution of Civil Engineers 49 (July 1971).

5. Sainsbury, R.N.: Private communication.

6. Wooton, L.R., Sainsbury, R.N., Warner, M.H., Cooper, D.H.: The Flow-Induced Oscillations of Piles. NPL Aero Special Report 025 (January 1969).

7. Stokes, G.G.: On the Effects of the Internal Friction of Fluids on the Motion of Pendulums. Mathematical and Physical Paper, 3, Cambridge 1901.

8. King, R., Prosser, M.J., Johns, D.J.: On Vortex Excitation of Model Piles in Water. J. Sound Vibration 29(2), 169-188 (1973).

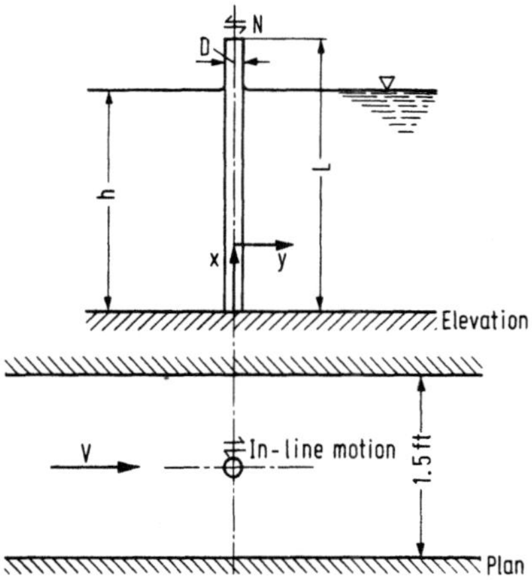
N
D
h
L
x
y
Elevation
V
In-line motion
1.5 ft
Plan

Fig.1

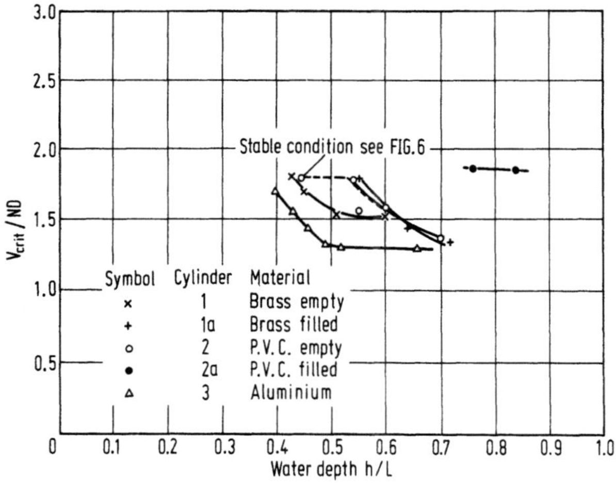
Stable condition see FIG.6
V_{crit}/ND
Water depth h/L
Symbol Cylinder Material
× 1 Brass empty
+ 1a Brass filled
○ 2 P.V.C. empty
● 2a P.V.C. filled
△ 3 Aluminium
3.0
2.5
2.0
1.5
1.0
0.5
0
0.1
0.2
0.3
0.4
0.5
0.6
0.7
0.8
0.9
1.0

Fig.2

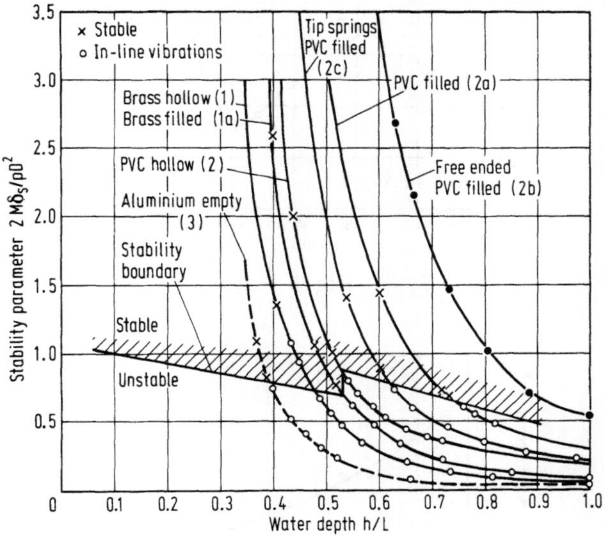
× Stable
o In-line vibrations
Tip springs PVC filled (2c)
PVC filled (2a)
Brass hollow (1)
Brass filled (1a)
PVC hollow (2)
Aluminium empty (3)
Stability boundary
Free ended PVC filled (2b)
Stable
Unstable
Stability parameter $2M\delta_s/\rho D^2$
Water depth h/L

Fig.3

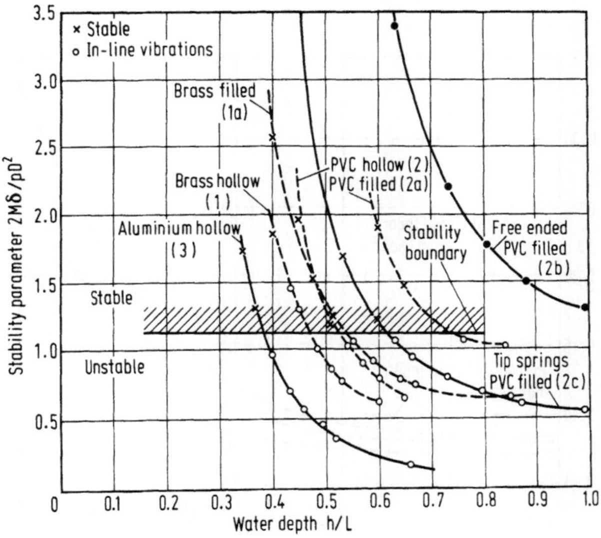
× Stable
o In-line vibrations
Brass filled (1a)
PVC hollow (2)
PVC filled (2a)
Brass hollow (1)
Aluminium hollow (3)
Stability boundary
Free ended PVC filled (2b)
Stable
Unstable
Tip springs PVC filled (2c)
Stability parameter $2M\delta/\rho D^2$
Water depth h/L

Fig.4

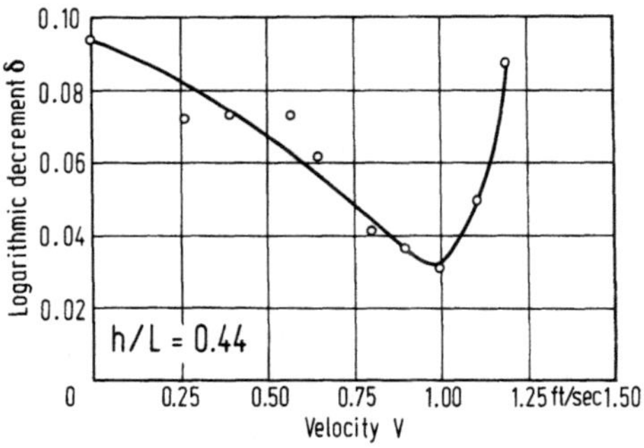

Fig.5

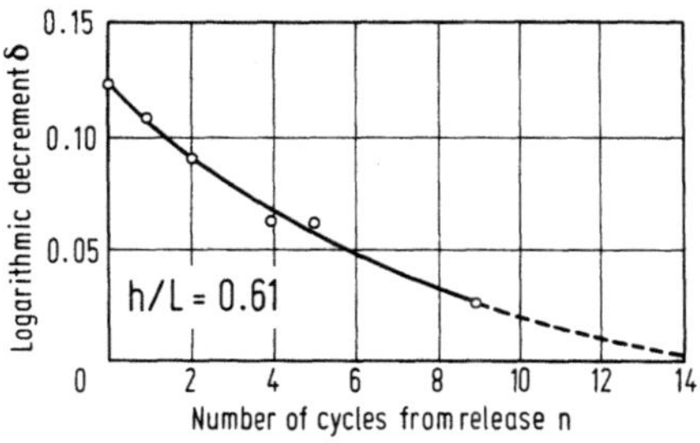

Fig.6

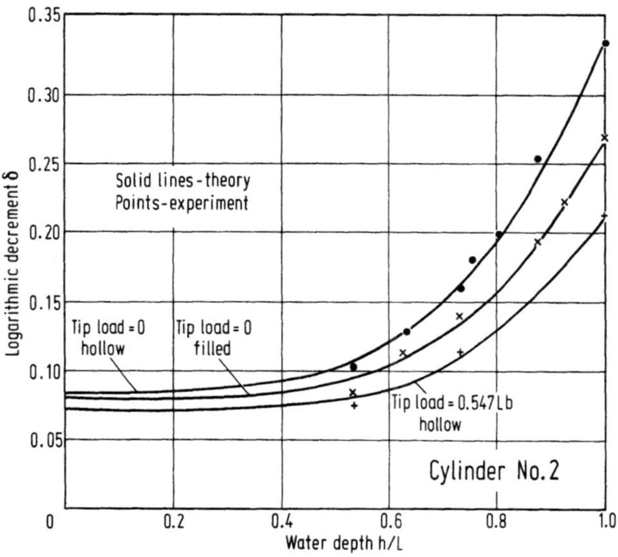

Fig.7

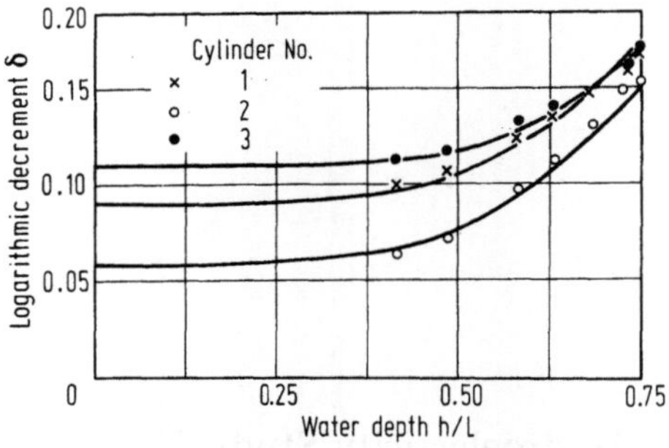
Cylinder No.
1
2
3
Logarithmic decrement δ
Water depth h/L
0.20
0.15
0.10
0.05
0
0.25
0.50
0.75

Fig.8

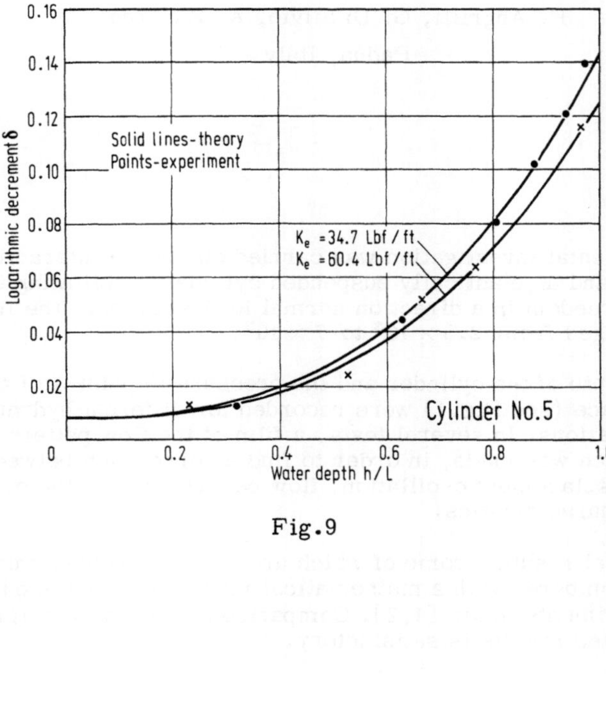
Solid lines-theory
Points-experiment
Ke =34.7 Lbf / ft
Ke =60.4 Lbf / ft
Cylinder No.5
Logarithmic decrement δ
Water depth h / L
0.16
0.14
0.12
0.10
0.08
0.06
0.04
0.02
0
0.2
0.4
0.6
0.8
1.0

Fig.9

Hydroelasticity Study of a Circular Cylinder in a Water Stream

By

F. Angrilli, G. Di Silvio, A. Zanardo

Padua, Italy

Summary

An experimental investigation was carried out on the interaction of a water flow and an elastically suspended cylinder, oscillating in one degree of freedom in a direction normal to the current. The Reynolds number ranged from 2.5×10^3 to 7×10^3.

Displacements of the cylinder and the pressure in point A of the cylinder surface (see Fig.1) were recorded for different hydraulic and elastic situations. In several tests, a film of the flow pattern at the water surface was made, in order to find a correlation between pressure and displacement oscillation, flow behaviour near the cylinder and wake characteristics.

Experimental results, some of which are reported below, were analized and compared with a mathematical model already tested with other experiments in air [1,2]. Comparison between experimental and calculated results is satisfactory.

Experimental Apparatus

The tests were carried out in a 11 m long flume with a square cross section of 0.5 m size (Fig.1). Special care was taken in order to reduce the turbulence of the flow upstream of the cylinder. The cylinder, made of PVC, is hollow and weighs about half a kilogram, equipment included. Its diameter is 40.5 mm, the lenght 500 mm. The elastic suspension is shown in Fig.1. Four different spring constants were tested, i.e., 0.72, 1.02, 1.38, and 1.78 kgf/m, corresponding to four natural frequencies of the system in still water, namely 0.38, 0.44, 0.50, 0.57 Hz.

The oscillations of the cylinder were detected by means of an angular displacement transducer. Pressure measurements were obtained using an extremely small-sized pressure transducer with a very low sensitivity to acceleration. The pressure sensitivity is of the order of 1 mm of water column, and the frequency response is very good in the range of the measured frequencies.

Experimental Results

Tests were performed with a stream velocity between 6 and 15 cm/s. As long as the stream velocity is below a certain value, both pressure and displacement oscillations are very small and rather irregular. They have, however, always the same frequency.

Pressure and displacement oscillations begin to become regular and remarkable with flow velocities rather close to resonance conditions (oscillation frequency equal to the natural frequency of the system). The maximum motion and pressure amplitudes are displayed with a velocity somewhat larger.

Experimental pressure and displacement frequencies are plotted versus stream velocity in Fig. 2, together with the Strouhal frequency of eddy shedding when the cylinder is at rest ($f = 0.2\ v/D$). It is interesting to note that the experimental frequencies are always in between the Strouhal frequency and the respective natural frequency of the system. In Fig. 3, the corresponding displacement amplitudes are given. The four curves have similar shapes and almost equal peaks.

For the natural frequency of $f_n = 0.50$ Hz, a complete photographic analysis was made. In Fig. 4, a series of 18 frames, covering an entire oscillation period, is given as an example of photographic visualization technique. The flow velocity in this test is 11 cm/s, corresponding to the maximum amplitude of the cylinder oscillation.

In Figs. 5a to 5d, displacement and pressure records are given, together with a schematization of the vortex pattern deduced from the photographic frames. The dots in a schematization represent the successive positions of the apparent center of the vortices as they result from the photographic records. Different marks are used for different vortices. Each photographic series, covering about an oscillation period, is formed by 18 frames. The figures marked on the vortex pattern associate the position of each vortex and the maximum displacement of the cylinder with the corresponding frame.

Separation giving rise to alternate vortices takes place on the cylinder surface in accordance with the relative motion between stream and cylinder. Following the motion, each vortex rolls along the cylinder surface until a suitable position is reached to be shed in the Von Kármán stable wake.

As long as the oscillation is small (Fig. 5a), vortex trails are not very different from those produced by a stationary bluff body, but for larger oscillations (Figs. 5b, c, d) the two vortex trajectories must cross (twice) each other to reach a stable configuration in the wake. It is interesting to note in Fig. 5d the short row of clockwise vortices (marked by a triangle) moving toward the wrong side of the wake and therefore doomed to be quickly destroyed.

Comparison with Mathematical Model

The mathematical model proposed in [2] was tested with the above mentioned experimental results. By using a constant lift coefficient C_L contained in the usual range of the experimental values for stationary cylinder, and a damping ζ deduced by free oscillations in still water, both the experimental frequencies and amplitudes are in very good agreement with the computation [3].

References

1. Ferguson, N., Parkinson, C.V.: Surface and Wake Flow Phenomena of the Vortex-Excited Oscillation of a Circular Cylinder, Paper 67, Vibr.31, Vibrations Conference, ASME, Boston 1967.

2. Di Silvio, G.: Self-Controlled Vibration of a Cylinder in a Fluid Stream, J. Eng. Mech. Div. ASCE 95, No. EM2, Proc. Paper 6498 (April 1969).

3. Angrilli, F., Di Silvio, G., Zanardo, A.: Fluidelastic Vibrations: Comparison between Experimental and Calculated Results. Submitted for publication to Meccanica.

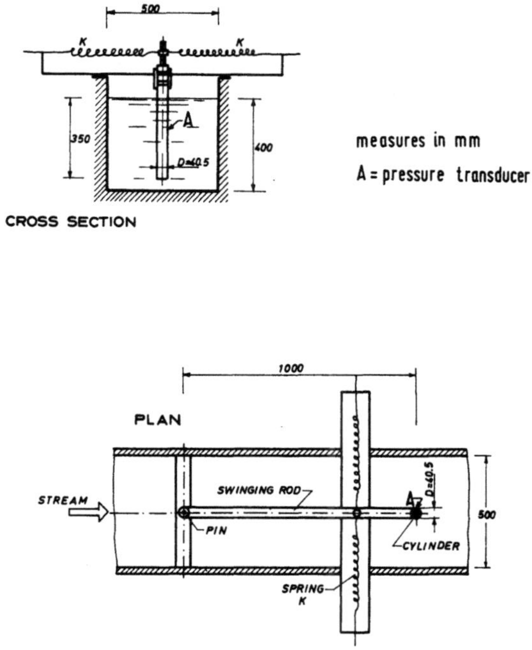

Fig.1. Oscillating cylinder and experimental flume.

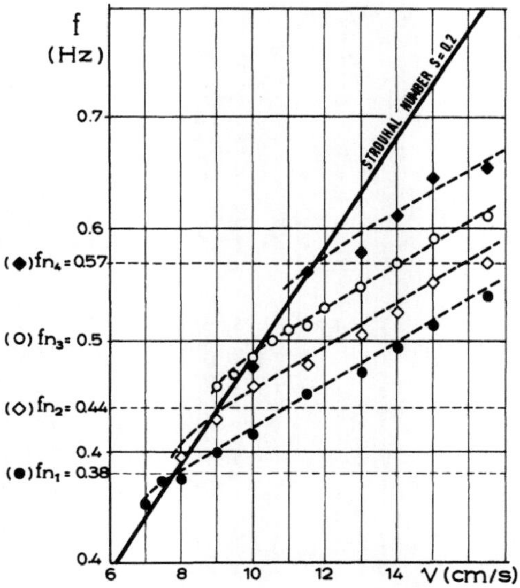

Fig.2. Pressure and displacement oscillation frequency vs. stream velocity with different natural frequencies.

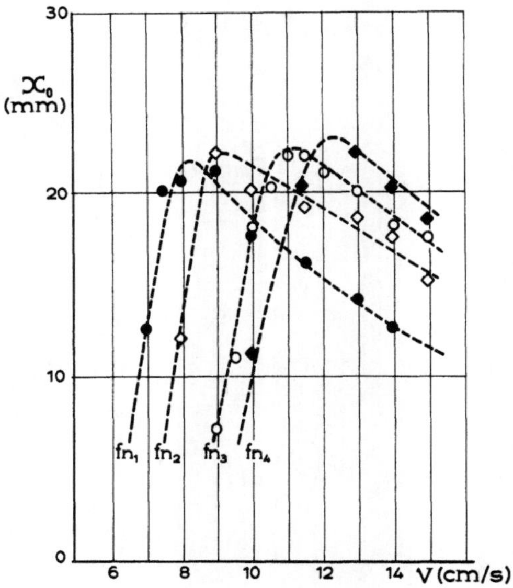

Fig.3. Displacement amplitude vs. stream velocity with different natural frequencies.

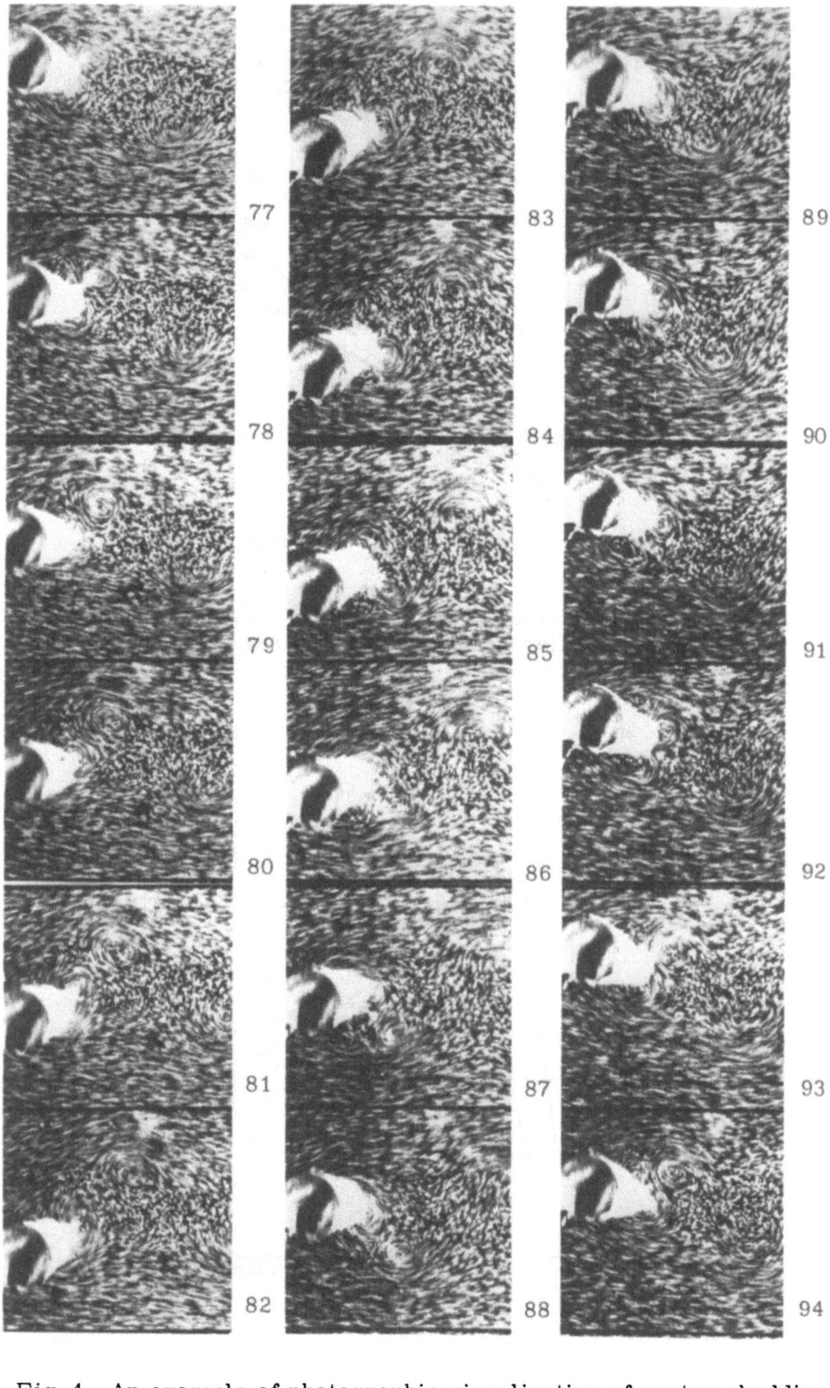

Fig.4. An example of photographic visualization of vortex shedding (f_n = 0.5 Hz, v = 11.0 cm/sec).

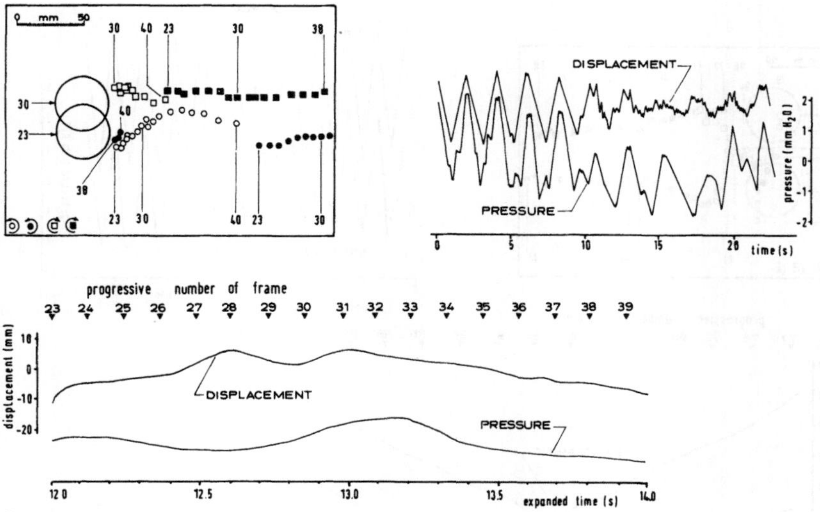

a

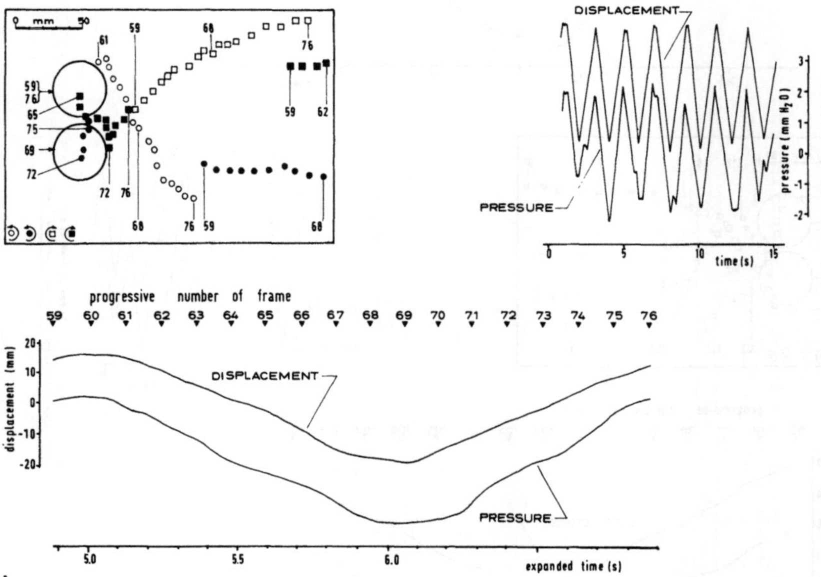

b

Fig.5. Correlation between vortex pattern, pressure and cylinder displacement. (Natural frequency $f_n = 0.5\,Hz$)

a) Stream velocity $v = 9.5\,cm/sec$. Oscillation and pressure frequency is below the natural frequency of the system: displacement curve precedes pressure curve.

b) Stream velocity: $10.5\,cm/sec$. Oscillation and pressure frequency is equal to the natural frequency (resonance conditions): displacement and pressure curves are in phase.

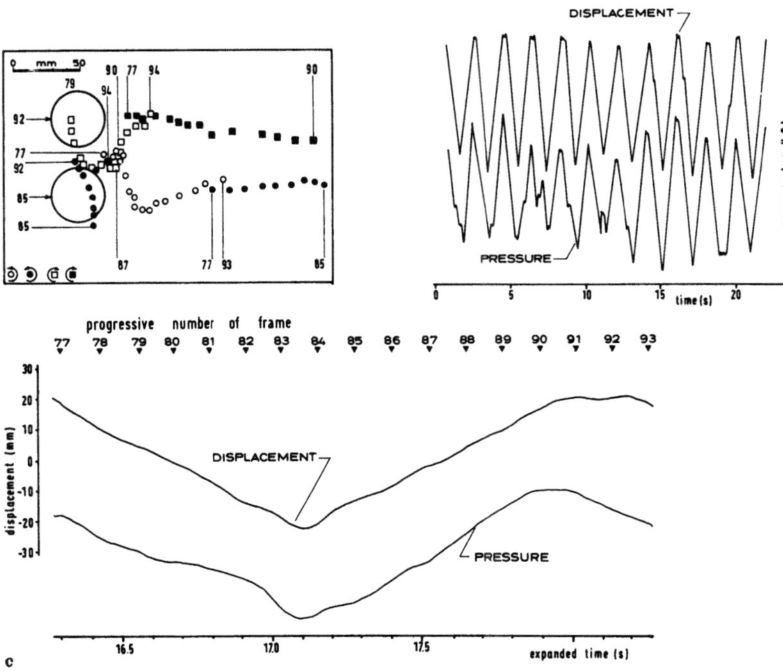

c

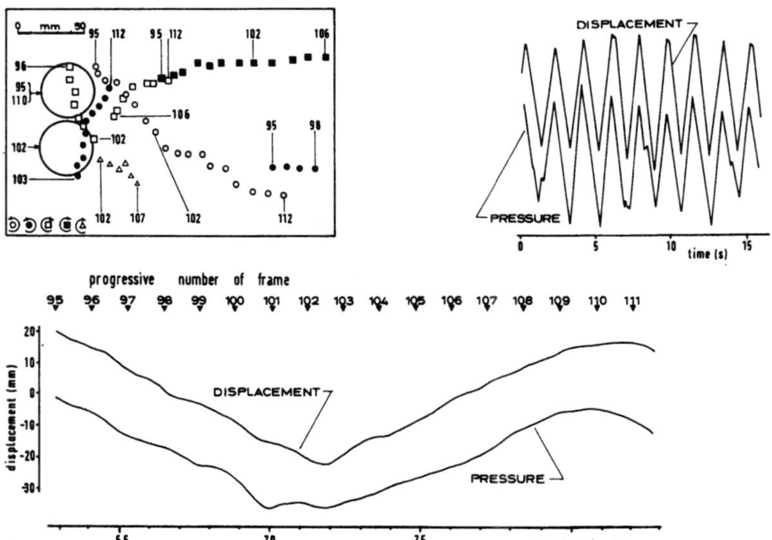

d

5c) Stream velocity: 11 cm/sec (see also Fig. 4). Oscillation and pressure frequency (maximum oscillation amplitude): displacement curve follows pressure curve a little.

5d) Stream velocity: 12 cm/sec. Oscillation and pressure frequency is above the natural frequency: displacement curve follows pressure curve considerably.

Correlations between Forces, Flow Field Features, and Confinement for Bluff Cylinders

By

G.H. Toebes

Lafayette, Indiana, U.S.A.

A movie was presented that shows in detail the unsteady flow field at and near the rear of bluff cylinders. The process of alternate vortex formation is clearly portrayed. The effect of non-circular cylinder orientation on fluidelastic stability is shown.

Flow field-force correlation is portrayed by simultaneous showing of the vortex formation and an X-Y recorder tracing the oscillatory lift.

Shown also are the effects of splitter plates and of flow confinement or "blocking" on vortex formation.

The essentially stochastic character of the phenomenon is illustrated by the lift-drag correlation traces or "Lissajous figures".

PANEL DISCUSSION

"Research Priorities in Flow-Induced Vibrations"

Chairman: E. Naudascher, Germany

PANEL 1

New Theoretical and Experimental Approaches

By

G.H. Toebes

U.S.A.

The workshop discussion started out by considering the following breakdown of its title topic:

I. Theory
(a) Basic Mechanisms, (b) Sample Calculations, (c) Model Laws.

II. Design
(a) Basic Mechanisms, (b) Sample Calculations, (c) Forcing Functions.

III. Experiment
(a) Flow Field, (b) Forces, (c) Flow Field-Force Correlations, (d) Model Laws, (e) Computer Simulation.

It was hoped that this catagorization would permit the group to arrive at some concensus of what is needed in terms of new theory and experimentation, based on recognized similarities among the great diversity of flow-induced vibration problems.

The ensuing exchanges showed that, whenever one divorces the discussion from its reference to a particular type of structure or a particular type of mechanism, it becomes as yet very difficult to discuss new approaches for the field of fluid-elasticity as a whole. In this overall sense, there are no new approaches because there are no old ones. In their absence one should perhaps expect the heterogeneity in theory, experimental facts, and ad hoc design solution that were evident in the discussion. Apparently, the time has not yet come to consider the subject in the framework of mechanisms[1] such as: (a) Divergence, (b) Flutter, (c) Vortex-Induced Vibrations, (d) Galloping, (e) Buffeting, Etc. Nevertheless, the heterogeneous collection of elements of theory, experimental facts, and trial and error design modification to get rid of problems is probably greater and more costly in terms of spent effort than it needs to be.

At a time when generalized-systems concepts are becoming more widely accepted, it may be convenient to remind ourselves of one of the factors that brought forth the systems approach, namely the relative decrease in the number of problems amenable to linear analysis. This, as much as anything else, strikes at a basic tenet in the practice of western science, which is that the worthwhileness of studying ever smaller details would be beyond question. This principle is being eroded successfully. Indeed, if a problem is decidedly non-linear, it becomes unpredictable how one is to integrate separate detail studies.

In our research planning, therefore, a better awareness appears to be needed of one of the fundamental systems concepts, namely that the value (worthwhileness, purpose, objective) of a subsystem (individual researches) is to be subservient to the total system (the field of fluid-elasticity). Another way of putting this is that we need to distinguish between research strategy and tactics.

The needed strategy may well be to consciously set out on a period of synthesis and consolidation of what we know presently. With a highly multivariable problem one should ask whether at this moment, an "everybody doing his own thing", is not all too prevalent. As a corresponding tactic we might adopt a quantitative design focus. Meant is not that the "schism" between theoretists and practitioners should be removed entirely (each has its own functions). Rather,

---

[1] Admittedly, in their "pure" form these mechanisms occur seldom in practical problems.

that for both the theoretical and the academic laboratory worker, a design focus is essential in keeping research from being pedantic and showing little return, now and later. At the same time, the practioner should convince his management that a longer-range view is needed by starting an in-house structural-fluid-mechanics group that gets and keeps itself educated through the years by regularly inviting a specialist.

From this point of view an urgently needed new approach is not first of all more attention to one or another detail of the field of flow-induced vibration. Rather, it is the will to find and engage in long-term, cooperative programs that engage groups of academic and industrial workers. It is the multi-disciplinary and sustained effort spent on learning all about a few recognized practical problems, that holds most promise for progress. Along this road there would appear to be enough new work for everybody.

PANEL 2

Air and Water Craft

(1) Aircraft Vibrations. G. Coupry, France.

When trying to decide on research priorities for flow-induced vibrations of aircraft, one has to keep in mind all the progress that has been already achieved in this field, and to try to understand why the success has been greater than could have been expected. It seems to me that one of the most important reasons for this success is the great amount of money that has been provided for decades to air and space agencies, and the fact that economics was of small importance for military projects.

Even with this financial support, the success would have been poorer, if most of the problems had not been, in some way, simpler than those faced in other areas, such as flow-induced vibrations of bridges, buildings, or hydraulic machinery. As a matter of fact, the shape of an airplane makes it possible to use a linear approach in predicting vibrational behaviour for most of the aerodynamic phenomena.

Starting from this point, it seems possible to split the future topics of research into two categories: problems for which the classical approach of the potential theory can be used, and phenomena for which a new mathematical tool is needed. In the first category, research priorities could be given to the following problems.

(a) Flight of aircraft in turbulence. This problem is of growing importance for the new commercial airplanes because of their size and of their increasing velocity. For both these reasons, they are concerned with much greater wave lengths of atmosphere, which correspond to much greater excitation. Progress should be made

- in getting a better knowledge of the statistics of gusts all around the world, and in trying to understand why the cumulative frequency distribution is nearly exponential and not Gaussian;
- in having a better understanding of what is really the scale of turbulence;
- in developing gust alleviation devices.

(b) Response of Aircraft Panels to jet noise, or response of spacecraft structures to boundary-layer excitation. These problems are of great industrial interest. Their solution relies on the use of statistical techniques, and needs an improved knowledge of the cross-power of the pressure field between different locations on the structure. On the other hand, the model representation of the structure can no longer be used in general, and the energy concepts are not always successful: a new approach to the problem is needed.

In the second category, I would list the following topics:

(a) Coupling between shock waves and boundary layer, which is responsible for buffet or buzz in the transsonic range. This problem, which was of importance at the time when the first transsonic flights were achieved in the 1950's, surfaces again with the advent of new commercial airplanes which fly at high subsonic speeds and have high relative wing thickness. This problem cannot be treated with linear techniques, and even its mechanism is not clearly understood. Visualization in wind tunnels could give some insight into the phenomena, with the Reynolds number correctly simulated.

(b) Vibrations of flaps, when fully extended. This will be a problem of increasing importance, with the development of V. STOL Aircraft and of Direct Lift-Control Systems. Here, again, experiments in highly pressurized wind tunnels will be necessary before deriving any mathematical model.

(c) Suppression or reduction of the wake of very big aircraft, such as the Boeing 767. This problem will be of greater importance due to the increase of the air traffic, especially in Europe.

To conclude, I want to emphasize the necessity of carrying out fundamental research on the wind-tunnel simulation of these problems. The cross-check of wind-tunnel and flight results is not always satisfactory, and one gets depressed when trying to simulate a control surface buffet on a model in a wind tunnel. Only a deep insight into the definition of Reynolds number could be of some help in this field.

(2) Water Craft Vibrations. H.N. Abramson, U.S.A.

High-performance marine craft experience many different types of dynamic excitation and response. Because such vehicles operate in a fluid medium, one might expect that much of our extensive experience in aeronautics could be carried over directly; to some extent this is true, but we need to understand fully the differences in environment. The marine vehicle operates at an interface between air and water and thus is subjected to wave actions and therefore loading conditions are complex and difficult to describe accurately. Secondly, a vehicle operating at this interface may entrain air in cavities formed behind those portions of the structure that penetrate the liquid surface ("ventilated flow") - in fact, the entire concept of "cavitated" flows is vital to high-speed marine craft. Finally, we find that the ranges of parameters of most importance to marine craft may be far outside of our normal aeronautical experience - this is particularly true of Reynolds number and the mass-density ratio. Of course, the important parameters Froude number and Weber number, usually neglected in aeronautical applications, must be considered.

Hydrofoil flutter remains the least understood problem of high-speed lifting surfaces as conventional unsteady-aerodynamics theory has not proven adequate. The flutter of flexible skirts or seals of surface-effect craft is known to occur with practical consequences for materials engineering. Water-jet propulsion systems, now being employed increasingly on high-speed marine vehicles, clearly require components that penetrate the air-water interface and are therefore subjected to very large dynamic forces from both the internal and external flows. Planning craft have also been observed to exhibit self-excited oscillations during high-speed operations. Towed bodies exhibit a variety of flow-excited dynamic responses, the combined system may prove dynamically unstable, or the towed body itself, particularly when towed at high speed by heliocopter or aircraft, may exhibit severe dynamic oscillations or instabilities. Towing cables themselves often exhibit dynamic responses of great significance. Of course, all forms of marine craft possess protuberances and appendages that often are subject to flow-induced vibration - typical of these are submarine periscopes, bow fins, cavities that resonate, etc. Displacement vehicles also exhibit a variety of dynamic responses to wave motion: elastic vibrations of the main hull (vertical, lateral, or torsional), slamming, springing, etc. Finally, we must mention the "singing" propeller as a perfect example of flow-induced vibration of a component of a marine vessel, and at the same time not forget that some of the most significant vibrations felt by the ship are the unsteady pressures on the hull provided by the propeller.

While research on all of these problems is important and to some extent is progressing, my personal opinion is that the highest research priorities are in the areas of developing new tools for accomplishing such research. Above all, our ability to predict and analyze these kinds of problems on marine vehicles is limited by our still inadequate knowledge of real fluid flow - new theoretical approaches must be sought and developed. New facilities that will provide for high test speeds with adequate simulation of cavitation conditions are urgently required. The development of proper modelling techniques, including model construction, will of course be required to utilize such facilities efficiently.

PANEL 3

Buildings, Towers, and Bridges

By
A.G. Davenport
Canada

Approximately 25 persons attended the work group discussion. From the long list of suggestions the following matters appeared to be of key importance.

1. Wind Tunnel Modelling of Atmospheric Flows

Several methods of modelling atmospheric flows for wind loading purposes are currently available including:

a) long boundary layer wind tunnels;

b) graduated grids and screens;

c) vortex generators;

d) combinations of the above;

e) conventional aeronautical wind tunnels.

Each of these methods has some limitations in its capability to model fully the atmospheric flow; this being the case, the procedure is one of matching as far as possible the salient properties of the atmosphere. It is important to establish some degree of confidence in these methods and the limitations produced by the directions and distortions of the full scale conditions, in particular the following:

a) turbulence scale and intensity;

b) mean velocity profile;

c) effect of horizontal directional wind shear due to Coriolis effects; and

d) unusual meteorological conditions such as thunder storms, low level jets, and tornadoes.

Modelling of tornadoes may require considerably different modelling procedures.

It was suggested that comparisons of the response of a "standard model" in different test conditions and wind tunnels could be valuable in the same way that the "standard sphere" was used to compare aeronautical low-speed wind tunnels. The Commonwealth Advisory Aeronautical Research Council standard model of a rectangular building model was suggested. Details of this are given in an appendix attached.

2. Problems of Reproducing Full-Scale Reynolds Numbers

Problems frequently arise in predicting aerodynamic behaviour at very high Reynolds numbers, in excess of 10^8, exemplifying full scale flow around large curved structures such as chimneys, cooling towers, and domes.

Apart from the possibility of full scale testing (the need for which is referred to below) such predictions must depend on either experimental modelling or mathematical modelling. Neither of these approaches at present is completely suitable for making these high Reynolds number predictions. Required is further work in either simulating high Reynolds number or in developing suitable theoretical understanding to enable adequate extrapolations from lower Reynolds numbers. The latter should incorporate the interactions of both surface roughness as well as turbulence in the flow.

3. Specification of Wind Loading and Other Effects

One yardstick of the success of engineering research is its successful translation into useful design procedures. This represents an area of research requiring attention. The progress in understanding of wind loading and other effects over the past few years has been considerable. However, the design methodology and quantitative data are still required in several areas. The mean and fluctuating pressures in turbulent boundary layers requires determination for a spectrum of typical shapes and flows.

4. Full Scale Monitoring of Structures

The active program of monitoring the response of full-scale buildings, towers, and bridges already begun should be introduced. Both long and short time scales should be considered; the former is related to the long-term effects of wind, estimates of the accumulation of fatigue damage, the sensitivity to different prevailing wind conditions and gradual changes in the structural response characteristics such as stiffness and damping. The short time scales studies are important to the understanding of the dynamic response characteristics. Without full scale testing, there is little real control over the effectiveness and suitability of the research carried out theoretically or at model scale.

Appendix: The C.A.A.R.C. Model

Reference: Wardlaw, R.L., Moss, G.F.: A Standard Tall Building Model for the Comparison of Simulated Natural Winds in Wind Tunnels. Proceedings of Wind Effects on Buildings and Structures, Tokyo, 1971, pp. 1245-1250.

The C.A.A.R.C. "standard model" project is described in detail in the above reference, which includes a complete description of the hypothetical building, the suggested standard measurements and the proposed methods of presentation.

Briefly, the hypothetical full-scale building should have the following properties:

exterior shape: prismatic, 100 ft by 150 ft by 600 ft high, smooth walls and flat roof,

density: 10 lbs/ft^3,
mode shape: linear first mode only of importance,
natural frequency: 0.2 Hz, in both directions,
damping: 1% of critical (secondary values of 0.33% and 3%).

The building is to be tested in a simulated 1200 ft deep boundary layer typical of a forest or urban environment (a = 0.28). The primary measurements to be compared are root-mean-square tip deflections for winds perpendicular to the faces. Secondary measurements would include other wind directions, power spectra of response, and surface pressures characteristics at the 400 ft and 200 ft level.

PANEL 4

Hydraulic Structures and Machinery

By
M.V. Morkovin
U.S.A.

This report attempts to summarize the discussion and especially the concerns of the Workshop on Hydraulic Structures and Machinery. We set ourselves two tasks: (1) to bring forth specific problems which currently cry out for solution for practical applications, (2) to stimulate formulation of more general principles and concepts of underlying mechanisms which could help us to unify the enormous heterogeneous bulk of specialized ad hoc problems and experiences.

Our group comprised the widest spectrum from designers to ultra researchers, who freely expressed their feelings about the state of the art and this Conference. Judging by the time spent on ideas, we felt overwhelmingly that the most burning problem was that of communication between the practicing engineers and the researchers. The vox populi - the cries for solutions of specific problems or categories were surprisingly feeble and, therefore, the following list cannot be considered more than a spotty starting point of a more systematic search.

Gate seals, generally (including the questions of whether quasisteady view is adequate and whether a gate needs to be watertight); gates, generally (curved tainter gates - a special area of ignorance); trash racks for pump storage intake; vanes associated with valves

(Mr. Douma's awesome photographs); the broad issue of simulation by scaled models (which probably depends on the type of sickness which the patient suffers - cavitation pains, resonant sickness, etc.); the broad issue of damping (within materials, solid friction, velocity dependence in oscillating systems and the challenge of Prof. C. Wood to seek geometries with fluid self-damping); pressure-wave and acoustical stimulation of dangerous coherences, etc. etc. Our Russian colleagues seem to be moving along the same paths, and a greater effort at cross-pollination with them was urged.

The role of theory and of the computer were discussed with little consensus. However, it was strongly suggested that the computer should be used for systematically delineating the *sensitivities* of the results to the many simplifying assumptions and to the parameters of the problem by "perturbing" them. Such sensitivities are all-important in ultimate applications. Similarly, the experimentalist could well "perturb" his conditions by adding roughness, flow disturbances, modifying edges of gates, etc., on purpose to obtain partial empirical sensitivities. We were also urged to consider normalization, even standardization of common units (e.g. gates, irrigation systems, especially data presentation) as a means for alleviating the burden of information pollution.

But time and again we returned to the problem of communication primarily between the engineer and the researcher and to the problem of making possible a more structured body of unifying understanding of the proliferating heterogeneous information. We agreed that the present organization of the engineering professions actually discourages such communication (even between colleagues at universities!) by the system of funding and professional reward mechanisms. Advances in salary and status come from ad hoc efforts and highly specialized or sophisticated publications rather than from *cross-discipline cooperation and teamwork between research and field applications*. These observations are not new, but we feel that our professional organizations (including IUTAM and IAHR), the governmental funding offices, the design and consulting firms, the universities, etc. should address themselves more directly to this *increasingly more urgent but delicate problem*. Reward carrot is preferable to the forcing stick. The planning, control, and reward system of a NASA Apollo Program (systems engineering) was of course unique in its broadly shared motivation and largesse of available funding. Adaptation of some of these procedures and especially innovation in motivation suitable to smaller and more loosely organized units appears highly desirable.

As a short-range example we urge our professional organizations and governmental and private funding officers to establish well-advertized special awards and prizes for technical expository writing - basic review articles and books. In our field, even a basic catalog of vibration-inducing mechanisms, describing technically the common characteristics, causes, and remedies of observed structural vibrations, would be highly beneficial for designers and students. As Mr. Douma stressed, the researchers would also be reminded of the need for simplicity and field practicability of the designs and cures. The professional and governmental organizations should also be opened to proposals for more focused workshops, which would differ from the usual gathering of narrow specialists, by the emphasis on ample dialog between the application - and research-minded engineers. The researcher is likely to derive special satisfaction through the wider applicability of his efforts and the designer through the deeper, more unified understanding of his own problems.

We feel that the engineer needs more than a handbook of correlated data, which rapidly becomes obsolete as the range of parameters broadens (energy, scale) and which may, therefore, even be dangerous. We believe that his engineering intuition would be well served through organizational techniques such as just described. We, therefore, look forward to:

cooperative application of fundamental knowledge across the various fields of Mechanics;

consolidation of current results and techniques of experimental and theoretical studies;

increased efforts at computer and experimental modelling with increased feedback from field experiences;

new, relevant research and design applications stimulated by the increased communication between the various groups.

PANEL 5

Marine Structures

By

E. Plate

Germany

The recent development of off-shore structures for varied purposes of ocean exploration and exploitation has increased the need for accurate methods of predicting steady and dynamic forces on such structures. The discussion in the working group has shown clearly that this need is at present only partially met. This is in part due to the complications which arise because in addition to vortex-induced motions of the kind considered in other sessions of the symposium there are the effects of the free surface and its unsteady deformation by water waves. Of the many problems in this area, the discussion group singled out only those associated with single cylinders either perpendicular or parallel to the surface.

The force on a vertical cylinder of diameter D acting over a height dy is optimistically described through what has become known as the Morison equation:

$$dF = C_M \rho_w \left(\frac{\pi D^2}{4}\right) \dot{u}_w \, dy + C_D \rho_w D u_w |u_w| dy , \qquad (1)$$

where the subscript w denotes the conditions of the water, u_w is the wave-induced velocity at the depth y from the surface in the absence of a cylinder and $\dot{u}_w$ is the corresponding acceleration. C_D is a drag coefficient, while C_M is a mass coefficient dependent not only on the incident wave but also on the diffraction pattern generated by the cylinder. Although C_M is often identified as a "virtual mass" coefficient, it is clear that it can be a virtual mass in the classical sense only, if the cylinder performs an oscillatory motion through a fluid which is otherwise at rest. Both coefficients are affected by the geometry of the pier and, as was pointed out, also on the wave dynamics. Professor Hino mentioned that in recent researches he has found that C_D is a function of a number $u_{w\,max}/(D \cdot f)$ where $u_{w\,max}$ is the maximum wave-induced velocity in the horizontal direction and f is the frequency (in Hz) of the wave. The number is called by Hino a Strouhal number, while others preferred to call it a Keulegan number. Professor Wiegel remarked that his group prefers to use Eq. (1) as given, but with u_w and $\dot{u}_w$ replaced by the relative velocities and accelerations between cylinder and waves, respectively.

In working with Eq. (1), it is found that the non-linearity of the drag term is probably not too important. However, it appears that the probability distribution of the stress at a point has a variance which exceeds that obtained by using Eq. (1) by as much as 40 to 100 %; this has important consequences in calculating the fatigue of off-shore structures. Material failure due to fatigue seems to be responsible for a considerable number of failures of structural elements, and it is significant to note that joints are particularly affected. It was recommended that stress measurements should be directly correlated with wave measurements to better assess the correspondence of the two.

Significant are lateral "lift" forces observed on vertical cylinders, which contribute to the motion of such elements. Professor Wiegel reports that whenever eddies at the surface similar to Kármán vortices are recognizable, no significant lateral displacements of the cylinders can be observed, while there are no eddies observed when strong lift forces are felt. An explanation for this is as yet not available. It was recognized that eddy pattern under waves may differ from those in a steady stream: long waves will cause symmetrical eddies, similar to starting vortices of cylinders set suddenly in motion, while buffeting of the structure might occur due to the return of the eddies generated by the structure itself in the return cycle of the wave motion.

Not well understood are also the forces on cylinders with axes parallel to the wave crest. Such cylinders suffer very large impact stresses when located in the region between wave crest and troughs, or significant downward and upward lifts when located below the water surface. Simultaneous upward lift on one cross member and downward lift on another might induce significant overturning moments into an off-shore structure. More important, however, appear to be the forces on pipelines on the ocean floor which, due to the shifting of the ocean-floor sediments, might at certain times be buried in the sediment while at others they are bridging gaps between ocean-floor ridges. The stresses on such pipes are of considerable magnitude, and research is needed to assess the forces on them.

The discussion closed with the clear indication that what is missing is field information on measured forces and stresses, which should be utilized in conjunction with improved theories and laboratory experiments for developing government-enforced standards for the design of such structures.

TECHNICAL SESSION F

"Flow-Induced Vibrations of Marine Structures"

Chairman: M. Hino, Japan
Co-Chairman: C. Zimmermann, Germany

GENERAL LECTURE

Ocean Wave Spectra, Eddies, and Structural Response

By
R. L. Wiegel
Berkeley, Cal., U.S.A.

Introduction

At 7:25 p.m. on 15 January 1961 during a North Atlantic storm, Texas Tower No. 4 collapsed and toppled into the sea, taking with it the lives of 28 persons. This 3-legged offshore platform, located in 185 ft of water about 80 miles east of Barnegat Inlet, New Jersey, was known as Old Shaky to those working on it. Tests showed that the platform oscillated with complex motions with frequencies (0.28 to 0.38 cps, periods of 2.5 to 3.5 sec) ranging about its fundamental natural frequency of about 0.31 cps (period of 3.2 sec). It suffered progressive damage, and had repairs made to it, from the time it was being towed to the site at the end of June and early July 1957 until it collapsed. It was subjected to large waves during a number of severe storms, including Hurricane Donna in September 1960. The collapse of this structure and the events leading up to it is one of the few cases of such an occurrence that has been documented. This information is available in great detail owing to an investigation by a committee of the U.S. Senate [69, 70].

Another example of vibration that has been described in the literature was in connection with the measurement of wave-induced forces on a vertical circular cylinder at the end of a pier (water depth of nearly 50 ft) at an exposed location in the Pacific Ocean 70 miles south of San Francisco, California [76]. During these tests, the entire end of the pier was felt to vibrate considerably when subjected to waves during storms, which in this case were 15 to 20 ft in height [74], p. 269). A 2-ft-diameter test section was mounted on a 0.5-ft-diameter supporting drill pipe. The test section was guyed by two wires attached at its bottom at an angle of 30 degrees with the direction of prevailing waves (3 wires were used in subsequent tests). Waves with a 12 ft "significant" height (average of the highest one-third of the waves) and an average period of 13 sec, caused a considerable lateral (normal to the mean direction of wave advance) vibration of the cylinder with a period of about 2.5 sec, for the length and mass of cylinder being tested. After nearly a week of continual storm conditions, the support pipe broke. An examination of the break indicated a fatigue failure. Large vortices were observed to form and move from the cylinder during this time.

The two examples described above are important to engineers who design structures which will be used in the ocean, as they illustrate what can happen to a structure or a structural element that is not stiff enough to have its fundamental frequency sufficiently higher than the energetic range of the fluid forcing (waves and wave-induced eddies) spectrum in the ocean. Furthermore, they are examples of fluid flow-structure interaction.

An analysis of these types of problems requires an understanding of the characteristics of eddies induced by an oscillating flow. A flow of this type may have a very large Reynolds number, but the period and distance traveled by the fluid particles may not be sufficiently great to permit the formation of eddies. In addition, when eddies are formed, and the flow reverses direction, the wake becomes the upstream flow, with the size of the eddies being of the same size as the pile diameters.

The term "flow-induced structural vibrations" is usually used in connection with interactions of a structural member and the flow, with the flow being non-reversing in the absence of the structure. In the ocean, however, one of the most noticeable and troublesome characteristics is its waves, with their oscillatory motions. In some ways they must be treated in a manner similar to what is done for turbulence, except that the scale of this "turbulence" is of the same magnitude as the scale of the structures. For example, waves in deep water are now being treated as the linear superposition of wave trains of many frequencies with random phase moving in a range of directions. Use has been made of the technique of obtaining the co-variance function from a wave time series, and from this calculating the one-dimensional energy spectral density distribution of the waves; now, use is made of the Fast Fourier Transform method to calculate the spectral distribution. Also, by making use of a wave gage array and a rather elaborate analysis, it is possible to obtain an estimate of the directional energy spectral density distribution of waves.

Owing to the oscillatory nature of ocean waves, much work has been done on developing theories to describe the wave-related motion of ships in a seaway, with their six degrees of freedom, and three natural periods-heave, pitch and roll. We are all aware of the difficulties involved in this problem, and of the great advances that have been made even for the case of large resonant pitching and heaving where the response of the ship becomes out of phase with the forcing function (waves) with the resulting "slamming" which has led to the destruction of many a ship.

Even more difficult to analyze is the problem of moored structures such as buoys, offshore floating and semi-submersible drilling structures, and moored ships and boats. These have six natural periods; the three associated with a freely floating body and three associated with the restoring force of the mooring lines (usually non-linear)-- surge, sway and yaw. Little work has been done on this problem compared with that done on the motion of ships under way.

Another type of problems that is encountered is that of a pipeline lying on the ocean bottom in shallow water. Owing to the shear flow characteristics of currents flowing past the pipe, a lift force is exerted on the pipe. If it is not adequately weighted or anchored it will be lifted a little, and then the force will be down [4]. This can easily result in a vibratory motion of the pipeline. This tendency is complicated by the presence of waves, with their oscillating flow, both horizontally and vertically, which amongst other things results directly in an oscillating vertical lift force.

In this paper the spectral characteristics of waves will be described. As the spectra are obtained using a theory that assumed the linear superposition of waves of infinitesimal amplitudes, an estimate will be made of the limits of the validity of the approximation. This will be followed by an exposition of the theory of wave forces exerted on a pile, and then how a linearization technique can be used to obtain practical results. The concept of relative velocities and relative accelerations will be used to show how one might approximate the analysis of the stresses induced in an offshore platform. Then, some information will be presented on how one might use a simulation technique to calculate forces exerted by waves without linearizing the wave force transfer equation.

Next, additional problems will be described which are associated with the formation of eddies in an oscillating flow. The effect of this phenomenon on the longitudinal drag and inertial forces will be developed together with information on lateral oscillating forces that are associated with the eddy formation. Experimental studies show these lateral forces to be as large as the longitudinal forces for certain conditions.

The case history of the design, construction, maintenance and collapse of Texas Tower No. 4 will be given as an example of some of the difficulties encountered in offshore operations.

Finally, some information will be presented on the motions of a highly non-linear vibration, that of a moored boat at a dock.

Ocean Wave Spectra

In analyzing the motions of marine structure, it is often useful to work with the energy spectral density of the surface waves of the ocean. For some problems a knowledge of the one dimensional spectra is sufficient, while for other problems (the motion of a moored floating drilling rig) a knowledge of directional spectra is necessary. These are the basic inputs to most of the vibration problems in the ocean.

One-Dimensional Wave Spectra

There have been a number of papers published on one-dimensional wave spectra (see, for example, [47]), and a large number of measured wave spectra have been published (see, for example, [44]). There are several possible ways of using actual spectra, one being a simulation technique [11] for a large number of spectra, or a large number of wave time histories reconstituted from spectra. Another way to use spectra is to develop a "standard" set of spectra. There have been a number of such standards suggested. One of these has been given by Scott [62], who re-examined the data of Darbyshire [19] and Moskowitz, Pierson and Mehr [44], and then recommended the following equation as being a better fit of the ocean data

$$S(\omega)/H_s^2 = 0.214 \exp - \left[\frac{(\omega - \omega_0)^2}{0.065\{(\omega - \omega_0) + 0.26\}} \right]^{\frac{1}{2}} \tag{1}$$

for $-0.26 < (\omega - \omega_0) < 1.65$, and, $= 0$, elsewhere. Here, $\omega = 2\pi f$ (in radians per second), ω_0 is the spectrum peak frequency, H_s is the "significant" wave height (in feet), and the energy density $S(\omega)$ is defined by

$$S(\omega) = \frac{1}{\pi} S_{\eta\eta}(f) \quad \text{or} \quad S(\omega) = \frac{1}{2} \sum^{\delta\omega} a_i^2/\delta\omega , \tag{2}$$

in which the summation is over the frequency interval $\omega, \omega + \delta\omega$, and a_i is the amplitude of the ith component, with

$$\eta(t) = \sum_{i=1}^{n} \eta_i(t) = \sum_{i=1}^{n} a_i \cos(\omega_i t + \Phi_i) . \tag{3}$$

Here Φ_i is the phase angle of the $i^{\underline{th}}$ component, $\eta(t)$ is the time history of the surface wave motion at a point, and t is time. The factor $\frac{1}{2}$ enters as $\sum^{n} a_i^2/2$ is the mean value of $\eta^2(t)$ during the motion. The term $a_i^2/\delta\omega$ is used, as the concept of a_i tends to lose physical significance (i.e., $a_i^2 \rightarrow 0$ as $n \rightarrow \infty$) whereas $a_i^2/\delta\omega$ does not; hence the value of using the energy density as a function of frequency. Scott found, using linear regression, that: $1/f_0 = 0.19\, H_s + 8.5$; $1/\omega_0 = 0.03\, H_s + 1.35$; and $T = 0.085\, H_s + 7.1$ where T is the average period (in seconds) of all waves in the record. There are many locations in the ocean where there are data on the significant wave height, although there may not be much data on spectra. Using the above relationships one may construct a spectra based upon the values of H_s for the location.

It can be shown that

$$T = 2\pi(m_0/m_2)^{\frac{1}{2}}, \tag{4}$$

where

$$m_k = \int_0^\infty \omega^k S_{\eta\eta}(\omega)d\omega . \tag{5}$$

For $k = 0$, we have the "variance", m_0, and for a narrow (i.e., "Rayleigh" spectrum) we have

$$H_s = 4\, m_0^{\frac{1}{2}} . \tag{6}$$

Using quadratic regression, Scott found: $f_0 = (0.501/T) + (1.43/T^2)$, and $\omega_0 = (3.15/T) + (8.98/T^2)$.

$S_{\eta\eta}(f)$ in Eq.(2) is defined by the Fourier transform

$$S_{\eta\eta}(f) = \int_{-\infty}^{\infty} R_{\eta\eta}(\tau)e^{-i2\pi f\tau}d\tau , \tag{7}$$

in which $R_{\eta\eta}(\tau)$ is the averaged lagged product of $\eta(t)$ (i.e., average of $\eta(t)\ \eta(t+\tau)$) and τ is the lag (seconds).

It is of considerable importance to the engineering profession to develop means by which the spectral approach can be studied in the laboratory. In studying some of the problems, it is necessary to know the relationship between the one-dimensional spectra in the ocean and the spectra generated in a wind-wave tank [53, 54]. Comparison of a number of wave spectra measured in the ocean, in lakes and in wave tanks have been made by Hess, Hidy and Plate [26]. Their results, shown in Fig. 1, are fully developed seas' wind-wave energy density spectra. The high frequency portion of the spectra all tend to lie close to a single curve, with the energy density being approximately proportional to ω^{-5} as predicted by the Phillips' equilibrium theory (see [77], for a physical explanation of this). A close inspection of these data by Plate and Nath [53] led them to conclude that the high frequency portion of the energy spectral density curve varies from the ω^{-5} "law", being proportional to ω^{-7} near the spectral peak, and being proportional to about ω^{-4} in the highest frequency range of the spectra. Figure 1 must be viewed with caution, however; other data indicate that the ω^{-5} portion of the graph is probably better represented by a reasonably wide band rather than a line. It would appear from the one example of Wiegel and Cross [77], in which they compared a normalized measured laboratory wind-wave energy density spectrum with one calculated by use of Miles' theory, together with other physical reasoning, that a theoretically sound basis exists for the development of a "standard" set of spectra.

It is evident from Fig. 1 that the width of the spectral density of waves generated by winds in the laboratory (labeled CSU wind-water tunnel) is much narrower than the width of the spectra density of waves generated by winds blowing over great areas of the ocean.

Directional Wave Spectra

Before directional spectra can be used in the design of structures in relatively deep water it is necessary to have measurements of such spectra, and to understand them sufficiently to be able to choose a "design" directional spectra. A few measurements have been made in the ocean [16, 35, 51], a few in a bay [65] and a few in the laboratory [41, 42, 22, 75]. An example of directional spectra measured in the ocean is shown in Fig. 2.

Mobarek [41] made use of simulated inputs to choose a reasonably reliable method to measure directional spectra, and to devise correction factors. He used this method to analyze data obtained in a hydraulic laboratory study. He found his laboratory results, when normalized, to be similar to normalized values of the measurements made in the ocean by Longuet-Higgins, et al. [35], as can be seen in Fig. 3. At the suggestion of Professor Leon E. Borgman, Dr. Mobarek compared the circular normal probability function (the solid curve in Fig. 3) with the normalized data and found the comparison to be good. Values in the ordinate are in terms of the normalized wave energy, E, rather than the energy density, $S_{\eta\eta}(f)$.

The probability density of the circular normal distribution function is given by [25, 17]:

$$P(\alpha,K) = \frac{1}{I_0(K)} \exp(K \cos \alpha) , \qquad (8)$$

where α is the angle measured from the mean $(\theta_m - \theta)$, K is measure of the concentration about the mean, and $I_0(K)$ involves an incomplete Bessel function of the first kind of zero order for an imaginary argument. The larger K, the greater the concentration of energy; it is analogous to the reciprocal of the standard deviation of the linear normal distribution.

It has been found that much useful information on directional spectra can be obtained from the outputs of two wave recorders, through use of the cospectra and quadrature spectra to calculate the linear coherence and the mean wave direction [45, 64]. It appeared to the author that if the directional spectra were represented by the circular normal distribution function it should be possible to obtain the necessary statistical parameters in a similar manner. It was believed that such a simplified approach could provide data of sufficient accuracy for many practical purposes. As a result of discusssions with Professor Leon Borgman, a theory was developed by Borgman [10] to do this, and tables were calculated to provide a practical means to obtain the required information.

Borgman [10] used s lightly different representation of the directional spectra

$$S_{\eta\eta 2}(f,\alpha) = S_{\eta\eta 1}(f) \exp[-K\cos(\theta - \theta_m)]/2\pi I_0(K), \qquad (9)$$

where the 2π in the denominator indicates an area under the curve of 2π rather than unity, f is the component wave frequency in cps, and $S_{\eta\eta 1}(f)$ is the one-dimensional spectral density. The estimation of the parameters $S_{\eta\eta 1}(f)$, $K(f)$ and $\theta_m(f)$ is achieved by cross-spectral analysis based on a sea surface record at two locations. $S_{\eta\eta 1}(f)$ and the co- and quadrature spectral densities for the two recordings are computed by the usual time series procedures. The theoretical relations between measured and unknown quantities is

$$\frac{C(f)}{S_{\eta\eta 1}(f)} = \int_0^{2\pi} \frac{\exp[K\cos(\theta - \theta_m)]}{2\pi I_0(K)} \cos[k\bar{D}\cos(\theta - \bar{\beta})]d\theta, \qquad (10)$$

$$\frac{Q(f)}{S_{\eta\eta 1}(f)} + \int_0^{2\pi} \frac{\exp[K\cos(\theta - \theta_m)]}{2\pi I_0(K)} \sin[k\bar{D}\cos(\theta - \bar{\beta})]d\theta, \qquad (11)$$

where $\bar{D}$ is the distance between the pair of recorders, k is the wave number $(2\pi/L)$ and $\bar{\beta}$ is the direction from wave recorder No. 1 to wave recorder No. 2. For a given frequency, all quantities are known except θ_m and K. Hence these two equations represent two nonlinear equations with two unknowns. Borgman has prepared tables which enable one to solve for θ_m and K, given $C(f)/S_{\eta\eta 1}(f)$ and $Q(f)/S_{\eta\eta 1}(f)$. Two solutions, symmetric about the direction between the pair of recorders result. This ambiguity may be eliminated by using three wave gages instead of two, or in many applications using other information regarding the main direction of the directional spectra. The relationship between the parameter K and the directional width of the spectrum can be seen in Fig. 4.

Using simulation techniques devised by Professor Leon Borgman, Dr. Fan [22], continuing the work of Mobarek, made an extensive study of the effects of different lengths of data, lag numbers, wave recorder spacings, filters, and different samples on the calculation of directional spectra, using several methods, using a known circular normal distribution input. He then used the "best" combination to obtain the directional spectra of waves generated in a model basin by wind blowing over the water surface. As a result of this study it appears that, for the case of waves being generated in a nearly stationary single storm, the directional spectra can be approximated by two parameters and should be tested for use in the design of an offshore structure. This work has been continued, for a number of arrays by Panicker [50].

Wave Induced Forces of a Pile

The most common type of structures in the ocean that are subject to vibratory motions are pile supported. Owing to this, a presentation will be made herein of the wave induced forces on a vertical circular cylindrical pile. Other orientations and cylinders of other cross sections are variations of this fundamental problem. Owing to space limitations only the approach using linear wave theory will be considered.

Linear Wave Theory

The coordinate system usually used is to take x in the plane of the undisturbed water surface and y as the vertical coordinate, measured positive up from the undisturbed water surface. The undisturbed water depth is designated as d. Sometimes the vertical coordinate is taken as measured positive up from the ocean floor, being designated by S.

The wave surface is given by

$$S_\eta = \eta + d = \frac{1}{2} H \cos 2\pi \left[\frac{x}{L} - \frac{t}{T} \right] + d , \qquad (12)$$

where H is the wave height, L is the wave length, T is the wave period, t is time, and η is the wave surface value of y. The wave length, L, and wave speed, C, are given by

$$L = \frac{gT^2}{2\pi} \tanh \frac{2\pi d}{L}, \qquad C = \frac{gT}{2\pi} \tanh \frac{2\pi d}{L}, \qquad (13)(14)$$

where g is the acceleration of gravity. The horizontal component of water particle velocity, u, the local acceleration, $\partial u/\partial t$, and the pressure, p, are given by

$$u = \frac{\pi H}{T} \frac{\cosh 2\pi S/L}{\sinh 2\pi d/L} \cos 2\pi \left[\frac{x}{L} - \frac{t}{T} \right], \qquad (15)$$

$$\frac{\partial u}{\partial t} = \frac{2\pi^2 H}{T^2} \frac{\cosh 2\pi S/L}{\sinh 2\pi d/L} \sin 2\pi \left[\frac{x}{L} - \frac{t}{T} \right], \qquad (16)$$

$$p + \rho g y = \frac{1}{2} \rho g H \frac{\cosh 2\pi S/L}{\cosh 2\pi d/L} \cos 2\pi \left[\frac{x}{L} - \frac{t}{T} \right], \qquad (17)$$

where ρ is the mass density of the water.

Similar expressions are available for the vertical components, and expressions are available of the water particle displacements [74].

Wave Forces on Circular Cylindrical Piles

In a frictionless, incompressible fluid the force exerted on a fixed rigid submerged body may be expressed as ([31], p. 93), in a fluid of infinite extent,

$$F_I = (M_0 + M_a) f_f = \rho B C_M f_f, \qquad (18)$$

where F_I is the inertia force, M_0 is the mass of the displaced fluid M_a is the so-called added mass which is dependent upon the shape of the body and the flow characteristics around the body, and f_f is the acceleration of the fluid at the center of the body, were no body present. C_M has been found theoretically to be equal to 2.0 for a rigid circular cylinder by several investigators (see, for example, [31]). The product of the coefficient of mass, C_M, the volume of a body, B, and the mass density of the fluid, ρ, is often called the "virtual mass" of a body (i.e., $M_0 + M_a$) in an unsteady flow ([20], p. 97). C_M is sometimes expressed as $C_M = 1 + C_a$, where C_a is the coefficient of added mass[1]. The mass of the fluid displaced by the body enters into Eq. (18) with one part of the inertial force being due to the pressure gradient in the fluid which causes the fluid acceleration (or deceleration). This force per unit length of cylinder, F_p, is given by

$$F_p = \oint p dy = \rho \frac{dU}{dt} \oint x dy = \rho A_0 \frac{dU}{dt}, \tag{19}$$

in which A_0 is the cross-sectional area of the cylinder and $\oint$ is a contour integral [39] which follows from the well-known relationship in fluid mechanics for irrotational flow $-(1/\rho)(dp/dx) = dU/dt$, where dp/dx is the pressure gradient in the fluid in the absence of the body. In many papers on aerodynamic studies using wind tunnels F_p is called the "horizontal buoyancy" (see, for example, [5]). The added-mass term, expressed by $C_a \rho A_0$ per unit length of cylinder, results from the acceleration of the flow around the body caused by the presence of the body. As the fluid is being accelerated around the body by the upstream face of the body (which requires a force exerted by the body on the fluid), the fluid decelerating around the downstream face of the body will exert a smaller or larger force on the downstream face, depending upon whether the flow is accelerating or decelerating. This concept can be seen more clearly for the case of a body being accelerated or decelerated, through a fluid. The force necessary to do this is proportional to the mass per unit length of the cylinder, M_c, plus the added mass, M_a,

$$F_I = (M_c + C_a \rho A_0) \frac{dU}{dt} = (M_c + M_a) \frac{dU}{dt} = C_M \rho A_0 \frac{dU}{dt}. \tag{20}$$

The leading face of the cylinder pushes on the fluid causing it to accelerate, and the fluid decelerating on the rear side of the cylinder

---

[1] In many papers the term virtual mass is used for the term added mass. Owing to this, care must be exercised in reading the literature on the subject.

pushes on the cylinder (with the equivalent reaction of the cylinder). In accelerated motion, the reaction at the front must be greater than the reaction at the rear as the fluid decelerating at the rear was not accelerated as much, when it was at the front, as the fluid in front is being accelerated at that instant.

It is unfortunate that the terms added mass and virtual mass have entered the literature as they tend to confuse our concept of the phenomenon. MacCamy and Fuchs ([36]; see also [74], p. 273) solved the diffraction problem of waves moving around a vertical right circular cylinder extending from the ocean bottom through the water surface, using linear wave theory. They solved for the potential, obtained the distorted pressure field from this potential, and integrated the x-component of force around the pile which resulted from this pressure field. In our coordinate system, their solution is

$$F_{Ih}(S) = \frac{\rho g H L}{\pi} \frac{\cosh 2\pi S/L}{\cosh 2\pi d/L} f_A(D/L) \sin\left(-\frac{2\pi t}{T} - \beta\right), \tag{21}$$

where

$$f_A(D/L) = \frac{1}{\left\{[J_1'(\pi D/L)]^2 + [Y_1'(\pi D/L)]^2\right\}^{\frac{1}{2}}}, \tag{22}$$

in which J_1 and Y_1 are Bessel functions of the first and second kinds, respectively, and the prime indicates differentiation. β is the angle of phase lag, and will not be shown here as $\beta < 5^{\circ}$ for values of $D/L < 1/10$, although it is very large for large values of D/L. When $D/L \to 0$, $f_A(D/L) \to \frac{1}{2}\pi(\pi D/L)^2$. Neglecting β for small values of D/L, it can be seen that Eq. (22) reduces to the commonly accepted equation for the inertial force, with $C_M = 2$.

In a real fluid, owing to viscosity, there is an additional force, known as the drag force, F_D. This force consists of two parts, one due to the shear stress of the fluid on the body, and the other due to pressure differential around the body caused by flow separation. The most common equation used in the design of pile supported structures is due to Morison, O'Brien, Johnson and Schaaf [43]. Consider the case of a pile installed vertically in water of depth d, extending from the bottom through the surface. The water particles move in an orbit due to the waves, with both horizontal and vertical components of velocity and acceleration, u, v, du/dt and dv/dt, respectively. The horizontal component of wave induced force, per unit length of pile, $F_h(S)$, is given by

$$F_h(S) = \frac{1}{2} C_D \rho_w D |u| u + C_M \rho_w \frac{\pi D^2}{4} \frac{du}{dt}, \qquad (23)$$

where du/dt is the total acceleration. If we consider only linear theory, the convective acceleration can be neglected, leaving only the local acceleration; i.e., $du/dt \approx \partial u/\partial t$. u and $\partial u/\partial t$ are given by Eqs. (16) and (17). It can be seen that the drag and inertia forces are in quadrature, so that the maximum total force "leads the crest" of the wave. The larger the drag force relative to the inertia force, the closer will be the maximum total force to the passage of the wave crest past the pile. As will be pointed out in a later section, there is a relationship between C_D and C_M, so that Eq. (23) is quite complicated, although it is not usually treated as such.

If a circular structure is placed at an angle to the waves, the vertical component of wave induced force can be treated in a similar manner, using v and $\partial v/\partial t$ as well as u and $\partial u/\partial t$.

If strictly linear theory is used the total horizontal component of wave force acting on a vertical circular pile can be obtained by integrating $F_h(S)\, dS$ from 0 to d. Very often in practice, one integrates $F_h(S)$ from 0 to S_η, obtaining results which are somewhere between the results for linear wave theory and those for second order wave theory. A digital computer program for this operation is available for this purpose, as are graphs and tables of results [18, 77].

Much time and money have been spent in obtaining prototype and laboratory values of C_D and C_M. Most of the work has been done by private companies and have just become available. The results of these long term prototype studies of wave forces on piles, by a consortium of oil companies, have been described [66, 1, 21, 73]. The results are given in a number of different ways, including distribution functions, and the reader should refer to the original publications. There was a considerable scatter of results, with C_D found to vary from 0.5 to 1.5, and C_M found to vary from 0.5 to 2.5.

Some of the first data which was published was done by Wiegel, Beebe and Moon [76]. There is a considerable scatter of C_D; this is also true for the values of C_M. One of the main reasons for this is that the analysis of the data was based upon two simplifications: First, that linear theory could be used to reduce the basic data, and second, that each wave (and force) of a series of irregular waves could be analyzed as one of a series of uniform waves having the height and period of the individual wave in the record. Agerschou and Edens [3] reanalyzed the published data of Wiegel, Beebe and Moon [76] and some unpublished data of Bretschneider, using both linear theory and Stokes Fifth Order theory. They concluded that for the range of variables covered, the fifth-order approach was not superior to the use of linear theory. They recommended for design purposes, if linear theory is used, that C_D should be between 1.0 and 1.4, and that C_M should be 2.0, these values being obtained for circular piles 6-5/8, 8-5/8, 12-3/4, 16 and 24 in. in diameter. (It should be noted here that the theoretical value of C_M for a circular cylinder in potential flow is 2.0). Wilson ([80]; see also, [84]) report average values of $C_D = 1.0$ and $C_M = 1.45$ for a 30-inch diameter pile.

At a recent conference, one design engineer stated he used values of C_D ranging from 0.5 to 1.5 and C_M from 1.3 to 2.0, depending upon his client [68]. The results reported above were obtained either as values of C_D and C_M at that portion of a wave cycle for which $F_D =$ = max and $F_I = 0$, and vice-versa, or for the best average values of C_D and C_M throughout a wave cycle, assuming C_D and C_M to be constant. Both of these methods of obtaining and reporting the coefficients should be refined, as the coefficients are dependent upon each other, and are also time dependent as well as dependent upon the flow conditions.

In the significant wave approach, the significant wave height, H_s, and significant wave period, T_s, are substituted for H and T in the above equations, treating the significant wave as one of a train of waves of uniform height and period. In the design wave approach, the chosen values of H_d and T_d are used in a similar manner.

One-Dimensional Wave Spectra Approach

Recently there have been several papers published on the study of wave forces exerted on circular piles, using probability theory. In these studies it was assumed that the continuous spectrum of component waves could be superimposed linearly, that the process was both stationary and ergodic, and that the phase relationship among the component waves was Gaussian.

Some years ago the author obtained both the wave and force spectral densities for a pile installed at the end of the pier at Davenport, California. It was not evident why the form of the two spectral densities should have been so similar considering the fact that the product $|u|\,u$ occurs in Eq. (23). Professor Leon E. Borgman [9] studied this problem in detail and developed the following theory.

The basic wave force equation is Eq. (23), which may be expressed as a function of time as

$$F(t) = C_1 |V(t)| V(t) + C_2 A(t) . \tag{24}$$

Here $F(t)$ is the time history of the horizontal component of force per unit length of circular pile at an elevation S above the ocean floor, and

$$C_1 = \frac{1}{2} \rho_w C_D D , \quad \text{and} \quad C_2 = \rho_w C_M \pi D^2/4 . \qquad (25a),(25b)$$

The theoretical covariance function for $F(t)$ using ensemble averaging with the Gaussian random wave model is

$$R_{FF}(\tau) = C_1^2 \sigma^4 G\left[R_{VV}(\tau)/\sigma^2 \right] + C_2^2 R_{AA}(\tau) , \tag{26}$$

where $R_{VV}(\tau)$ and $R_{AA}(\tau)$ are the covariance functions of the horizontal component water particle velocity, $V(t)$, and local acceleration A(t) (i.e., u and $\partial u/\partial t$), σ^2 is given by

$$\sigma^2 = 2\int_0^{\infty} S_{VV}(f)\,df\,, \tag{27a}$$

where f is the frequency of the component wave (f = 1/T), and

$$G(r) = \left[(2+4r^2)^2 \arcsin r + 6r\sqrt{1-r^2}\right]/\pi\,, \tag{27b}$$

in which $G(r) = G(R_{VV}(\tau)/\sigma^2)$. The covariance functions $R_{VV}(\tau)$ and $R_{AA}(\tau)$ are calculated from the spectral densities $S_{VV}(f)$ and $S_{AA}(f)$ by use of the Fourier transforms

$$R_{VV}(\tau) = \int_{-\infty}^{\infty} S_{VV}(f)e^{i2\pi f\tau}\,df \quad \text{and} \quad R_{AA}(\tau) = \int_{-\infty}^{\infty} S_{AA}(f)e^{i2\pi f\tau}\,df\,,$$

(28a),(28b)

where

$$S_{VV}(f) = \frac{(2\pi f)^2 \cosh^2 2\pi S/L}{\sinh^2 2\pi d/L} S_{\eta\eta}(f) = T_V(f)S_{\eta\eta}(f)\,, \tag{29a}$$

$$S_{AA}(f) = \frac{(2\pi f)^4 \cosh^2 2\pi S/L}{\sinh^2 2\pi d/L} S_{\eta\eta}(f) = T_A(f)S_{\eta\eta}(f)\,. \tag{29b}$$

The functions $T_V(f)$ and $T_A(f)$ are called transfer functions. The fundamental quantity $S_{\eta\eta}(f)$ is the spectral density of the water waves, and is obtained from Eq. (7).

Borgman found that Eq. (28) could be expressed in series form as

$$G(r) = \frac{1}{\pi}\left[8r + \frac{4r^3}{3} + \frac{r^5}{15} + \frac{r^7}{70} + \frac{5r^9}{1008} + \ldots\right], \tag{30}$$

and that the series converges quite rapidly for $0 \leq r \leq 1$. He found that for r = 1, the first term $G_1(r) = 8r/\pi$ differed from G(r) by only 15 % and that the cubic approximation $G_3(r) = (8r + 4r^3/3)/\pi$

differed from $G(r)$ by only 1.1 %. Substituting the first term of the series into Eq. (26) results in

$$R_{FF}(\tau) = \frac{C_1^2 \sigma^4}{\pi} \left[\frac{8 R_{VV}(\tau)}{\sigma^2} + \dots \right] + C_2^2 R_{AA}(\tau) \qquad (31)$$

and its Fourier transform

$$S_{FF}(f) = \frac{C_1^2 \sigma^4}{\pi} \left[\frac{8 S_{VV}(f)}{\sigma^2} + \dots \right] + C_2^2 S_{AA}(f) , \qquad (32)$$

which is the desired force spectral density.

Borgman made a numerical analysis of one situation. The numerical integration of $S_{VV}(f)$ gave $\sigma^2 = 1.203\ ft^2/sec^2$ and a least square fitting of the theoretical covariance of $F(t)$ against the measured force covariance gave estimates of $C_D = 1.88$ and $C_M = 1.73$. The transfer functions $T_V(f)$ and $T_A(f)$ were calculated and plotted; it could be seen that $T_A(f)$ was nearly constant in the range of circular frequencies $(2\pi/T)$ for which most of the wave energy was associated. The calculated and measured force spectral densities are shown in Fig. 5. The reason for the excellent fit is that for the conditions of the experiment $T_V(f)$ was nearly constant and the linear approximation to $G(r)$, $G_1(r)$, was a reliable approximation.

Jen [28] made a model study of the forces exerted by waves on a 6-in diameter pile in the 200 ft long by 8 ft wide by 6 ft deep wave tank at the University of California, Berkeley. In addition to using periodic waves, irregular waves were generated by a special wave generator using as an input the magnetic tape recording of waves measured in the ocean. The dimensions of the waves relative to the diameter of the pile were such that the forces were largely inertial. Jen found for the regular waves that $C_M \approx 2.0$, and using Borgman's method to analyze the results of the irregular waves tests found $C_M \approx 2.1$ to 2.2. The reason for this close agreement between theory and measurement of C_M is probably due to the small value of H/D, which resulted in quasi-potential flow. (This will be discussed in a subsequent section.)

Equation (32) permits the calculation of the force spectral density at a point. This is useful but the design engineer usually needs the total force on a pile, and the total moment about the bottom. In addition, the total force and the total moment on an entire structure is needed. These problems have been considered by Borgman [9, 10, 11, 12], Foster [23], and Malhotra and Penzien [37]. In obtaining a solution to this problem, the integration of the force distribution is performed from the ocean bottom to the still water level as this is in keeping with linear wave theory. There is no difficulty in obtaining the solution for the inertial force, but cross product terms appear in the solution for the drag force. Borgman made use of the linearization of G(r) by restricting it to the first term of the series given by Eq. (31) to obtain the approximate solution for the total force spectral density $S_{QQ}(f)$.

$$S_{QQ}(f) \approx S_{\eta\eta}(f)\left\{\frac{8}{\pi}\left[\frac{2\pi f C_1}{\sinh 2\pi d/L}\int_0^d \sigma(S)\cosh(2\pi S/L\,dS\right]^2\right.$$

$$\left.+\left[\frac{(2\pi f)^2 C_2}{\sinh 2\pi d/L}\int_0^d \cosh(2\pi S/L)dS\right]^2\right\}, \qquad (33)$$

in which

$$\int_0^d \cosh(2\pi S/L)dS = \frac{\sinh 2\pi d/L}{2\pi/L}. \qquad (34)$$

The first integral in Eq. (33) cannot by pre-evaluated, but must be calculated for each sea-surface spectral density used. The total moment about the bottom can be obtained in a similar manner.

Borgman [10, 11, 12] has found this linearization of the drag term to be the equivalent of using $(V_{RMS}\sqrt{8/\pi}V(t)$ in place of $|V(t)|V(t)$ in Eq. (24); the physical reason for this is not clear, however. It should be pointed out here, that another linearization has been used by nearly every investigator in the past, with essentially no discussion;

that is, the use of $\partial u/\partial t$ rather than du/dt. Work is needed to determine the size of error introduced by this linearization of the drag term.

A relatively simple transfer function has been obtained by Borgman [9, 10, 12] to calculate the total force and overturning moment the pile array of an offshore platform, and the reader is referred to the original work for information thereof.

A solution to the problem of the stress distribution in a real structure with a number of legs and cross bracings has been obtained by A. Malhotra and J. Penzien [37, 38]. They considered the relative velocities and relative acceleration of a structure which was forced to vibrate in a random manner by water waves. This work was a continuation of the work done by Foster [23]. The methods and computer programs developed by these engineers are too complex and long to be able to summarize herein. They obtained a linearized drag term similar to that of Borgman (see above), but with the relative velocity between pile and water particle motions being used; e.g., $(V_R)_{RMS}\sqrt{8/\pi}\, V_R(t)$, with an iterative method being given to calculate $(V_R)_{RMS}$ and $V_R(t)$. Although there was little difference between the drag terms calculated by the two methods, there was an appreciable difference in the calculated horizontal displacement of the deck for large storms. The authors made calculations for structures in 400, 600, 800, 1,000 and 1,120 ft water depths. The computer program prepared by the authors for use with a CDC 6400 digital computer permitted them to obtain a complete solution of one tower to a single sea state, using seven degrees-of-freedom, in about one minute. They presented results on the effect of tower stiffness and leg spacing on the deck motion for a large range of wave spectra. They found that the larger the waves the more important the drag term, with the drag predominating for fully developed seas associated with winds in excess of about 50 miles per hour, making use of the Pierson-Moscowitz-Bretschneider model. This model is

$$S_{\eta\eta}(\omega) = \frac{\alpha g^2}{\omega^5} \exp[-\beta(g/\omega W)^4], \tag{35}$$

in which α and β are dimensionless constants having the values of 8.1×10^{-3} and 0.74, respectively, and W is the mean windspeed at a height of 19.5 m above the sea surface. Owing to a mistake in one term, the results are only qualitatively correct [14], but corrected calculations have been made which will be published shortly.

A simpler theory was developed by Nath and Harleman [46] for a pile supported platform in which the inertial forces predominated (owing to their choice of pile diameters, water depths, wave periods and wave heights). This permitted a linear solution as they dropped the drag term; furthermore they did not use relative motions. They constructed an elastic model and tested it in a wave tank. They found results of the theoretical analysis compared rather favorably with the experimental results, in tests using waves of single frequency and in tests using a wave spectrum. The effect of pile spacing was investigated, both theoretically and experimentally. The results showed a large response occurred when the wave frequency was such that it matched both the time frequency of the structure and the space frequency of the structure (that is, when the wave length was the same as the pile spacing as measured in the direction of wave advance; conversely, a minimum response occurred when one leg spacing was half a wave length - see also [9]).

Directional Wave Spectra Approach

Professor Joseph Penzien and Mr. B. Berge, University of California, have applied the general ideas developed by Foster and by Malhotra and Penzien, for the case of one dimensional spectra to the case of directional spectra. The studies are still under way.

Eddies and Associated Forces

Introduction

The formation of eddies in the lee of a circular cylinder in uniform steady flow has been studied by a number of persons. It has been found that the relationship among the frequency (cps) of the eddies, f_e, the diameter of the cylinder, D, and the flow velocity, V, is given by the Strouhal number, S,

$$S\left(1 - \frac{19.7}{Re}\right) = \frac{f_e D}{V} \approx S , \qquad (36)$$

where Re is the Reynolds number. Except in the range of laminar flow, the Reynolds number effect in this equation can be neglected. For flow in the sub-critical range (Re less than about 2.0×10^5), $S \approx 0.2$. For $Re > 2.0 \times 10^5$, there is a considerable variation of S; in fact, it is most likely that a spectrum of eddy frequencies exists (see [74], p.268 for a discussion of this). Extensive data on S at very high Reynolds numbers, as well as data on C_D (Fig. 6) and the pressure distribution around a circular cylinder with its axis oriented normal to a steady flow, has been given by Rosko [57]. Few data are available on the resulting oscillating transverse forces.

What is the significance of S for the type of oscillating flow that exists in wave motion? For deep water the horizontal component of water particle velocity (Eq. 15) is approximately $u = (\pi H/T)\cos 2\pi t/T$ at $x = 0$. An average of u can be used to represent V; i.e., $V_w \approx u_{avg} \approx \pi H/2T$, where V_w is the "average" horizontal component of water particle velocity due to a train of waves of height H and period T. For at least one eddy to have time to form it is necessary for $T > 1/f_e \approx 2DT/\pi HS$; if $S \approx 0.2$, $H > 10\,D/\pi$.

Keulegan and Carpenter [29] studied both experimentally and theorerically the problem of the forces exerted on bodies by an oscillating flow. The oscillations were of the standing water wave type in which the wave length was long compared with the water depth so that the horizontal component of water particle velocity was nearly uniform from top to bottom. Furthermore, the body was placed with its center in the node of the standing wave so that the water motion was simply back and forth in a horizontal plane. They found that C_M and C_D depended upon the number $u_{max}T/D$ where $u = u_{max}\cos 2\pi t/T$. They observed that when $u_{max}T/D$ was relatively small no eddy formed, that a single eddy formed when $u_{max}T/D$ was about 15, and that numerous eddies formed for large values of the parameter. It is useful to note that this leads to a conclusion similar to the one above. For example, if one used the deep water wave equation for $u_{max} = \pi H/T$, then $u_{max}T/D > \pi H/D > 15$, and $H > 15D/\pi$.

It appears from the work described above that a high Reynolds number oscillating flow can exist which is quite different from high Reynolds number in steady rectilinear flow unless the wave heights are larger than the diameter of the circular cylinder. It is likely that the Keulegan-Carpenter number is of greater significance in correlating C_D and C_M with flow conditions than is Reynolds number ([74], p. 259), and that the ratio H/D should be held constant to correlate model and prototype results, or at least should be the appropriate value to indicate the prototype and model flows are in the same "eddy regime" (see [49], for similar results for a cylinder oscillating in water).

When the Keulegan-Carpenter number is sufficiently large that eddies form, an oscillating "lift" force will occur. For a vertical pile the "lift" (transverse) force will be in the horizontal plane normal to the direction of the drag force. Few data have been published on the coefficient of lift, C_L, for water wave type of flow [15, 6, 7]. In uniform rectilinear flows it can be almost as large as C_D, although there are few results available.

Photographs taken of flow starting from rest, in the vicinity of a circular cylinder for the simpler case of a non-reversing flow, show that it takes time (the fluid particles must have time to travel a sufficient distance) for separation to occur and eddies to form. The effect of time on the flow, and hence on C_D and C_M has been studied by Sarpkaya and Garrison [60, 61]. A theory was developed which was used as a guide in analyzing laboratory data taken of the uniform acceleration of a circular cylinder in one direction. Figure 7 shows the relationship they found between C_D and C_M, which was dependent upon ℓ/d, where ℓ is the distance traveled by the cylinder from its rest position and D is the cylinder diameter. They indicated the "steady state" (i.e., for large value of ℓ/D) values of $C_D = 1.2$ and $C_M = 1.3$.

The results shown in Fig. 7 are different than those found by McNown and Keulegan [40] for the relationship between C_D and C_M in oscillatory flow. They measured the horizontal force exerted on horizontal circular cylinder places in a standing water wave, with the cylinder being parallel to the bottom, far from both the free surface and the bottom, and with the axis of the cylinder normal to the direction of motion of the water particles. The axis of the cylinder was placed at the node of the standing wave so that the water particle motion was only horizontal (in the absence of the cylinder). Their results are shown in Fig. 8. Here, T is the wave period and T_e is the period of a pair of eddies shedding in steady flow at a velocity characteristic of the unsteady flow. The characteristic velocity was taken as the maximum velocity. They found that if T/T_e was 0.1 or less, separation and eddy formation were relatively unimportant, with the inertial effects being approximately those for the classical unseparate flow, and if T/T was greater than 10, the motion was quasi-steady.

"Lift" Forces Exerted by Progressive Water Waves

Studies in a hydraulic laboratory have been made by Bidde [6, 7] for the case of "deep water" and "transitional water" waves acting on a vertical "rigid" circular cylinder which extended from near the bottom through the water surface. For this case the undisturbed water particle motion was not simply a rectilinear back and forth motion, but the water particles moved in an elliptical orbit in a vertical plane, so that they were never at rest. Furthermore, any eddies that formed were affected by the free surface at the interface between the air and water. One of the most crucial factors in oscillating flow of this type is the fact that the wake formed during one portion of the cycle becomes the upstream flow in another portion of the cycle.

During the first stages of the study, immiscible fluid particles with the same specific gravity as the water were made of a mixture of carbon tetrachloride and xylene, with some zinc oxide paste added to make the particles easily visible. The fluid was injected into the water by means of a long glass tube which had a rubber bulb mounted at one end. The other end of the tube was heated and drawn to make the tip opening the desired size. Stereophotographic sets were taken of the trajectories of these tracer particles, and a computer program [24] was used to calculate the space position of them. However, it was found too difficult and lengthy a job to pursue.

Owing to the difficulty described briefly above, a description of the wake regime was developed which was based upon its surface characteristics, using magnesium powder sprinkled on the water surface in the vicinity of the pile. The relationship between the wake characteristic and the wave height, with the wave period being held constant is given in Fig. 9 and Table 1, together with the values of Reynolds number and Keulegan-Carpenter number. Similar tables were constructed for a number of wave periods. It was found that the Keulegan-Carpenter number correlated reasonably well with the different regimes of the surface wake characteristics. When the Keulegan-Carpenter number was about 3, one or two eddies formed, when its value was about 4 several eddies formed and shed, having the appearance of a von Kármán vortex street, when it was 5 to 7 the wake started to become turbulent, and when it was larger than 7, the wake became quite turbulent, and the turbulent mass of water swept back and forth past the pile. The Reynolds number was somewhere between 4,000 and 7,000 when the wake became quite turbulent with no detectable von Kármán vortex street. A similar phenomenon occurs in steady flow for Reynolds numbers greater than 2,500 [59].

It can be seen from this study that the von Kármán vortex street formed when the wave height was about equal to the pile diameter. A few experiments made with a pile with a diameter nearly four times the size showed similar results when the wave height was about the same size as the pile diameter.

When eddies form, in addition to their effect on the longitudinal drag and inertial forces, lift forces are also exerted on the cylinder. For a vertical cylinder these lift forces act horizontally, but normal to the longitudinal forces (longitudinal being in the direction of wave motion). Examples of waves, lift forces and longitudinal forces are shown in Fig. 10 for three different values of the Keulegan-Carpenter number (3.2, 6.2 and 10.2). The terms top and bottom associated with the lift and longitudinal forces refer to the forces measured by the top and bottom strain gages on the transducer; the total lift and total longitudinal forces are the sums of the top and bottom values.

There is considerable agreement between the visual observations described previously and the force measurements. Figure 10a shows a set of records for a Keulegan-Carpenter number of 3.2. The lift force has just begun to be non-zero. For this value of the Keulegan-Carpenter number the first eddies develop and shed. The eddy strength is probably very small to that the lift force recorded is negligible. The lift forces for this case have a frequency which is about the same as the wave frequency. This might be due to the fact that the flow is not perfectly symmetrical. The horizontal component of velocity in one direction (wave crest) are slightly different from those in the opposite direction (wave trough), and for the threshold condition the eddies only shed for one direction of the flow. The Keulegan-Carpenter number is 6.2 for the run shown in Fig. 10b. The eddy shedding of the waves. This shows that there is time only for two eddies to shed in each direction. The lift forces are about 25 % of the longitudinal force. The wake is not yet completely turbulent, and the left force records show a more or less regular pattern. The Keulegan-Carpenter number for the run shown in Fig. 10c is 10.2. The wake is fully turbulent. The transverse ("lift") force record appears to be random. The ratio of maximum lift to maximum longitudinal force is about 40 %.

An equation for lift forces is

$$\text{Lift force} = F_L = C_L \rho u^2 A/2 \,, \tag{37}$$

where C_L is the coefficient of lift, u is the horizontal component of water particle velocity, ρ is the mass density of water, and A is the projected area of the cylinder. Use of this equation leads to difficulties as the time history of the force does not necessarily vanish when u goes through zero. Owing to this, very large values of C_L can be calculated from the laboratory measurements. This difficulty can be largely overcome by defining the relationship of Eq. (37) only for maximum values of the force, $F_{L_{max}} = C_{L_{max}} \rho u^2_{max} A/2$. Chang [15] found values of $C_{L_{max}}$ between 1.0 and 1.5 for values of the Keulegan-Carpenter number greater than about 10.

In the study by Bidde the ratio of lift to longitudinal force was used as a basic parameter rather than C_L as this parameter is comparatively less sensitive to any systematic errors in the instrumentation used to measure the forces, as similar errors would be present in both lift and longitudinal force measurements, and these errors would have a certain tendency to cancel out. Some of the data are shown in Fig. 11 on the relationship between the wave height and the ratio to lift force to longitudinal force.

Table 1. Observation of Surface Characteristics of Eddies (from Bidde, [6, 7].
Water Depth = 2.0 ft, Cylinder Diameter = 1-5/8 in., Wave Period = 2.0 sec.

Run Number	Wave Height [feet]	Surface Reynolds Number	Surface Keulegan-Carpenter Number	Observations
1	0.028	850	0.9	No separation, no eddies (Amplitude of motion does not reach cylinder diameter)
2	0.04	1,220	1.3	
3	0.055	1,680	1.8	Small separation
4	0.07	2,140	2.3	Very weak von Kármán street
5	0.08	2,450	2.7	Clear von Kármán street
6	0.095	2,920	3.2	Wake of prior semicycle, when swept back gives rise to additional eddies
7	0.105	3,230	3.5	
8	0.120	3,700	4.0	
9	0.135	4,180	4.6	
10	0.155	4,810	5.2	Eddies swept back by the time they are formed
11	0.180	5,610	6.1	
12	0.20	6,250	6.8	
13	0.22	6,900	7.5	Becoming highly turbulent
14	0.24	7,550	8.2	
15	0.27	8,530	9.3	Extremely turbulent, no more eddies visible
16	0.30			
17	0.32			
18	0.34			
19	0.35	11,200	12.2	

The relationships between the Keulegan-Carpenter number and the Reynolds number and the ratio of lift force to longitudinal force are shown in Fig. 12. This graph indicates that the lift forces start at a Keulegan-Carpenter number of 3 to 5, and then increase steadily with increasing values. At a Keulegan-Carpenter number of about 15 the ratio of lift to longitudinal force shows a slight tendency to stop increasing.

The work described above was continued by Wiegel and Delmonte [78] to determine the effect of large values of Keulegan-Carpenter number on the ratio of "lift" forces to longitudinal forces. These data have also been shown on Fig. 12 (labeled UCB 1972). As can be seen in Fig. 10c the amplitude and frequency of the "lift" forces become irregular for larger values of Keulegan-Carpenter number, and it is necessary to specify what is measured. Bidde presented the ratios of average maximum lift forces to average maximum longitudinal forces. The longitudinal forces were uniform so that no problem existed in measuring and reporting them. Bidde drew a line by eye through the crests of the larger lift forces and another line through the troughs, and reported the lift force as the distance between the two lines.

Wiegel and Delmonte measured the crest to trough distance for thirty consecutive lift forces and measured the crest to trough distance for four consecutive longitudinal forces. The ratio of the averages of these two sets of data were plotted on Fig. 12.

The Coastal Engineering Research Center (U.S. Army, Corps of Engineers) kindly lent to Wiegel and Delmonte the original wave force records obtained by Ross [58] in their large wave tank (635 ft long by 15 ft wide by 20 ft deep). Longitudinal and "lift" forces were measured for waves with periods as long as 16 sec and heights as great as 8 ft. The two sets of data labeled "BEB" in Fig. 12 are ratios of forces per unit length of pile, using the average of four consecutive waves and the maximum of the four waves.

It would be expected that Bidde's data would lie between the maximum and mean values reported by Wiegel and Delmonte for Keulegan-Carpenter numbers that are not too far apart, and this appears to be the case.

Bidde found for the smaller diameter pile, that Reynolds number appears to be as good a parameter for correlation purposes as Keulegan-

Carpenter number. However, Reynolds number fails to correlate the ratio of lift to longitudinal force when the values of the larger pile are compared with those of the smaller pile. For the smaller pile the value of the Reynolds number for eddies to form is about 0.5×10^4, whereas it goes up to 2.5×10^4 for the larger pile. For the same conditions the Keulegan-Carpenter number is 3 to 5 for both the piles. The UCB (1972) and BEB (1972) mean value data given by Wiegel and Delmonte also show an effect of Reynolds number.

When the oscillating flow is such that the Keulegan-Carpenter number becomes large, one might expect the ratio of lift to longitudinal forces would approach the values obtained for steady rectilinear flows. Wiegel and Delmonte were not able to find many data, however. The results they found have been plotted in Fig. 12 [8. 27].(The curves labeled by Fung and Macovsky were drawn from their results as they appeared in Humphrey's paper).

Although it is evident that much work remains to be done before the problem is solved, it is clear that an engineer must consider a rather large "lift" force as well as a longitudinal force in his design. Owing to the fact that the frequencies of the "lift" forces are higher than the wave frequencies (from 2 to 5 times as high in the laboratory studies) they may be more important than the waves in exciting a resonant response of some structures.

<u>Failure of Texas Tower No. 4</u>

Some details of the collapse of Texas Tower No. 4 were given in the Introduction. Information on the events leading to its ultimate failure will be given below, mostly by a series of quotations from one of the two reports of the U.S. Senate Investigating Committee [69, 70].

During fabrication of Texas Tower No. 4 several change orders were approved, with the following consequences:

"(a) In permitting the substitution of a permanent platform for the originally contemplated temporary one, it meant that the permanent platform would be jacked up above the water before the legs had been embedded into the ocean floor and before any concrete stiffening had been placed in the legs.

(b) Without the legs first being embedded, there was insufficient draft (water depth) above the upper panels of bracing to float the platform into position between them. For this reason, the upper panels of bracing had to be folded down in the initial stages of construction to be connected up later underwater.

(c) In order to fold down the upper panel of bracings, an increase in the tolerance between the size of the pin and the holes into which they were to be inserted was granted. Difficulty in fabrication had required an increase in tolerance from the 1/64 in. called for in the design to 1/16 in. For the upper panels of bracing this was further increased to 1/8 in."

While towing the tower to the site a storm was encountered which, although it did not exceed the criteria for the towing operation, delayed the tip-up operation.

"It was discovered that the two diagonals in the upper panel of braces on the A-B side had broken loose from their lashings and were damaged. During the tip-up process these diagonals sheared off their connecting pin plates and were lost."

It was decided to attempt repairs at sea rather than to tow the tower back to port. Next:

"The platform was towed into position between the three legs (now in a vertical position) and sea swells of about 3 ft in height caused it to dent the three legs, the indentations being an average of 10 ft high, 6 to 8 feet wide, and about 10 to 12 in. deep. The plating of the legs exposed to the crushing action of the platform framework was 13/16 in. thick which was too thin, without being supported by the concrete stiffening, to resist the force. It was agreed that the contractor's operations were carried out in accordance with the erection procedure contemplated by the plans and specifications. It appears that adequate provision was not made to stiffen the steel shells of the tower legs in the critical region. On July 12, 1957, the Bureau of Yards and Docks agreed to finance the direct costs of strengthening the legs at their indentations which ultimately would end up about 10 ft underwater after embedment of the footings."

"The heavy permanent platform was jacked up on the three legs prior to any concrete stiffening being placed in the legs, being jacked up clear of the water on July 8, 1957."

"The water depth, as noted previously, turned out to be 185 ft instead of 180 ft as originally contemplated in the design. Also, the total range of tide was measured at 3.5 ft instead of the original estimate of 1 ft. Thus, in order to achieve a safe clearance of the platform above mean sea level, it was necessary to modify the design by:

(a) Reducing the embedment of the footings from 20 ft to 18 ft;
(b) raising the platform to the maximum extent permittet by the design.

These resulted in achieving a platform elevation of 66.5 ft instead of the 67 ft called for in the design."

"Repairs to Texas Tower No. 4"

"Agreement apparently having been reached to attempt repairs of the tower at sea, the design engineers, under contract with the builder, designed a collar connection encircling legs A and B as a means by which to secure the replacement diagonals to the legs. Dardelet bolts having a serrated shank were inserted through the collars into the legs to keep the collars from moving vertically on the legs. The installation of these bolts required underwater cutting and fitting to close tolerances and the repair was only as good as the capability and integrity of the divers working under adverse conditions at an underwater depth of 65 ft. Moreover, the repair was made even more difficult by the reactions of the legs to the forces of the sea in that, even at this time, the tower's foundation was in motion.

By the summer of 1958, the Air Force personnel operating the tower had complained of sensations of considerable movement of the platform with frequencies of 15 to 18 cpm. Although those who observed the motion had no means of properly measuring its extent, such motions did not occur during severe weather conditions as the maximum wind velocity and wave height during that period were about 30 knots and 15 ft, respectively. However, the frequency of the horizontal oscillations gave some clue as to the stiffness of the tower (design frequency was 37 to 46 cpm) and the design engineers made a special analysis to determine whether or not the replaced diagonals on the A-B side were functioning. This analysis led to the conclusion that the upper tier of bracing on the A-B side was not functioning and this, in turn, prompted an analysis of the strength of the tower under this condition and the amplitude of motion which might indicate danger. It was estimated that, if the tower leg steel were stressed to the yield point, the tower could stand a 125 mile per hour wind combined with a 36 foot high wave or an 87 mile per hour wind combinded with a 67 ft wave. The horizontal deflection at the same time would be about 6 in.. Because of this motion and the fact that the tower had not yet experienced a hurricane, the tower was totally evacuated of personnel in advance of Hurricane Daisy in August of 1958."

At a conference during the winter of 1958-1959 it was revealed that

"... the 8-in. LD clevis holes were bored one-eighth inch oversize to facilitate assembly. Moreover, two clevis brackets were broken off in tower transit to the installation site. A repair at sea was attempted by replacing the broken brackets with eared collars. These collars attempted to anchor the lower end of the diagonals in the upper frame of the A-B plane to caissons A and B, respectively. Considerable doubt was expressed regarding the success of this repair."

It was also indicated that tower motions of plus or minus 2 in. could occur without taking up known clearances in the first truss beneath the ocean surface.

A study was made by Brewer Engineering Laboratories Inc. [71] during a 5 month interval; winds to 65 knots and waves to 30 ft were experienced. The U.S. Senate report makes the following statements on their findings:

> "(a) That the observed natural tower frequencies (17 to 23 cpm translational and 23 to 24 cpm rotational) were approximately one-third of those predicted by the designer's theoretical calculations.
>
> (b) That the subsea truss work was essentially ineffective for excursions up to 3 in. and rotations to 0.1°. It was expected that the clearances in the pin connections would be taken up with increasing deflections of the tower platform.
>
> (c) That positive evidence of the fact that relative motion between members of the underwater truss system occurred during the ever-present tower oscillations, was provided by hydrophones. The metallic rumbling noises heard beneath the tower were coincident with the frequency of tower motion. They were interpreted to result from the movement of very heavy metal objects.
>
> (d) That 10-ft waves produced the greatest tower motions and therefore stresses over the range of waves 0 to 30 ft in height experienced during the study.
>
> (e) That hydrodynamic forces (waves) were by far the more important over aerodynamic forces (wind)."

An underwater survey was made by Marine Contractors, Inc., and they found (November 1958):

> "(a) The pins in the horizontal brace at -23 ft at midpoint in the A-B plane were loose (A diagonal) and withdrawn 9 in. (B diagonal).
>
> (b) The Dardelet keeper plates on the B leg were loose and several of the studs and nuts were missing from them.
>
> (c) None of the Dardelet bolts on either side of the collar on the A leg were in place, either having sheared off or fallen out and there was evidence of vertical motion of the collar on the caisson.

(d) In tightening the collar bolts to a certain torque, they would be found looser within a day or two.

(e) The A leg developed oil leaks which could not be repaired."

Next, the U.S. Senate report states:

"The work to replace the Dardelet bolts with T-bolts began in November 1958 but was discontinued when about 50 percent complete because of weather. None of the T-bolts were installed until the following year and the entire repair was completed in May 1959. In the meantime, with the tower in this condition, it was exposed to five severe storms with maximum waves of 33 ft in height and winds up to 90 miles per hour."

"By January of 1960, less than a year after the collars were fixed, the operating personnel again complained of excessive platform motion. Marine Contractors, Inc., performed another underwater inspection in February 1960. In the report of Mr. Alan Crockett, general manager for Marine Contractors, Inc., there appears the following statement:

'This concern did a similar survey on tower No. 4 last October 1959 [sic; 1958] and the results did not show the magnitude of clearance to be found in the pins that we have appreciated during this survey. We feel that there is approximately three-fourth-inc. increase in clearance between the surveys ***. The tower movement is very erratic in an oscillatory direction ***. The noise factor heard on the tower in the vicinity of A caisson is resulting from the motion of the tower taking up total clearances in the pins and flanges on one side or the other to bringing the two metal surfaces together at the extremity of motion causing the metallic bang.'"

In a letter dated 1 April 1960, the design engineer stated:

"The loose pin connections are a very serious matter since there seems to be no way of satisfactorily remedying the condition. Furthermore, the condition is one which will tend to worsen at an increasing rate with time. This is because the looseness induces impact stresses in the pins and pin plates which are greater than for the nondynamic design assumptions and will become increasingly greater as the play in the joint enlarges."

As a result of these findings and additional engineering, repairs were made to the tower.

"The installation of the X-bracing above water was a matter of emergency because a condition existed which would result in the probable loss of the tower if it was not corrected. This bracing was installed at elevations plus 9 ft to plus 58 ft above water in the area presenting maximum resistance to the passage of waves, and represented a scheme which was diametrically opposed to the original concept of keeping resistance to wave passage to a minimum. No effort was made to rectify the admittedly serious conditions of loose pins and worn connections underwater but, nonetheless, the design engineers on August 10, 1960, certified that the above-water X-bracing had restored the tower to its original design strength."

"About a month later, on September 12, 1960, Hurricane Donna passed through the area. The actual maximum wind velocities and wave heights experienced at tower No. 4 from the effects of Hurricane Donna have not been clearly substantiated, other than admittedly having exceeded the original design criteria of 125 miles per hour for wind velocity and 35-ft breaking wave height. Some sources claim winds of 132 miles per hour and breaking waves of a least 50 ft in height. Other claim winds of 115 miles per hour velocity and breaking waves of 65 ft in height, while others claim waves of 75 ft in height, the latter purportedly being a measurement above mean sea level and not from trough to crest. Light structural steel for the exhaust vents, 8 ft above the base of the platform which was 66.5 ft above mean sea level, was dented from wave action. Shown below is a photograph of wave action against the above-water X-bracing taken at some time during Hurricane Donna." (Writer's note - not included herein.)

During this hurricane the "flying bridge" beneath the tower had been torn loose and had to be repaired.

"J. Rich Steers, Inc., the original builders under contract with the Air Force, completed repair to the "flying bridge" on November 1, 1960, and subsequent inspections revealed the following damage:

(a) The above-water X bracing was cracked and fractured in its primary and secondary members.

(b) The upper tier A diagonal was fractured.

(c) The two diagonals in the second tier of bracing at the midpoint of the horizontal at elevation -75 ft had torn loose from their attachment and were moving freely. These damages were all on the A-B side."

Other studies were made, and designs completed for additional structural work. It was decided to repair, through installation of a sleeve, the attachment of the two diagonals on the horizontal brace at the -75 ft level, and this was completed on January 1, 1961, after much difficulty from wind and waves. Other work was also planned; however:

"On December 12, 1960, the tower was subjected to another severe storm with high seas and winds of 87 knots. Then, on January 7, 1961, the divers discovered that the B diagonal in the lowest panel of bracing was broken."

"As a result of the meeting of January 12, 1961, it was decided to completely evacuate the tower by February 1, 1961. This was the date at which Steers would have used the grouting (sand, gravel, and cement) supplies, rewelded the X bracing above water, and this would also have allowed the Air Force time in which to winterize and preserve equipment which would be left on board pending a return in the spring to resume repair under more favorable weather conditions."

"The cable bracing had not been installed."

"Three days later, January 15, 1961, at 7:25 p.m., on Sunday evening, during a winter storm, Texas tower No. 4 collapsed. There were no survivors of the 28 men on board."

"The maximum prevailing weather at the time the tower disappeared, from the radar scope of the supply ship, consisted of winds of force 11 (approximately 55 knots) and waves of 35 to 40 ft in height."

"The tower platform rests on the ocean floor 200 yards from its original location on a southwest bearing of 242°. It rotated in a counterclockwise direction through 35° so that the A-B plane is now on a bearing of N. 9°W. rather than N. 26°E."

Moored Ships

The problem of mooring so as to prevent damage to boat, mooring lines or dock is extremely complex, depending upon wave heights, periods and direction (directional wave spectra), and upon the weight, shape, natural periods of the boat and upon the characteristics of its moorings. Some of the first theoretical studies of a moord ship were made by Wilson [81, 83] and Abramson and Wilson [2]. Using these studies as a guide, together with theoretical studies by a number of investigators of the motions of freely floating ships, the author and his colleagues made a series of hydraulic model studies of using geometrically and dynamically similar models in waves. Tests were made for a number of conventional ships, moored in the open ocean and also alongside docks. Most of this work has been summarized by the author [74]. Tests were also made for special types of bodies such as floating drilling rigs [67], and submerged oil storage tanks [79].

A theoretical study has been made recently by Raichlen [56] for the simplified case of the surging motion of several classes of small boats (ranging from 2 to 8 tons on 20 to 40 ft in lengths) subjected to uniform periodic standing waves with crests normal to the longitudinal axis of the moored boats, with two bow lines and two stern lines. The restoring force versus displacement of the moorings were nonlinear, as is apparently the normal case. A detailed analysis was made for one of the boats (Harbor Boat No.3), which had a length of 26 ft, beam of 9 ft -2 in., maximum draft of 2 ft -4 in. and an approximate displacement (unloaded) of 5200 lbs. Details of the mooring configuration and mooring line characteristics were also presented. Measurements of the period of free oscillation of surge for three mooring line conditions (zero slack, 4 in. and 8 in. slack) for several different metal displacements were made and compared with theory. The comparisons, shown in Fig. 15, are quite good.

Some of the complexities of the problem can be seen from Fig. 16 which compares the maximum motion (in one direction only) of the boat as a function of wave period and the forcing function ζ for taut mooring lines and for 8 in. slack. ζ is a rather complicated function, and Raichlen describes it as the maximum with respect to time of the water particle velocity averaged over the displaced volume of the moored body. All other things being usual, ζ is directly proportional to the standing wave amplitude. It is evident that a boat moored with slack lines at one tide stage may have taut lines at another stage of the tide, so that its response will vary with tide stage, all other conditions being equal.

Two other examples have been chosen from Raichlen's report, and are shown in Fig. 17. The maximum positive displacement from rest is shown as a function of wave period and ζ for two boats, one of 3700 lbs with a length of 22 ft -5 in., and the other of 17,000 lbs and a length of 38 ft. First it appears peculiar that the smaller boat should have larger "natural periods" than the larger boat. The reason for this was that the mooring lines of the larger boat were much stiffer compared with its weight than was the case for the smaller boat. This emphasizes again that the mooring is extremely important to the problem and there can be no simple wave height criteria for a harbor.

References

1. Aagaard, P.M., Dean, R.G.: Wave Forces: Data Analysis and Engineering Calculation Method. Preprints of 1969 Offshore Technology Conference, Houston, Texas, Vol. 1, Paper No. OTC 1008, May 1969, pp. 194-206.

2. Abramson, H.N., Wilson, B.W.: A Further Analysis of the Longitudinal Response of Moored Vessels to Sea Oscillations. Proc. Midwestern Conf. Fluid and Solid Mechanics, Purdue Univ., Indiana, September 1955, pp. 236-251.

3. Agerschou, H.A., Edens, J.J.: Fifth and First Order Wave-force Coefficients for Cylindrical piles. Coastal Engineering: Santa Barbara Specialty Conference, October 1965, Amer. Soc. Civil Engrs., 219-248 (1966).

4. Aria, M., Kiya, M.: Lift of a Cylinder in Shear Flow. Proc. U.S.-Japan Seminar on Similitude in Fluid Mechanics, Cambridge, Mass., Minneapolis, Minn., Iowa City, Iowa, Palo Alto, Calif., September 21-28, 1967, prepared by the Japanese Delegation (Organizer, Dr. Tokio Uematsu), pp. 33-60.

5. Bairstow, L.: Applied Aerodynamics, London: Longman, Green 1939, pp. 390-396.

6. Bidde, D.D.: Wave Forces on a Circular Pile due to Eddy Shedding. Ph. D. Thesis, Department of Civil Engineering; also Tech. Rept. HEL 9-16, Hydraulic Engineering Laboratory, Univ. of California, Berkeley, Calif., June 1970, 741 pp.

7. Bidde, D.D.: Laboratory Study of Lift Forces on Circular Piles. J. of the Waterways, Harbors and Coastal Engineering Division, Proc. ASCE 97, No. WW4, 595-614 (November 1971).

8. Bishop, R.E.D., Hassan, A.Y.: The Lift and Drag on a Circular Cylinder in a Flowing Fluid. Proc. Roy. Soc. London A, 277, 32-50 (1964).

9. Borgman, L.E.: The Spectral Density for Ocean Wave Forces. Coastal Engineering: Santa Barbara Specialty Conference, October 1965, Amer. Soc. Civil Engrs., 147-182 (1966).

10. Borgman, L.E.: The Estimation of Parameters in a Circular-Normal 2-D Wave Spectrum. Univ. of Calif., Berkeley, Calif., Inst. Eng. Res., Tech. Rept. HEL 1-9, 1967.

11. Borgman, L.E.: Ocean Wave Simulation for Engineering Design. Proc. Conf. on Civil Engineering in the Oceans, Amer. Soc. Civil Engrs., 31-74 (1968).

12. Borgman, L.E.: Spectral Analysis of Ocean Wave Forces on Piling. J. Waterways and Harbors Div. Proc. ASCE 93, No. WW2, 129-156 (May 1967).

13. Burling, R.W.: The Spectrum of Waves at Short Fetch. Dtsch. Hydrogr. Z. 12, 45-64, 96-117 (1959).

14. Chakrabarti, S.K.: Discussion of Nondeterministic Analysis of Offshore Structures. J. of the Engineering Mechanics Division, Proc. ASCE 97, No. EM3, 1028-1029 (June 1971).

15. Chang, K.S.: Transverse Forces on Cylinders due to Vortex Shedding in Waves. M.S. Thesis, Massachusetts Institute of Technology, Department of Civil Engineering, January 1964.

16. Chase, J., Cote, L.J., Marks, W., Mehr, E., Willard, J., Pierson, Jr., Rönne, F.C., Stephenson, G., Vetter, R.C., Walden, R.G.: The Directional Spectrum of a Wind Generated Sea as Determined from Data Obtained by the Stereo Wave Observation Project. New York University, College of Engineering, under Contract to the Office of Naval Research (NONR 285(03)), July 1957, 267 pp.

17. Court, A.: Some New Statistical Techniques in Geophysics, Advances in Geophysics, Vol. 1, Academic Press 1952, pp. 45-85.

18. Cross, R.H.: Wave Force Program. Techn. Rept. HEL 9-4, Hyd. Eng. Lab., Univ. of Calif., Berkeley, September 1964, 26 pp.

19. Cross, R.H., Wiegel, R.L.: Wave Forces on Piles: Tables and Graphs. Tech. Rept. HEL 9-5, Hyd. Eng. Lab., Univ. of Calif., Berkeley July 1965, pp. 57.

20. Dryden, H.L., Murnaghan, F.D., Bateman, H.: Hydrodynamics, Dover Publications 1956, pp. 634.

21. Evans, D.J.: Analysis of Wave Force Data. Preprints of 1969 Offshore Technology Conference, Houston, Texas, Vol. 1, Paper No. OTC 1005, May 1969, pp. 151-170.

22. Fan, Shou-shan: Diffraction of Wind Waves. Tech. Rept. HEL 1-10, Hyd. Eng. Lab., Univ. of Calif., Berkeley, August 1968, 175 pp.

23. Foster, E.T.: Predicting Wave Response of Deep-Ocean Towers. Proc. of the Conference on Civil Engineering in the Oceans, ASCE, Conference, San Francisco, Calif., September 6-8, 1967, ASCE, 1968, pp. 75-98

24. Glaser, G.H.: Determination of Space Coordinates of Particles in Water by Stereophotogrammetry. M.S. Thesis, Dept. of Civil Engineering, University of California, Berkeley, Calif. 1966.

25. Gumbel, E.J.: The Circular Distribution. Am. Math. Stat. 21, No. 143 (1952) (abstract).

26. Hess, G.D., Hidy, G.M., Plate, E.J.: Comparison between Wind Waves at Sea and in the Laboratory. National Center for Atmospheric Research, Ms No. 68-80, Boulder, Colorado, 1968, 13 pp.

27. Humphreys, J.S.: On a Circular Cylinder in a Steady Wind at Transition Reynolds Numbers. J. Fluid Mech. 9, Part 4, 603-612 (1960).

28. Jen, Yuan: Laboratory Study of Inertia Forces on a Pile. J. Waterways and Harbors Div., Proc. ASCE 94, No. WW1, 59-76 (February 1968).

29. Keulegan, G.H., Carpenter, L.H.: Forces on Cylinders and Plates in an Oscillating Fluid. J. Res. Nat. Bureau of Standards 60, No. 5, 423-440 (May 1958).

30. Kipper, J.M., Jr., Joseph, E.J.: A Study of Wave Persistence for Selected Locations in the North Atlantic Ocean, North Sea, and Baltic Sea, U.S. Naval Oceanographic Office, TR-149, 1963.

31. Lamb, Sir H.: Hydrodynamics, 6th ed., New York: Dover Publications 1945.

32. Larras, J.: Probabilité d'apparition des houles dont l'amplitude dépasse une valeur doneé. Comptes Rendus, Acad. Sc. Paris 260 (Groupe 10), 3125-3128 (March 1965).

33. Larras, J.: Nouvelles recherches sur la probabilité d'apparition des houles dont l'amplitude dépasse une valeur donneé. Comptes Rendus, Acad. Sc. Paris 265 (Série B), 434-436 (August 1967).

34. Larras, J.: Les phénomènes aléatoires et l'intenieur: propriétés additives et lois de probabilités. Travaux, pp. 119-120 February 1969).

35. Longuet-Higgins, M.S., Cartwright, D.E., Smith, N.D.: Observations of the Direction Spectrum of Sea Waves Using the Motions of a Floating Buoy, Ocean Wave Spectra. Proceedings of a Conference, Prentice-Hall 1963, pp. 111-132.

36. MacCamy, R.C., Fuchs, R.A.: Wave Forces on Piles: a Diffraction Theory. U.S. Army, Corps of Engineers, Beach Erosion Board, Tech. Memo. No. 69, December 1954, 17 pp.

37. Malhotra, A.K., Penzien, J.: Nondeterministic Analysis of Offshore Structures. J. Engng. Mech. Div., Proc. ASCE 96, No. EM6, Proc. Paper No. 7777, 985-1003 (December 1970).

38. Malhotra, A.K., Penzien, J.: Response of Offshore Structures to Random Wave Forces. J. of the Structural Div., Proc. Amer. Soc. Civil Engrs. 96, ST10, 2166-2173 (October 1969).

39. McNown, J.S.: Drag in Unsteady Flow, IX Congrès International de Mécanique Appliqueé, Actes, Tome III, 1957, pp. 124-134.

40. McNown, J.S., Keulegan, G.H.: Vortex Formation and Resistance in Periodic Motion. J. Engng. Mech. Div., Proc. ASCE 85, No. EM 1, 1-6 (January 1959).

41. Mobarek, Ismail El-Sayed: Directional Spectra of Laboratory Wind Waves. J. Waterways and Harbors Div., Proc. ASCE 91, No. WW3, pp. 91-119 (August 1965).

42. Mobarek, I.E., Wiegel, R.L.: Diffraction of Wind Generated Waves. Proc. Tenth Conf. on Coastal Engng. I, ASCE 185-206 (1967).

43. Morison, J.R., Johnson, J.W., O'Brien, M.P., Schaaf, S.A.: The Forces Exerted by Surface Waves on Piles. Petroleum Trans. 189, TP 2846, 149-154 (1950).

44. Moskowitz, L., Pierson, W.J., Jr., Mehr, E.: Wave Spectra Estimated from Wave Recors Obtained by OWS WEATHER EXPLORER and the OWS WEATHER REPORTER (II). Rept. No. 63-5, Geophysical Science Lab., New York University, New York, N.Y., March 1963.

45. Munk, W.H., Miller, G.R., Snodgrass, F.E., Barber, N.F.: Directional Recording of Swell from Distant Storms. Phil. Trans., Roy. Soc. London A, 255, No. 1062, 505-584 (18 April 1963).

46. Nath. J.H., Harleman, D.R.F.: Dynamics of Fixed Towers in Deep-Water Random Waves. J. Waterways and Harbors Div., Proc. ASCE 95, No. WW4, 539-556 (November 1969).

47. National Academy of Science, Ocean Wave Spectra. Proc. of a Conference, Easton, Maryland, May 1-4, 1961, Prentice-Hall 1963, 357 pp.

48. Paape, A.: Wave Forces on Piles in Relation to Wave Energy Spectra, Proc. of the Eleventh Conference on Coastal Engineering, London 1968, ASCE, 1969, pp. 940-953.

49. Paape, A., Breusers, H.N.C.: The Influence of Pile Dimension on Forces Exerted by Waves. Proc. Tenth Conf. Coastal Engng., Vol. II. ASCE, 1967, pp. 840-847.

50. Panicker, N.N.: Determination of Directional Spectra of Ocean Waves from Gage Arrays. Ph. D. Thesis in Civ. Engng., University of California, Berkeley, California, 1971. Also, Tech. Rep. HEL 1-18, Hyd. Engng. Lab., August 1971, 315 pp.

51. Panicker, N.N., Borgman, L.E.: Directional Spectra from Wave Gage Arrays. Proceedings of the 12th International Conference on Coastal Engineering, Washington, D.C., September 13-18, 1970, Amer. Soc. Civ. Engrs., 1971, pp. 117-136.

52. Pierson, W.J.: The Directional Spectrum of Wind Generated Sea as Determined from Data Obtained by the Stereo Wave Observation Project. Coll. Engr., N.Y.U., Meteor, Papers, 2, No. 6, 1962.

53. Plate, E.J., Nath. J.H.: Modeling of Structures Subjected to Wind Generated Waves. Colorado State University, Civil Engineering Department, CER 67-68 EJP-JN68, Fort Collins, Colorado, 1968, 29 pp.

54. Plate, E.J., Nath, J.H.: Modeling of Structures Subjected to Wind Waves. J. Waterways and Harbors Div., Proc. Amer. Soc. Engrs. 95, WW4, 491-522 (November 1969).

55. Powers, W.H., Jr., Draper, L., Briggs, P.M.: Waves at Camp Pendleton, California. Proc. of the Eleventh Conference on Coastal Engineering, London, England, September 1968, Vol. 1, Amer. Soc. Civil Engineers, New York, N.Y., 1969, pp. 1-8.

56. Raichlen, F.: Motion of Small Boats Moored in Standing Waves, California Institute of Technology, W.M. Keck Laboratory of Hydraulic and Water Resources, Report No. KH-R-17, August 1968, 158 pp.

57. Roshko, A.: Experiments on the Flow Past a Circular Cylinder at Very High Reynolds Numbers. J. Fluid Mech. 10, Part 3, 345-356 (May 1961).

58. Ross, C.W.: Large-Scale Tests of Wave Forces on Piling (Preliminary Report). U.S. Army, Corps of Engineers, Beach Erosion Board, Washington, D.D., Tech. Memo No. 111, May 1959, 25 pp.

59. Rouse, H.: On the Role of Eddies in Fluid Motion. Amer. Scientist 51, No. 3, 285-314 (September 1963).

60. Sarpkaya, T.: Lift, Drag, and Added-Mass Coefficients for a Circular Cylinder Immersed in a Time-Dependent Flow: J. Appl. Mech. 30, Series E, No. 1, 13-15 (March 1963).

61. Sarpkaya, T., Garrison, C.J.: Vortex Formation and Resistance in Unsteady Flow. J. Appl. Mech. 30, Series E, No. 1, 16-24, (March 1963).

62. Scott, J.R.: A Sea Spectrum for Model Tests and Long-Term Ship Prediction. J. Ship Res. 9, No. 3, 145-152 (December 1965).

63. Singh, K.Y., Draper, L.: Waves off Benghazi Harbour, Libya. Proc. of the Eleventh Conference on Coastal Engineering London, England, September 1968, Vol. 1, Amer. Soc. Civil Engineers, New York, N.Y., 1969, pp. 9-18.

64. Snodgrass, F.E., Groves, G.W., Hasselmann, K.F., Miller, G.R., Munk, W.H., Powers, W.H.: Propagation of Ocean Swell across the Pacific. Phil. Trans., Roy. Soc. London A, 259, No. 1103, 431-497 (1966).

65. Stevens, R.G.: On the Measurement of the Directional Spectra of Wind Generated Waves Using a Linear Array of Surface Elevation Detectors. Woods Hole Oceanographic Institution, Ref. No. 65-20, April 1965, 118 pp. (unpublished manuscript).

66. Thrasher, L.W., Aagaard, P.M.: Measured Wave Force Data on Offshore Platforms. Preprints of 1969 Offshore Technology Conference, Houston, Texas, Vol. 1, Paper No. OTC 1007, May 1969, pp. 183-194.

67. Tickner, E.G., Wiegel, R.L., Swanstrom, R.D.: Model Study of the Dynamics of the Shell Model "Donut" Moored in Water Gravity Waves. University of California, Berkeley, Calif., IER Tech. Rept. No. 122-1, August 1958 (unpublished).

68. University of California, Berkely: Design and Analysis of Offshore Drilling Structures. Continuing Education in Engineering, Short course, 16-21, September 1968.

69. U.S. Senate, Inquiry into the Collapse of Texas Tower No. 4: Hearings before the Preparedness Investigating Subcommittee of the Committee on Armed Services, United States Senate, 87th Congress, First Session, May 3, 4, 10, 11 and 17, U.S. Government Printing Office, Washington, D.C., 1961, (a), 288 pp.

70. U.S. Senate, Investigation of the Preparedness Program: Report by the Preparedness Investigating Subcommittee of the Committee on Armed Services, United States Senate, under the Authority of S. Res. 43 (87th Congress, First Session) on the Collapse of Texas Tower No. 4, U.S. Government Printing Office, Washington, D.D., 1961, (b), 41 pp.

71. Vanstone, R.A.: Texas Tower No. 4 Platform Motion Study. Brewer Engineering Laboratories Incorporated Report No. 173, Report to The Hallicrafters Co., 4401 West 5th Ave., Chicago, Ill., 10 June 1959, 95 pp.

72. Walden, H.: Comparison of One-Dimensional Wave Spectra Recorded in the German Bight with Various 'Theoretical' Spectra, Ocean Wave Spectra, Englewood Cliffs: Prentice-Hall 1963, pp. 67-80.

73. Wheeler, J.D.: Method for Calculating Forces Produced by Irregular Waves. Preprints of 1969 Offshore Technology Conference, Houston, Texas, Vol. 1, Paper No. OTC 1006, May 1969, pp. 171-182.

74. Wiegel, R.L.: Oceanographic Engineering, Englewood Cliffs: Prentice-Hall, 1964.

75. Wiegel, R.L., Al-Kazily, M.F., Raissi, H.: Wind Generated Wave Diffraction by a Breakwater. Univ. of Calif., Berkeley, Calif., Hyd. Eng. Lab., Tech. Rept. No. HEL 1-19, April 1972.

76. Wiegel, R.L., Beebe, K.E., Moon, J.: Ocean Waves Forces on Circular Cylindrical Piles. J. Hyd. Div., Proc. ASCE 83, No. HY2, Paper No. 1199 (April 1957).

77. Wiegel, R.L., Cross, R.H.: Generation of Wind Waves, J. Waterways and Harbors Div., Proc. ASCE 92, No. WW2, 1-26 (May 1965).

78. Wiegel, R.L., Delmonte, R.A.: Wave-Induced Eddies and Lift Forces on Circular Cylinders, Univ. of Calif., Berkely, Calif., Hyd. Eng. Lab., Tech. Rept. No. HEL 9-19, April 1972.

79. Wiegel, R.L., Dilley, R.A., Whisenand, S.F., Williams, J.B.: Model Study of a Submerged Buoyant Tank in Waves, 1969 Offshore Technology Conference, May 18-21, Houston, Texas, Preprints, Vol. 1, Paper No. OTC 1067, pp. 695-712.

80. Wilson, B.W.: Analysis of Wave Forces on a 30-inch Diameter Pile under Confused Sea Conditions, U.S. Army Corps of Engineers, Coastal Engineering Research Center, T.M. 15, October 1965, 83 pp.

81. Wilson, B.W.: A Case of Critical Surging of a Moored Ship. J. of the Waterways and Harbors Division, Proc. Amer. Soc. Civ. Engrs. 85, No. WW3, Paper No. 2318, December 1959 (b), pp. 157-176.

82. Wilson, B.W.: The Energy Problem in the Mooring of Ships Exposed to Waves. Bull. Perm. Intern. Assoc. of Navigation Congresses, No. 50, 1959 (a), pp. 7-71.

83. Wilson, B.W.: Ship Response to Range Action in Harbor Basins. Proc. Amer. Soc. Civ. Engrs. 76, Separate No. 41, (November 1950).

84. Wilson, B.W., Reid, R.O.: Discussion of Wave Force Coefficients for Offshore Pipelines (by Beckmann and Thibodeaux). J. Waterways and Harbors Div., Proc. ASCE 88, No. WW1, 61-65 (February 1963).

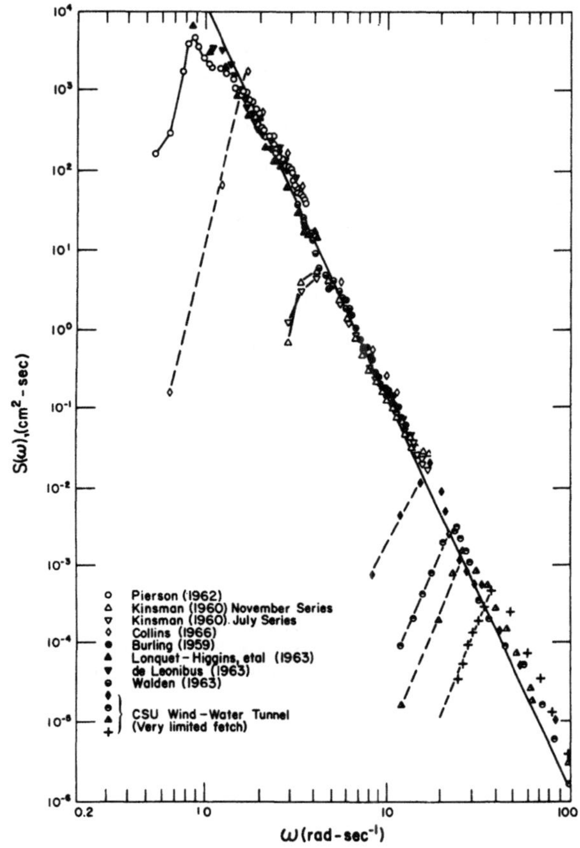

Fig.1. (From Hess, Hidy and Plate [26].)

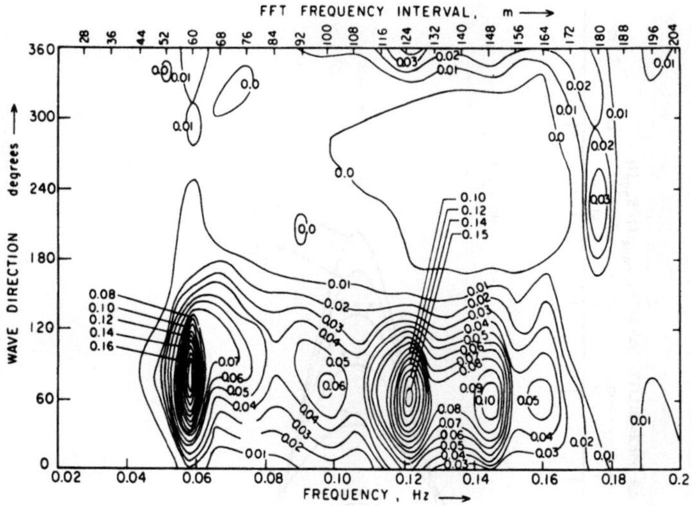

Fig.2. Contour plot of directional spectrum at Point MUGU, Calif., March 27, 1970, 1623 hours (from Panicker, [50]).

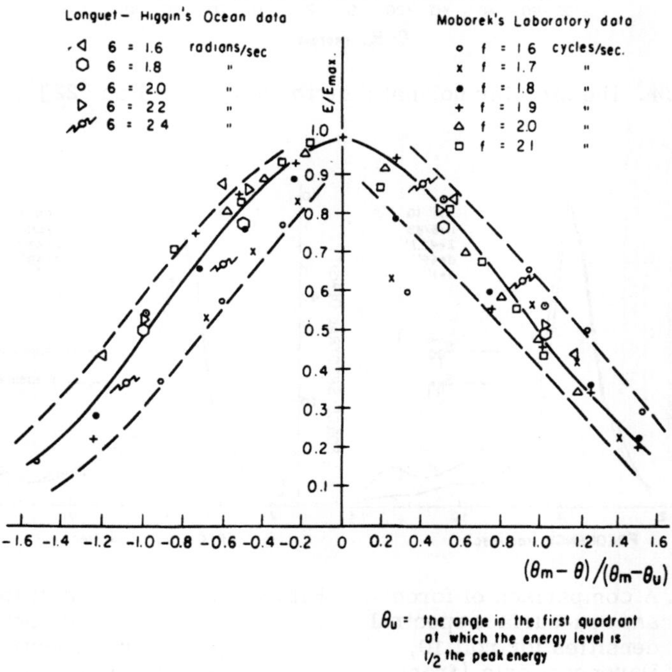

Fig.3. Normalized plot of directional spectra (from Mobarek [41]).

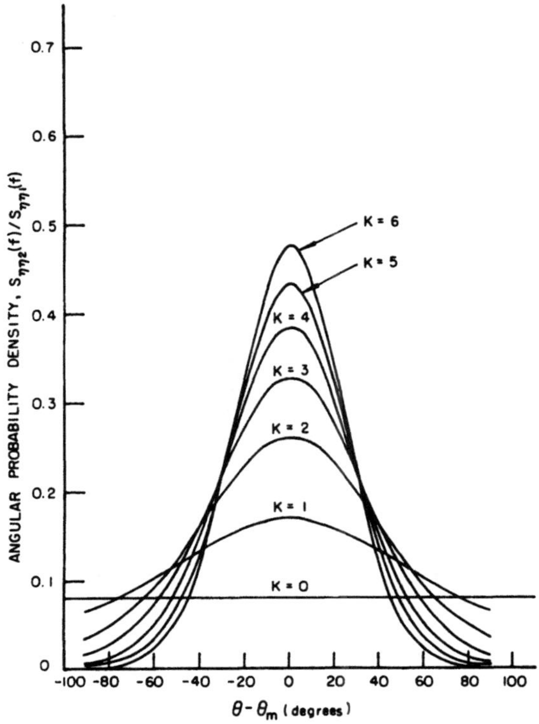

Fig.4. The circular normal distribution (from Fan [22]).

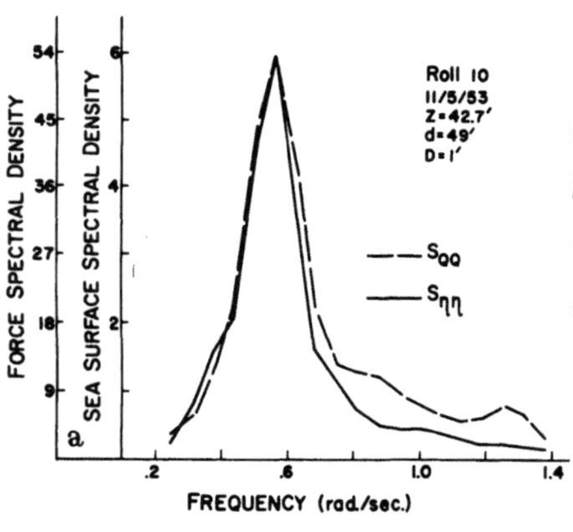

Fig.5a. A comparison of force and sea-surface spectral densities for roll 10, Davenport data (from Borgman [9]).

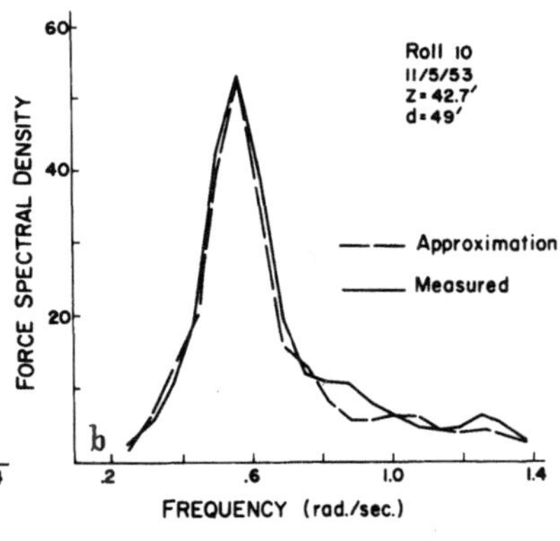

Fig.5b. A comparison of the measured and computed force spectral density for roll 10, Davenport data (from Borgman [9]).

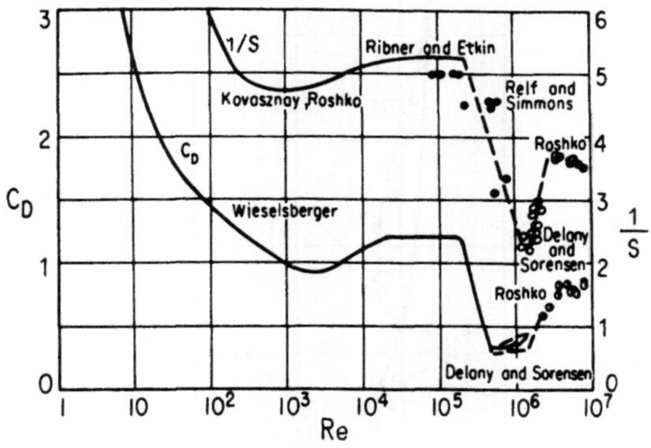

Fig.6. Drag coefficient and reciprocal of Strouhal number versus Reynolds number (from Roshko, [57]).

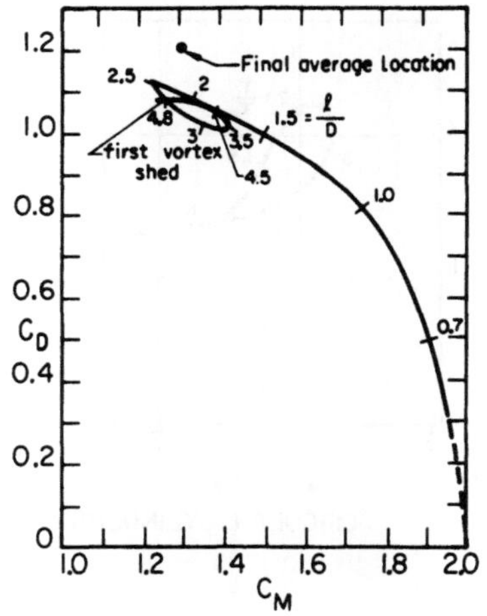

Fig.7. Correlation of drag and inertia coefficients (from Sarpkaya and Garrison, [61]).

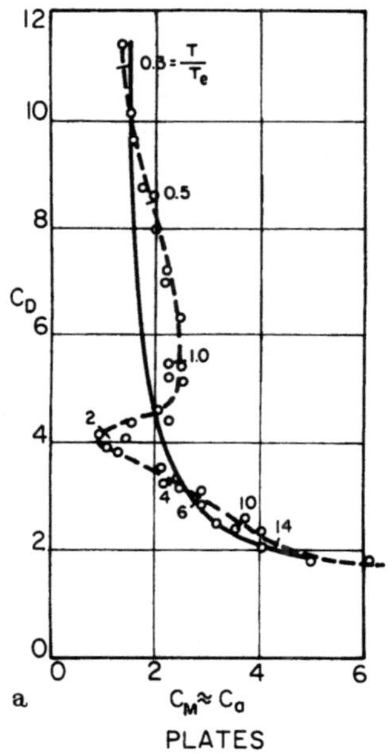

a

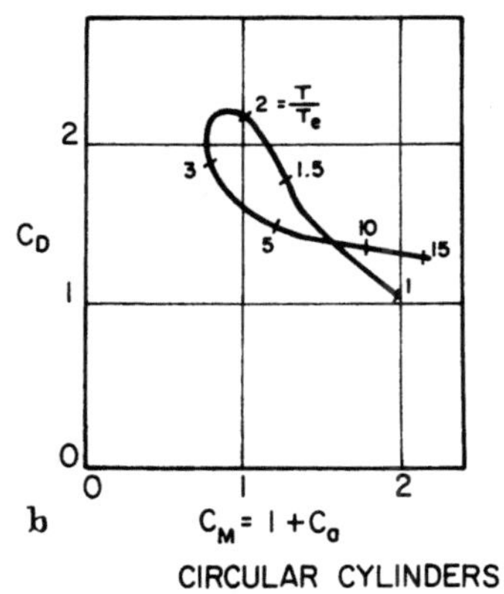

b

Fig.8. Inter-relationship between coefficients of coefficients of drag and of virtual mass for (a) flat plates and (b) circular cylinders (from McNown and Keulegan, [40]).

a) Plates; b) Circular cylinders.

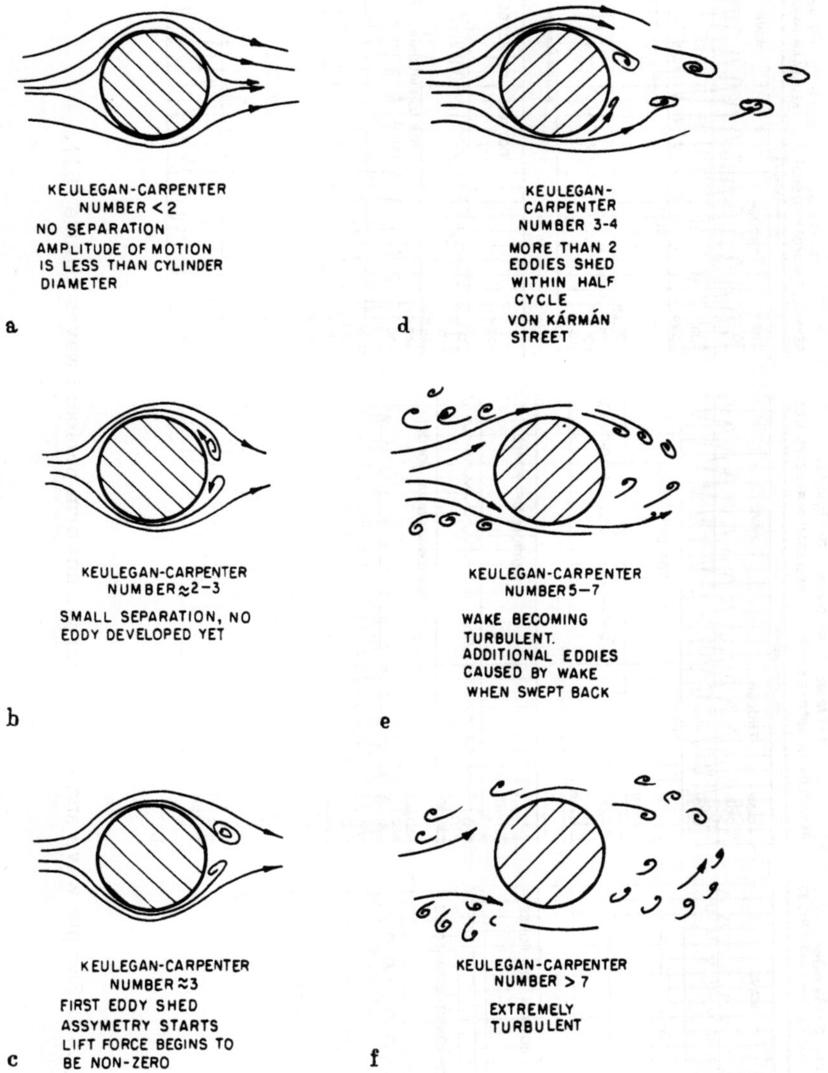

Fig.9. Wake characteristics as a function of the Keulegan-Carpenter number (Bidde, [6]).

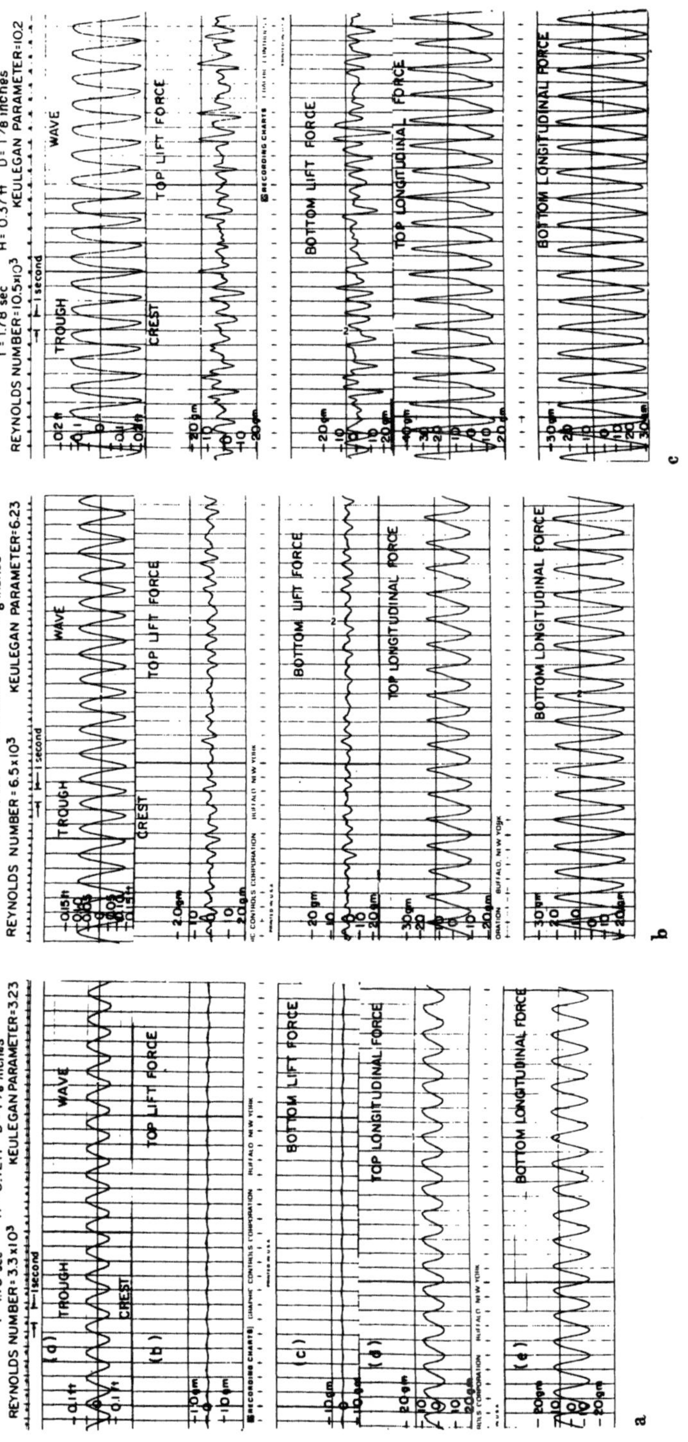

Fig. 10. Sample wave and force records, uniform periodic waves (Bidde [6]).

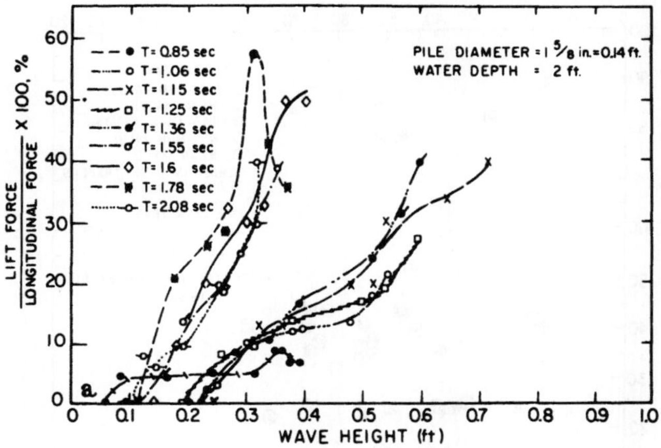

Fig.11a. Ratio of lift to longitudinal forces vs wave height, runs 69 to 244 (Bidde, [6]).

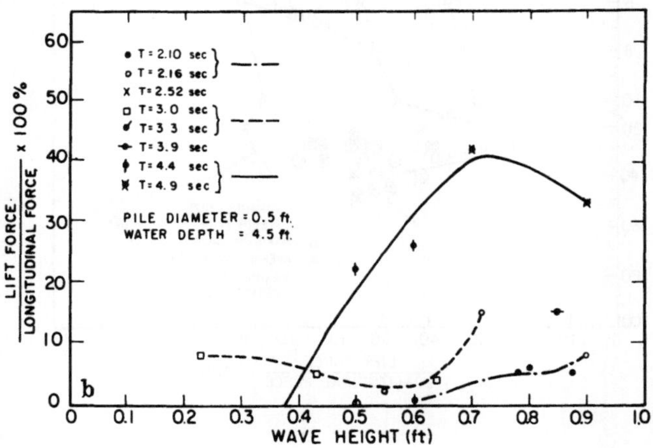

Fig.11b. Ratio of lift to longitudinal force vs wave height, runs 313 to 348 (Bidde, [6]).

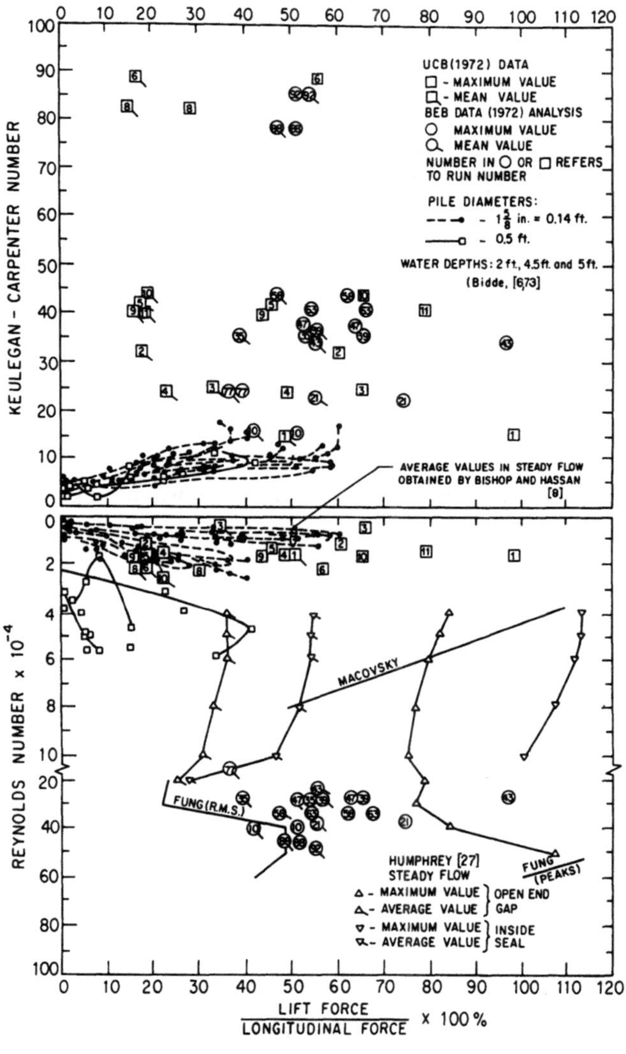

Fig.12. Ratio of lift force to longitudinal force (%) versus Keulegan-Carpenter number and Reynolds number.

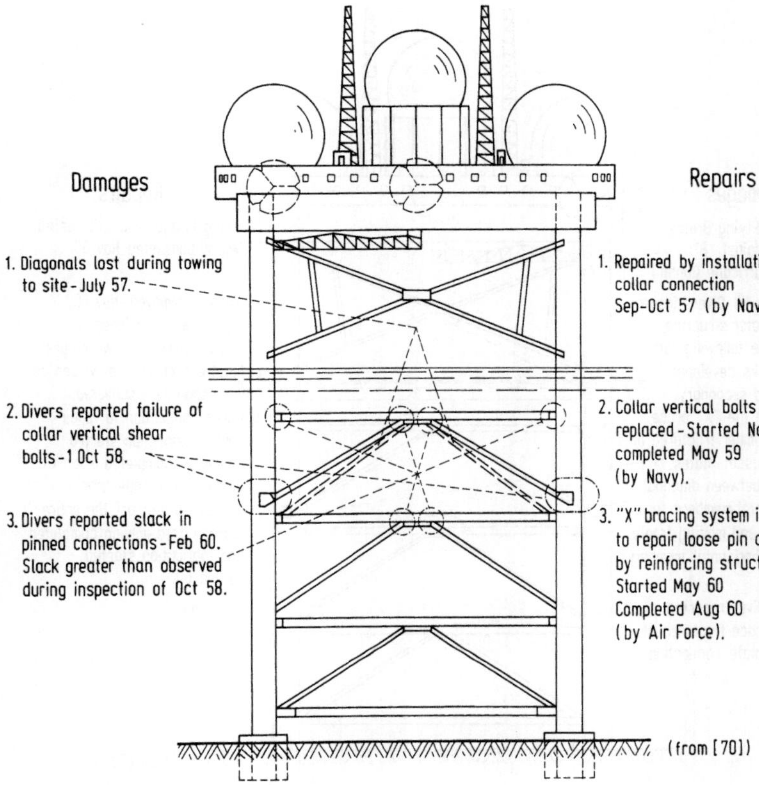

Fig.13. Texas Tower No. 4.
History of damages and repairs 2 november 1957 to 12 September 1960.

(From [70]).

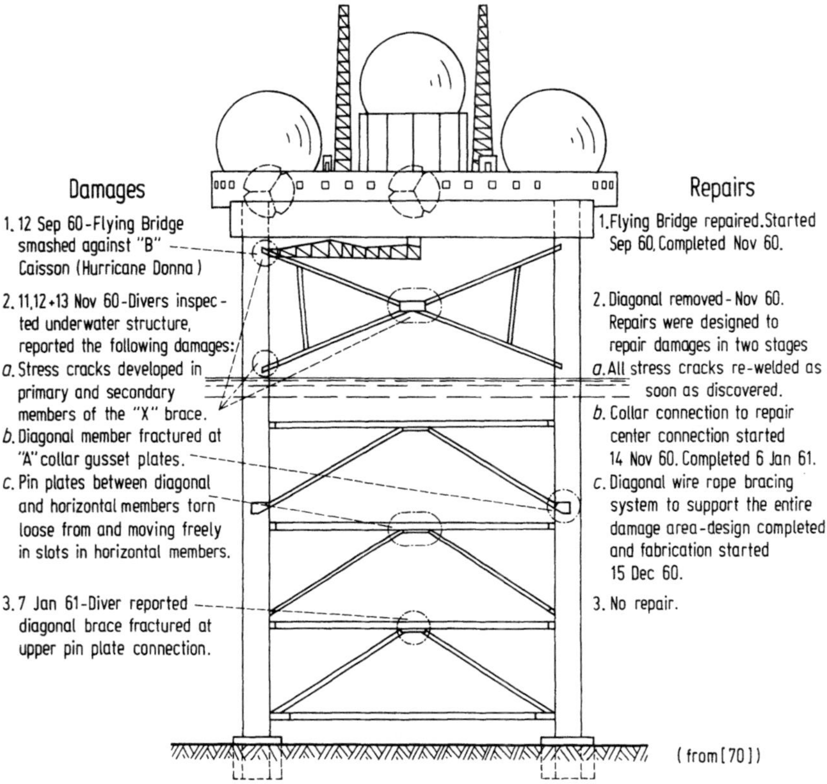

Fig.14. Texas Tower No. 4.
History of damages and repairs 12 September 1960 to 7 January 1961.

(From [70]).

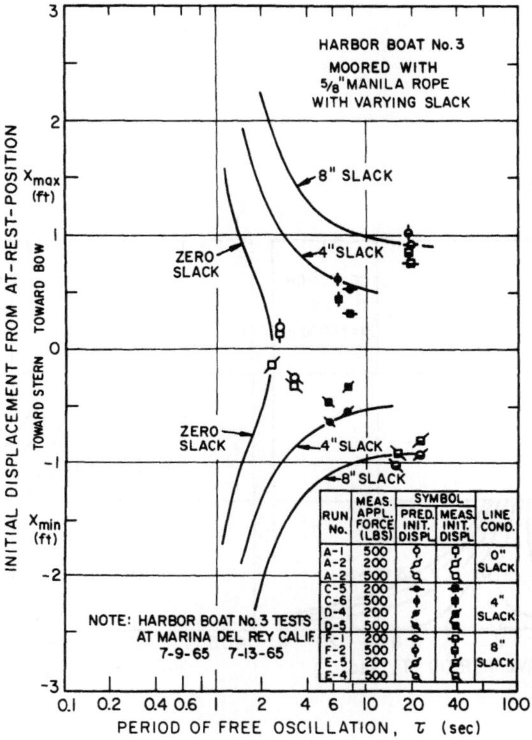

Fig. 15. Measured and predicted periods of free oscillation. Harbor Boat No. 3. (From Raichlen, [56].)

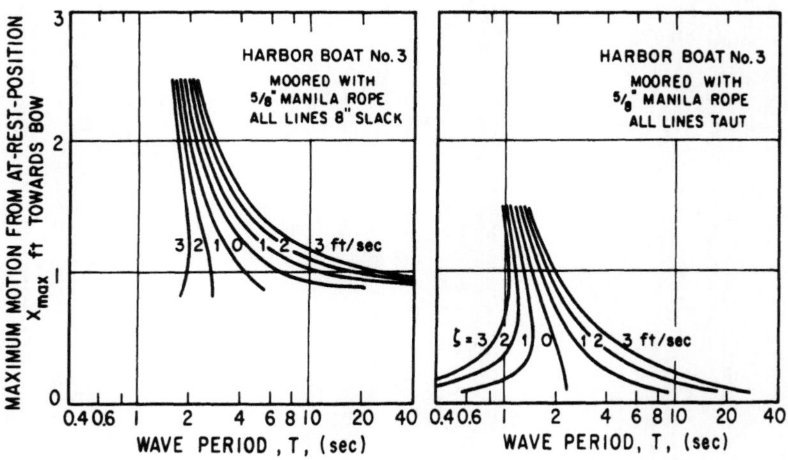

Fig. 16. Response curves: Harbor Boat No. 3. (From Raichlen [56].)

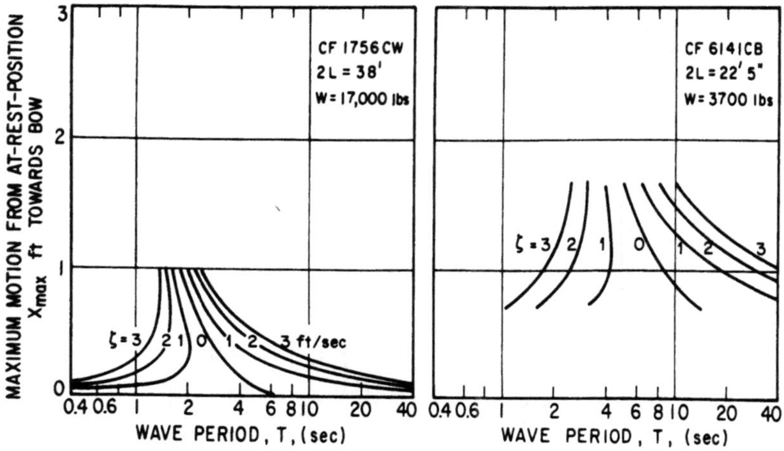

Fig.17. Response curves: 17,000 lb Boat and 3,700 lb Boat. (From Raichlen, [56].)

CONTRIBUTIONS

Some Aspects of the Oscillations of Full-Scale Piles

By

L.R. Wootton
Epsom,
Surrey, U.K.

M.H. Warner
Southall,
Middlesex, U.K.

D.H. Cooper
Southall,
Middlesex, U.K.

Summary

A full-scale experiment has been carried out to study the vortex-induced motion of circular steel piles in flow. The response of three test piles was studied in detail but this present paper deals mainly with the fluid dynamics aspects of the investigation. A yawmeter probe was placed behind one of the piles and has yielded much useful information. The response is thought to be due to both symmetrical and alternate shedding of vortices, depending on the approaching flow speed and other parameters.

Introduction

In 1967/68 severe problems were encountered during the construction of a deep water oil terminal at Immingham in the Humber Estuary on the English North Sea coast. These problems were due to the in-line oscillation, in the fast ebb tide flow, of the cylindrical steel piles used to support the jetty structures. Following the safe completion of the jetty [1], a full scale investigation was carried out at Immingham into the cause of the flow induced motion [2, 3]. A device for preventing these oscillations was developed from this work [4].

The Design of the Experiment

Three test piles (of 457 mm, 610 mm and 762 mm diameter and 40 m in length) were specially driven in the undisturbed flow upstream of the jetty approach-way in water with a mean depth of about 23 m. The tops of the piles were connected by short capping beams to a frame that was welded to the jetty structure at a strong point under an expansion loop (Fig. 1).

The capping beams, which provided a form of encastré fixity in one direction and pinned end fixity in the orthogonal direction, could be rotated and re-attached to the piles and frame at 11.25° intervals. The fixity of the piles in the bed was found to be encastré in all directions.

The motion of the piles was measured using accelerometers mounted in the in-line and cross-flow directions, although the results showed that the motion was best described using the components along and across the capping beam direction. That is, the predominant direction of the motion was in either the encastré-encastré or the encastré-pinned mode.

The flow speed at Immingham was fastest on the ebb tide and at the test rig site, reached well over 2 m/sec on spring tides. The flow speed built up rapidly to a maximum as the tidal level fell from high water but then the speed decreased more slowly until the slack water condition was reached. At this site, slack water was up to one hour after low water. Thus the ebb tide condition, during which all dynamic tests took place, lasted for nearly $7\frac{1}{2}$ hours (Fig. 3). The combination of physical conditions present meant that pile oscillations occurred near the critical Reynolds numbers and this was considered to be one of the most important parameters to be studied.

Measurements were made in the wake of the largest pile to obtain information about the cause of the motion, subsequently confirmed to be a form of vortex shedding. The 762 mm pile was used for this work since it provided data to the greatest Reynolds number. A yawmeter probe was mounted on an adjustable traverse gear behind the pile (Fig. 6). A yawmeter was used rather than a speed measuring device since, although the flow speed at a point in the wake fixed relative to the pile may fluctuate at multiple frequencies of the shedding, flow direction is unambiguous. The traverse gear was designed with minimum drag and to be as rigid as possible to prevent vibration being transmitted to the yawmeter. The position of the probe relative to the rear of the pile could be changed at low slack water using the traverse gear; during the test runs the probe was kept stationary. The optimum position of the probe is shown in Fig. 6. The flow velocity was monitored continuously at six depths ahead of the test piles using current flowmeters.

The signal from the differential pressure transducer mounted on the probe was recorded on magnetic tape. The recordings were replayed through a digital computer and their spectra analysed. Since the spectra were fairly narrow band and only the predominant frequencies (and not the quantitative value of the power spectral density) were required, short sample length could be used. By careful selection these samples were chosen to cover times when the flow velocity had been reasonably constant.

The Results

Prior to the observation of the Immingham piles almost the only form of vortex excitation that had been studied was that in which the motion occurs across the flow direction when the frequency of the shedding of pairs of vortices coincides with the natural frequency of the pile. If the Strouhal number is 0.2, then the peak amplitude occurs approximately when $V = 5\ ND$. However, excitation also occurs in the direction of the flow when the shedding of individual vortices coincides with natural frequency of the pile. The flow speed for this motion is half that of the cross flow response, i.e. at $V = 2.5\ ND$. Because the cross flow component of the force is much greater than that in line with the flow, the cross flow amplitude is much greater. For structures in air, the in-line force is generally insufficient to cause significant motion of a structure and this is why this type of excitation has been little studied in the past. The greater density of water, however, makes the forces large enough for in-line response to occur, and to be a very serious problem.

Although the cross flow response did occur on a few of the piles during the construction of the jetty at Immingham, the experimental work concentrated on the in-line motion since this is produced at a lower flow speed and, for reasons explained in Reference 1, is more likely to occur on completed structures. The response of the 457 mm pile is shown in Fig. 2. The results are plotted in the form of non-dimensional amplitude against reduced flow velocity, V/ND. The peak at V/ND = 2.5 occurs as expected but, additionally and of greater importance to the civil engineering designer, there is a peak at lower flow speeds. The response shown in Fig. 2 was obtained as the tidal level fell from high water to mean water and the flow speed increased; a similar curve was obtained as the flow speed decreased and the level fell to slack low water. The response over a complete tidal cycle (for the case where the capping beam is at an angle to the flow) is shown in Fig. 3.

A further unexpected fact which emerged was that the motion was not completely in-line with the flow. This can be seen in Fig. 4 which shows typical motion in plan of the pile. For example, with the capping beam at right angles to the flow, the motion of the 610 mm pile between V/ND = 1.20 (where the movement commenced) and 1.50 was a Lissajous figure with a frequency ratio in the orthogonal directions associated with the beam of 5:4 (Fig.4a). At V/ND = 1.50, this became pure in-line motion commensurate with a small hump in the response curve as shown in Fig. 2.

The frequency of vortex shedding behind the stationary 762 mm pile (Fig.5) shows that there is no discernable Reynolds number effect over the range studied, a result that is significantly different to those of previous studies behind stationary cylinders over the critical Reynolds-number regime. The Immingham results are most likely due to the very rough surface finish of the piles, a similar effect of surface roughness is reported in Reference 5. Equally important was the fact that the vortex shedding was strongly periodic throughout the critical Reynolds number range. The spectra of the wake behind the stationary 762 mm pile showed a strong peak at $nD/V \simeq 0.22$ (Fig.7 is a typical spectrum).

However, it was the spectra obtained beyond the pile when it was responding to the flow that presented the most interesting results (Fig.8). At low values of V/ND (i.e., those below about 1.7) the frequency of oscillation of the piles was three times that of the predominant shedding frequency, not twice as had previously been expected. It was oscillation of this type that was responsible for most of the problems that occurred during the construction of the jetty at Immingham and would cause similar problems at any similar future jetty. The two curves shown in Fig. 8 refer to different values of reduced velocity. It is significant that the basic Strouhal frequency of about 0.22 is unaffected by the value of V/ND whereas the reduced frequency of the pile depends directly on the flow speed.

Visual observation were also made of the vortex shedding at the water surface. Under certain conditions, when the pile was oscillating, vortices could be shed simultaneously in pairs from the two sides of the pile. Sometimes there was a sequential mix of simultaneous and alternate shedding and under other conditions the two types would occur purely. No simultaneous shedding was observed from a stationary cylinder. Attempts to photograph the simultaneous shedding generally were not very successful but one result is shown in Fig. 9. It was observed that the symmetrical shedding always formed on alternate street well downstream of the oscillating pile.

Because this photograph is rather unsatisfactory, a clearer indication of symmetrical shedding is shown in Fig. 10 for a model pile. It is not intended to discuss this model work in any detail in this paper; it will suffice to say oscillations in-line with the flow were observed with pairs of vortices being shed symmetrically at the frequency of oscillation.

The Cause of the Oscillations

The peak in the response at V/ND = 2.5 (Fig.2) fitted well with the explanation of a single vortex being shed every cycle of oscillating of the pile. This motion is shown schematically in Fig. 12d. It is clear that if this is the correct explanation for this part of the response, then the peak at the lower speed must be due to some other cause. What follows is a tentative attempt to explain the kind of mechanism causing these oscillations; a full account appears in Ref. [3].

The shedding from a stationary pile is regularly periodic and at a Strouhal number of 0.2. This produces a fluctuating force that has a predominant component across the flow direction with a reduced frequency of $nD/V = 0.2$ and a much smaller component in line with the flow at a reduced frequency of 0.4. If the flow speed is very low, neither of these are close to the pile frequency and there is no appreciable response.

The problem of developing a rational explanation for the low speed peak is shown clearly by considering the possibility of the single vortices being shed every one and a half cycles of oscillation (Fig.12a). This situation clearly satisfies the peaks of $nD/V = 0.2$ and a pile frequency of $ND/V = 0.6$ (though not the vortex shedding peak at $nD/V = 0.4$) as shown in the spectra, Fig. 8. However, it can be seen that there is no energy input into the pile in the flow direction. These are several alternative explanations. First, it is possible that the alternative and symmetrical shedding occur in blocks, that is, a series of, for example, 50 cycles of symmetric shedding followed by a similar series of alternate shedding. The spectra cannot distinguish between this and the missed shedding shown in Fig. 12b, for example. This block shedding explanation suffers from two defects. If the force is experiencing long term changes, then the response would be expected to follow suit and this did not occur. Also there is no obvious reason why the type of shedding should change so radically in this way.

One pattern of vortex shedding that fits the spectra evidence completely can be hypothesised. This is shown in Fig. 12b. The period between 1,2 and 4,5 produces the strong peak in the spectra at a third of the pile frequency, i.e. at $nD/V = 0.2$. The period between vortices 1, 2 and 3 for example, produces the peak at $nD/V = 0.4$. Also, the shedding can easily form an alternative street downstream of the pile with a Strouhal number of 0.2. There is positive energy input from the flow to the pile, albeit somewhat small, and since the pattern repeats itself every six cycles of oscillation the force and response pattern are repetitive. There are, however, two points which make this hypothesis difficult to accept. First, visual observations always indicated that the symmetrical vortex shedding occurred every single cycle of oscillation of the pile. Secondly, it is difficult to understand why the vortex shedding should develop such a complex pattern as shown in Fig. 12b, when simpler alternatives are available.

An alternative possibility is that the development with increasing flow speed of the flow pattern round a flexible cylinder that is free to oscillate is similar to the flow at Reynolds numbers below a few hundred. Sequentially the development may be as follows. At low flow speeds, the pile does not respond to the alternate vortex shedding. At a certain value of V/ND, a small perturbation causes the pile to oscillate and vortices to be shed symmetrically from the cylinder. The wake instability, however, is relatively unaffected by whether the shedding is alternate or symmetric and still oscillates at a Strouhal number of 0.2 with the vortices passing along the wake. In other words, the wake instability is a separate phenomenon to the vortex shedding. This is shown schematically in Fig. 11. The pressure probe sensed both the shedding frequency at nD/V = 0.6 and the wake instability at nD/V = 0.2 with a subsidiary peak at nD/V = 0.4.

At a certain value of V/ND, the pattern of symmetric shedding broke down and the conventional alternate shedding occurred instead. The result of this breakdown is shown clearly in the response curve (e.g. at V/ND = 2.0 in. Fig.2). The peak of this type of motion is then at V/ND = 2.5. It might be considered possible that the flow speed for the onset of symmetric shedding and the change to alternate shedding are functions not of reduced flow speed but, instead, of some alternative parameter. One such parameter could be amplitude as suggested above, another could be based on the acceleration of the motion of the pile and on the gravitational parameter, V^2/gD. The evidence suggested that the last of these did not produce the required correlation of the results.

These results lead to an important conclusion. The Strouhal number of the wake remains almost constant, at about 0.23, though the shedding from the pile may occur symmetrically. The dominant feature of the flow is the stability of the wake far downstream of the pile. Whether or not the pile responds to the vortex-induced forces and, even, modifies the process of the shedding, the final wake layout must have a

Strouhal number of 0.23. It is interesting to compare this result with the theoretical work of Birkoff (reviewed by Marris [6]). His work shows that, for a two-dimensional vortex trail in a non-viscous fluid the ratio b/a of mean transverse spacing to mean longitudinal spacing is invariant, so that all mean spacing ratios are stable. It is the periodicity that is determined by stability criteria. An excellent example of this is in the flow behind the oscillating piles. The wake directly downstream of the piles does not satisfy the Von Kármán spacing stability criteria. Well downstream of the pile it does, however, satisfy the concept of the universal Strouhal number; that is a Strouhal number based on wake width not the pile diameter. On a few exceptionally calm days it was possible to observe the progress of the vortices on the water surface. It was noticeable that the vortices that were shed symmetrically remained in that pattern for a large number of pile diameters and only gradually formed an alternate shedding in the wake.

Conclusions

The shedding of vortices from a stationary pile was found to be regularly periodic up to the highest Reynolds numbers studied ($Re \simeq 1.5 \times 10^6$) with a shedding frequency of $nD/V = 0.22$. As the flow speed increased from zero, the piles started to oscillate when the shedding frequency was close to 1/3 of their natural frequency. Symmetrical shedding could be observed from behind the pile when it was responding in the flow direction under these conditions. At a reduced velocity of about 2.0, the cause of the response changed radically with vortices being shed alternately from the sides of the cylinder. At much higher flow speed still, the well known cross flow response occurred, though this is not discussed in this paper.

Acknowledgements

This investigation was initiated by the British Transport Docks Board and was carried out jointly by their Research Station, the National Physical Laboratory, John Mowlem and Co. Ltd. and Atkins Research & Development. It was partly financed through special contributions to the Construction Industry Research and Information Association. This paper is published by permission of the Director of CIRIA.

Nomenclature

D pile diameter
g acceleration due to gravity
N natural frequency of pile
n frequency of shedding of pairs of vortices
Re Reynolds number
V flow speed
η reduced amplitude; displacement divided by D

References

1. Sainsbury, R.N., King, D.: The Flow Induced Oscillations of Marine Structures. Proc. Inst. Civ. Engrs. 49, 269, 302 (July 1971).

2. Wootton, L.R., Warner, M.H., Sainsbury, R.N., Cooper, D.H.: Oscillations of Piles in Marine Structures. A Resume of the full-scale experiments at Immingham CIRIA Technical Rep. 41, April 1972.

3. Wootton, L.R., Warner, M.H., Sainsbury, R.N., Cooper, D.H.: Oscillations of Piles in Marine Structures. A Report of the full-scale experiments at Immingham. CIRIA Technical Note 40, April 1972.

4. Wootton, L.R., Warner, M.H.: Reducing Oscillations of Bodies in Flowing Fluids. British Patent No. 55904/70, November 1970.

5. Wootton, L.R.: The Oscillations of Large Circular Stacks in Wind. Proc. Inst. Civ. Engrs. 43, 573-598 (August 1969).

6. Marris, A.W.: A Review on Vortex Streets, Periodic Wakes and Induced Vibration Phenomena. J. Basic Eng. Trans. ASME 185-196 (June 1964).

Fig.1a. The experimental test rig at Immingham.

Fig.1b. The yawmeter probe and traverse gear.

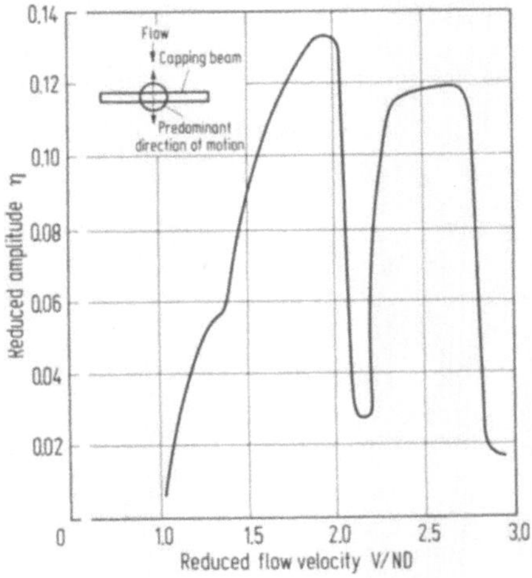

Fig.2. The response of the 457 mm diameter pile in the flow direction.

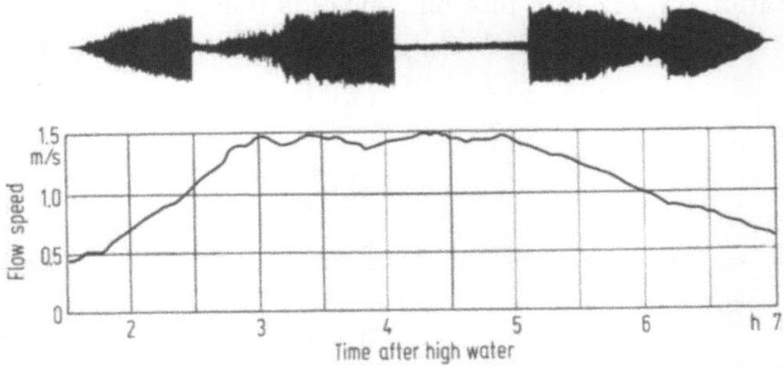

Fig.3. The response of the 457 mm pile over a complete tidal cycle: capping beam at 57° to the flow.

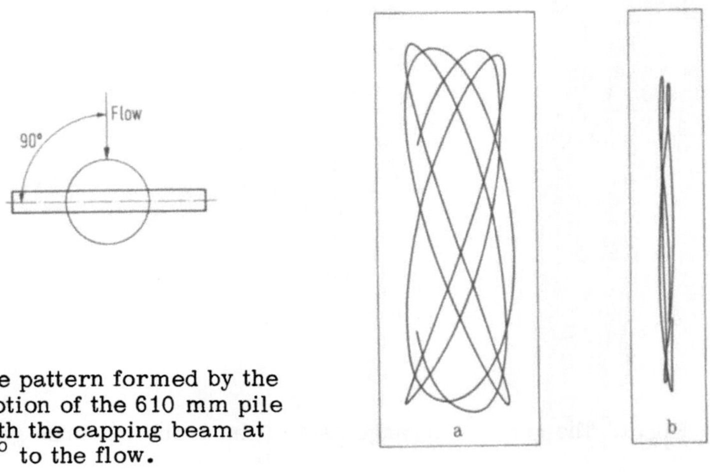

Fig.4. The pattern formed by the motion of the 610 mm pile with the capping beam at 90° to the flow.

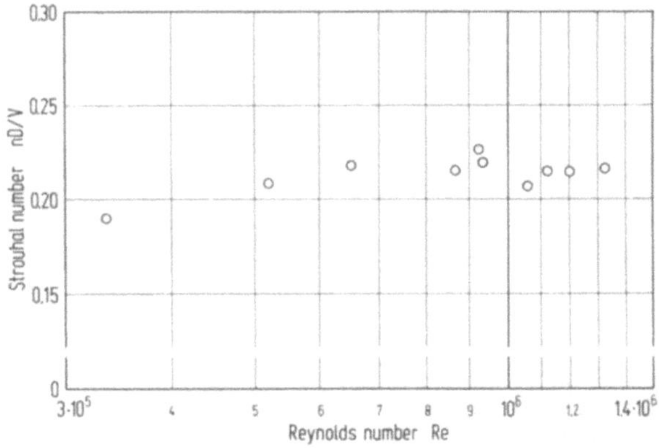

Fig.5. The dependence of the frequency of vortex shedding from the stationary 762 mm pile on Reynolds number.

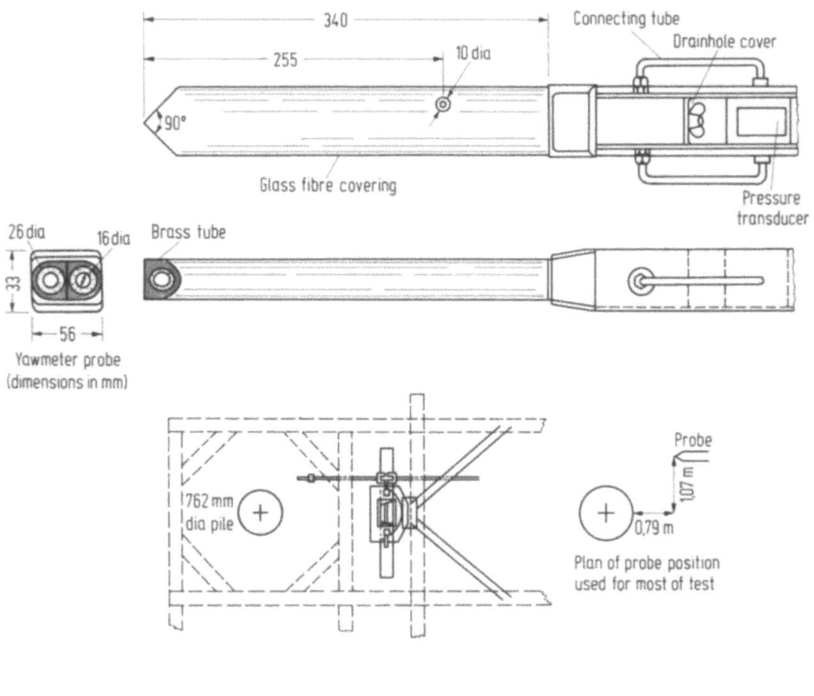

Fig.6. The general arrangement of the probe traverse gear.

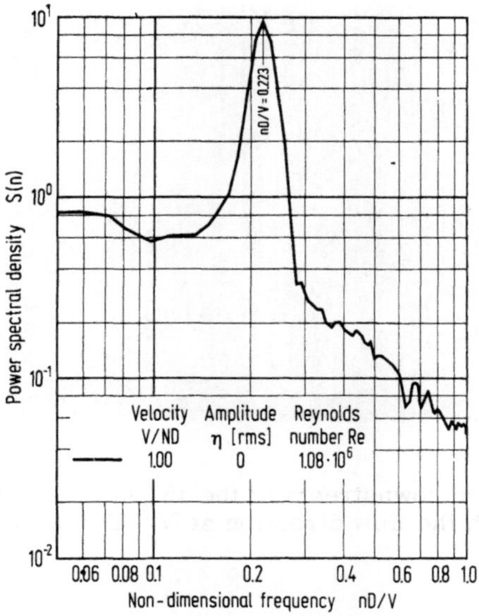

Fig.7. The power spectrum of the velocity fluctuations behind the stationary 762 mm pile.

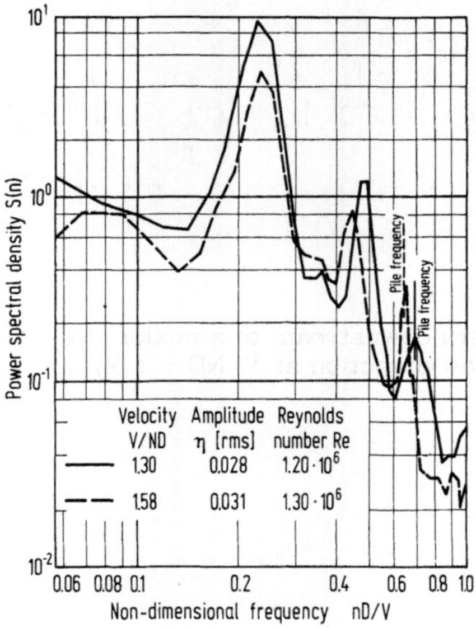

Fig.8. The power spectra of the velocity fluctuations in the wake behind the 762 mm pile.

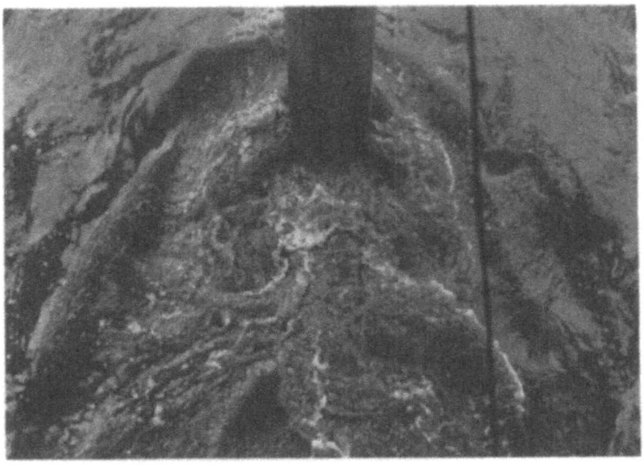

Fig.9. Flow pattern downstream of the 457 mm pile oscillating in-line with the flow direction at V/ND = 1.4.

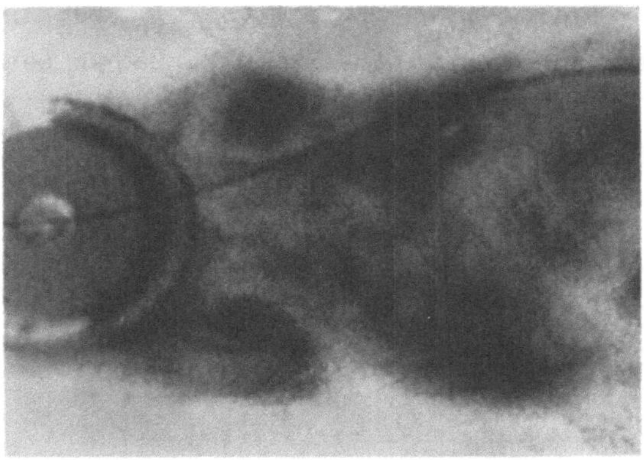

Fig.10. Flow pattern downstream of a model pile oscillating in-line with the flow direction at V/ND = 1.4.

Fig.11. Schematic diagram of one possible type of shedding, in which the rate of symmetrical shedding is not directly related to the wake instability.

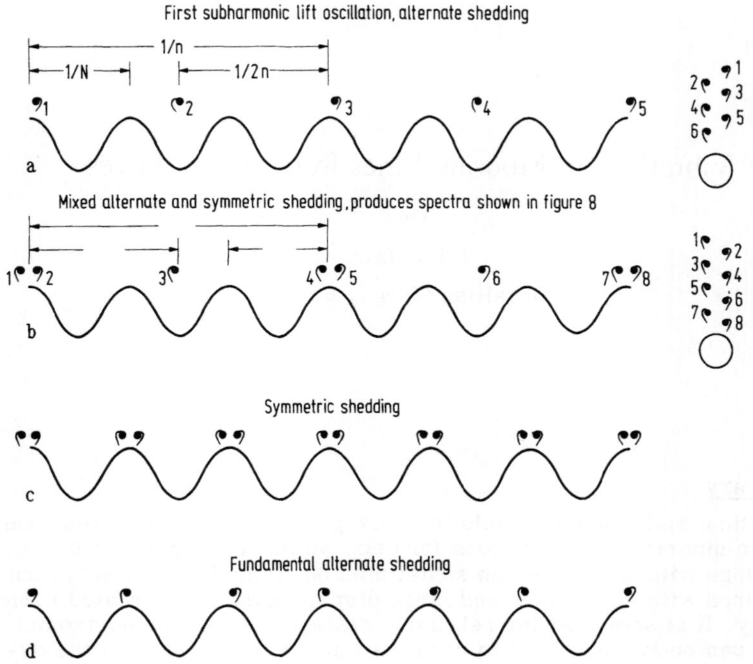

Fig.12.

(a) The pattern of shedding such that n = N/3. Note that there is no energy input in-line with the flow. V/ND = 1.67.

(b) The pattern of shedding such that n = N/3 and there is positive energy input to the pile V/ND = 1.67.

(c) The shedding frequency is controlled by the motion. The vortices form an alternate street downstream.

(d) The pattern such that n = N/2. This is the fundamental drag motion. V/ND = 2.5.

Vibrations in Mooring Lines from Ocean Waves

By

J.H. Nath

Corvallis, Oreg., U.S.A.

Summary

Analytical and numerical solutions are presented for some relatively simple mooring configurations for deep water oceanographic buoys. Moorings with taut lines can sometimes be treated analytically, but moorings with slack lines and large displacements are treated numerically. It is seen that for relatively slack moorings the horizontal loads can be large enough to create a taut condition and that the dynamic response then increases while keeping the dynamic loading constant. A dimensionless means of presentation is developed so that the dynamic responses of moorings with different geometry can be directly compared.

Introduction

The mooring of buoys, drill rigs, ships and other objects at sea presents a complex design problem for ocean engineers. Such moorings can be classified according to the number and type of mooring lines and position of the object moored. Each class of mooring has distinctive problems and this paper is concerned with single line moorings for oceanographic research buoys in water depths that range from 1,000 to 10,000 meters.

The moorings in question are sometimes called deep water single point moorings. The terminology refers to the single anchor and should not be confused with the single point large-buoy moorings used in shallow water to unload tankers, which have several anchors but only one point for attachment of the ship.

The loads on the deep water system are generated from gravity, wind, current, waves and the geometry of the system. The response of the system (line tensions and motions, etc.) is determined from the geometry and the constitutive relationship for the line. This paper is mostly concerned with the mooring line response when the system is excited by ocean waves. The waves act on the buoy and on the extreme upper portion of the line. For very long lines it is often sufficient to consider the wave action on the buoy alone.

The wind and current also play important roles in determining the magnitudes of dynamic responses. The wind is usually gusty, which creates a time varying force on the buoy and the current can create the shedding of vortices from the line which may interact with the mooring line motion to create a classical hydro-elastic response motion. It will be seen that the wind and current will have another influence on the dynamic motion of the mooring line. If the horizontal drag forces on the buoy and mooring line are strong, even a slack mooring will be pulled taut and the influence of waves on the dynamic line tensions, for example, will be more pronounced.

Before proceeding, a few definitions should be made. Scope is a dimensionless term used to designate the unstressed line length divided by the water depth. A taut mooring is usually one where the scope is less than one. For such a mooring the line is very elastic and is usually nylon. However, some moorings which have scopes greater than one, but still small, will behave similarly to taut moorings if the horizontal forces from wind and current are sufficiently large. Thus, in this paper the word taut will refer to any mooring where the dynamic response can be determined from tight line concepts. A slack mooring is one where a considerable sag exists in the mooring line and taut mooring concepts cannot predict the dynamic line action. Presently no clear-cut way exists to define the dividing region between taut and slack moorings but the topic remains a point for research.

The major purpose for the research that made the base for this paper was to at least estimate the manner in which horizontal loads influence the dynamic action of the mooring line from waves. Thus a few representative scopes were selected with two different wind and current conditions for some typical deep ocean moorings. A major problem was how to represent the results so that only the dynamic action was compared in a way that was not unduly influenced by the fact that different line scopes required different diameters, because it is usually required to maintain stresses from the constant loads at a given percentage of the breaking strength. A dimensionless transfer function was developed that allowed the various different moorings to be directly compared.

Some Dynamics of Taut Moorings

It is useful at first to consider some closed solutions to simple taut moorings such as the one shown in Fig. 1. Consider that the only forces acting on the system are from the wave action, which will be assumed to move the buoy in the vertical direction only; from the prestress in the line; and from gravity, which creates the buoyancy force acting on the buoy. The line is assumed to be neutrally buoyant in order to shorten the equations and illustrate only the important points. It is a simple matter to add the submerged weight of the line, and nylon, for that matter, is nearly neutrally buoyant. Let the displacement of the line in the z direction be denoted as u. Partial differentiation is noted by subscript. Thus the strain is $u_z(z,t)$. At first consider all damping to be negligible and that the modulus of elasticity, E, is constant. The depth is so large that L denotes both the water depth and the stressed length of the line. The horizontal wind and current will displace the buoy horizontally and add curvature to the line. However, for initial simplicity it will be assumed that the line is mostly straight and that the buoy moves primarily in the z direction. The motion of the line is governed by the wave equation,

$$u_{tt} = C^2 u_{zz}, \tag{1}$$

where C is the celerity of an elastic wave in the line, $(E/\rho)^{\frac{1}{2}}$, with ρ the saturated volumetric mass density. Smith and Nath [11] show that the upper boundary condition is determined by the equation of motion of the buoy,

$$M u_{tt}(L,t) + EA\, u_z(L,t) + K u(L,t) = K a_0 \cos \omega t \,, \tag{2}$$

where M is the effective mass of the buoy (including hydrodynamic added mass), A is the cross-sectional area of the line and K is the effective spring constant, or restoring force, representing the buoyancy on the buoy. The lower boundary condition is

$$u_z(0,t) = 0 \,. \tag{3}$$

By solving the boundary value problem with the right side of Eq. (2) set to zero, the undamped modal frequencies of the system can be readily obtained. The resulting equation is:

$$\beta \cos \beta + (k - m\beta^2) \sin \beta = 0 \,, \tag{4}$$

where $\beta = \omega_n L/C$ (n = 1,2,..., the modal indicator), $k = KL/EA$ (the ratio of the hydrostatic restoring force at the buoy to the effective spring constant of the line) and $m = M/\rho LA$ (the ratio of the effective buoy mass to the total line mass). It is seen that the modal frequencies depend on k and m. However, some buoy systems have a very large k and m/k ratio. For example, the 12.2 m diameter ONR/General Dynamics Ocean Data Station, moored in 6000 m of water with 38 mm diameter plaited nylon rope, would have a k value of about 10,000 to 13,500 or more. The m value would be about 20. Thus by dividing through by k and letting it get very large, it is seen that Eq. (4) reduces to

$$\sin \beta = 0 \tag{5}$$

and the modal frequencies will be

$$\omega_n = \frac{n\pi C}{L} \qquad n = 1,2,\ldots, \tag{6}$$

where the value n = 0 is ignored as not useful.

The solution for line displacement is:

$$u(z,t) = \frac{k a_0 \sin \frac{\omega z}{C} \cos \omega t}{\alpha \cos \alpha + (k - m\alpha^2) \sin \alpha}, \qquad (7)$$

where $\alpha = \omega L/C$.

The solution for line tensions can be readily obtained from

$$T = AEu_z. \qquad (8)$$

For the example of the large buoy cited above with the 38 mm nylon line the first mode frequency would be about 0.055 Hz. Thus the danger of resonance with the ocean waves of one or more modes of vibration cannot be avoided for many deep ocean moorings. Therefore, it is important to consider damping when one wants to compute the dynamic tensions at the modal frequencies. That is, Eqs. (7) and (8) are adequate only when the geometry is such that resonance cannot occur with ocean wave frequencies. However, Eqs. (4) and (6) are very useful for estimating modal frequencies if damping is not close to the critical damping.

Equations similar to Eqs. (1) through (8) can be developed for the transverse modes of vibration, which are important for the more inelastic types of lines like steel wire rope and chain in less water depth. The submerged weights of such lines are much too great for deep ocean moorings such that they cannot support their own weight.

When damping is considered, a three parameter model as proposed by Reid [10] and utilized in Smith and Nath [12] will allow deformation to occur for high rates of strain. A schematic of the model is shown in Fig. 2. It is far from a perfect model, but it is linear and is possibly better than the usual two-parameter model which excludes E_1. Longitudinal damping is due to structural damping or energy dissipation from rubbing together of the individual rope fibers. In addition, the hydrodynamic drag force in the longitudinal direction can be approximated as linear and included in parallel with the shown model. Test information is sorely needed in order to obtain the proper damping coefficients for submerged lines.

For a taut mooring as indicated in Fig. 1 with damping as indicated in Fig. 2 the constitutive relationship becomes

$$\sigma + \tau\sigma_t = E_0(u_z + \lambda u_{zt}), \tag{9}$$

where σ is stress, $\tau = q/E_1$ and $\lambda = q(E_0 + E_1)/E_0E_1$.

The damping coefficient, q and the supplemental modulus of elasticity, E_1, are related to both the area of the hysteresis loop, which is obtained from testing, and the dynamic modulus of elasticity. A summary of the complete relationships is given in Smith and Nath [12]. For this type of damping one should not investigate the influence of changes in the damping coefficient, q, alone, but instead should consider changes in τ.

The solution for the line displacement becomes

$$u(z,t) = R\left[\frac{(1 + i\gamma)\sin\left(p\alpha\frac{z}{L}\right)a_0 e^{i\omega t}}{\frac{\alpha}{kp}\cos p\alpha + \left(1 - \frac{m\alpha^2}{k} + i\gamma\right)\sin p\alpha}\right], \tag{10}$$

where $\gamma = Q\omega/K$ with Q the effective linear hydrodynamic damping on the buoy and $p = \pm[(1 + i\omega\tau)/(1 + i\omega\lambda)]^{1/2}$. For many, if not most, moorings, the damping at the buoy with respect to other types of damping is small and can be neglected. When Eq. (10) is expanded, a lengthy solution results, which is reviewed in Smith and Nath [11].

Some Dynamics of Slack Moorings

Some analytical solutions exist for slack moorings, or catenaries for special cases. Examples are those of Davenport [1] and Smith and Thompson [13]. Davenport's solution depends on small motion, and the assumption that the cable sag takes on a parabolic shape. The solution was in terms of the "guy modulus" or the effective spring constant type restoring force on a tower offered by a heavy guy. Although line elasticity was considered for determining the guy modulus, only the transverse vibrations were considered in determining the frequency response, because longitudinal frequencies were considerably out of the range of excitation frequencies. The second citation depends on a catenary chain being inextensible.

However, due to recent emphasis on designing competent deep ocean moorings using synthetic lines with considerable sag and which are very elastic, numerical solutions have been developed. One is explained by Nath [5] which utilizes a derivation by Langer [4] and the method of characteristics, to solve, by numerical means, the partial differential equations that arise from considering a small element of the line in motion for the two dimensional mooring as shown in Fig. 3. Four equations are developed in four unknowns from applying concepts of the conservation of mass, conservation of momentum and continuity of the line filament. A two parameter damping model was assumed for the internal line damping where E_1 is zero in Fig. 2. Non-linear hydrodynamic damping was included as well as a non-linear modulus of elasticity, E_0. The upper boundary condition was established as the motion of the attachment point at the buoy. The motion of the buoy was also modeled numerically as a rigid body in motion, acted upon by gravity, wind, waves, surface current and mooring line tension. For the line, the static position due to steady wind and current was first computed and then the dynamic conditions due to sinusoidal surface waves were superimposed. Thus a very slack mooring could be investigated with a fairly high degree of accuracy. A brief description of the numerical solution follows.

The four dependent unknown variables for the line motion are the axial velocity, v_a, the normal velocity, v_n, the line tension, T, and the angle, θ, from the horizontal to the line tangent. The independent variables are time, t, and the position on the line, s. The solution for the dependent variables is given by,

$$\begin{bmatrix} 1 & 0 & -v_n^{(1)} & -\frac{1}{C_a^{(1)}pa} \\ 1 & 0 & -v_n^{(1)} & +\frac{1}{C_a^{(1)}pa} \\ 0 & 1 & \left(v_a^{(1)} - c_n^{(1)}\right) & 0 \\ 0 & 1 & \left(v_a^{(1)} - c_n^{(1)}\right) & 0 \end{bmatrix} \begin{Bmatrix} v_a^{(2)} \\ v_n^{(2)} \\ \theta^{(2)} \\ T^{(2)} \end{Bmatrix} = \begin{Bmatrix} f_1 \\ f_2 \\ f_3 \\ f_4 \end{Bmatrix} \quad \begin{matrix} \text{on } \frac{ds}{dt} = \\ +\left(\frac{E}{p}\right)^{\frac{1}{2}} \\ -\left(\frac{E}{p}\right)^{\frac{1}{2}} \\ +\left(\frac{T}{M}\right)^{\frac{1}{2}} \\ -\left(\frac{T}{M}\right)^{\frac{1}{2}} \end{matrix} \tag{11}$$

The right hand side of Eq. (11) describes the forcing functions, which include the hydrodynamic loads, internal damping, and some coupling terms from the dependent variables which are in terms of the values at a previous time increment. As noted, the solution is in the time domain and the superscript (1) refers to a value at the previous time step and (2) refers to a current time step. The solution proceeds along the characteristic curves represented by ds/dt.

The advantages of the numerical solution are that mooring lines with all degrees of "slackness" can be analyzed; the coupling between axial and normal motions due to line curvature is included; some non-linear aspects are considered; and added utility is gained from having the numerical program in the time domain because solutions for transient motions are obtainable. Some disadvantages are that sometimes numerical instabilities occur from the damping terms, requiring small time increments, and some problems become fairly expensive to run. However, usually such problems have no other possibilities for solution.

Yet to be accomplished with the above program is the consideration of motion in three dimensions so that problems such as vortex shedding and the consequent strumming of the line can be investigated. Nath [6] shows that results from the numerical program give a good check with analytical results for taut line buoy moorings.

Some Comparative Results of Analyses and Tests

In 1969 the United States Coast Guard conducted on in-situ test of mooring lines on Aid-to-navigation buoys and one large discus buoy of 12.2 m diameter. The testing was done in 1100 meters of water near San Clemente Island, California and is reviewed in General Dynamics [3]. Buoy motions and line tensions were measured as well as wind and current velocities. Waves were measured by means of the motion of the discus buoy, which is a very close surface follower, as shown in Gaul and Brown [2]. For some comparisons, field conditions with fairly sinusoidal waves were selected. These conditions were reproduced with the numerical program. An example of results of mooring line tensions measured at the discus buoy and those computed with the numerical program are shown in Fig. 4. In order to compare the dynamic loads closely, the mean tension for the measurement was displaced slightly from the computed value.

The purpose of the work reported in Nath [6] was to investigate, with the numerical model, the influence of line scope, wind and current magnitudes on the dynamic action of the line. Two environmental conditions were investigated - one had a wind of 24.8 m/sec with a surface current of 0.77 m/sec, and the other had a wind of 74.4 m/sec with a surface current of 2.21 m/sec. Two scopes were investigated - one of 0.87 and the other of 1.18. (The larger scope yielded an "in-place", or "stretched" scope of about 1.5 for the high wind.) A non-dimensional representation of results was needed. The method was to divide the mooring line tension by a characteristic mass of the line times the acceleration of the buoy. Thus the dimensionaless tension, T', became

$$T' = \frac{T}{\rho A L H \omega^2}, \qquad (12)$$

The dimensionless frequency was developed as the wave frequency divided by the frequency of the first mode for the line, from Eq. (6). Note that since n = 1,2,... in Eq. (6), that by using the first mode frequency for non-dimensionalizing, the higher modes are represented on the frequency response curve by 2,3,... .

Some results of the study are shown in Fig. (5). More points for plotting were desired but were not obtained because of a shortage of funds. The figure clearly shows the influence from the modal shapes and that much larger relative dynamic tensions occur for the smaller scopes and/or higher wind magnitudes. One disadvantage of this representation is that for high frequencies the ordinates are forced to being small. The first mode responses were not obtained because the wave periods were so large as to create large computer running times and cost.

A frequency response curve was developed by the General Dynamics Corporation for six moorings in the Caribbean Sea. All the moorings had a scope of 1.3 which produced a "stretched" scope of about 1.7. The water depths ranged from 2740 m to 3660 m. The mooring line tension at the buoy are shown in Fig. 6. The figure shows that all computed points fall on the same smooth curve, thus for design work one really only needs to compute the peak values. Which modes will be excited depends on the line dimensions and types and the full range of the expected wave spectrum.

In the test program conducted by the U.S. Coast Guard/General Dynamics Corporation, some long records were obtained of line tension at the discus buoy and of the waves. A spectrum analysis was made and an experimental transfer function was obtained (T'). The wave spectrum (actually vertical motion spectrum of the discus buoy) was obtained both optically and by means of an accelerometer record. These transfer functions were compared with computed transfer functions using the numerical model at discrete wave frequencies. One result is shown in Fig. 7. It is seen that for this particular case, the mooring line had a greater degree of damping than that illustrated with the model, at low frequencies. The figure shows that different wave heights were used in the computations with a fairly linear response.

For conditions where the relative spring constant, k, is not large, the first modal frequency can possibly be computed from Eq. (4) and used as the non-dimensionalizing frequency. However, in such a case the higher modes will not necessarily occur near $\omega/\omega_1 = 2,3,\ldots$. Similar conditions will exist for very slack lines.

Conclusions

The analytical solutions presented can be expected to yield good results for engineering design purposes providing the mooring can be considered to be taut. Of course, steady state loads from horizontal wind and current must be added to the analysis. Experimental data is needed on the damping characteristics for longitudinal motion of submerged ropes.

For semi-slack moorings a convenient non-dimensionalizing parameter is expressed as ω_1 from Eq. (6). For very slack lines, some other procedure will need to be developed.

The numerical model yields good results for engineering purposes. If expanded to include the third dimension, it could possibly be used for vortex shedding and strumming problems.

Future work will be devoted to classifying the designs for various types of moorings. Of particular interest is being able to distinguish between taut conditions and slack conditions when the scope is greater than one.

References

1. Davenport, A.G., Steels, G.N.: Dynamic Behavior of Massive Guy Cables. J. Structural Div. ASCE (April 1965).

2. Gaul, R.D., Brown, N.L.: A Comparison of Wave Measurements from a Free-Floating Wave Meter and the Monster Buoy. Transations of the Second International Buoy Technology Symposium, MTS, September 1967.

3. General Dynamics Corporation Convair Division: Sea Lanes Test Program Final Report. Report No. GDC-AAX70-020 for the U.S. Coast Guard, National Data Buoy Development Project Office, October 1970.

4. Langer, R.M.: The Catenary in Space - Free Motions of Flexible Lines. J.R.M. Bege Company, Arlington, Mass, or AD 611429, Defense Documentations Center, December 1964.

5. Nath, J.H.: Dynamics of Single Point Ocean Moorings of a Buoy - A Numerical Model for Solution by Computer. Progress Report for the Office of Naval Research, U.S.A., Oregon State University, Ref. 69-10, July 1969.

6. Nath, J.H.: Analysis of Deep Water Single Point Moorings. Technical Report for the Office of Naval Research, U.S.A., Colorado State University, CER 70-71 JHN4, August 1970.

7. Nath, J.H., Felix, M.D.: Dynamics of Single Point Mooring in Deep Water. J. of the Waterways, Harbors, and Coastal Engineering Division of ASCE, (November 1970).

8. Nath, J.H.: Dynamic Response of Taut Lines for Buoys. MTS J. (July/August 1971).

9. Peller, R.C.: Summary of EEP Dynamic Moor Analyses. Internal Rep., General Dynamics Corp., DF-Buoy-42, September 8, 1971.

10. Reid, R.O.: Dynamics of Deep Sea Mooring Lines. A and M Project 204, Reference 68-11F, Department of Oceanography, Texas A and M University, Texas, July 1968.

11. Smith, C.E., Nath. J.H.: Some dynamics of Taut-Moored Buoy and Systems. Submitted to MTS Conference in Washington, D.C., September 1972.

12. Smith, C.E., Nath. J.H.: Parameters Affecting the Natural Frequencies of Buoy/Taut Line Systems. Submitted to MTS Journal.

13. Smith, C.R., Thompson, R.S.: On the Small Oscillations of Suspended Chain. Submitted to ASME, J. Appl. Mech.

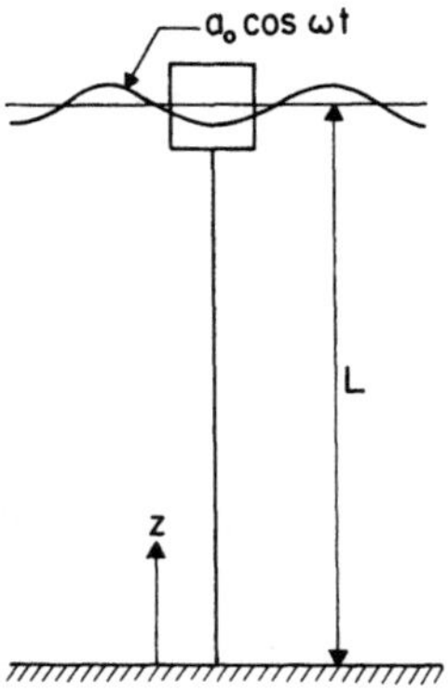

Fig.1. Taut line buoy mooring.

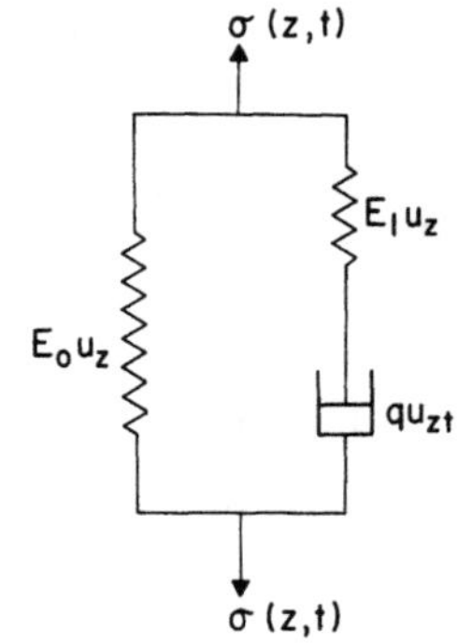

Fig.2. Three parameter constitutive model.

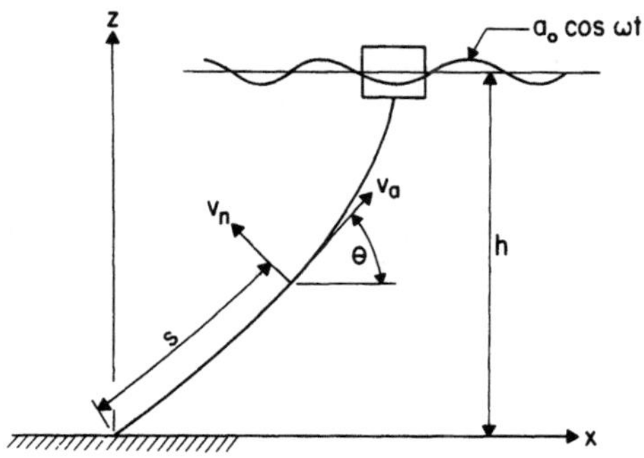

Fig.3. Slack line buoy mooring.

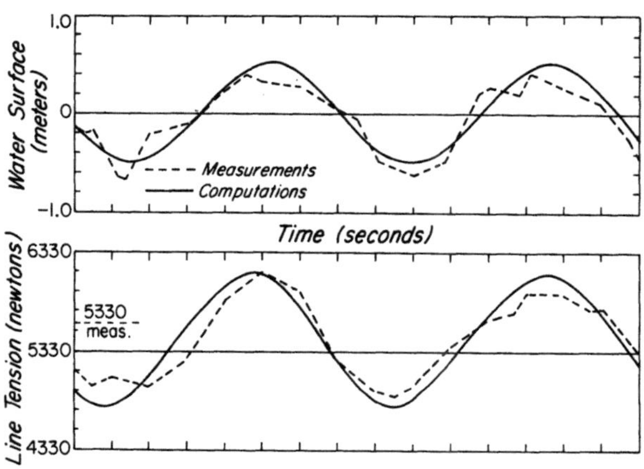

Fig.4. Waves and line tensions, USCG test.

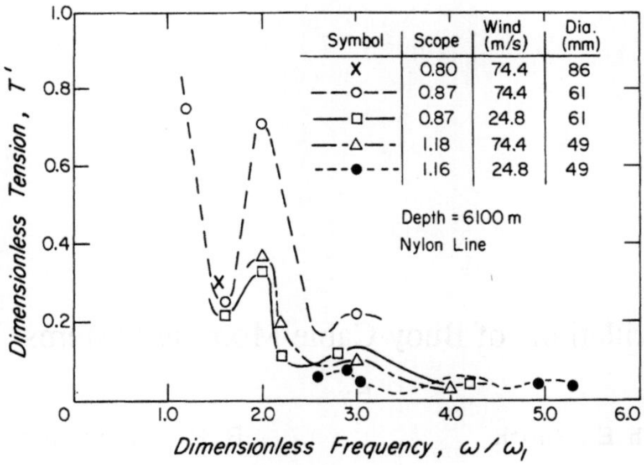

Fig.5. Line tension at the anchor.

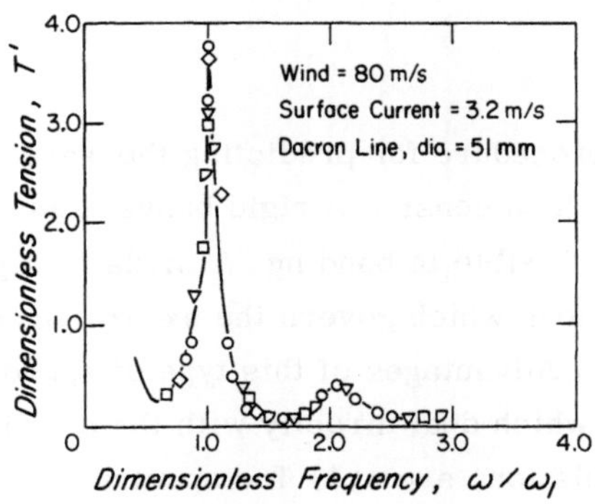

Fig.6. Line tension at the buoy.

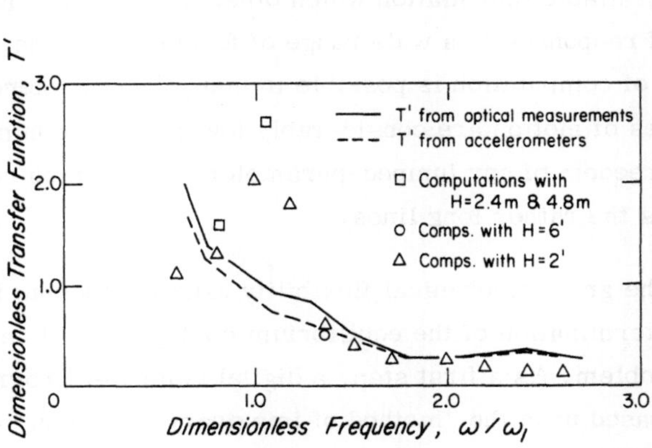

Fig.7. Line tensions at the buoy. Sea lanes test program.

Oscillations of Buoy-Cable Mooring Systems

By

Ch.E. Smith
Corvallis, Oreg. U.S.A.

R.F. Dominguez
College Station, Tex. U.S.A.

Outlined here is a procedure for predicting the wave-excited motions in ocean moorings which consist of rigid buoys interconnected by lines which are perfectly flexible in bending. A modal analysis is made of the linearized equations which govern the excursions about an equilibrium configuration. Advantages of this type of approach, over integration procedures which deal directly with the coordinates introduced for the initial formulation, are: (1) The spectrum of natural frequencies of vibration and associated damping decay rates is obtained, making readily available information which otherwise would require consideration of responses to a wide range of forcing functions, and (2) An economy of computation is possible in many cases, where the important modes of motion are considerably fewer than the number of degrees of freedom of any lumped-parameter model which reasonably approximates the rather long lines.

Because of the great mechanical flexibility usually inherent in such systems, determination of the equilibrium configuration is a highly nonlinear problem. As a first step, a digital computer program was developed, based upon the "method of imaginary reactions" as described in Ref. [1], for determining the configuration resulting from an arbitrary set of forces at the selected mass points. In addition to determining the equilibrium configuration, the procedure is also used to evaluate the matrix of flexibility influence coefficients for small deviations about the equilibrium configuration.

The dynamic behavior is described by a set of equations of the form

$$[a][m]\{\ddot{q}\} + [a][c]\{\dot{q}\} + \{q\} = [a]\{f(t)\}. \qquad (1)$$

Here, [a] is the square matrix of flexibility influence coefficients mentioned above, $\{q\}$ is a column matrix of displacement coordinates, [m] and [c] are matrices of inertia and viscous damping parameters, respectively. The column $\{f\}$ represents the set of unsteady force components acting upon the various points in the system; these may arise as variations of buoyant force acting upon a surface buoy or through fluctuations in drag force at submerged points.

The unforced system ($\{f\} = \{0\}$) admits "natural" motions represented by

$$\{q(t)\} = \{\Phi\} e^{\lambda t}. \qquad (2)$$

The characteristic roots λ_r ($r = 1,2,\ldots$) and associated amplitude configurations $\{\Phi\}_r$ are the eigenvalues and associated eigenvectors satisfying

$$[\lambda^2[a][m] + \lambda[a][c] + [1]]\{\Phi\} = \{0\}. \qquad (3)$$

In the analysis being described here, these eigenvalues and eigenvectors are evaluated by means of an iterative procedure outlined in Ref. [2]. A real λ and corresponding real $\{\Phi\}$ represent a natural motion which is readily interpreted from Eq. (2). The natural motion represented by a complex conjugate pair of roots,

$$\lambda_r = \alpha_r + i\omega_r; \quad \lambda_r^* = \alpha_r - i\omega_r, \qquad (4)$$

and members of the associated modal columns,

$$\Phi_{rj} = |\Phi_{rj}| e^{i\theta_{rj}}; \quad \Phi_{rj}^* = |\Phi_{rj}| e^{-i\theta_{rj}}, \qquad (5)$$

takes the form

$$q_{rj} = |\Phi_{rj}| e^{\alpha_r t} \cos(\omega_r t + \theta_{rj}) . \tag{6}$$

Here, j denotes the typical member of the column $\{q\}$ or $\{\Phi\}$.

The modal analysis of forced motion centers around a set of normal coordinates $\xi_r(t)$, which measure the components of the displacement vector in the directions defined by the modal columns; this is expressed by the transformation

$$\{q(t)\} = \sum_r \xi_r(t) \{\Phi\}_r . \tag{7}$$

In terms of these coordinates, the equations governing forced motion take the uncoupled form

$$\dot{\xi}_r - \lambda_r \xi_r = \Xi_r(t) . \tag{8}$$

The Ξ_r represent components of the applied force vector, each of which excites a motion only in the direction defined by the corresponding modal column. Evaluation of these force components is greatly simplified by a set of orthogonality relationships pointed out in Ref. [3]. These relationships,

$$(\lambda_r + \lambda_s) \{\Phi\}_r^T [m] \{\Phi\}_s + \{\Phi\}_r^T [c] \{\Phi\}_s = C_r \delta_{rs} , \tag{9}$$

are valid when the matrices [m], [c], and [a] are symmetric. (Here, T denotes the transpose of a matrix and δ_{rs} the Kronecker delta symbol.) Use of these relationships leads to the simple formula

$$\Xi_r(t) = \frac{\{f(t)\}^T \{\Phi\}_r}{C_r} . \tag{10}$$

The contribution to the displacement from a complex conjugate pair of terms in Eq. (7) is denoted by

$$\{q\}_r = \xi_r \{\Phi\}_r + \xi_r^* \{\Phi^*\}_r . \tag{11}$$

This combination may be formed after the solutions to Eq. (8) have been obtained. However, it is more instructive to rearrange the differential equations which govern ξ and ξ^*, to form the alternative uncoupled form

$$\{\ddot{q}\}_r - 2\alpha_r\{\dot{q}\}_r + |\lambda_r|^2\{q\}_r = \{Q(t)\}_r , \tag{12}$$

in which

$$\{Q\}_r = 2\,\mathrm{Re}\,\{(\dot{\Xi}_r - \lambda_r^*\Xi_r)\{\Phi\}_r\} . \tag{13}$$

For a given forcing function, the relative importance of a particular mode may be estimated by the relative magnitude of Ξ_r or Q_{rj}. In the case of an oscillatory mode, the possibility of a resonance condition is present, so that the variation of Q_{rj} with time must also be considered. Typically, the number of terms required for reasonable accuracy in the sum of Eq. (7) is considerably lower than the number of degrees of freedom of the system.

To illustrate this point, a resonance diagram has been generated by this method, for a two-point mooring depicted in Fig. 1. The excitation considered was a simple harmonic fluctuation in water surface height at the center spar buoy. The steady-state output shown in Fig. 2 is the vertical displacement of that same buoy. As a minimal representation of the physical system, three lumped masses were used to represent each of the four sections of line; with an additional lumped mass at each buoy, and constraints which state that the interconnections between mass points are inextensible, the model was one with 14 degrees of freedom. Contributions from the first 4 modes are shown on the diagram. Although a proper superposition of the mode contributions would include an account for phase differences not shown on the diagram, it is clear from the amplitude plot that far fewer than 14 modes will be needed to accurately estimate the buoy displacement through a relatively wide range of frequencies.

References

1. Skop, R.A., O'Hara, G.J.: The Static Equilibrium Configuration of Cable Arrays by Use of the Method of Imaginary Reactions. 36 Numb. Leaves. Naval Research Laboratory, Ocean Structures Branch, Ocean Technology Div., Washington, D.C. NRL Report 6819.

2. Grad, J., Brebner, M.A.: Eigenvalues and Eigenvectors of a Real General Matrix. Communications of the ACM 11, No. 12, 820-826 (December 1968).

3. Foss, K.A.: Coordinates Which Uncouple the Equations of Motion of Damped Linear Dynamic Systems. The American Society of Mechanical Engineers. Applied Mechanics Div., September 1958, pp. 361-364.

4. Dominguez, R.F.: The Static and Dynamic Analysis of Discretely Represented Moorings and Cables by Numerical Means. Ph. D. Dissertation, Oregon State University, June 1971.

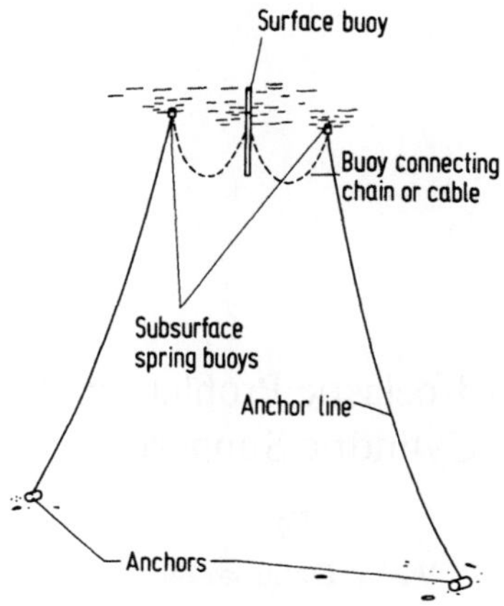

Fig.1. Two point mooring system.

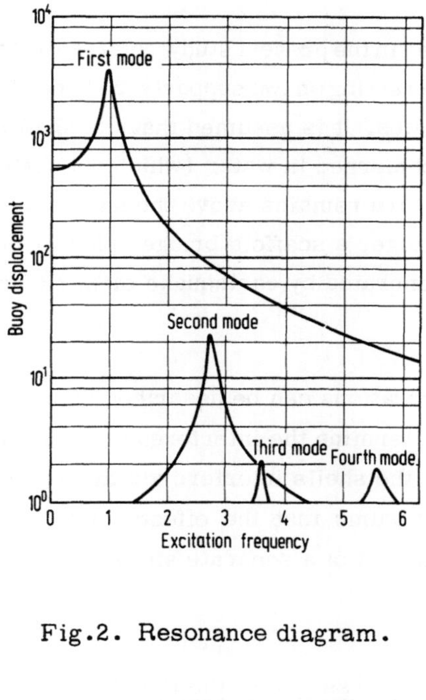

Fig.2. Resonance diagram.

An External Hydroelastic Problem of Two Circular Cylindric Supports

By

V.A. Dzhupanov

Sofia, Bulgaria

Introduction

The problem discussed in the present study relates to the cases when some superstructure resting on two supports in shape of circular cylindrical shells vibrates. It is assumed that the shell supports are partly or entirely submerged in water (which we shall call fluid), while the superstructure remains above the surface. The structure can be a part of a bridge, a scaffold bridge, pier or quay, and the vibrations may be generated by earthquake or any other (e.g. technological) reasons.

Assuming that the vibrations can be described by a known harmonic law, the task is to determine the interference of the supports through the fluid. Certainly, the shells interfere via the superstructure and the ground, but we assume that the effects of this interaction are well known or are subject of a separate study.

Main, Prerequisites and Essence of the Problem

1. Two elastic circular cylindrical shells are defined by means of their middle surfaces with radii R_1 and R_2.

2. The fluid in which the shells are vibrating is ideal, incompressible, and heavy. Its depth h is equal to the length of the supports.

3. The process of shell vibration has a stationary character and the small radial displacements w_1 and w_2 can be described by

$$w_j = Y_j(x)\cdot\cos n_j\theta_j\cdot T(t)\,, \quad j = 1 \quad n_j = n\,, \quad j = 2 \quad n_j = m\,, \tag{1}$$

where $Y_j(x)$ are forms of the vibrations of the mid-surfaces in sections $\theta_j = 0$. If ω is the circular frequency, then

$$T(t) = \exp(i\omega t)\,. \tag{2}$$

4. It is assumed that the small displacements w_j generate gravitational waves in the fluid with small amplitudes.

5. The vibration forms of the shells in vacuum and in fluid are either identical or the differences are admissibly small.

6. Before the movement of the "shells-water" system was started, its members were at rest.

The aim is to determine the hydrodynamic pressure on the shell surfaces and the added masses of fluid, taking into account their interference.

Given the above conditions, the movement of the fluid should be potential. The potential of the fluid velocities Φ should satisfy the following boundary-value problem (see Fig.1).

$$\frac{\partial^2\Phi}{\partial r^2} + \frac{1}{r}\frac{\partial\Phi}{\partial r} + \frac{1}{r^2}\frac{\partial^2\Phi}{\partial\theta^2} + \frac{\partial^2\Phi}{\partial x^2} = 0\,, \tag{3}$$

$$\left.\frac{\partial\Phi}{\partial x}\right|_{x=0} = 0\,, \qquad \left[\frac{\partial^2\Phi}{\partial t^2} + g\frac{\partial\Phi}{\partial x}\right]_{x=h} = 0\,, \tag{4)(5}$$

$$\lim \sqrt{r_j}\left[\frac{\partial\Phi}{\partial r_j} - ik\,\Phi\right]_{r_j\to\infty} = 0\,, \tag{6}$$

$$\left.\frac{\partial\Phi}{\partial r_j}\right|_{r_j=R_j} = \frac{\partial w_j}{\partial t}\,, \quad (j = 1,2)\,. \tag{7}$$

The hydrodynamic pressure is determined from the linearized integral of Cauchy-Lagrange [4]

$$p_j = -\rho \left.\frac{\partial\Phi}{\partial t}\right|_{r_j=R_j}\,, \tag{8}$$

where ρ is the density of the fluid.

Velocity Potential

1. We take

$$\Phi = \Phi_1(x, r_1, \theta_1; t) + \Phi_2(x, r_2, \theta_2; t)\,, \tag{9}$$

where

$$\Phi_j = \frac{\operatorname{ch} k x}{\operatorname{ch} k h}\,\dot{T}(t)\sum_{\lambda=0}^{\infty}\Delta_\lambda A_{j\lambda} H_\lambda^{(2)}(k r_j)\cos\lambda\theta_j\,, \quad (j = 1,2)\,. \tag{10}$$

$A_{j\lambda}$ are unknown constants and $H_\lambda^{(2)}$ is Hankel's function of the second kind. Obviously, the potentials (10) are solutions to Eq. (3) and satisfy Eqs. (4), (5) and (6). The wave number k is being determined from Eq. (5), i.e., from Eq. (11), where g is the acceleration of gravity.

$$gk \cdot \operatorname{th}(kh) = \omega^2\,. \tag{11}$$

$A_{j\lambda}$ are determined from Eq. (7); to this end, the known addition formulae of the Hankel's functions are used. We define

$$\overline{H}_\lambda^{(2)}(kR_j) = \frac{1}{k}\frac{\partial}{\partial r_j}\left[H_\lambda^{(2)}(k r_j)\right]_{r_j=R_j}\,; \tag{12}$$

$$\left.\begin{aligned}
&Q_\lambda = \int_0^h \frac{ch^2 kx}{ch\ kh}\,dx, \quad L_{js} = \int_0^h Y_j(x)\cdot ch\,kx\,dx\,; \\
&\delta_{sn} = \begin{cases} 1, & s = n, \\ 0, & s \neq n, \end{cases} \quad \delta_{sm} = \begin{cases} 1, & s = m, \\ 0, & s \neq m\,; \end{cases} \quad s = 0,1,2,3,\ldots\,;\ j = 1,2\,; \\
&P_{sn} = L_{1n}\left[k\cdot Q\cdot\Delta_s\cdot\overline{H}_s^{(2)}(kR_1)\right]^{-1}, \\
&P_{sm} = L_{2m}\left[k\cdot Q\cdot\Delta_s\cdot\overline{H}_s^{(2)}(kR_2)\right]^{-1}; \\
&\Delta_s \equiv \Delta_\lambda = 0{,}5 \rightarrow \lambda \equiv s = 0\,, \quad \Delta_s \equiv \Delta_\lambda = 1 \rightarrow \lambda \equiv s \neq 0\,.
\end{aligned}\right\} \tag{13}$$

The following infinite system of simultaneous algebraic equations can be obtained if Eq. (7) is satisfied:

$$\left.\begin{aligned}
&A_{1s} + \sum_{\lambda=0}^{\infty} A_{2\lambda}\cdot b^*_{s\lambda}\cdot G_s^1 = \delta_{sm}\cdot P_{sm}, \\
&\sum_{\lambda=0}^{\infty} A_{1\lambda}\cdot c^*_{s\lambda}\cdot G_s^2 + A_{2s} = \delta_{sm}\cdot P_{sm},
\end{aligned}\right. \tag{14}$$

where

$$G_s^j = J_s'(kR_j)/\Delta_s\cdot\overline{H}_s^{(2)}(kR_j)\,, \tag{15}$$

$$\left.\begin{aligned}
&b^*_{s\lambda} = \Delta_s\Delta_\lambda(-1)^\lambda\left[(-1)^s\cdot H_{\lambda-s}^{(2)}(ka) + H_{\lambda+s}^{(2)}(ka)\right], \\
&c^*_{s\lambda} = \Delta_s\Delta_\lambda\left[H_{\lambda-s}^{(2)}(ka) + (-1)^s\cdot H_{\lambda+s}^{(2)}(ka)\right].
\end{aligned}\right\} \tag{16}$$

From Eq. (16) it may be seen that the block of coefficients $\|c^*\|$ is obtained from the block $\|b^*\|$ by means of transposition. This makes it easier to determine the elements of the matrix after introducing the relations:

$$b_{s\lambda} = b^*_{s\lambda}\cdot G_s^1, \quad c_{s\lambda} = c^*_{s\lambda}\cdot G_s^2. \tag{17}$$

2. The factor G_s plays an important role in proving that the systems of the type (14) can be solved. We shall not dwell on this particular problem, but we should note its engineering implication, namely, that with the increase of the index s the module of G_s relatively quickly tends to 0. This makes it possible from the infinite system (14) to "cut out" (substitute by zero) thos coefficients before the unknowns whose modules are close to zero. In this way, the solution of Eq. (14) is reduced to the solution of a system of simultaneous algebraic equations with a finite number of unknowns. The details are given in Ref. [1,3].

It appears that the number of unknowns which you should leave should be equal to $s_1 + s_2$, where s_1 and s_2 are determined from the corresponding inequalities

$$c_{s\lambda} \leqslant \varepsilon, \quad b_{s\lambda} \leqslant \varepsilon, \tag{18}$$

where ε is small and is determined by engineering considerations.

3. Let us take a practical example. Two reinforced concrete cylindrical shells have the same radius of the mid-surface $R = 0,4$ m and vibrate with a frequency of $f \sim 1$ Hz. The distance between their axes is $a = 8R$, and their height is $h = 10$ m. If $1\ N = 1/9806,65$ t, then $E = 2.10^6\ t/m^2$ and $\gamma = 2,4\,t/m^3$. Vibrations - with forms of their first tone

$$Y_j(x) = (T_h \cdot U_x - S_h \cdot V_x)/(T_h \cdot U_h - S_h \cdot V_h), \tag{19}$$

where

$$\begin{aligned} T_h &= (\mathrm{sh}\,\beta h + \sin \beta h)/2, \quad V_h = (\mathrm{sh}\,\beta h - \sin \beta h)/2; \\ S_h &= (\mathrm{ch}\,\beta h + \cos \beta h)/2, \quad U_h = (\mathrm{ch}\,\beta h - \cos \beta h))/2; \end{aligned} \tag{20}$$

$$\beta = \sqrt[4]{\omega^2 \cdot m_0/EJ^*}, \quad \omega = 2\pi f, \tag{21}$$

m_0 = length-mass and J^* = inertia moment of the transversal section.

It turns out that one can take into account only ten unknown constants (five for each shell), if $\varepsilon = 0{,}0001$. The results are given in the following Table 1.

Table 1

λ	$A_{1\lambda}$	$A_{2\lambda}$
0	$-0{,}29750 + i \cdot 0{,}19067$	$-0{,}10721 + i \cdot 0{,}10341$
1	$0{,}56714 + i \cdot 0{,}76524$	$0{,}64073 + i \cdot 0{,}72809$
2	$0{,}06277 + i \cdot 0{,}11247$	$-0{,}09012 - i \cdot 0{,}16755$
3	$-0{,}01833 + i \cdot 0{,}00661$	$-0{,}02577 + i \cdot 0{,}01143$
4	$-0{,}00066 - i \cdot 0{,}00082$	$0{,}00119 + i \cdot 0{,}00114$

Hydrodynamic Pressure

According to Eq. (8) it is obtained

$$p_I = \rho\omega^2 \cdot T(t)\,\frac{\operatorname{ch} kx}{\operatorname{ch} kh} \sum_{\lambda=0}^{4} \left[\Delta_\lambda A_{1\lambda} H_\lambda^{(2)}(kR) \cos\lambda\theta_1 + A_{2\lambda} \sum_{s=0}^{4} b_{s\lambda}^{*} \cdot J_s(kR) \cdot \cos s\theta_1 \right],$$

$$p_{II} = \rho\omega^2 \cdot T(t)\,\frac{\operatorname{ch} kx}{\operatorname{ch} kh} \sum_{\lambda=0}^{4} \left[\Delta_\lambda A_{2\lambda} H_\lambda^{(2)}(kR) \cos\lambda\theta_2 + A_{1\lambda} \sum_{s=0}^{4} c_{s\lambda}^{*} \cdot J_s(kR) \cdot \cos s\theta_2 \right]. \qquad (22)$$

The values of the hydrodynamic pressure for $x = h$ are given in Table 2 (the factor for all values is $T(t)$). There, p_0 = hydrodynamic pressure without the effect of the interference; p_I, p_{II} = hydrodynamic pressure for each respective shell with the interference being taken into account (the values are given in t/m^2, where $1\,t = 9806{,}65$ N).

Table 2

θ_j	p_0		p_I		p_{II}	
	Re p_0	Im p_0	Re p_I	Im p_I	Re p_{II}	Im p_{II}
0	-4,6424	39,304	3,3975	38,304	8,2780	28,304
$\pi/8$	-4,2890	36,312	3,4267	36,776	8,0756	28,316
$2\pi/8$	-3,2827	27,792	2,7583	31,166	7,7530	26,049
$3\pi/8$	-1,7766	15,041	0,1566	19,991	4,4488	18,304
$4\pi/8$	0	0	-3,0946	4,408	-0,3186	4,500
$5\pi/8$	1,7766	-15,041	-7,7530	-10,879	-3,8702	-11,332
$6\pi/8$	3,2827	-27,792	-9,0145	-21,251	-4,6623	-23,676
$7\pi/8$	4,2890	-36,312	-8,2012	-25,732	-3,6240	-29,965
π	4,6424	-39,304	-7,5127	-26,390	-2,9387	-31,655

Added Masses, Kinetic Energy of the Interference

For the first shell we receive

$$M_I = \frac{\operatorname{Re} E_{kI} + i \operatorname{Im} E_{kI}}{\dot{T}^2(t)\pi R \int_0^h Y_1^2(x)\,dx} = M_{I1} + iM_{I2}, \tag{23}$$

where

$$E_{kI} = -\frac{\rho R}{2} \int_0^{2\pi} \int_0^h \left[\Phi_I \frac{\partial \Phi_I}{\partial r_1} \right]_{r_1=R} d\theta_1 \cdot dx. \tag{24}$$

The kinetic energy, Equation (24), can be taken as a criterion for the dynamic interference of the shells through the fluid. In this specific example, for $n = m = 1$

$$E_{kI} = \rho\omega^2 \pi R \frac{L}{\operatorname{ch} kh} T^2(t) \left[A_{11} H_1^{(2)}(kR) + \sum_{\lambda=0}^{4} A_{2\lambda} b_{1\lambda}^* J_1(kR) \right],$$
$$E_{kII} = \rho\omega^2 \pi R \frac{L}{\operatorname{ch} kh} T^2(t) \left[A_{21} H_1^{(2)}(kR) + \sum_{\lambda=0}^{4} A_{1\lambda} c_{1\lambda}^* J_1(kR) \right], \tag{25}$$

or numerically:

$$E_k^0 = (-1,2359 + i \cdot 10,463)\,T^2(t)\,,$$
$$E_{kI} = (\ 0,7550 + i \cdot \ 9,104)\,T^2(t)\,, \quad [E] = t \cdot m\,, \qquad (26)$$
$$E_{kII} = (\ 1,2876 + i \cdot \ 9,469)\,T^2(t)\,.$$

It is obvious that when the interference through the fluid is taken into account, the inertial terms in both the pressure and the energy are subject to considerable changes.

Conclusion

The above considerations prove that the effect of the interference of the supports through the fluid may be significant in some cases and therefore should be taken into account.

References

1. Guz A.N., Tchernishenko, I.S., Shnerenko, K.I.: Spherical Bottoms Weakened by Perforations, Kiev: Naukova Dumka 1970.

2. Dzhupanov V.A.: On the Analysis of the Dynamic Interaction of Circular Cylindrical Shell with Liquid. Sofia, Theoretical and Applied Mechanics 2, No 3 (1971).

3. Ivanov E.A.: Diffraction of Electromagnetic Waves upon Two Bodies, Minsk: Nauka i Technika 1968.

4. Sheinin, I.S.: Vibrations of Hydraulic Structures in Liquid. Handbook on Dynamics of Hydraulic Structures, Pt 1, Leningrad: Energia 1967.

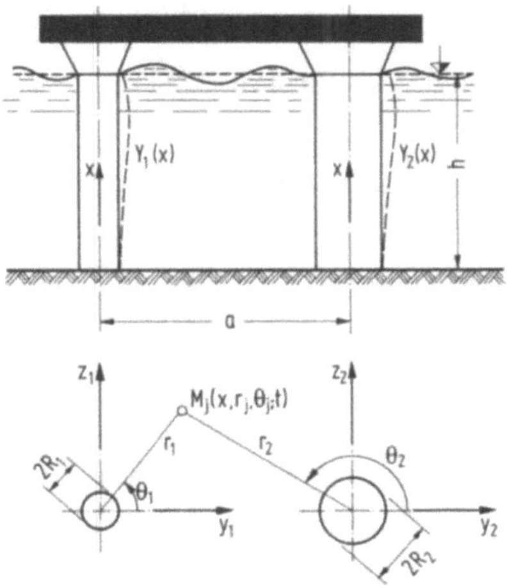

Fig. 1

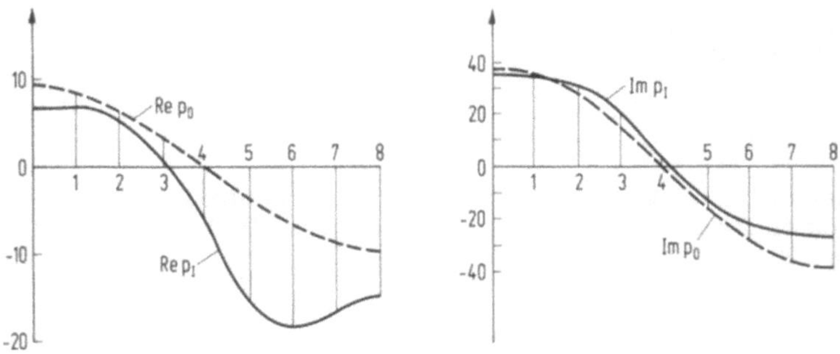

Fig. 2. Hydrodynamic pressure according to the values in Table 2.

Flow-Induced Vibrations of Two Cylinders in Tandem, and their Suppression

By

M. M. Zdravkovich

Salford, U.K.

Introduction

When a flexible rear cylinder is displaced from the center line of the front cylinder wake, a lateral force is produced. The rear cylinder is then inside the shearing velocity profile of the front cylinder wake, and potential flow theory predicts that the force acting on it is directed away from the center line, i.e., there should be a lift force. Experiments, however, show that, depending on the lateral displacement and longitudinal spacing, the force is usually directed towards the wake center, so it is a restoring force. Mair and Maull [1] offered a likely explanation that the flow entrainment into the first wake produces a local mean velocity directed towards the wake center line and, hence, produces the restoring force. The present experiments showed that a hysteresis in the lift and restoring forces may cause or maintain low-frequency large-amplitude aeroelastic vibrations.

Experimental Arrangements

The experiments were carried out in a 1.14 by 0.84 m open-circuit wind tunnel in the Aerodynamics Laboratory at the University of Salford. Both cylinders were made of PVC tubing 11.4 cm outer diameter. The rigid cylinder was fixed vertically to the floor and ceiling of the working section. A metal ring built into the center of the fixed cylinder was provided with 36 circumferential holes for the measurement of static pressure. The vibrating cylinder, 12 mm shorter, was fitted with a cap at each end, through which was passed a rod, 12 mm diameter, extending outside the tunnel through the slots. For some tests the ends of the rod were attached to the free ends of two horizontal arms which were free to swing about a vertical axis. This axis was 1.26 m upstream of the moving cylinder. For other tests, the rod was attached at its ends to cords anchored on the laboratory ceiling and floor. The former arrangement allowed only transverse motion, but the latter allowed both transverse and streamwise motions. For both arrangements, different sets of 4 springs could be laterally attached to the ends of the rod, enabling the natural frequency to be changed. Vibrational motions were simultaneously recorded with a linear displacement transducer on a UV recorder.

Vibrational Tests

The first set of experiments was performed with the front cylinder fixed and the rear cylinder attached to the swing arms without springs, Fig.1(i). This arrangement produced a large-amplitude sinusoidal vibration with a low frequency (Strouhal number was 0.01) for all gap-to-diameter ratios, G/D, up to 2.5. The effect of G/D on the peak-to-peak amplitude, 2A, and frequency, f, is shown in Fig.2 as a function of the Reynolds number, Re. When G/D was increased beyond 2.5, an abrupt reduction in amplitude occurred of more than 10 : 1. These vibrations were random with no distinct frequency and continued so up to the biggest tested G/D = 6.

When the springs were fitted, only small-amplitude random vibrations occurred in the whole range of G/D and Re. It might be that the initial instability was too weak to displace the rear cylinder sufficiently far sideways into the shear layers of the front one. To check this, the rear cylinder, fitted with springs, was pushed transversely; the typical oscillogram is shown in Fig.4. For the first three initial displacements the vibrations, at a frequency 20% below the natural frequency, f_n, were damped out. The fourth displacement, however, built up large amplitudes at a frequency which was above f_n. They were maintained while the flow was accelerated up to the upper speed limit and also during the deceleration down to the self-damping limit ($Re = 10^5$). All efforts to start large-amplitude vibration when $G/D > 2.5$ were unsuccessful. The amplitude and frequency ratios are presented in Fig.3. The similarity between Figs.2 and 3 is obvious. The amplitudes are less in the latter, presumably due to the damping factor of the spring system which was 0.24.

The second set of experiments was performed with the front cylinder attached to the swing-arms and the rear one was fixed, Fig.1(ii). This arrangement did not produce large-amplitude vibrations, either with or without springs. The only exception was for $G/D = 0.5$, where self-excited large-amplitude vibrations were observed when the front cylinder was fitted with springs. Without springs, the flow always tended to move the front cylinder to one side or the other so as to be non-aligned with the rear, fixed one. This is in agreement with Hori's [2] measurements of the lift force on the front cylinder, which was always directed away from the wake center line. With springs, the small-amplitude random vibrations which occurred were larger than those for the arrangement (i), for all values of G/D.

The third set of experiments was carried out with two cylinders fixed together at various G/D, both attached to the swing arms, see Fig.1(iii). This arrangement was least prone to vibration. The large-amplitude mode was impossible to start, either with or without springs, in the whole ranges of G/D and Re, while the small-amplitudes were reduced in comparison with the arrangement (i). It seems that the cause of vibration lies in the gap between the cylinders where flow is likely to be affected by the relative transverse displacements of cylinders.

Origin of Instability

The mean static pressure distribution was measured around the rear cylinder at subcritical and transitional Reynolds numbers. Both cylinders were fixed and the ratio of transverse displacement relative to the cylinder diameter, T/D, was taken as a parameter. In Fig.5a, G/D = 1/2 and T/D = 1/4, the pressure coefficient curves indicate that the drag is negative and the lift is directed away from the wake center line. The assumed flow patterns and the resulting aerodynamic forces are shown above the graphs in Fig.5. Fig.5b for T/D = 1/2, shows a switch in the direction of drag and lift, and the appearance of a stagnation point suggests a flow through the gap. The magnitude of the negative lift, i.e. the restoring force, is greater for transitional flow than for the subcritical one. Further displacement to T/D = = 3/4, Fig.5c, shows that the pressure distribution is almost symmetrical, with a flat part of the curve around the rear of the cylinder typical of subcritical flow, and two suction peaks and a decrease in base suction for transitional Reynolds numbers.

The rear cylinder was displaced in the opposite direction in Fig.6, G/D = 1, but lift and drag at T/D = 1/8 (Fig.6a) are again directed away from the wake and towards the gap. The position of the suction peaks in Fig.6b, T/D = 1/4, shows a force switch for subcritical flow while the transitional flow retains the same pattern. Fig.6c shows both flows having a similar pattern, but the restoring force much greater for higher Re.

A distinct trend to symmetry in pressure distribution around the rear cylinder may be seen in Fig.7 where G/D = 3. There is neither switch of the aerodynamic forces nor sudden flow through the gap, judging from the gradual build up of the stagnation pressure in graphs 7a, b and c.

Mechanism of Vibration

The resulting mean aerodynamic forces on the rear cylinder indicate that an initial instability from the neutral position may lead to vibration but the restoring force caused by the sudden change of the flow pattern tends to suppress it. Therefore, the observed large-amplitude mode cannot be classified as a galloping oscillation. The feeding force which maintained the large-amplitude mode was found to be due to a hysteresis effect. The switch from the subcritical to transitional pressure distribution occurred at a higher Reynolds number when the velocity was increasing, but the reverse switch was postponed to a lower Reynolds number when the velocity was decreasing. When the rear cylinder crosses the wake of the front one during vibration, it enters the higher velocity flow of the shear layer and the switch of the lateral force occurs at a higher T/D. On the way back, the restoring force is maintained due to hysteresis, because the rear cylinder enters now the lower velocity flow. Thus, the aerodynamic forces are in phase with the elastic forces for a longer time feeding the large amplitude vibration. The minimum displacement necessary to start this mode supports the proposed mechanism (see Fig.4). A similar mechanism based on the hysteresis of the jet switch behind a cascade of cylinders was observed by Roberts [3] and this also resulted in low-frequency large-amplitude vibrations, but in the streamwise direction.

Suppression of Vibration

The axial-rod shroud which has been developed at Salford [4] was found to be effective in suppressing vibrations of a single cylinder immersed in water flow [5]. The application to the tandem cylinder arrangement proved to be equally effective. The large-amplitude vibrations were impossible to excite for a whole range of G/D and Re not only in the arrangements (i), (ii) and (iii), but also when the front cylinder was on swing-arms and the rear one on the cords, see Fig.1(iv). The typical small-amplitude transverse vibrations with additional vibration of the rear cylinder in a streamwise direction are presented in Fig.8a. When the rear cylinder was shrouded, lateral vibrations of both cylinders are reduced as seen in Fig.8b. Only when both cylinders were shrouded was the rear cylinder fully stabilised, as seen in Fig.8c.

Conclusion

The observed large-amplitude low-frequency vibration of the rear cylinder may be attributed to the hysteresis of the flow switch through and around the gap between the cylinders. The axial-rod shroud totally suppressed this mode of vibration.

References

1. Mair, W.A., Maull, D.J.: Aerodynamic Behaviour of Bodies in the Wakes of other Bodies. Phil. Trans. Roy. Soc. London A, 269, 425-437 (1971).

2. Hori: Experiments on Flow Around a Pair of Parallel Circular Cylinders. Proc. 9th Japan Nat. Congr. Appl. Mech., paper III-11, pp 231-235 (1959).

3. Roberts, B.W.: Low Frequency Aeroelastic Vibrations in a Cascade of Circular Cylinders. Mech. Engng. Sci. Monograph No 4, September 1966.

4. UK Patent Appl 40768/71, Improvements in and Relating to Flow Stabilisation.

5. Zdravkovich, M.M.: Circular Cylinder Enclosed in Various Shrouds. ASME, Vibrations Conf, Toronto, Paper No 71-VIBR-28 (1971).

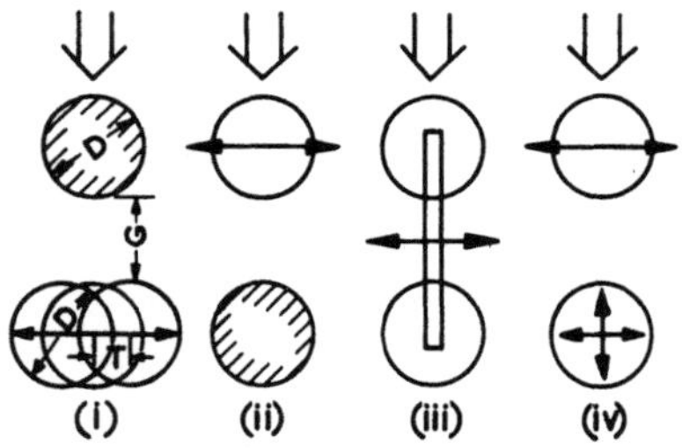

Fig.1. Test arrangements of cylinders.

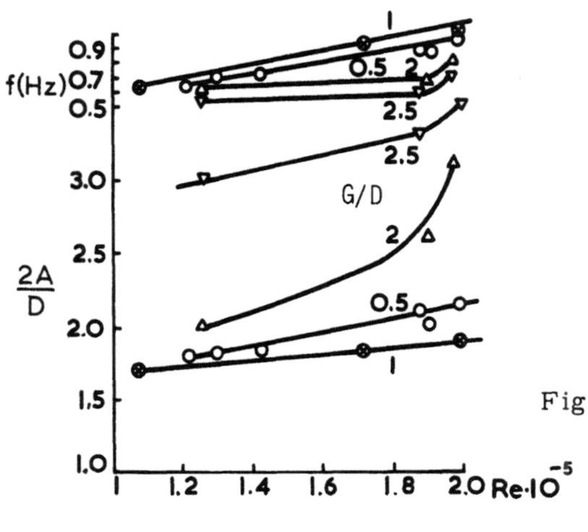

Fig.2. Frequency and double-amplitude ratio vs Reynolds number without springs.

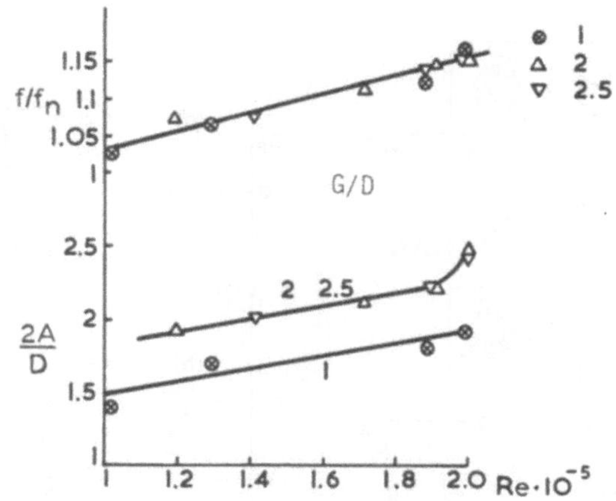

Fig. 3. Frequency and double-amplitude ratios vs Reynolds number with springs. (Natural frequency 1.87 Hz.)

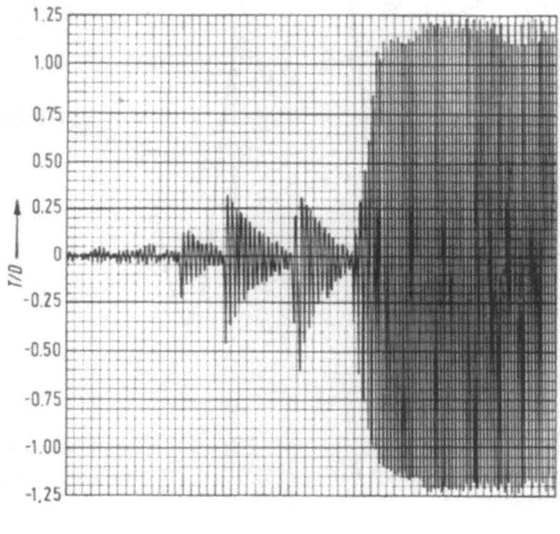

Fig. 4. Artificial excitation of large amplitude mode of vibration.

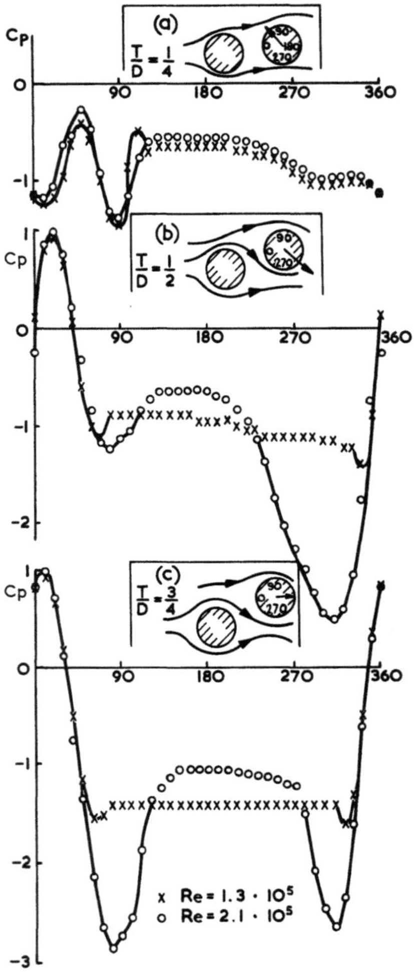

Fig.5. Pressure coefficient around rear cylinder for G/D = 0.5.

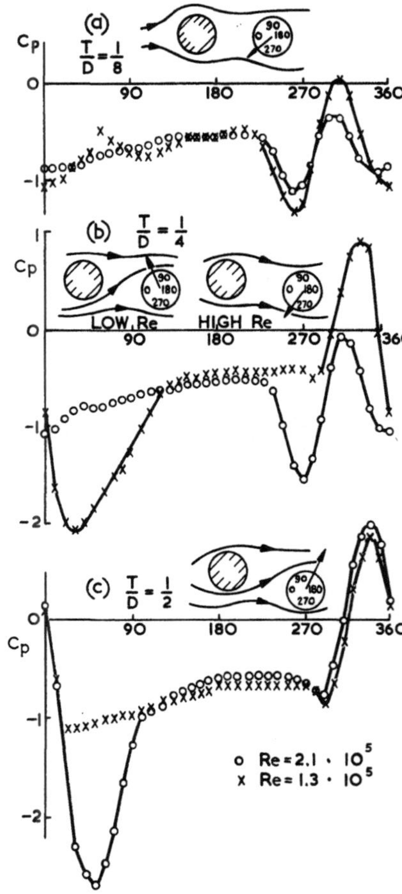

Fig.6. Pressure coefficient around rear cylinder for G/D = 1.

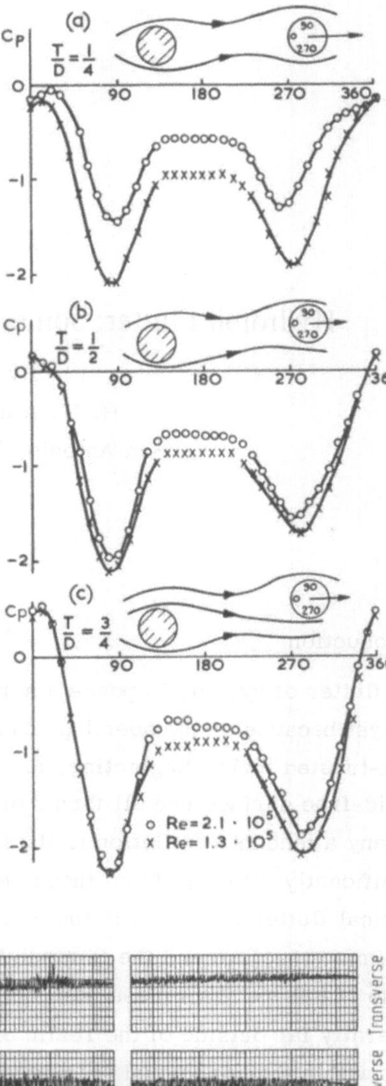

Fig.7. Pressure coefficient around rear cylinder for G/D = 3.

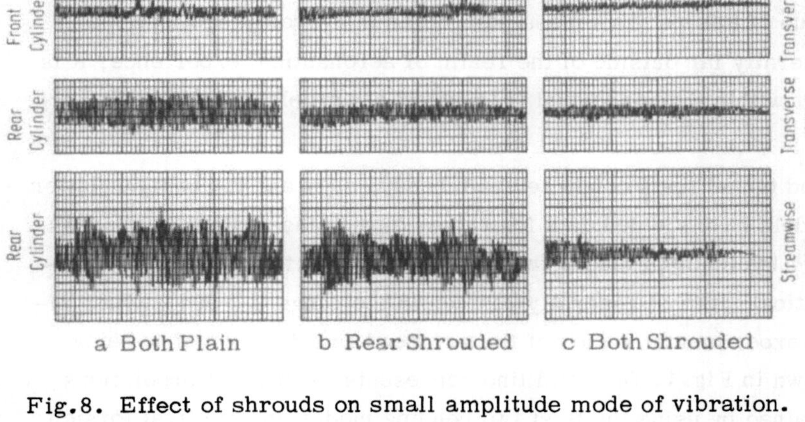

Fig.8. Effect of shrouds on small amplitude mode of vibration.

Hydrofoil Flutter: Some Recent Developments

By

H. N. Abramson

San Antonio, Tex., USA

Introduction

The flutter of hydrofoils poses a somewhat perplexing problem to the analyst because of the seeming inadequacy of his conventional and well-trusted tools. Neglecting, for the moment, the presence of the liquid-free surface and all forms of cavitation (either with or without any attendant ventilation to the free surface), the only factor significantly different from those pertaining to conventional aeronautical flutter problems is that defining the relative masses of the dynamic structure and the immediately surrounding fluid. While the value of this relative mass parameter in hydrofoil applications is generally far outside of the realm of aeronautical experience, it is well within the conceptual framework of aeroelasticity.

Wind tunnel tests conducted with heavy gases and lightweight flutter models have yielded data for values of the mass ratio μ as low as 1.44 (which approaches the range of interest for hydrofoils; aeronautical interest generally resides in the range $\mu > 20$), with typical experimental values of flutter speed as a function of μ being shown in Fig. 1. The solid line represents theoretical predictions obtained by using the first two bending modes and the first torsion mode in a modal flutter analysis employing conventional aerodynamic strip theory. This analysis indicates the existence of a minimum mass ratio below which flutter will not occur: the main feature of this data, however, resides in the generally unconservative nature of flutter prediction near μ_{min}. Despite rather extended discussions by many investigators,[1] the basic reasons for this general inability of conventional flutter theory (which has been so successful in aeronautical applications for many years) to yield if not accurate at least conservative predictions, at low values of μ, still remain elusive. Subsequent experimental data with representative hydrofoil models tested in water have served only to compound the problem by continuing to demonstrate the fundamental inadequacy of the theoretical basis of the classical formulation of the unsteady hydrodynamic loads.

This rather fundamental problem of flutter prediction of a fully-wetted, subcavitating hydrofoil is the central subject of this brief review of certain recent developments. The content, after some initial and brief discussion of the role of real fluid effects, is directed toward recent developments in the rather pragmatic questions of calculation techniques and model testing.

Some Speculation on Real Fluid Effects

Some investigators (the present author being one of the principals) have held to the belief that the observed discrepancies between calculated and measured flutter speeds cannot be attributed solely to various secondary factors (such as finite aspect ratio, tunnel wall corrections, and effects of hydrodynamic loads on structural mode shapes), but rather are a consequence of basic inadequacies of the theoretical values of the oscillatory lift and moment coefficients at low values of the reduced velocity. Real fluid effects, involving both the boundary layer and the wake, are suggested as the basic factor involved. Independent measurements of such oscillatory coefficients

---

[1] The very extensive background of this entire field may be reviewed in the monograph of Ref. 1 and the later literature survey of Ref. 2.

do indeed reveal that the classical analyses appear to be somewhat deficient.

The possible role of real fluid effects in modifying oscillatory loads has been discussed in detail by several investigators, who have noted a number of examples of experimental evidence regarding the effects of motions of the boundary layer transition point. In particular, mechanically fixing the transition point caused deviations between measured pressure distributions and those predicted by theory to be practically eliminated. Encouraged by these kinds of observations, researchers have continued to explore this avenue by developing analyses of the problem of foils in unsteady viscous flow and the problem of unsteady boundary layers on oscillating foils [2]. It seems, at least to this author, that the carry-over from classical steady potential flow theory of the concept of a fixed rear stagnation point by the arbitrary imposition of the Kutta condition is not justifiable. In fact, the rather intriguing question of the theoretical implications of an oscillatory value of the circulation on flow conditions at the trailing edge appears not to have been much explored.

While intellectually challenging, these kinds of speculations unfortunately have so far not been very helpful in the practical problems of predicting hydrofoil flutter. Instead, more recent efforts have been directed toward empirical modifications to the circulation terms (e.g., variations in magnitude and phase) in the theoretical expressions for the oscillatory coefficients and their incorporation into general analysis methods [3].

Recent Applications of Modified Strip Theory

Perhaps the most rewarding of all recent developments for subcavitating hydrofoil flutter relates to the application of Yates' modified strip theory [4]. This analysis method employs arbitrary spanwise distributions of lift-curve slope and center of pressure location, and these can be derived either from measured data or from calculations using lifting surface theory. Application of this analysis to a now-classical SwRI hydrofoil flutter model test yielded rather conservative flutter speed values and so it was suggested that some additional improvement might be achieved if the noncirculatory terms in the hydrodynamic analysis were extended from two-dimensional to the three-dimensional form as had been done with the circulatory terms. For low values of the mass ratio parameter the noncirculatory terms, which include virtual mass, would seem to be of relatively greater importance and therefore these further modifications were undertaken [5], with the results presented in Fig.2.

The SwRI flutter model underwent a catastrophic structural failure at a speed of 48.1 knots and at a frequency of 17.5 Hz. Measured values of the exponential decay factor and oscillation frequency for the unstable mode are shown in Fig.2. together with representative calculations. Theories 1 to 4 contain no empirical corrections of any kind, and all are drastically unconservative in predicting flutter speed. Theory 5 represents the original Yates analysis [4] and Theory 6 represents that obtained with the additional modifications to the noncirculatory terms [5], both of which are overly conservative. (Both sets of calculations assume a value of structural damping of 0.02.)

As mentioned earlier, empirical modifications to the circulatory terms in the theoretical hydrodynamic coefficients have also been proposed in the form of a generalized Kutta condition which itself is formulated in terms of a factor $\delta_1 = 1 - \eta e^{i\Phi_0}$ applied to the trailing-edge tangential velocity, where $\eta = (C_{1\alpha}/2\pi)(1/\cos\Phi_0)$. Employing this modification in the Yates analysis leads to the results [3] shown in Fig.3. There can be no question that variations in lift-curve slope and in center of pressure location have a most profound influence on flutter speed and frequency in the range of parameters representative of hydrofoils, but further modifications in the theoretical oscillatory hydrodynamic terms are also required if the theory is ultimately to be utilized reliably and routinely.

Figure 4 shows comparisons of the several methods of analysis [3, 4, 5] with yet another set of flutter model test data [6], with similar conclusions.

The Use of Small Models in Hydrofoil Flutter Model Research

While flutter theory applied in the range of parameters pertaining to hydrofoils has met with a notable lack of success in yielding predicted values of flutter speed in reasonably close agreement with measured values, experiments with hydrofoil flutter models also have all too often failed to produce usable results in the form of actual flutter occurrences. The design of the flutter models themselves is a difficult compromise between adequate flexibility to ensure flutter within the relatively restricted speed range of most of the existing test facilities and yet sufficient strength to withstand the static bending or torsional divergence loads. In fact, the usually simple divergence calculations take on new proportions as the spanwise center of pressure location again turns out to be a critical factor in making predictions for stiffness requirements [6].

The usual types of built-up construction commonly employed in aeroelastic models generally becomes prohibitively expensive in the hydroelastic case, and only a very few such models have been fabricated. The upper boundary on mass ratio for a given model chord length is limited by the materials that can be employed in the model, and the static load requirements preclude all but high strength materials for the main spar. The mass ratio can then be reduced by selective hollowing of the interior of the model, which will also permit some variation of static unbalance and radius of gyration. Plugging of these hollowed-out portions with dense materials can be done, but at significant cost, so that only some limited variations in mass and inertial properties are possible. Finally, since the elastic axis locations for flexible models must necessarily be forward of the quarter chord to prevent divergence, little variation in this parameter is available either. Thus, hydrofoil flutter model design and fabrication for research purposes is both very difficult and expensive.

Flutter testing with such models is itself a demanding effort as special care must be taken by the investigator to analyze and interpret his test conditions very carefully and objectively. Besides the usual problems associated with flutter testing (careful control of excitation conditions and test velocity, for example), particular attention has to be given to flow conditions. The simple fact that the foil is oscillating may result in intermittent incipient cavitation, particularly as the vibration amplitude increases - the effect of such flow anomalies on center of pressure location and $C_{l\alpha}$ are unknown. Thus, it is entirely possible that a foil nearing its flutter speed but with only a gradually decreasing damping could, under these kinds of ill-defined flow conditions, experience a sudden and large decrease in damping with consequent large-amplitude flutter oscillations or even catastrophic failure. Such a drastic change in flow conditions could conceivably reduce the fully-wetted flutter speed by 20%, or more. The limited data presently available seems to lend considerable credence to these kinds of speculations and so any evaluations of flutter prediction methods or analyses by comparison with model flutter points must be done with caution.

References

1. Abramson, H.N., Chu, W.H., Irick, J.T.: Hydroelasticity: With Special Reference to Hydrofoil Craft. NSRDC Report 2557, U.S. Navy, September 1967.

2. Abramson, H.N.: Hydroelasticity: A Review of Hydrofoil Flutter. Appl. Mech. Rev. 22, 2, 115-121 (February 1969).

3. Chu, W.H., Abramson, H.N.: Further Calculations of the Flutter. Speed of a Fully Submerged Subcavitating Hydrofoil. AIAA J. Hydronautics 3, 4 168-174 (October 1969).

4. Yates, E.C., Jr.: Flutter Prediction at Low Mass-Density Ratios with Application to the Finite-Span Subcavitating Hydrofoil. AIAA Paper 68-472.

5. Liu, Y.N., Besch, P.K.: Hydrofoil Flutter Analysis Using a Modified Strip Theory. NSRDC Report 3624, U.S. Navy, July 1971.

6. Besch, P.K., Liu, Y.N.: Flutter and Divergence Characteristics of Four Low Mass Ratio Hydrofoils. NSRDC Report 3410, U.S. Navy, January 1971.

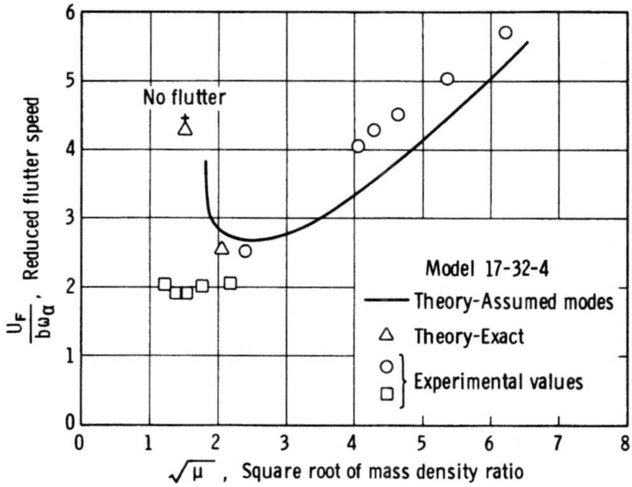

Fig.1. Comparison of flutter results obtained by "exact method," "assumed mode method" and experiments.

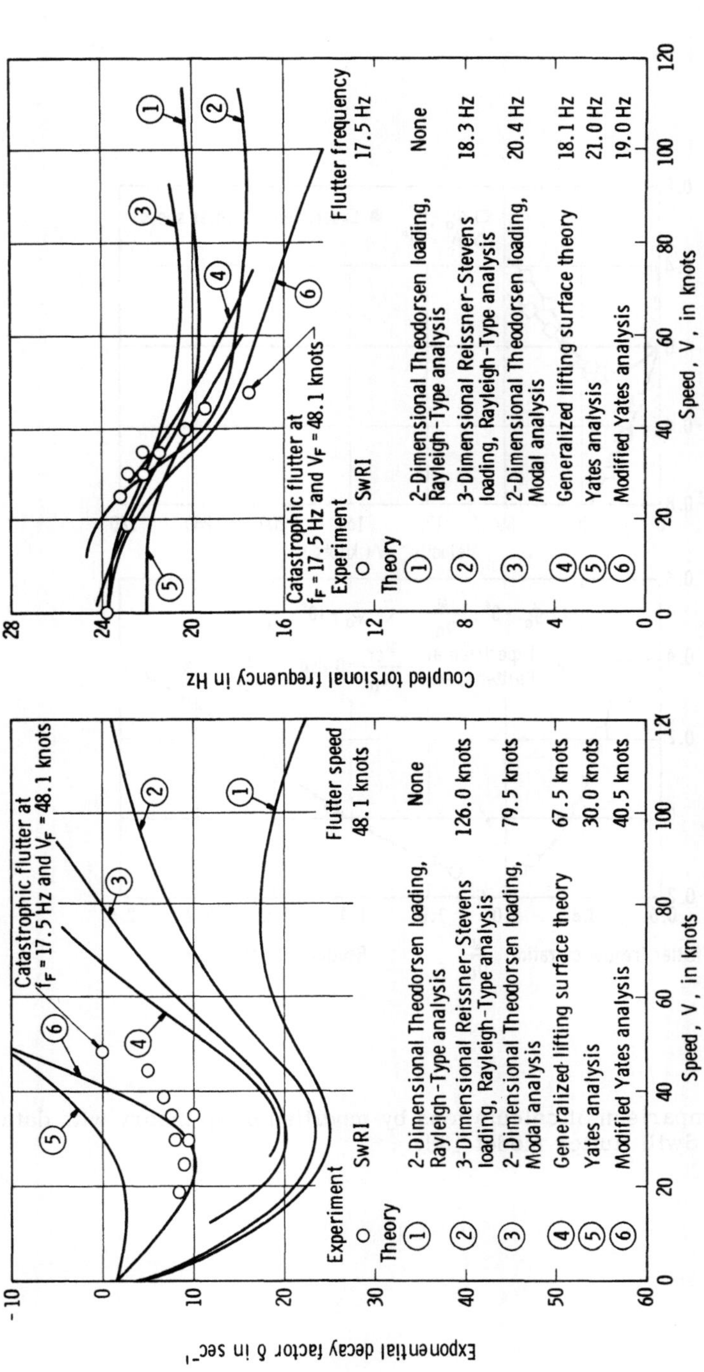

Fig.2. Comparison of various calculations with data for SwRI flutter model [5].

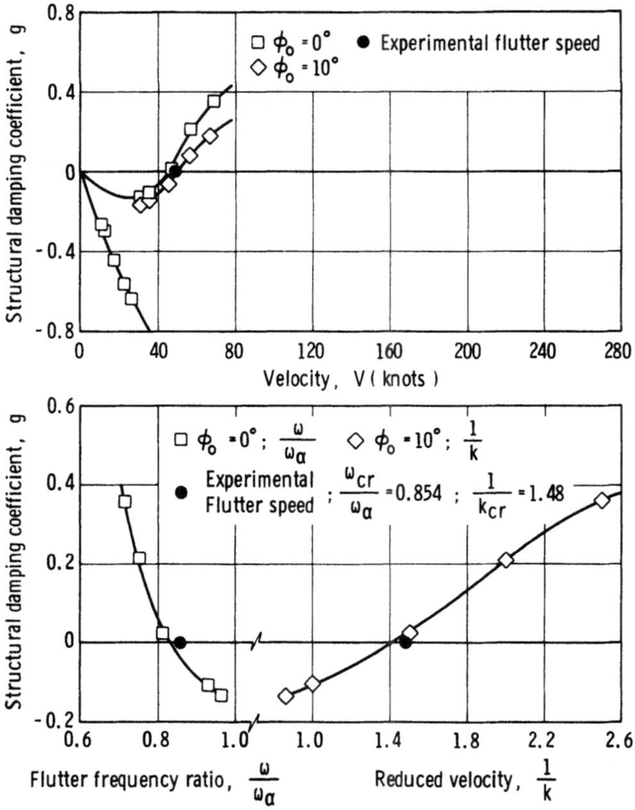

Fig.3. Comparison of calculations by modified strip theory with data for SwRI flutter model [3].

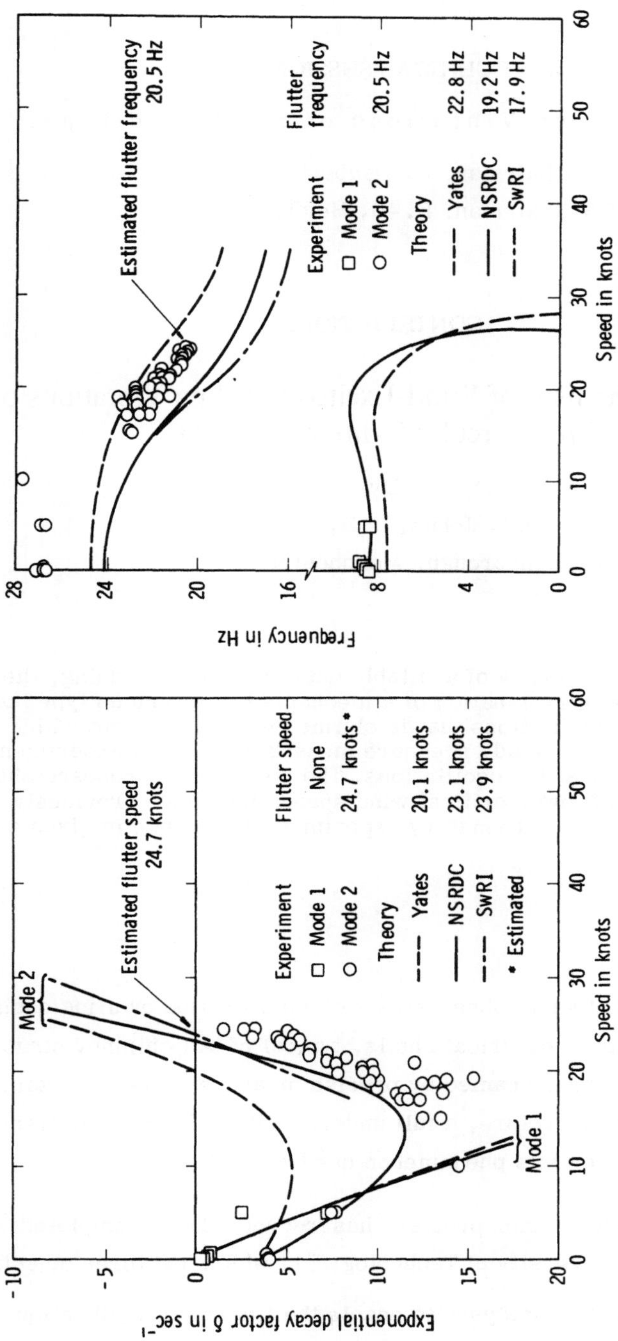

Fig. 4. Comparison of various calculations with flutter data from model [6].

TECHNICAL SESSION G

"Flow-Induced Vibrations of Shells and Pipes"

Chairman: A. Daubert, France

Co-Chairman: F.-D. Heidt, Germany

CONTRIBUTIONS

On the Mechanism of Wind-Excited Ovalling Vibrations of Thin Circular Cylindrical Shells

By

D.J. Johns, C.B. Sharma

Loughborough, Manchester, U.K.

Summary

Following a brief review of available data on vortex shedding, the paper presents an explanation of a mechanism for various types of vortex-excited oscillations that is claimed to be consistent with available data on unsteady pressure measurements and experiments on swaying and ovalling oscillations. The mechanism proposed shows the possibility of lower critical wind speeds than have previously been observed and confirmatory experimental evidence is given and discussed.

Introduction

The earliest, full scale observations of wind-excited ovalling oscillations of circular, cylindrical shells, typical of tall chimney structures, were not accompanied by precise measurements of certain relevant data and, therefore, a full understanding of the mechanism of wind excitation for this phenomenon has been delayed.

A detailed study of this problem has recently been completed at Loughborough University of Technology [1] which had three objectives:

(a) Development of analyses to enable the linear, free vibration characteristics of such shells with clamped-free boundary conditions to be predicted. This has been reported also in Ref. [2].

(b) Experimental study of these vibration characteristics under harmonic excitation and correlation with the theory from (a). This has been reported also in Ref. [3].

(c) Experimental study of the wind-excited vibrations to assess the validity of a postulated mechanism of wind-excited ovalling oscillations. The study, the mechanism, and other relevant data are reported in this paper.

Vortex Shedding Data

It is well known that the frequency of vortex shedding (N) and the associated Strouhal number, $S_N = ND/V$ is strongly dependent on Reynolds number $R_N = \rho DV/\mu$, particularly for rigid circular cylinders at values of $R_N > 2 \times 10^5$.

For an "infinite" aspect ratio, rigid circular cylinder at subcritical values of R_N ($<2 \times 10^5$) the value obtained for S_N is about 0.2. The corresponding periodic forces due to vortex shedding have been observed to cause transverse bending oscillations and, in some cases, in line bending oscillations in the flow direction. The oscillatory in-line force attains a maximum every time an individual vortex is shed and, hence, has the same frequency as the shedding of a single vortex; whereas the fluctuating transverse force has the same frequency as the shedding of a pair of vortices from alternate sides. Therefore it may be presumed that the unsteady pressure distribution due to vortex shedding has a resultant which is not orthogonal to the wind direction.

It is also known for sub-critical flow that the cylinder response to the above forces is not strictly a resonance effect but "aeroelastic", since the cylinder motion alters the flow field (and hence the response) significantly. Papers 8 and 37 of Ref. [4] have presented relevant data and in the latter paper it was concluded that:

(1) Cylinder motions increase the circulatory strength of developing vortices.

(2) Cylinder motions increase the two-dimensionality of the flow field (and, hence, the lengthwise correlation of the periodic forces).

(3) The dynamic lift contains higher harmonics of the Strouhal frequency, N.

Unfortunately, most practical structures of interest operate in the transition or supercritical regions of R_N where the data available is more scanty and less conclusive.

It is believed that the probable main causes for the disagreements between the data of various research workers are differences in end effects and other three-dimensional effects, and the fact that many of the cylinders tested have been rigid thus precluding aeroelastic coupling as referred to above and often termed "synchronism". This belief is reinforced by the data of Ref. [5] which, in particular, shows a wider spectral peak to the oscillatory forces and a lower "effective" value of $S_N \simeq 0.16$. (See also Paper 25 of Ref. [6].)

Vortex-Induced Oscillations - A Possible Mechanism

Two basic types of vortex induced oscillations of thin-walled circular shell structures have been experienced in practice viz "sway" or cantilever type bending in line with or transverse to the flow direction, and "ovalling". If the shell radial motion has the form

$$\omega(x,\theta) = F(x)\cos n\theta , \qquad (1)$$

the sway motion corresponds to $n = 1$ and the general ovalling or "breathing", motions correspond to $n \geqslant 2$. There is no evidence of $n = 0$ motions ever being excited by vortex shedding and the structural frequencies associated with $n = 0$ are so high as to practically preclude the possibility.

Transverse, sway oscillations at a natural frequency Ω can occur when $V = \Omega D/S_N$, i.e., when $N = \Omega$ and the instability will then persist for a range of flow speeds dependent on the amount of structural damping present.

The assumed 1 : 1 relationship between transverse oscillations and vortex shedding is given by Fig. 1a but it should be noted that Fig. 1b presents a possible alternative relationship of 3 : 1 which would result in a lower critical flow speed. Note also that in-line oscillations would imply a 2 : 1 relationship.

The earliest observations of ovalling oscillations prompted the suggestion that a 2 : 1 relationship existed between the ovalling natural frequency and the vortex shedding frequency as in Fig. 2a. However this assumption was questioned in Paper 28 of Ref. [7] and Fig. 2b shows an alternative 4 : 1 relationship if the ovalling axes remain orthogonal to the flow direction. Figures 2c and 2d show possible 1 : 1 and 3 : 1 relationships, if the ovalling axes are oriented at 45° to the flow direction. If r is the value of the appropriate relationship, the corresponding critical flow speed is

$$\left.\begin{aligned} V &= 5\Omega D/r \quad \text{if} \quad S_N = 0.2 \\ \text{or} \quad V &\simeq 6\Omega D/r \quad \text{if} \quad S_N \simeq 0.16 \end{aligned}\right\} \qquad (2)$$

with $r = 1, 2, 3, \ldots$ signifying progressively lower critical speeds. It is further suggested that the lowest perceived critical flow speed would depend upon the structural damping present (i.e., on some parameter such as $2m\,\delta/\rho D^2$ which is important for sway motions) and possibly on some other parameter such as the ratio of aerodynamic to structural stiffness, since the tendency of any motion to increase synchronism would depend on such a ratio in the case of wind excitation.

The ovalling data for the full-scale chimney mentioned in Paper 28 of Ref. [7] suggests that a value of $r = 1$ applied rather than the value $r = 2$.

Figures 2c and 2d are based on the assumption that the resultant of the unsteady pressure distribution due to vortex shedding is not orthogonal to the wind direction. Evidence for this possibility was presented in the preceding section. Further evidence is contained below.

Unsteady Pressure Distributions During Vortex Shedding

It should be possible to infer the preferred mode of ovalling oscillation from the measured unsteady pressure distribution due to vortex shedding. Thus, in Paper 28 of Ref. [7] it was shown that the effective center of pressure of the circumferential distribution acts on a line close to 135° from the windward generator at a $R_N = 0.8 \times 10^5$. This would tend to support the r = 1 mechanism of ovalling suggested above. Figure 3 presents the results referred to above for C_p^* and corresponding data from Ref. [8] also, for subcritical Reynolds numbers in the range $0.36 - 0.47 \times 10^5$.

From the data of Ref. [7] (paper 28) and Case 0 of Ref. [8] which both correspond to low levels of free stream turbulence intensity, it is clearly seen that the effective center of pressure of the unsteady pressure distribution is well aft of $\theta = 90°$. As the turbulence level increases to case 4 (Ref. [8]) the center of pressure moves forward towards $\theta = 90°$.

The above data suggest that the ovalling mode of Fig.2c or 2d (r = 1 or 3) may be more easily excited in conditions of low Reynolds number or low turbulence level but that at high turbulence levels and possibly high Reynolds number the modes of Fig.2a or 2b (r = 2 or 4) may be more easily excited.

A detailed examination of Fig.4 (from Ref. [8]) of the unsteady pressures measured at various circumferential positions as recorded in Ref. [8] showed that the effect of increasing turbulence is to broaden and lower the pressure peak at the Strouhal frequency. Evidence of the second harmonic of the vortex shedding was clearly seen at $\theta = 90°$, 120°, 150° and 180°, and of the third harmonic at 120° and 150°, for some but not all of the four cases of turbulence.

The existence of higher harmonics of the Strouhal frequency could also cause oscillations of a structure at speeds in Eq. (2) corresponding to $r > 1$, and this may be an alternative mechanism to explain such critical speeds. However, it is believed that it would not explain the existence of critical conditions corresponding to $r > 3$ which are reported in the following section, but it would reinforce the possibility of the mechanism previously described for r = 1, 2 and 3.

Wind-Tunnel Test Results

The shells described in Table 1 have been designed to be excited in various modes according to Eq. (2). They were constructed with one or two axial lap joint seams rivetted and/or bonded. Previous tests [3] had shown these seams to have negligible effect on the prediction of the natural frequencies.

Table 1. Properties of Shells Tested.

Shell	Length L(in.)	Radius a(in.)	Thickness h(in.)	L/a	a/h	Material
1	20	2.4	0.01	8.3	240	Aluminium
2	36	2.4	0.01	15.0	240	"
3	60	2.4	0.01	25.0	240	"
4	71	2.4	0.01	29.6	240	"
5	36	4.0	0.01	9.0	400	"
6	46	4.0	0.01	11.5	400	"
7	71	4.0	0.01	17.75	400	"
8	70	6.0	0.01	11.67	600	"
9	36	4.0	0.01	9.0	400	Steel

Table 2 contains the detailed wind tunnel test results together with estimates of predicted natural frequency and predicted critical wind speed based on the two Eqs. (2) which probably represent upper and lower limits of the critical speed. Comparison of the actual and predicted critical speeds enables the value of r to be determined.

Shell 1 was designed such that the frequencies of modes n = 2,3 were nearly identical. It was observed to oscillate in the two modes simultaneously corresponding to a value of r = 3.

Shell 2 was designed to vibrate in the mode n = 2 only. The shell behaved as predicted with a value of r = 3. Oscillations began at 40 feet per second (fps) and the maximum amplitude occurred at 49 fps with a value of r = 3.

Shell 3 was designed such that the frequencies of modes n = 1,2 were nearly identical. Ovalling (n = 2) commenced first at 39 fps (r = 3); at 59 fps the swaying mode (n = 1) was excited but with some evidence

Table 2. Wind Tunnel Test Results - Estimated Values of r.

Shell	n	Predicted Frequency Ω(Hz)	V_{crit} = 5ΩD (fps)	V_{crit} = 6ΩD (fps)	Measured Frequency f(Hz)	Measured V_{crit} (fps)	r
1	1	410.4	820.8	984.0	-	-	-
	2	142.1	284.2	342.0	138	110	3
	3	140.6	281.2	337.0	137	110	3
	4	242.2	484.4	582.0	-	-	-
2	1	136.0	272.0	326.4	-	-	-
	2	62.0	124.0	148.8	60	40 ÷ 49	3
	3	126.3	252.6	303.1	-	-	-
3	1	49.5	99.0	118.8	50	59	2
	2	46.8	93.5	112.3	45	39	3
	3	124.7	249.3	299.3	119	64	4 or 5
	4	238.6	477.2	572.6	-	-	-
4	1	35.4	70.8	85.0	33	75	1
	2	45.4	90.9	109.0	43(41)	33(66)	3(2)
	3	124.5	249.0	299.0	-	-	-
5	1	220.7	735.6	882.8	-	-	-
	2	74.0	246.7	296.0	77	60	5
	3	56.6	188.6	226.4	54	40	5
	4	88.4	294.5	353.6	-	-	-
6	1	137.3	457.8	549.2	-	-	-
	2	47.2	157.2	188.8	46	59(65)	3
	3	49.6	165.3	198.4	48	38(44)	5
	4	86.7	289.5	346.8	-	-	-
7	1	58.5	195.1	234.0	-	-	-
	2	24.5	81.7	98.0	23	30	3
	3	45.7	152.3	182.8	43	52	3
	4	86.1	286.9	344.0	-	-	-
8	1	89.0	445.1	534.0	-	-	-
	2	29.6	148.1	177.6	29	35	4
	3	24.2	120.8	145.2	22(19)	26(45)	5(3)
	4	39.0	195.1	234.0	36	38	6
	5	62.0	310.0	372.0	-	-	-
9	1	227.2	757.4	908.0	-	-	-
	2	76.2	254.0	304.8	-	-	-
	3	58.2	194.2	232.8	58	72	3
	4	91.0	303.3	364.0	-	-	-

N.B. For bracketed values see text.

also of ovalling ($r = 2$); at 64 fps the $n = 3$ mode alone was excited with a value of $r = 4,5$. The addition of an aluminium square section ring (0.125" × 0.125") to the top of Shell 3 eliminated the $n = 2$ oscillations. It lowered the sway frequency slightly but the sway critical speed was increased (65 fps) as was the $n = 3$ critical speed (72 fps) and the $n = 3$ frequency (125 Hz).

Shell 4 behaved in a very interesting manner. Although its swaying ($n = 1$) frequency was the lowest, it started ovalling at 33 fps ($r = 3$) and reached maximum amplitude at 35 fps with the ovalling axes at 45° to the wind direction as shown in Fig.5. As the wind speed was increased a gradual change in the position of the ovalling axes occurred and at 66 fps the axes were orthogonal to the wind direction and with a slightly reduced frequency of oscillation. The corresponding value of r was approximately 2.

Some supplementary tests were made on Shell 4 to investigate the effects of added distributed "structural" damping in the form of (a) 0.005" thick polyvinyl coating (b) 0.005" thick Fablon. The coated shell commenced ovalling at an increased wind speed of 40 fps and slightly reduced frequency. At 66 fps swaying commenced, rather than ovalling with the axes orthogonal to the wind. The main reason for the change is believed to be due to the added mass of coating lowering the sway mode frequency sufficiently for it to "swamp" any tendency to oval with axes orthogonal to the wind. The latter tendency would also be somewhat removed by the inclusion of damping. The addition of the fablon sheet to the coated shell had a more dramatic effect. No ovalling was observed at all and the swaying oscillations commenced at about 78 fps. The sway oscillations were much reduced in amplitude compared to either the coated shell or plain shell. From these results it might be concluded that added damping has been more effective in eliminating the ovalling than in affecting the swaying oscillations. This might explain why so few full-scale ovalling incidents have been reported.

Shell 5 was designed to oscillate in the $n = 2,3$ modes. At a wind speed of 40 fps, the $n = 3$ mode was excited corresponding to $r = 5$, and at 56 fps it had stopped. At 60 fps ovalling occurred $(n = 2)$, again with a value of $r = 5$.

Shell 6 was designed to have almost equal frequencies in the $n = 2, 3$ modes and, hopefully, almost equal critical speeds. However the $n = 3$ mode was excited first at 38 fps $(r = 5)$; as wind speed increased the $n = 2$ mode appeared and at 59 fps $(r = 3)$ it was dominant. These tests were conducted with the two axial seams on the windward and leeward generators; subsequent tests with the seams at 90° to the wind direction showed no change in measured frequencies but slight increases in the critical speeds, e.g., from 38 to 44 fps for $n = 3$ and from 59 to 65 fps for the $n = 2$ mode. This cannot be explained, neither can the fact that in both sets of tests the $n = 3$ mode was excited well before the $n = 2$ mode even though the latter had the lower predicted critical speed.

Shell 7 behaved as expected with the ovalling commencing first at 30 fps $(r = 3)$. Maximum amplitude occurred at 36 fps and it then decreased to zero as wind speed increased. At 52 fps the $n = 3$ mode appeared $(r = 3)$ reaching a maximum amplitude at 60 fps. At this speed there was significant distortion of the shell due to the high static pressures acting.

Shell 8 also behaved as expected and each of the modes $n = 2, 3, 4$ was excited. At 26 fps $n = 3$ oscillation commenced $(r = 5)$ and then ceased at 33 fps. Ovalling commenced at 35 fps with the axes orthogonal to wind and very low amplitude $(r = 4)$. This motion ceased and a strong $n = 4$ motion started at 38 fps $(r = 6)$. At 45 fps $n = 3$ motion recommenced $(r = 3)$ but with a much reduced frequency of 19 Hz; the shell was almost collapsed at the windward face at this speed.

Shell 9 might be considered as the steel equivalent of shell 5 but it behaved quite differently. The $n = 3$ mode commenced at the much higher speed of 72 fps $(r = 3)$ instead of 40 fps $(r = 5)$ which suggests that the formula of Eq. (2) is not the sole criterion for instability. Likewise the $n = 2$ mode could not be excited at all on the steel shell 9.

For Shell 9 and most of the others it was found that once instability had occurred and as the oscillation amplitude increased with increasing wind speed the frequency decreased. This indicated a possible "softening" effect of structural non-linearity.

The results of comparable wind-tunnel tests on much shorter shells having values of $L/a = 2$, 2.4 and typical of open-ended oil storage tanks have shown similar results. Wind excitation occurred in the modes of lowest frequency in the range $n = 6$ to 8 and at wind speeds which suggested values of r in the range 7 to 14 for the different modes. Unfortunately, the significance of r in these cases and the existence of a well-correlated Strouhal frequency is less certain since the aerodynamic effect of the free, open end is likely to be more significant than for the longer shells described earlier. In all these cases the difference between predicted natural frequency and measured value was less than 10%.

Conclusions

(1) A mechanism for wind excitation due to vortex shedding of the swaying ($n = 1$) and circumferential bending ($n \geq 2$) modes of tall circular shells has been proposed which suggests that critical wind speeds might be experienced which are one - rth those previously considered (Eq. 2).

(2) Wind tunnel tests have confirmed this possibility and shown values of r from 1 to 6 for various modes.

(3) Significant differences in results between comparable aluminium and steel shells (Shells 5, 9) and for aluminium shells with differing amounts of added damping (Shell 4) have supported the view that the inherent structural damping, through some parameter such as $2m\delta/\rho D^2$ known to be important for swaying oscillations, and/or a structural-aerodynamic stiffness ratio (involving $E/\rho V^2$), are influential in affecting instability.

(4) Data were presented for the unsteady pressure distributions during vortex shedding which would confirm the possibility of the excitation mechanism postulated.

Notation

a	Shell radius	r	Critical parameter Eq. (2)
C_p^*	Unsteady pressure coefficient	R_N	Reynolds number $\rho DV/\mu$
D	Shell diameter	S_N	Strouhal number ND/V
E	Youngs Modulus	V	Wind velocity
f	Frequency - measured	x	Axial coordinate
h	Shell thickness	δ	Logarithmic decrement
L	Shell length	θ	Circumferential coordinate
m	Shell mass per unit length	$\bar{\Phi}$	Non-dimensional unsteady pressure spectra (Fig. 4)
n	Circumferential wave number		
N	Strouhal frequency	ρ	Air density
q	Dynamic pressure = $\rho V^2/2$	Ω	Natural frequency - predicted

References

1. Sharma, C.B., Johns, D.J.: Wind-Induced Oscillations of Circular Cylindrical Shells: an Experimental Investigation. Loughborough University of Technology, Report TT7008, July 1970.

2. Sharma, C.B., Johns, D.J.: Vibration Characteristics of a Clamped-Free and Clamped-Ring Stiffened Circular Cylindrical Shell. J. Sound Vibration 14 (4), 459-474 (1971).

3. Sharma, C.B., Johns, D.J.: Natural Frequencies of Clamped-Free Circular Cylindrical Shells. J. Sound Vibration.

4. Anon: Wind Effects on Buildings and Structures. International Res. Seminar, N.R.C. Ottawa, Canada. Published by University of Toronto, September 1967.

5. Wootton, L.R.: The Oscillation of Model Circular Stacks due to Vortex Shedding at Reynolds Numbers from 10^5 to 3×10^6. N.P.L. Rep. 1267 (June 1968).

6. Anon: Aerodynamics of Atmospheric Shear Flows. Conference in Munich. Published by AGARD as CP 48, September 1969.

7. Johns, D.J., Scruton, C., Ballantyne, A. (Editors): Wind Effects on Buildings and Structures. Symposium held at Loughborough. Published by Loughborough University of Technology, April 1968.

8. Surry, D.: The Effect of High Intensity Turbulence on the Aerodynamics of a Rigid Circular Cylinder at Subcritical Reynolds Number. UTIAS Report (Toronto), October 1969.

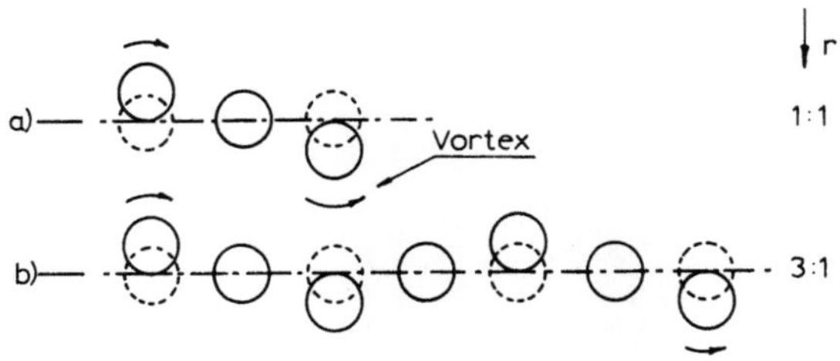

Fig.1. Alternative relationships between bending oscillations and vortex shedding.

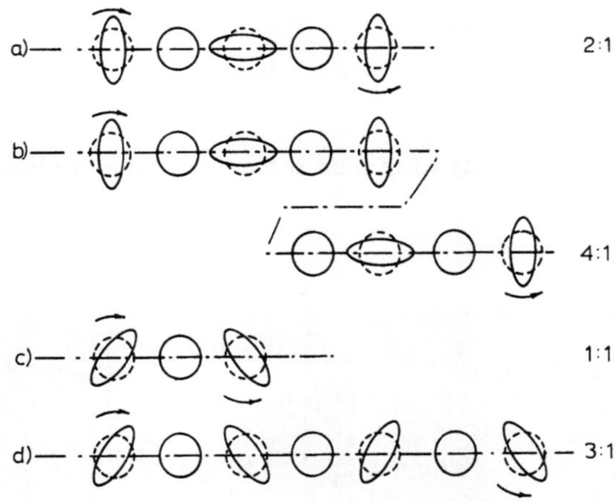

Fig.2. Alternative relationships between ovalling oscillations and vortex shedding.

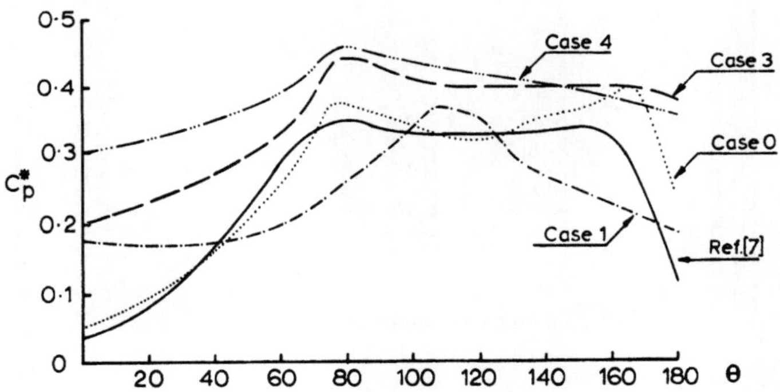

Fig.3. r m s pressure distributions for various turbulence inputs (Ref. [7] and [8]).

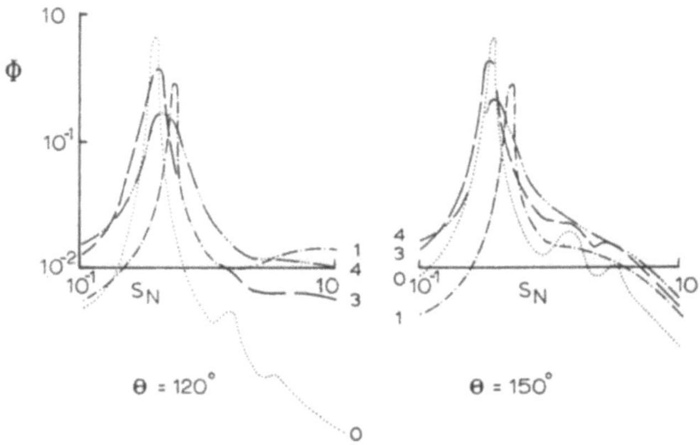

Fig. 4. Pressure spectra for four turbulence inputs (Ref. [8]).

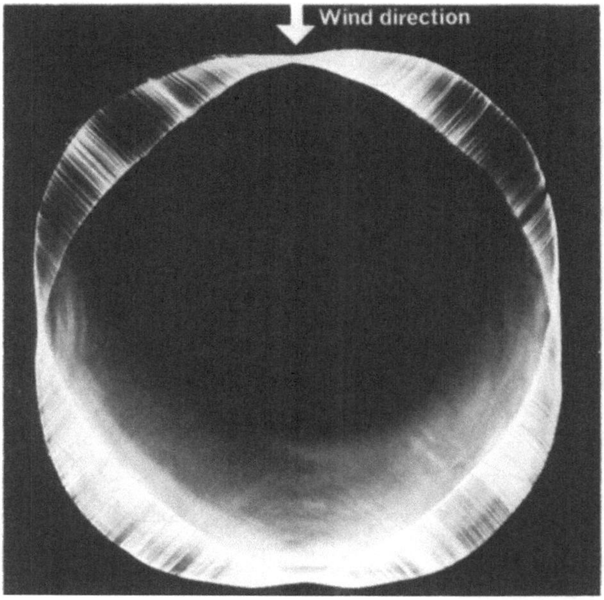

Fig. 5. Ovalling oscillations of shell 4.

Vibrations of Continuous Pipes Conveying Fluid

By

S.S. Chen

Argonne, Ill., U.S.A.

Summary

This paper presents a theoretical investigation of the transverse vibrations of continuous pipes conveying fluid. A general dynamic three-moment equation is derived including the effects of fluid forces and axial force. The three-moment equation is then used to study the free waves and the response to harmonic pressure in a periodically supported pipe. Finally, the free and forced vibrations of multiple span pipes are analyzed; the effects of various parameters on natural frequencies are studied.

Introduction

The transverse vibration of pipes conveying fluid has been investigated extensively because of its importance in the design of oil pipelines, propellant lines, pump-discharge lines, heat-exchanger tubes, nuclear fuel rods, and other system piping [1 to 8]. However, all these investigations have been limited to single span pipes. Therefore, a study is made of the dynamic characteristics of continuous pipes on multiple supports.

A dynamic three-moment equation is derived for a continuous pipe conveying fluid. The equation incorporates the effects of fluid centrifugal force, fluid Coriolis force, fluid pressure, and external axial compression. The three-moment equation is first used to study the free waves in a periodically supported pipe. Next, the response of continuous pipe to convected harmonic pressures is considered. Finally, the three-moment equation is employed to study the free vibration and forced vibration of multiple span pipes. The effects of various parameters on natural frequencies are investigated and the mode shapes and energy-trapping modes are also discussed.

The General Dynamic Three-Moment Equation

The governing differential equation for the transverse vibration of straight pipes conveying fluid is

$$EI\,\frac{\partial^4 y}{\partial x^4}+(MV^2+pA+P)\,\frac{\partial^2 y}{\partial x^2}+2MV\,\frac{\partial^2 y}{\partial x\,\partial t}+(m+M)\,\frac{\partial^2 y}{\partial t^2}=q\psi(x)e^{i\omega t}\,, \quad (1)$$

where E is the modulus of elasticity, I is the moment of inertia of the pipe, x is the axial coordinate, y is the transverse displacement, t is the time, p is the fluid pressure, A is the internal flow area, P is the axial compression, V is the flow velocity, and m and M are the masses per unit length of the pipe and fluid, respectively. The system is postulated to be excited by the sinusoidal force per unit length $q\psi(x)e^{i\omega t}$, where ω is the forcing frequency, q is a constant, and $\psi(x)$ describes the axial variation of the loading.

Assume that the solution of Eq. (1) is

$$y = l\eta(x)e^{i\omega t}\,, \quad (2)$$

where l is the length of the span under consideration. Further, we define the following dimensionless parameters:

$$\xi = x/l\,, \quad , \quad \beta = \left(\frac{M}{m+M}\right)^{1/2}, \quad Q = \frac{ql^3}{EI}\,,$$

$$u = \left(\frac{M}{EI}\right)^{1/2} Vl\,, \quad \Omega = \left(\frac{m+M}{EI}\right)^{1/2} l^2\omega\,, \quad \Gamma = \frac{Pl^2}{EI}\,, \quad (3)$$

$$\nu = \frac{Apl^2}{EI}\,.$$

Equation (1) becomes

$$\frac{d^4\eta}{d\xi^4} + (u^2 + \nu + \Gamma)\frac{d^2\eta}{d\xi^2} + i2\beta u\Omega\frac{d\eta}{d\xi} - \Omega^2\eta = Q\psi(\xi)\,. \qquad (4)$$

The general solution of Eq. (4) may be written in terms of four arbitrary constants C_n $(n = 1, 2, 3, 4)$ as

$$\eta(\xi) = \sum_{n=1}^{4} C_n e^{i\lambda_n\xi} + \Phi(\xi)\,, \qquad (5)$$

where the λ_n are solutions of the following equation:

$$\lambda^4 - (u^2 + \nu + \Gamma)\lambda^2 - 2\beta u\Omega\lambda - \Omega^2 = 0 \qquad (6)$$

and $\Phi(\xi)$ is the particular solution of Eq. (4).

Consider two neighboring spans of continuous pipes as shown in Fig. 1b. The notations without bar refer to the variables of the left-hand-side span, while the notations with bar refer to those of the right-hand-side span of the jth support. The pipe displacements must satisfy the following conditions:

$$\left.\begin{aligned} &\eta(0) = 0\,, \quad \frac{d^2\eta(0)}{d\xi^2} = -\alpha_{j-1}\,, \quad \left(\alpha_{j-1} = \frac{1}{EI}M_{j-1}\right)\,, \\ &\eta(1) = 0\,, \quad \frac{d^2\eta(1)}{d\xi^2} = -\alpha_j\,, \quad \left(\alpha_j = \frac{1}{EI}M_j\right)\,, \end{aligned}\right\} \qquad (7)$$

$$\left.\begin{aligned} &\bar\eta(0) = 0\,, \quad \frac{d^2\bar\eta(0)}{d\bar\xi^2} = -\bar\alpha_j\,, \quad \left(\bar\alpha_j = \frac{\bar I}{\bar{EI}}M_j\right)\,, \\ &\bar\eta(1) = 0\,, \quad \frac{d^2\bar\eta(1)}{d\bar\xi^2} = -\bar\alpha_{j+1}\,, \quad \left(\bar\alpha_{j+1} = \frac{\bar I}{\bar{EI}}M_{j+1}\right) \end{aligned}\right\} \qquad (8)$$

and

$$\frac{d\eta(1)}{d\xi} = \frac{d\bar\eta(0)}{d\bar\xi}\,, \qquad (9)$$

where M_{j-1}, M_j, and M_{j+1} are the bending moments at the supports j-1, j, and j+1, respectively. Substitution of Eq. (5) into Eqs. (7) yields

$$\begin{bmatrix} 1 & 1 & 1 & 1 \\ e^{i\lambda_1} & e^{i\lambda_2} & e^{i\lambda_3} & e^{i\lambda_4} \\ \lambda_1^2 & \lambda_2^2 & \lambda_3^2 & \lambda_4^2 \\ \lambda_1^2 e^{i\lambda_1} & \lambda_2^2 e^{i\lambda_2} & \lambda_3^2 e^{i\lambda_3} & \lambda_4^2 e^{i\lambda_4} \end{bmatrix} \begin{bmatrix} C_1 \\ C_2 \\ C_3 \\ C_4 \end{bmatrix} = \begin{bmatrix} -\Phi(0) \\ -\Phi(1) \\ \alpha_{j-1} + \dfrac{d^2\Phi(0)}{d\xi^2} \\ \alpha_j + \dfrac{d^2\Phi(1)}{d\xi^2} \end{bmatrix} . \tag{10}$$

From Equations (10), we obtain

$$C_n = a_n \left[\alpha_{j-1} + \frac{d^2\Phi(0)}{d\xi^2}\right] + b_n \left[\alpha_j + \frac{d^2\Phi(1)}{d\xi^2}\right] + d_n\Phi(0) + e_n\Phi(1) , \tag{11}$$

where a_n, b_n, d_n, and e_n are functions of λ_i. For conciseness, they are not given here. Therefore, the solution for $\eta(\xi)$ between j-1 and j supports is

$$\eta(\xi) = \sum_{n=1}^{4} \left\{ a_n \left[\alpha_{j-1} + \frac{d^2\Phi(0)}{d\xi^2}\right] + b_n \left[\alpha_j + \frac{d^2\Phi(1)}{d\xi^2}\right] + d_n\Phi(0) + e_n\Phi(1) \right\} e^{i\lambda_n\xi} + \Phi(\xi) . \tag{12}$$

Similarly, using the equation of motion and Eqs. (8), we obtain the expression for $\bar{\eta}(\xi)$ between j and j+1 supports:

$$\bar{\eta}(\xi) = \sum_{n=1}^{4} \left\{ \bar{a}_n \left[\bar{\alpha}_j + \frac{d^2\bar{\Phi}(0)}{d\bar{\xi}^2}\right] + \bar{b}_n \left[\bar{\alpha}_{j+1} + \frac{d^2\bar{\Phi}(1)}{d\bar{\xi}^2}\right] + \bar{d}_n\bar{\Phi}(0) + \bar{e}_n\bar{\Phi}(1) \right\} e^{i\bar{\lambda}_n\bar{\xi}} + \bar{\Phi}(\bar{\xi}) , \tag{13}$$

where the variables with bar are similarly defined as those without bar. Finally, the condition of continuity at the joint j as given by Eq. (9) yields the following general three-moment equation:

$$\sum_{n=1}^{4}\left(a_n\lambda_n e^{i\lambda_n}\right)\frac{1}{EI}M_{j-1}+\left\{\sum_{n=1}^{4}\left(b_n\lambda_n e^{i\lambda_n}\right)\frac{1}{EI}-\sum_{n=1}^{4}(\bar{a}_n\bar{\lambda}_n)\frac{\bar{l}}{\overline{EI}}\right\}M_j$$

$$-\sum_{n=1}^{4}(\bar{b}_n\bar{\lambda}_n)\frac{\bar{l}}{\overline{EI}}M_{j+1}$$

$$=-\sum_{n=1}^{4}\left(a_n\lambda_n e^{i\lambda_n}\right)\frac{d^2\Phi(0)}{d\xi^2}+\sum_{n=1}^{4}(\bar{a}_n\bar{\lambda}_n)\frac{d^2\bar{\Phi}(0)}{d\bar{\xi}^2}-\sum_{n=1}^{4}\left(b_n\lambda_n e^{i\lambda_n}\right)\frac{d^2\Phi(1)}{d\xi^2}$$

$$+\sum_{n=1}^{4}(\bar{b}_n\bar{\lambda}_n)\frac{d^2\bar{\Phi}(1)}{d\bar{\xi}^2}-\sum_{n=1}^{4}\left(d_n\lambda_n e^{i\lambda_n}\right)\Phi(0)+\sum_{n=1}^{4}(\bar{d}_n\bar{\lambda}_n)\bar{\Phi}(0)$$

$$-\sum_{n=1}^{4}\left(e_n\lambda_n e^{i\lambda_n}\right)\Phi(1)+\sum_{n=1}^{4}(\bar{e}_n\bar{\lambda}_n)\bar{\Phi}(1)+i\frac{d\Phi(1)}{d\xi}-i\frac{d\bar{\Phi}(0)}{d\bar{\xi}}\,. \qquad (14)$$

Equation (14) is the three-moment equation for the system under consideration.

Free Waves in Periodically Supported Pipes

The three-moment equation is used to study the free waves in pipes. We consider a fluid-conveying pipe which rests on identical supports equally spaced at distance l. In this case, the three-moment equation for free vibration is

$$\sum_{n=1}^{4}\left(a_n\lambda_n e^{i\lambda_n}\right)M_{j-1}+\sum_{n=1}^{4}\left(b_n\lambda_n e^{i\lambda_n}-a_n\lambda_n\right)M_j-\sum_{n=1}^{4}(b_n\lambda_n)M_{j+1}=0. \qquad (15)$$

Equation (15) applies at each support; therefore, the set of equations for all supports constitutes a recurrence relationship between successive support moments. For an infinite, periodically supported pipe, the solution of Eq. (15) is

$$M_j = M_{j-1} e^{i\mu} , \tag{16}$$

where μ is seen to be the propagation constant. Substitution of Eq. (16) into Eq. (15) yields the characteristic equation:

$$\sum_{n=1}^{4} \left(a_n \lambda_n e^{i\lambda_n} \right) e^{-i\mu} + \sum_{n=1}^{4} \left(b_n \lambda_n e^{i\lambda_n} - a_n \lambda_n \right) - \sum_{n=1}^{4} (b_n \lambda_n) e^{i\mu} = 0 . \tag{17}$$

In general, μ is complex. The real part $Re(\mu)$ represents the phase difference between moments at adjacent supports. The imaginary part $Im(\mu)$ represents the exponential decay rate of the moment wave as it propagates from one support to the next.

μ depends on the parameters Γ, ν, β, u, and Ω; therefore, we may write

$$\mu = \mu(\Gamma, \nu, \beta, u, \Omega) . \tag{18}$$

The propagation constants of two examples are shown in Fig. 2. When $Im(\mu)$ is zero, free waves can propagate without attenuation; these regions are called propagation bands. When $Im(\mu)$ is not zero, free waves cannot propagate; these regions are called stop bands. There exist alternate bands of free propagation and attenuation; this property is intrinsic to a periodic structure.

The effects of flow velocity, fluid pressure, and axial compression have been investigated. Increasing the flow velocity is to shift the position of each band. Also, the propagation bands become wider as flow velocity increases. At $u = \pi$, free wave with zero frequency can propagate. In fact, π is the first buckling flow velocity.

The effects of fluid pressure and axial compression are similar to those of flow velocity. Increasing the fluid pressure and axial compression can also make the system unstable. However, the role of β is different. Note that β is proportional to the Coriolis acceleration. Due to the Coriolis force effect, the free waves propagate with different speeds in the downstream and upstream directions. This can be seen from Fig.2b; the positive-going and negative-going waves have different magnitudes in μ. Since the wave velocities are different in two directions, the system does not possess classical normal modes.

The Response of an Infinite Periodically Supported Pipe to a Convected Pressure

Consider an infinite periodically supported pipe which has uniform cross-section and is excited by a convected pressure field

$$f(x,t) = q e^{i(\omega t - kx)} . \tag{19}$$

The convection velocity of the pressure is ω/k, and the phase difference between pressures at points with l apart is -kl. The motion of the system can be studied using Eqs. (4) and (14). The axial variation of the loading ψ is given by

$$\psi(\xi) = e^{-i\varkappa\xi}, \quad \varkappa = kl . \tag{20}$$

Using Eq. (20) and solving Eq. (4) yield the particular solution $x(\xi)$;

$$\Phi(\xi) = \frac{Qe^{-i\varkappa\xi}}{\varkappa^4 - (u^2 + \nu + \Gamma)\varkappa^2 + 2\beta u\Omega\varkappa - \Omega^2} . \tag{21}$$

Since the system properties are uniform, substituting Eq. (21) into Eq. (14), and assuming the solution to be

$$M_j = M_{j-1} e^{i\varkappa} , \tag{22}$$

then, we obtain

$$M_j = (ql^2) F(\Gamma, \nu, \beta, u, \Omega, \varkappa) , \tag{23}$$

where

$$F = \frac{\left[\sum_{n=1}^{4} d_n\lambda_n\left(1-e^{-i\lambda_n}\right)-\sum_{n=1}^{4} a_n\lambda_n\left(1-e^{-i\lambda_n}\right)\varkappa^2-\varkappa\right]+\left[\sum_{n=1}^{4} e_n\lambda_n\left(1-e^{-i\lambda_n}\right)-\sum_{n=1}^{4} b_n\lambda_n\left(1-e^{-i\lambda_n}\right)\varkappa^2+\varkappa\right]e^{-i\varkappa}}{\left[\sum_{n=1}^{4} a_n\lambda_n e^{i(\lambda_n-\varkappa)}+\sum_{n=1}^{4}\left(b_n\lambda_n e^{i\lambda_n}-a_n\lambda_n\right)-\sum_{n=1}^{4} b_n\lambda_n e^{i\varkappa}\right]\left[\varkappa^4-(u^2+\nu+\Gamma)\varkappa^2+2\beta u\Omega\varkappa-\Omega^2\right]} . \tag{24}$$

The response of moment M_j depends on the parameters: Γ, ν, β, u, Ω, and $\varkappa$. M_j will become infinite for certain ranges of parameters. From Eqs. (23) and (24) it is seen that resonance occurs if

$$\sum_{n=1}^{4} a_n\lambda_n e^{i\lambda_n}e^{-i\varkappa}+\sum_{n=1}^{4}\left(b_n\lambda_n e^{i\lambda_n}-a_n\lambda_n\right)-\sum_{n=1}^{4} b_n\lambda_n e^{i\varkappa}=0 \tag{25}$$

or

$$\varkappa^4-(u^2+\nu+\Gamma)\varkappa^2+2\beta u\Omega\varkappa-\Omega^2=0 . \tag{26}$$

Comparing Eqs. (25) with (17), we see that resonance occurs if the phase difference between pressures is equal to that between free waves. This is, the convection velocity of the pressure field is equal to the phase velocity of one of the free wave components of the periodically supported pipe. On the other hand, Eq. (26) indicates the normal coincidence of the unsupported pipe and pressure field frequencies. These two kinds of resonances exist in the infinite, periodically supported pipe.

Multiple Span Pipes

The natural frequency and response of continuous pipes are difficult to obtain by conventional method. With the three-moment equation, the problem is greatly simplified. For the purpose of illustration, we consider an N-span pipe which is simply-supported at the ends (Fig.1a). Applying the three-moment equation to each of two consecutive spans, we obtain

$$\begin{bmatrix} B_1 & C_1 & 0 & 0 & 0 & - & - & - & - & - & - & - \\ A_2 & B_2 & C_2 & 0 & 0 & - & - & - & - & - & - & - \\ 0 & A_3 & B_3 & C_3 & 0 & - & - & - & - & - & - & - \\ 0 & 0 & A_4 & B_4 & C_4 & - & - & - & - & - & - & - \\ - & - & - & - & - & - & - & - & - & - & - & - \\ - & - & - & - & - & - & - & A_{N-4} & B_{N-4} & C_{N-4} & 0 & 0 \\ - & - & - & - & - & - & - & 0 & A_{N-3} & B_{N-3} & C_{N-3} & 0 \\ - & - & - & - & - & - & - & 0 & 0 & A_{N-2} & B_{N-2} & C_{N-2} \\ - & - & - & - & - & - & - & 0 & 0 & 0 & A_{N-1} & B_{N-1} \end{bmatrix} \begin{bmatrix} M_1 \\ M_2 \\ M_3 \\ M_4 \\ \vdots \\ M_{N-4} \\ M_{N-3} \\ M_{N-2} \\ M_{N-1} \end{bmatrix} = \begin{bmatrix} F_1 \\ F_2 \\ F_3 \\ F_4 \\ \vdots \\ F_{N-4} \\ F_{N-3} \\ F_{N-2} \\ F_{N-1} \end{bmatrix} \tag{27}$$

where A_j, B_j, C_j are the coefficients of M_{j-1}, M_j, and M_{j+1} in Eq. (14), respectively, and F_j represents the right-hand-side of Eq. (14) of the jth support.

For a given system under external excitations, Eqs. (27) are used to solve for the bending moment M_j. Having the bending moment M_j, we can calculate the displacement from Eqs. (12) and (13).

Equations (27) are also useful to find the natural frequency. For free vibration, the F_j's are zero and Eqs. (27) become a system of homogeneous linear equations. A necessary condition for the existence of a nontrivial solution is that the determinant of the coefficient matrix equals zero; thus, the frequency equation may be written as

$$f(\Omega, u, \Gamma, \beta, \nu) = 0 . \tag{28}$$

With the aid of a digital computer, the natural frequency can be obtained numerically.

If the pipe is uniform, we have

$$\begin{aligned} A_1 &= A_2 = A_3 = \dots = A_{N-1} = A , \\ B_1 &= B_2 = B_3 = \dots = B_{N-1} = B , \\ C_1 &= C_2 = C_3 = \dots = C_{N-1} = C . \end{aligned} \tag{29}$$

Then, the frequency equation can be shown as follows:

$$(AC)^{\frac{N-1}{2}} (z^{N-1} + z^{N-3} + z^{N-5} + \dots + z^{5-N} + z^{3-N} + z^{1-N}) = 0\,, \quad (30)$$

where z is defined by

$$z + \frac{1}{z} = B/\sqrt{AC}\,. \quad (31)$$

Multiplying Eq. (30) by $(z^2 - 1)$, we obtain

$$(AC)^{\frac{N-1}{2}} (z^{2N} - 1)/z^{N-1} = 0\,. \quad (32)$$

The solutions of Eq. (32) are

$$\begin{aligned} &A = 0\,, \quad C = 0\,, \\ &z = \cos\frac{\gamma\pi}{N} + i \sin\frac{\gamma\pi}{N}\,, \quad \gamma = 1, 2, 3, \dots N\,. \end{aligned} \quad (33)$$

In general, A and C cannot be zero; therefore, the frequency equations are given by the last equation of Eqs. (33). Using Eq. (31), we obtain

$$B = 2\sqrt{AC}\cos\frac{\gamma\pi}{N}\,, \quad \gamma = 1, 2, 3, \dots N \quad (34)$$

or

$$\sum_{n=1}^{4}\left(b_n\lambda_n e^{i\lambda_n} - a_n\lambda_n\right) = 2\left[-\sum_{n=1}^{4}\left(a_n\lambda_n e^{i\lambda_n}\right)\sum_{n=1}^{4}(b_n\lambda_n)\right]^{1/2}\cos\frac{\gamma\pi}{N}\,, \quad (35)$$

$$\gamma = 1, 2, 3, \dots N\,.$$

Next, we consider two specific numerical examples to show the effects of various parameters. First, the natural frequencies of an eight-span uniform pipe conveying fluid are computed; the results show:

(1) There are eight natural frequencies in each propagation band for the eight-span pipe under consideration. In general, for a continuous pipe with N spans, there are N frequencies in each propagation band.

(2) The flow velocity tends to reduce the natural frequencies; this is attributed to the fluid centrifugal force.

(3) The effects of axial compression and fluid pressure are similar to that of flow velocity.

(4) The Coriolis force, which is proportional to β, tends to reduce the natural frequencies in the first propagation band but to increase the frequencies in the second band. The Coriolis force also has the effect of increasing the natural frequencies in the other propagation bands.

Secondly, we calculate the natural frequency of an eight-span nonuniform pipe (i.e., the dimensionless axial force is taken to be 4.0 in the fourth and fifth spans and zero in the others). The results are given in Table 1, where the frequency of the corresponding uniform pipe (i.e., $u = 0$, $\nu = 0$, $\beta = 0.5$ and $\Gamma = 0.0$) are also included for comparison.

Table 1. Dimensionless Natural Frequencies of Continuous Pipes

First Propagation Band		Second Propagation Band	
Uniform Pipe	Nonuniform Pipe	Uniform Pipe	Nonuniform Pipe
9.87	9.00	39.48	38.81
10.30	10.23	40.41	40.34
11.51	10.94	42.82	42.16
13.29	12.94	46.17	45.81
15.42	15.09	49.96	49.57
17.71	17.24	53.91	53.34
19.92	19.76	57.63	57.46
21.67	21.19	60.52	59.87

In order to understand the difference between the uniform and nonuniform pipes, the propagation constants for the two cases, i.e., $u = 0$, $\nu = 0$, $\beta = 0.5$ and $\Gamma = 0$ and 4.0, have been computed. From Table 1 and the propagation constants, it is seen that no natural frequency exists in the stop bands for a uniform pipe. However, certain frequencies do exist in the stop band for a nonuniform pipe; in this case, the stop band of one portion of the structure may be the propagation band of another portion of the structure.

For example, for $7.51 < \Omega < 9.91$ (from propagation constants not shown here) the flexural wave propagates freely for $\Gamma = 4.0$, but is attenuated for $\Gamma = 0$. If natural frequencies exist in this range, the shape of the modes will differ markedly from that for a uniform pipe, these modes are called energy-trapping modes. There are two energy-trapping modes for the nonuniform pipe given in Table 1: $\Omega = 9.00$ and $\Omega = 38.81$. The mode shape for $\Omega = 9.00$ has been investigated; the absolute amplitude of the uniform tube is the same in each span, while for the nonuniform pipe, the amplitudes in the fourth and fifth spans are much larger than the others; a large portion of the energy is confined in the center two spans. Thus if the excitations approach this particular frequency, the vibration energy is trapped in the center portion of the pipe.

References

1. Housner, G.W.: Bending Vibrations of a Pipe Line Containing Flowing Fluid. J. Appl. Mech. 19, 205-208 (1952).

2. Niordson, F.I.N.: Vibrations of a Cylindrical Tube Containing Flowing Fluid. Trans. Roy. Inst. Techn., Stockholm No. 73 (1953).

3. Benjamin, T.B.: Dynamics of a System of Articulated Pipes Conveying Fluid, Parts I and II. Proc. Roy. Soc. London A, 261, 457-499 (1961).

4. Gregory, R.W., Paidoussis, M.P.: Unstable Oscillation of Tubular Cantilevers Conveying Fluid, Part I -- Theory, Part II -- Experiment. Proc. Roy. Soc. London A, 293, 512-542 (1966).

5. Chen, S.S.: Flow-Induced Instability of an Elastic Tube. Presented at ASME Vibration Conference, September 8-10, 1971, Toronto, Canada; ASME Paper No. 71-Vibr-39.

6. Chen, S.S., Wambsganss, M.W.: Parallel-Flow-Induced Vibration of Fuel Rods. Presented at the First International Conference on Structural Mechanics in Reactor Technology, September 20-24, 1971, Berlin, Germany; J. Nucl. Engng. Design 18, 253-278 (1972).

7. Chen, S.S.: Dynamic Stability of Tube Conveying Fluid. J. Engng. Mech. Div. ASCE 97, No. EM5, 1469-1485 (October 1971).

8. Chen, S.S.: Vibration and Stability of a Uniformly Curved Tube Conveying Fluid. J. Acoustical Soc. Amer. 51, 223-232 (1972).

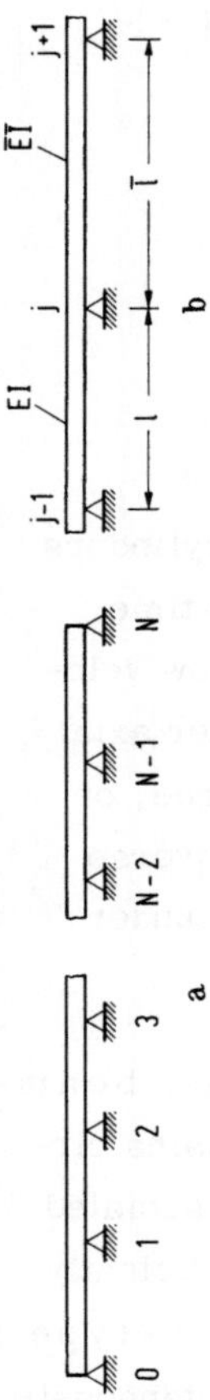

Fig.1. The continuous pipe.

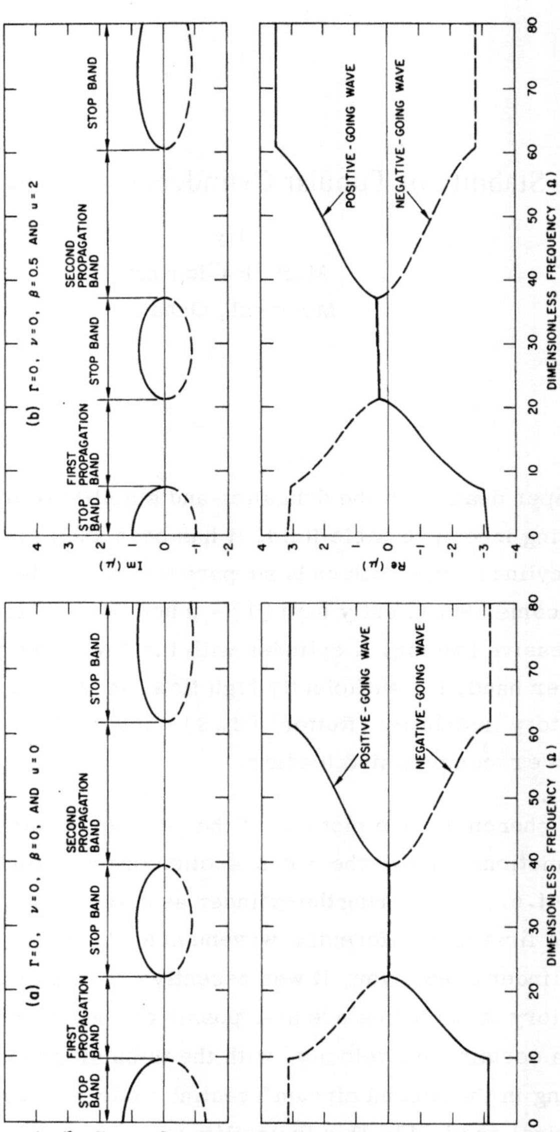

Fig.2. Propagation of flexural waves in an infinite, periodically supported pipe conveying fluid.

Stability of Tubular Cylinders Conveying Fluid

By

M.P. Païdoussis

Montreal, Canada

This paper deals with the dynamics and stability of tubular cylinders conveying incompressible fluid. It has been known for some time that a cylinder with both ends supported may buckle if the flow velocity becomes sufficiently high [1] - much like a column under axial compressive loading. A cylinder with the downstream end free, on the other hand, for sufficiently high flow, is subject to transverse oscillatory instability (flutter) [2, 3] - much like a column under follower compressive loading.

In both phenomena the motions of the cylinder are transverse, beam-type motions, where the cross-section of the cylinder remains circular - i.e., considering the cylinder as a shell, they are associated with the first circumferential wavenumber $(n = 1)$. For sufficiently thin cylinders, however, it was recently observed that shell-type oscillatory instabilities are also possible; they develop spontaneously above a certain flow velocity, with the tutbular cylinder in this case vibrating in the second circumferential mode $(n = 2)$ of a circular cylindrical shell [4]. This instability was observed both in cantilevered cylinders and cylinders with both ends supported. In both cases the instability developed into a limit-cycle of large amplitude, the largest amplitudes occurring in the vicinity of the downstream end. Depending on the characteristics of the cylinder the shell-type flutter either preceded or was superimposed on the beam-type flutter.

In this paper we shall examine the dynamics of the system by means of a linear theory in which the cylinder is taken to be a thin shell and the aerodynamic forces are obtained by potential-flow theory. The equations of motion [5] involving the axial, circumferential and radial motions of the middle surface of the shell, u, v and w respectively, are given by

$$L_1(u,v,w) = \gamma\partial^2 u/\partial t^2 ,$$

$$L_2(u,v,w) = \gamma\partial^2 v/\partial t^2 , \qquad (1)$$

$$L_3(u,v,w) = \gamma(\partial^2 w/\partial t^2 - P/\rho_s h) ,$$

where L_1, L_2 and L_3 are linear differential operators involving derivatives in the axial and circumferential co-ordinates x and θ; $\gamma = \rho_s a^2(1 - \nu^2)/E$, where ρ_s is the density of the shell and a its radius, ν is Poisson's ratio and E Young's modulus; h is the thickness of the shell and P is the difference between internal and external pressures measured at the wall.

We express the velocity field by $\underline{V} = \underline{\nabla}\Psi$ where $\Psi = Ux + \Phi$, Φ being the perturbation potential, and the pressure by $P = P_0 + p$, where P_0 is due to the steady flow involving the mean flow velocity U, and p is the perturbation pressure. Then using Bernoulli's equation $\partial\Psi/\partial t + V^2/2 + P/\rho = 0$ we obtain the following expression for p:

$$p = -\rho(\partial\Phi/\partial t + U\partial\Phi/\partial x)\big|_{r=a} . \qquad (2)$$

We next seek solutions of the form

$$\begin{Bmatrix} u \\ v \\ w \end{Bmatrix} = \begin{Bmatrix} A \\ B \\ C \end{Bmatrix} \exp[i(\lambda x/a + n\theta + \omega t)] \qquad (3)$$

and

$$\Phi = R(r)\exp[i(\lambda x/a + n\theta + \omega t)] . \qquad (4)$$

Using Laplace's equation $\nabla^2\Psi = 0$, the conditions of impermeability of the shell $\partial\Phi/\partial r|_{r=a} = (\partial w/\partial t + U\partial w/\partial x)$ and Eq. (4) we obtain

$$p = \frac{\rho}{n + \lambda I_{n+1}(\lambda)/I_n(\lambda)} \left(\frac{\partial^2 w}{\partial t^2} + 2U \frac{\partial^2 w}{\partial x \partial t} + U^2 \frac{\partial^2 w}{\partial x^2} \right). \qquad (5)$$

Substituting Eq. (5) into Eq. (1) and using Eq. (3), we obtain three linear algebraic equations in which the determinant of the coefficients of A, B, C must vanish for non-trivial solution; this yields a transcendental equation in λ in which ω also appears, i.e.

$$F(\lambda, \omega, n, h/a, \ell/a, \varepsilon, \bar{U}) = 0\,, \qquad (6)$$

where $\varepsilon = \rho a/\rho_s h$, and $\bar{U} = U[\rho_s(1 - \nu^2)/E]^{1/2}$ is the dimensionless flow velocity. If an ω is assumed, then the corresponding λ's may be obtained such that the complete solution becomes

$$\begin{Bmatrix} u \\ v \\ w \end{Bmatrix} = \sum_j \begin{Bmatrix} A_j \\ B_j \\ C_j \end{Bmatrix} \exp[i(\lambda_j x/a + n\theta + \omega t)]\,. \qquad (7)$$

Substitution of Eq. (7) into the eight boundary conditions (e.g. $u = v = $ $= w = 0$, $\partial w/\partial x = 0$ for a clamped edge) should lead to another vanishing determinant

$$G(\lambda, n) = 0 \qquad (8)$$

if the assumed ω is a natural frequency of the system.

One unfortunate aspect of this classical method of solution is that Eq. (6) being transcendental in λ will yield an infinite number of λ's, and hence in Eq. (7) $j = 1, 2, \ldots \infty$; this is then not consistent with Eq. (8) which is obtained on the assumption that $j = 1, \ldots 8$ equalling the number of boundary conditions. Fortunately, it was found that truncating the solution to eight terms - or even to four terms for clamped-clamped boundary conditions - yields adequately accurate results. A Ritz-Galerkin type of solution which by-passes this difficulty is currently under way.

[There is no inconsistency in that the method employed here yields an infinite number of independent solutions. If a Laplace transform method of solution in the x-coordinate is utilized, the general solution may be expressed in terms of only six independent functions. The difference, let us say, is equivalent to solving a dynamical problem by modal analysis, where the solution involves an infinite set of admissible functions, yet the closed-form solution, if it can be found, involves only a finite number of independent solutions.]

The complex natural frequencies, ω, were calculated [6] as functions of the flow velocity for systems similar to those of the experiments, i.e. for thin rubber tubes conveying low pressure air. For each set of parameters $\bar{U}$, ℓ/a, h/a, ε, there is a doubly infinite set ω_{nm} where n is the circumferential wavenumber and m is the number of axial half-waves.

It was found that with fixed ℓ/a, h/a and ε and increasing $\bar{U}$, the natural frequencies of cylinders with both ends supported decrease but remain real. For sufficiently high $\bar{U}$ the ω_{nm} vanish in turn indicating multiple buckling instabilities. For some n, the $m = 1$ and $m = 2$ modes buckle at almost identical $\bar{U}$ and, immediately after, their mode-loci coalesce giving rise to coupled-mode flutter. Cantilevered cylinders, on the other hand, lose stability by flutter, but do not buckle; at least one mode (for each n) crosses from positive to negative $\mathrm{Im}(\omega)$ while $\mathrm{Re}(\omega) \neq 0$.

Next, the effect of ℓ/a and of ε on stability was investigated, for systems similar to those used in the experiments: i.e. thin rubber ($h/a = 2.27 \times 10^{-2}$, $\nu = 0.5$) conveying low pressure air. We note that for both clamped-clamped and cantilevered cylinders (Figs.1 and 2) the system loses stability in $n = 3$ for small ℓ/a, then in $n = 2$ for larger ℓ/a, and very slender cylinders lose stability in $n = 1$. This is easily explained by considering the amount of strain energy involved in the beam-type motions ($n = 1$) and shell-type motions ($n > 1$) for small and large ℓ/a.

In Figs. 3 and 4 we see the effect of mass ratio ε; the heavier the fluid the lower the critical point for instability; this makes sense if we note that the fluid forces are associated with changes in fluid momentum.

We next consider the mechanism of instability. The onset of buckling is associated with the term $U^2\partial^2 w/\partial x^2$ of Eq. (5). The mechanism associated with $n = 1$ is well known. For $n > 1$ we note that any essentially inextensible movement of the shell reduces the flow area and, hence, the internal pressure; this effect increases with flow velocity, and at a certain $\bar{U}$, it exceeds the effect of restoring flexural forces, leading to buckling. This mechanism is capable of producing buckling for any boundary conditions. In the case of cantilevered shells, however, the system loses stability only by flutter. The mechanism of flutter may be examined [3] by considering the rate of work done by the non-conservative forces of Eq. (5), i.e.

$$dW/dt \propto \int_0^{\ell} \int_0^{2\pi} (\ddot{w} + 2U\dot{w}' + U^2 w'')\, \dot{w}\, a\, d\theta\, dx ,$$

which, for $w \propto \cos n\theta$, leads to an equation

$$dW/dt \propto -(\dot{w}^2 + U\dot{w}w')\Big|_0^{\ell} . \tag{9}$$

Clearly for a cantilever, for large enough U and for $\overline{(\dot{w}w')}_{\ell} < 0$, dW/dt will be positive, i.e. the system will absorb energy and flutter will occur; incidentally, $\overline{(\dot{w}w')}_{\ell} < 0$ means a sort of dragging motion of the free lips of the cantilevered shell, which was the type of motion actually observed. Equation (9) paradoxically indicates that a cylinder supported at both ends cannot lose stability by flutter.

References

1. Niordson, F.I.: K. tek. Högsk. Handl. No. 73 (1953).

2. Gregory, R.W., Paidoussis, M.P.: Proc. Roy. Soc. London A, 293, 512 (1966).

3. Paidoussis, M.P.: J. Mech. Eng. Sci. 12, 85 (1970).

4. Paidoussis, M.P., Denise, J.-P.: J. Sound Vib. 14, 433 (1971).

5. Flügge, W.: Statik und Dynamik der Schalen, 2. Aufl., Berlin-Göttingen-Heidelberg: Springer 1957.

6. Paidoussis, M.P., Denise, J.-P.: J. Sound Vib. 20, 9 (1972).

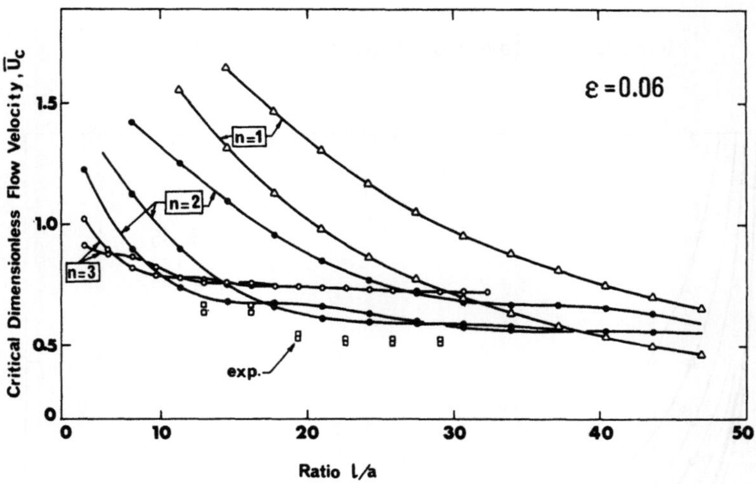

Fig.1. Clamped-clamped cylinders: effect of l/a.

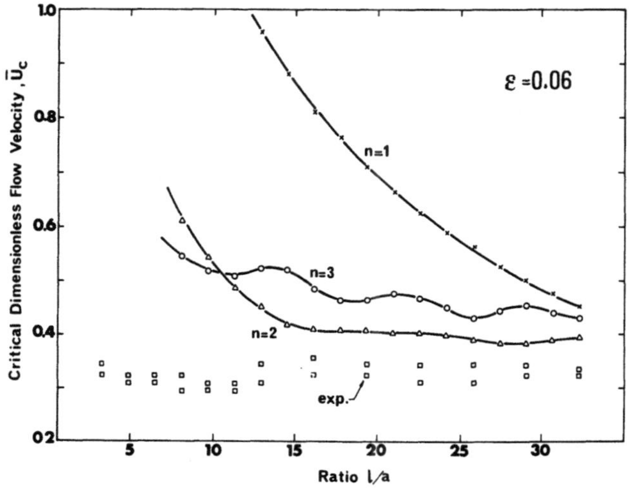

Fig.2. Cantilevered cylinders: effect of l/a.

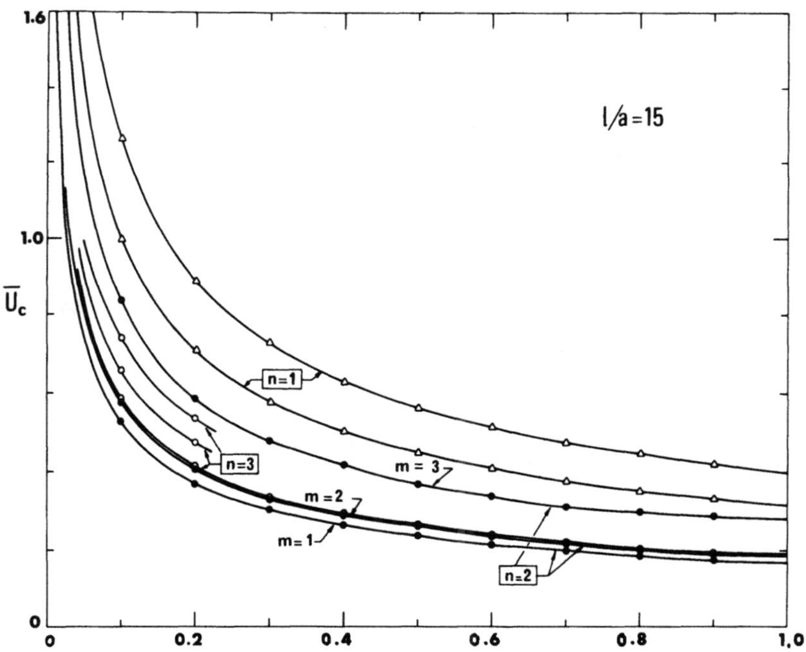

Fig.3. Clamped-clamped cylinders: effect of ε.

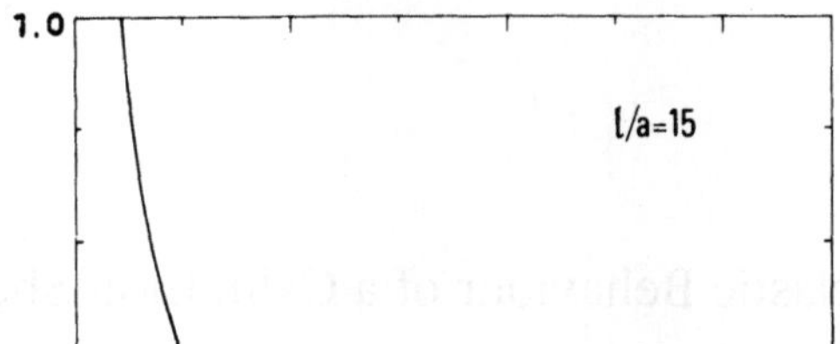

Fig. 4. Cantilevered cylinders: effect of ε.

Hydroelastic Behaviour of a Cylindrical Shell

By

D.S. Weaver
Hamilton, Canada

T.E. Unny
Waterloo, Canada

Introduction

The dynamics of flexible pipes containing flowing fluids has received considerable attention over the last twenty years (see, for example, Refs. [1], [2]). Whether cantilevers or fixed-ended pipes were considered, the majority of the analyses have been concerned only with beam-like modes of deformation and suitably simplified fluid loading terms. However, in many modern applications, the pipes may be relatively short and thin, and one might expect circumferential as well as longitudinal deformation to occur. In this case, the behaviour would be similar to that of the cylindrical shell flutter problem which has been studied extensively be aeroelasticians, mostly for an external supersonic flow. (For review, see Ref. [3].)

This paper discusses the general stability behaviour of a simply supported cylindrical shell conveying an incompressible fluid as shown in Fig. 1. The equation of motion used for the shell is that form of the Flügge equations given by Kempner [4] and the pressure loading on the shell was derived using classical potential theory. The resulting non-self-adjoint differential equation was solved using assumed solution of the form

$$w(x,\theta,t) = \sum_{m=1}^{\infty} A_m \sin \frac{m\pi x}{\ell} \cos n\theta e^{i\omega t} \tag{1}$$

and the method of Galerkin. The details of the derivation and solution are given in Ref. [5].

A two term approximation of Eq. (1) yields a biquadratic characteristic equation in the complex frequency, C, which is easily solved exactly for various values of flow velocity, V, length ratio, ℓ/R, thickness ratio, h/R, and circumferential mode number, n. A comparison of solutions for two, three and four terms in Eq. (1) shows that two terms give excellent results.

General Behaviour

Typical behaviour of the shell for increasing flow velocity and fixed ℓ/R, h/R and n is shown in Fig. 2. For zero flow velocity, the natural frequencies of the liquid filled shell in the first two longitudinal modes $m = 1, 2$ are found. As the flow velocity is increased there is a gradual reduction in natural frequency until static divergence or buckling occurs in the first mode. Such instability is similar to that found by previous authors for beam modes $(n = 1)$ using the appropriate simplified approach. However, if the flow velocity is increased further, the divergence disappears, the first mode frequency reappears and coalesces with that of the second mode to produce coupled mode flutter. The recent paper of Paidoussis and Denise [6] reports similar findings and obtained experimental corroboration.

Effect of Circumferential Mode, n

For a given cylinder the flow velocity at which instability occurs will depend on the number of circumferential waves, n. The lowest of these velocities is called the critical flow velocity and is found by holding ℓ/R and h/R fixed while examining the stability behaviour for various values of n. Typical results for the divergence boundaries are shown in Fig. 3 for a length ratio $\ell/R = 20$.

As suggested by Holt and Strack [7] for the case of an exterior supersonic flow, the critical flow velocity generally occurs for some shell mode $(n > 1)$. To avoid confusion the flutter boundaries are not shown. However, in every case the critical flutter boundary is above the corresponding divergence boundary and the critical value of n is either the same or slightly larger.

Effect of ℓ/R and h/R

The locus of minima of plots such as Fig. 3 gives the critical divergence and flutter boundaries as functions of thickness ratio for a given length ratio as shown in Figs. 4 and 5. These curves show clearly that the instability of short, thin shells is associated with a large number of circumferential waves. The longer and/or thicker the cylinder, the smaller the number of these waves. Finally, for sufficiently long, thick cylinders the unstable mode is the beam mode $(n = 1)$. Calculations show that the transition from beam modes to shell modes occurs roughly at $\ell/R = R/h$, i.e. for $\ell/R > R/h$, the beam mode will be critical and the simpler approach to the problem will suffice. In fact, it is easily shown that in such cases, the theory used here reduces to that of Ref. (1) and verified experimentally in Ref. [2].

Nomenclature

Complex frequency $C = \omega\left(\frac{\ell}{\pi}\right)^2\left(\frac{\rho_m h}{D}\right)^{1/2}$, flow velocity

$V = \dfrac{U\rho_m 12(1-\nu^2)}{\pi^2 E}$, ρ_m = shell density, D = flexural stiffness =

$= \dfrac{Eh^3}{12(1-\nu^2)}$.

References

1. Niordson, F.I.N.: Vibrations of a Cylindrical Tube Containing Flowing Fluid. Trans. Roy. Inst. of Stockholm 73, (1953).

2. Dodds, H.L., Runyan, H.L.: Effect of High Velocity Fluid Flow on the Bending Vibrations and Static Divergence of a Simply Supported Pipe. NASA TN D-2870 (June 1965).

3. Johns, D.J., Parthan, S.: Flutter of Circular Cylindrical Shells, A Review. Loughborough University of Technology, TT-6917, November 1969.

4. Kempner, J.: Remarks on Donnell's Equations. J. Appl. Mech. 22, No. 1, 117-118 (1955).

5. Weaver, D.S., Unny, T.E.: On the Dynamic Stability of Fluid-Conveying Pipes. J. Appl. Mech. 40, No. 1, 48-52 (1973).

6. Paidoussis, M.P., Denise, J.P.: Flutter of Thin Cylindrical Shells Conveying Fluid. J. Sound Vibration 20, no. 1, 9-26 (1972).

7. Holt, M., Strack, S.L.: Supersonic Panel Flutter of a Cylindrical Shell of Finite Length. J. Aerospace Sci. 28, 197-208.

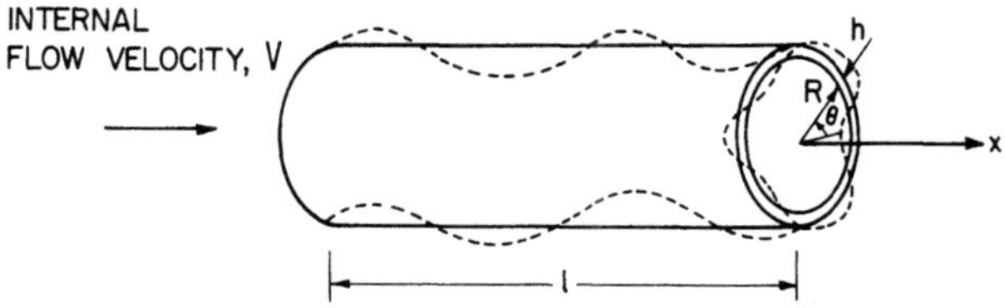

Fig.1. Configuration

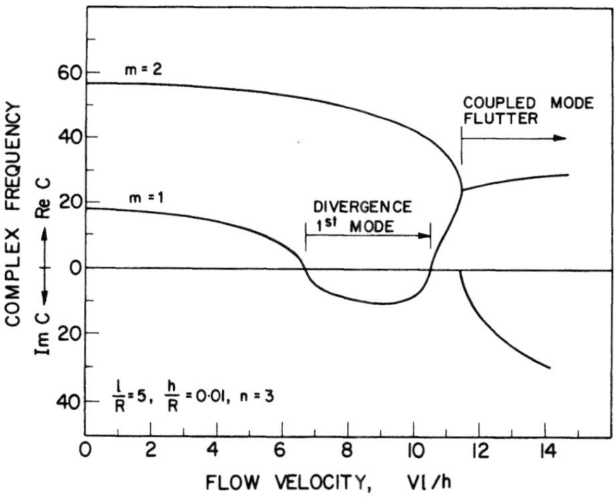

Fig.2. Effect of flow velocity

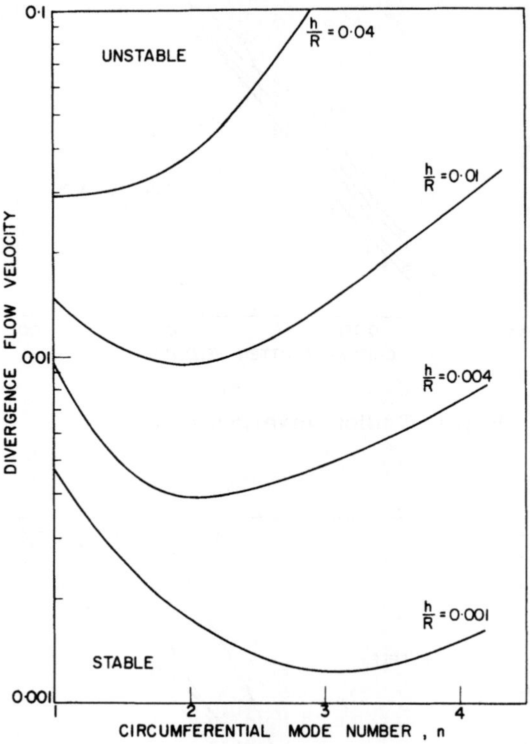

Fig.3. Critical circumferential mode

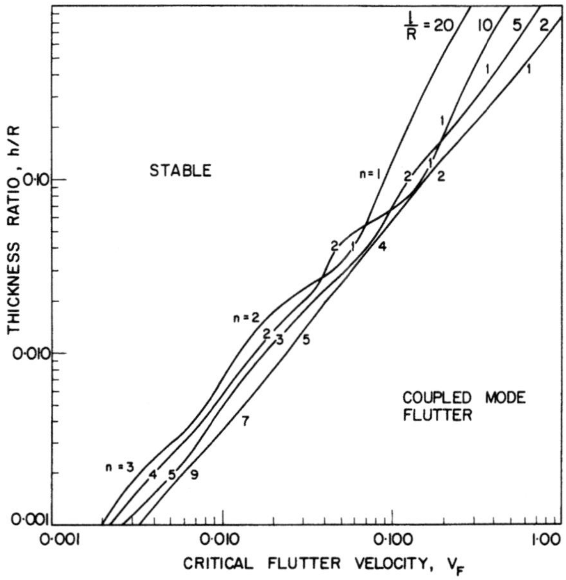

Fig. 4. Critical divergence velocity

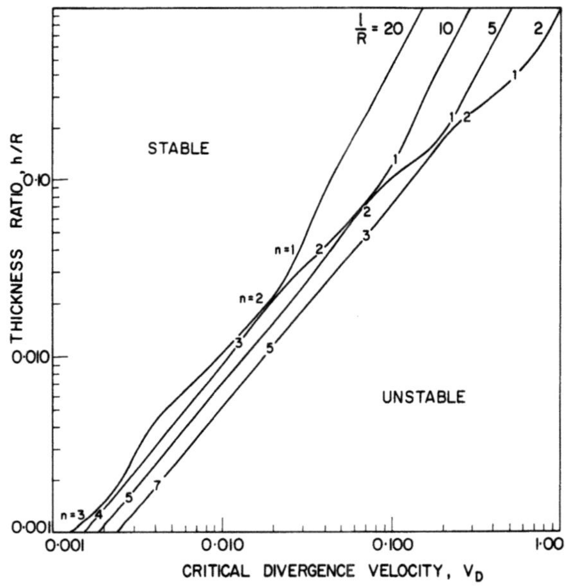

Fig. 5. Critical flutter velocity

Two-Phase Flow by Induced Random Vibrations

By

F. Hara, T. Shigeta, H. Shibata

Tokyo, Japan

Introduction

Oscillations which have a certain periodicity and a relatively large amplitude are often observed in pipe systems of electric power plants, chemical plants, or nuclear plants. It is thought that the gasliquid two-phase pipe flow plays an important role as an excitation source in these oscillations. To prevent such oscillations, it is necessary to make clear the influence of the two-phase pipe flow on the vibrations of the pipe systems. This paper deals with random vibration, of a L-shaped pipe system, induced by an air-water two-phase pipe flow, involving:

(1) Classification of flow patterns of an air-water two-phase pipe flow,

(2) Spectral analysis of void-signals of the two-phase flow and of strain signals of random vibration,

(3) Clarification of excitational characteristics of the two-phase flow, and

(4) Theoretical considerations about the random vibrations induced by the two-phase flow.

Experiments

Figure 1 shows a schematic diagram of the air-water pipe flow loop from which the experimental data referred in this paper were obtained. Air was supplied to the air-water mixer by a rotary air compressor through a surge tank, which reduced the fluctuations of air pressure, and a flowmeter. Water was supplied through an outlet of the city water-works. The air-water mixture flowed upwards through the vertical pipe and was turned into a horizontal direction by a 90° bend, then ran through the horizontal pipe to a riser and was finally released.

The L-shaped pipe system consisted of aluminium annular tubes (L_1, L_2 in Fig. 1) and an acrylic 90° bend, connected as shown in Fig. 1; both ends of the pipe system were rigidly fastened. The inside diameter of these annular tubes was 10 mm. At the outlet of the air-water mixer, steel wire mesh (495 meshes per square inch) was attached for homogenization of the air-water mixture.

During experimentation, air pressure inside the surge tank was kept in the range of from 345 to 354 mm Hg by means of a by-pass circuit.

(A) Pictures of the two-phase flow pattern were taken at both points just before the entrance of the bend and after its exit. The two-phase flow conditions for this observation are shown in terms of flow rates of air and water, that is, $Q_a(l/m)$ and $Q_w(l/m)$, respectively, in Table 1.

(B) The authors detected air and water by using platinum electro-probes at the entrance and exit of the 90° bend and obtained void signals which indicate whether there exists air or water at the points under observation. They used two electro-probes located at points 6 and 7 in Fig. 1, the distance between them was 95 mm. Void signals transformed into electric signals were recorded on magnetic tape.

(C) Strains at both ends of the L-shaped pipe in the random vibrations induced by the two-phase flow were detected by strain gauges attached at 1 and 3 for out-of-plane vibration and 2 and 4 for in-plane vibration. These electric signals of strain were recorded on magnetic tape.

Flow Patterns of Two-Phase Pipe Flow

The result of classification of the photographs obtained from experiment A is shown in Table 1.

The typical flow patterns are shown in Photo 1. In Table 1, the following can be discerned: When the flow rate of air is low, the bubble flow is dominant in spite of the change in the water flow rate. However, the size of the air bubbles depends heavily on the water flow rate, that is, the higher the flow rate of water, the smaller the air-bubble size. Along with the increase in the air flow rate, the two-phase flow pattern changes from a bubble flow to a sausage flow; a piston flow to an annular + piston flow. However, the transition of flow patterns depends on the water flow rate for each flow rate of air.

Table 1. Classification of Two-Phase Flow Patterns just before the Entrance of the 90° Bend.

Q_a \ Q_w	0.6	1.2	3.6	8.4
0.5	B	B	B	b
1.0	B	B	B	B
2.5	S_0	S_0	S_0	S_0 + P
5.0	S_0	S_0 + P	S_0 + P	P
10.0	P	P	P	P
20.0	P	P	P	P
30.0	A + P	P	P	

B---Bubble Flow, b---Small Bubble Flow, S_0---Sausage Flow, P---Piston Flow, A---Annular Flow.

The sign + means that both flow patterns are co-existent.

From the above observations, it can be concluded that the significant measures which determine air-water two-phase flow patterns are the flow rates of air and water, in other words, void fraction and total flow rate of the two-phase flow.

Spectral Analyses of Void Signals

Void-signals recorded on magnetic tape were translated into digital form by an analog-digital data converter and, then, their power spectral densities (psd) were calculated by the Akaike's method. The size of data used in these calculations was 1,500 digits and the sampling interval was 0.01 sec.

The results for the typical flow patterns are shown in Fig. 2. The slug flow which was observed in the range of low flow rates of each fluid has a high peak at a very low frequency in its psd distribution. The bubble flow and sausage flow show a considerable flatter psd distribution. The piston flow has a large peak at a certain frequency in psd distribution and its location changes owing to the flow rates of air and water. The psd for the coexisting annular and piston flow has a predominant peak in the low-frequency range and decreases monotonously with frequency.

From the above observations, it can be concluded that there exists a strong relationship between the psd distributions of void signals and the flow patterns of the two-phase pipe flow.

Spectral Analyses of Strain Signals

Strain signals of random vibrations of the L-shaped pipe system were translated into digital form in the same way as in the case of void signals and their psd's corresponding to the typical flow patterns were calculated. The results are shown in Fig. 3. The cross spectral densities (csd) between the void signals and strain signals are also described in Fig. 3.

For the slug flow, the psd of the strain-signals has no dominant peaks. However, there can be seen small peaks at the first, second and third natural frequencies of the out-of-plane vibration of this L-shaped pipe system. The csd distribution is quite similar to the psd distribution of the slug flow void signal. The bubble flow and sausage flow cause the very small peaks at the first and second natural frequencies in the psd distribution. The csd changes randomly along the whole range of frequency considered. For the piston flow, the psd has three sharp peaks at the first, second, and third natural frequencies, and it can be easily seen from (c) and (d) in Fig. 3 that the location of a dominant peak depends on the void-fraction of the piston flow. The same can be said for the csd distributions. For the annular + piston flow, a peak is located at the first natural frequency in the psd distribution and the csd is similar to the psd distribution of the correspondent void signal.

The psd's for the piston flow and the annular + piston flow have a relatively large value at a very low frequency below the first natural frequency. This means that the vibrations induced in these flow patterns include beats. Finally, Fig. 3 shows that the sulg, bubble or sausage flows do not so much influence the random vibrations of the L-shaped pipe system, but the piston flow plays an important role as an excitation source to the random vibrations.

Intensity of Random Vibrations

The authors adopted the square root of the auto-correlation function of strain-signals at $\tau = 0$ as the intensity of the random vibrations of their pipe system and the velocity of the two-phase flow as its intensity. That velocity was calculated as

$$V = \Lambda/\tau_d , \tag{1}$$

where Λ is the distance between two points 6 and 7 in Fig. 1 and τ_d is the time-lag of the corss-correlation function between two void signals obtained at 6 and 7. The relationship between these intensities is shown in Fig. 4. This figure shows that the intensity of random vibrations to the bubble, sausage and annular + piston flows is very low. On the other hand, the piston flow gives a considerably stronger intensity to the random vibrations.

Theoretical Considerations

In order to investigate the reason why the piston flow induces strong random vibration in the L-shaped pipe system, the authors have formulated the equation of motion for the turbulent two-phase flow model, namely,

$$EI\frac{\partial^4 y}{\partial x^4} + \{M + m(x,t)\}\frac{\partial^2 y}{\partial t^2} + 2m(x,t)U\frac{\partial^2 y}{\partial t \partial x} + m(x,t)U^2\frac{\partial^2 y}{\partial x^2} = f, \quad (2)$$

where it should be noted that a straight pipe with both ends simply supported is adopted as a vibration model; $M, m(x,t)$ are masses per unit length of pipe and fluid, respectively; U is the average velocity of the two-phase flow; $f(x,t)$ is an excitation force caused by turbulence; other variables have the usual meaning found in vibration theory. Applying the modal analysis method to Eq. (2), the following equation can be obtained for the first mode within the limit of first approximation,

$$(M_0 + \varepsilon \cos \omega t)\ddot{Y}_1 + \frac{2\pi U}{l}\eta \sin \omega t \dot{Y}_1 +$$

$$+ \left(\frac{\pi}{l}\right)^2 \left\{EI\left(\frac{\pi}{l}\right)^2 - m_{11}^0 - \varepsilon \cos \omega t\right\} Y_1 = \delta \cos \omega t, \quad (3)$$

where Y_1 is the amplitude of the first mode, l is the pipe length, ω is a dominant frequency of the void signal of the piston flow and $\varepsilon, \eta, \delta$ are small parameters. Denoting $\sqrt{(\pi/l)^2\{EI(\pi/l)^2 - m_{11}^0\}/M_0}$ by ω_0, an approximate solution of Eq. (3) yields

$$\exp\left\{-\frac{A}{\Delta}\cos \Delta t\right\}\sin(\omega_0 t + \varphi), \quad \frac{B}{\Delta}\cos \Delta t \sin(\omega_0 t + \psi). \quad (4)$$

Here A and B are constants and $\Delta = \omega - \omega_0$. From Eqs. (3) and (4), the following can be concluded: The piston flow induces parametric excitation as a reaction of time-varying Coriolis' force and centrifugal force acting on running water pistons, the turbulence of the piston flow plays the role of random external force, and the vibrations take the form of beats.

Conclusions

The following conclusions can be obtained from the experimental results and theoretical considerations:

(1) The power spectral densities of the void signals and the strain signals have a strong correspondence to the two-phase flow patterns,

(2) The intensity of the random vibrations depends on the two-phase flow patterns and its velocity and it is strongest in the piston flow, and

(3) The turbulent piston flow plays the role of parametric excitation and external force to the vibrations of the pipe conveying it.

Acknowledgements

The authors wish to express their appreciation to Profs. A. Watari, T. Ishihara, H. Sato, and S. Ono of the Institute of Industrial Science, University of Tokyo, for their valuable discussions. They also thank Messrs. K. Chitoshi and K. Sogabe for assisting their experimental works and Misses K.L. Maxwell and F. Ogino for correcting the English expressions in this paper and typing it.

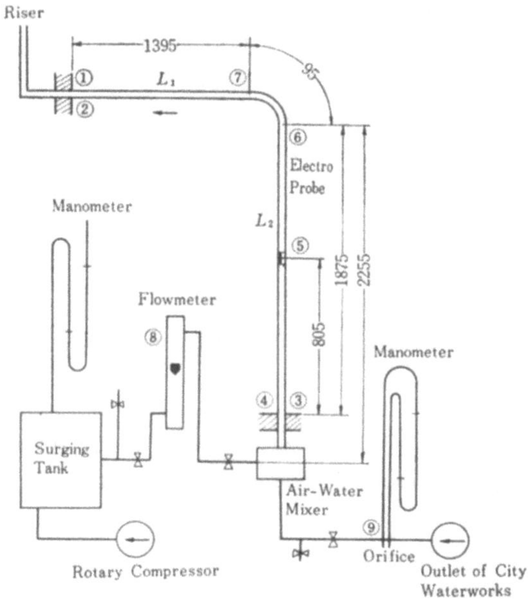

Fig.1. Schematic diagram of the test set-up.

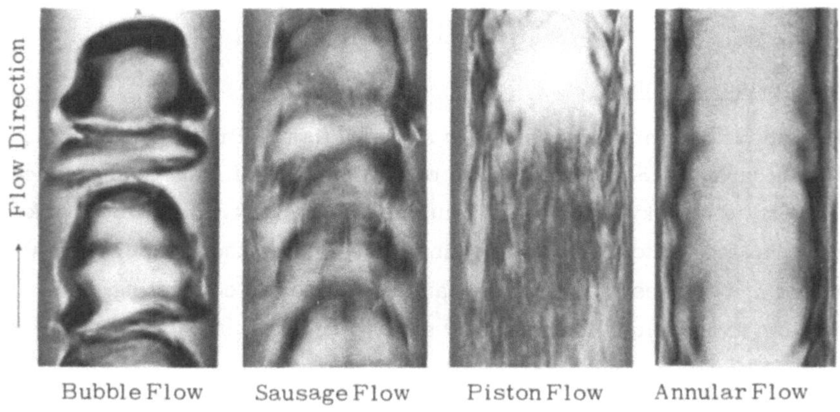

Photo 1. Photographs of typical two-phase flow patterns.

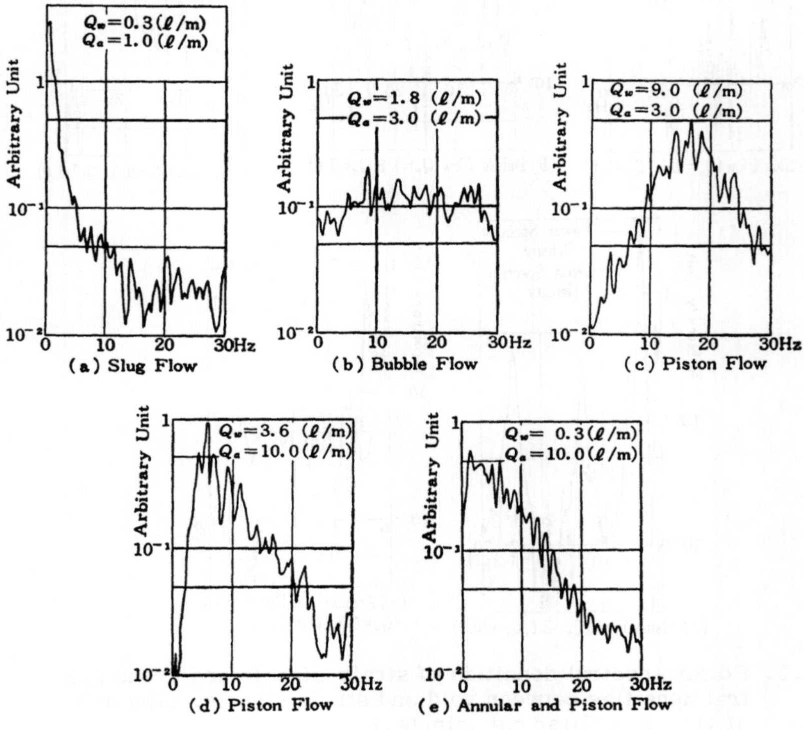

Fig.2. Power spectral densities of two-phase flow void signals.

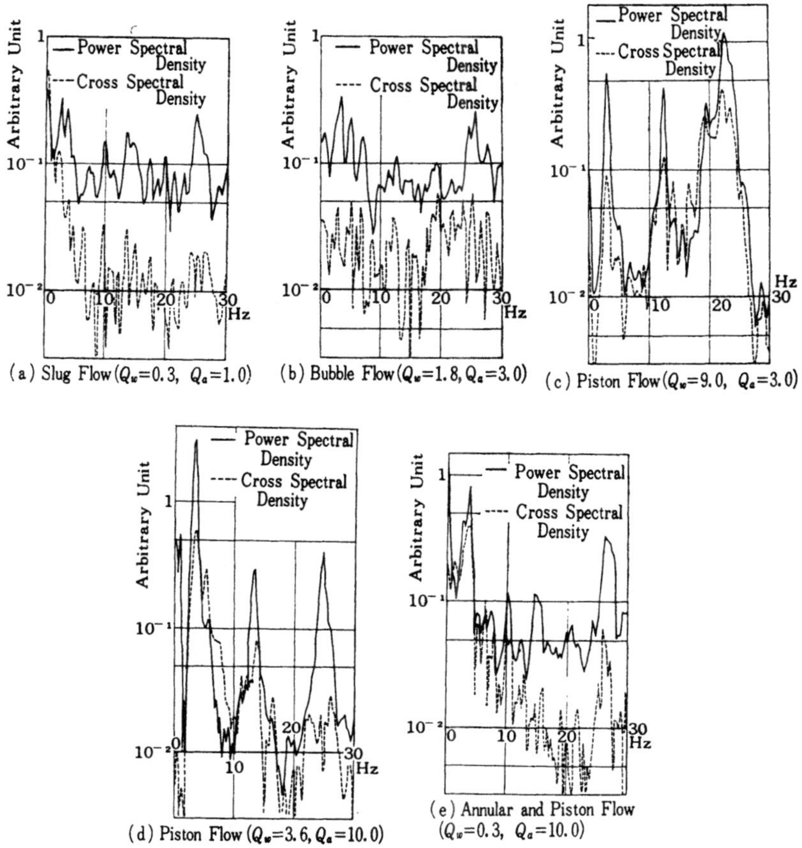

(a) Slug Flow (Q_w=0.3, Q_a=1.0) (b) Bubble Flow (Q_w=1.8, Q_a=3.0) (c) Piston Flow (Q_w=9.0, Q_a=3.0)

(d) Piston Flow (Q_w=3.6, Q_a=10.0) (e) Annular and Piston Flow (Q_w=0.3, Q_a=10.0)

Fig.3. Power spectral densities of strain-signals and cross spectral densities between void and strain signals. (The unit of Q_W, Q_a is liter per minute.)

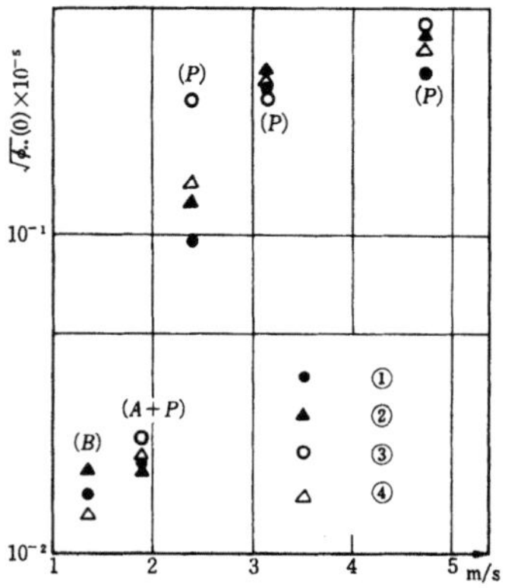

Fig.4. Relationship between intensities of random vibrations and two-phase flow patterns.

TECHNICAL SESSION H

"Flow-Induced Vibrations Relating to Buildings"

Chairman: C. Scruton, U.K.
Co-Chairman: E. Plate, Germany

CONTRIBUTIONS

Turbulence-Induced Vibrations of Bluff Bodies

By
P.W. Bearman
London, U.K.

Summary

This paper describes an experimental investigation into the response of a bluff body to a turbulent stream. In the first part of the paper, measurements are presented of the fluctuating pressure recorded at the stagnation point on a stationary, two-dimensional flat plate mounted normal to the flow. The effects of the distortion of the approaching turbulence are examined. This is followed by a report on some measurements of the response of a flexibly mounted bluff body of finite aspect ratio where the scale of turbulence was three times the width of the body. It is shown that the response in the drag direction, at the resonant frequency, can be predicted to good accuracy by the method of Davenport over a range of values of the structural damping. There is evidence, however, of a complex interaction between the stream turbulence and the wake flow at low frequencies.

Introduction

Within recent years increased interest has been generated in bluff body flows due to the recognition of their engineering importance in various non-aeronautical branches of aerodynamics. In no field is this more true than that of 'Architectural Aerodynamics' where the building of taller, more slender and more lightly damped structures has led to buildings having a greater susceptibility to vibration due to the action of the natural wind. The three major causes of flow-induced vibrations of buildings are galloping, regular vortex shedding and oscillations forced by the turbulent fluctuations produced and convected within the Earth's boundary layer. Galloping and vortex shedding cause oscillations predominantly in a plane transverse to that of the mean wind; turbulence, however, can induce both strong fluctuating lift and drag forces. Very little, generally applicable, theoretical or experimental work has been carried out to determine the magnitude of the lift forces that are generated. Also comparatively little is known about the interaction between turbulence and vortex shedding or galloping. On the other hand, the fluctuating drag forces produced by a turbulent flow have been investigated more extensively (Vickery [1] and Bearman [2]) and it is possible to predict their magnitude, to a limited accuracy, by surprisingly simple calculation procedures.

A method for calculating the response of a flexible bluff body in the along wind direction was proposed by Davenport [3] who introduced the concept of a frequency dependent aerodynamic admittance function to relate the upstream velocity fluctuations to the resulting fluctuating forces. Vickery [1] suggested a method for predicting the magnitude of this function, for simple shapes, based on the assumptions that the instantaneous drag force on a small element of the body is the same as if the element were in the flow in isolation and that the correlation pattern of the resulting fluctuating forces over the body surface is identical to the correlation pattern within the approaching turbulence. The aerodynamic admittance, as calculated by the above authors, contains two terms: one is a drag term depending directly on the fluctuating velocity and the other is a virtual mass or inertia term related to the derivative of the velocity. In most applications of the theory the virtual mass term is conveniently dropped on the argument that, at the high wavenumbers when this term might become important, it will be negligible because of its very small correlation over the body surface. Aerodynamic admittances calculated by this method have shown surprisingly good agreement with experiment [1,2], especially when the scale of turbulence is large compared with the size of the body, although of course gross assumptions have been made about the nature of turbulence and about the flow around bluff bodies and the interaction between the two. Hunt [4] has formulated a theory, based on turbulence rapid distortion theory, to analyse the turbulence in a flow sweeping past a body. He shows how the turbulence will be distorted by the stretching and rotating of vortex line filaments as they are convected past the body. Using his theory it is possible to obtain a clearer understanding of the mechanisms that determine the aerodynamic admittance.

The successful use of the admittance calculation methods to predict response depends on there being a linear relationship between the approaching velocity fluctuations and the resulting force fluctuations; fluctuating pressures on the body having an identical correlation pattern to the fluctuations in the approaching velocity field; negligible influence of any effect of the stream turbulence on the near wake structure; no influence of any distortion of the turbulence structure approaching the body and no interaction between body movements and the turbulence-induced fluctuating forces. In order to investigate some of these effects two experimental investigations were initiated: one on a rigid two-dimensional bluff body on which fluctuating pressures could be measured and the second on a linear mode, cantilevered bluff body, the amplitude of which could be controlled by varying the damping.

Experiments on a Stationary Bluff Body

Experimental Arrangement

The experiments were conducted in a wind tunnel with a 3 ft (0.91 m) by 3 ft by 15 ft (4.57 m) long working section. Highly turbulent flow was generated by the installation of square-mesh grids at the beginning of the working section. Measurements were made at a wind speed of 18 m/sec. The bluff body used was a flat plate spanning the tunnel and mounted normal to the flow direction. The prime interest was the distortion of the approaching turbulence and in order to remove any unsteadiness in the flow, generated by vortex shedding and turbulence in the wake of the plate, the wake was filled in along the theoretical free streamlines. The profile shape of the resulting body was designed according to Roshko's [5] notched hodograph method. The model cross-section is shown in Fig. 1 and the size of the equivalent flat plate, D, was 2.54 cm. Turbulence measurements were made with a Disa constant-temperature, linearized hot-wire anemometer. A traverse gear was embedded in the model and a hot wire could be traversed out along the mean stagnation streamline. Fluctuating pressure measurements were made with a microphone, connected by a short length of tubing to a surface hole on the stagnation line. Further details of the experimental arrangement are given in a previous paper [6].

Experimental Results

The turbulence structure behind each of four grids was investigated on the centre line of the working section, in the absence of the model, at a distance from the grids corresponding to the distance to the stagnation point. The measurements of the intensity of the along-wind component of turbulence $\sqrt{\overline{u^2}}/U$ and the longitudinal integral scale L_x are given in Table 1.

Table 1

Grid	$\frac{\sqrt{\overline{u^2}}}{U} \times 10^2$	L_x cm	$\frac{L_x}{D}$
A	2·09	3·05	1·2
B	2·93	4·57	1·8
C	5·71	6·54	2·57
D	6·40	6·05	2·38

Table 2

Grid	Maximum $\sqrt{\overline{u^2}}/U$	Amplification of intensity
A	0·18	8·6
B	0·21	7·43
C	0·32	5·31
D	0·34	5·6

The intensities of all three components of turbulence were measured ahead of the body along the stagnation streamline. It is difficult to interpret their meaning if they are simply plotted as a variation of local turbulence intensity because the changes in intensity are dominated by the changes in the mean velocity. Instead, the local root mean square value of the turbulence component has been divided by its value recorded at x/D = 4.25 (denoted by suffix 0). Figure 2 shows the variation of $\sqrt{\overline{u^2}}/\sqrt{\overline{u_0^2}}$ ahead of stagnation for the four values of scale examined. Very near the stagnation point the hot wire can only give a rough indication of the level of the fluctuating velocities because of the very high local turbulence intensities. The hot wire output was linearized but further errors are likely to arise at high levels of turbulence due to the non-linear yaw response. Table 2 shows the highest values of local intensity recorded and in all cases this occurred at about $x/D = 0.1$.

The other component of turbulence in the plane containing the cross-section of the model, $\sqrt{\overline{v^2}}/\sqrt{\overline{v_0^2}}$, showed a generally opposite effect to $\sqrt{\overline{u^2}}/\sqrt{\overline{u_0^2}}$ with the larger scales showing amplification and the smaller attenuation as x/D tends to zero. The third component $\sqrt{\overline{w^2}}/\sqrt{\overline{w_0^2}}$, which is in a direction parallel to the stagnation line along the model, shows only amplification. At the surface all three components must be reduced to zero.

Power spectral density measurements of the fluctuating pressure at the stagnation point were made for each of the grids and the spectra for grids A and C, representing the smallest and largest values of L_x/D examined, are shown in Figs. 3 and 4. The power spectral density of the pressure, $F(p)(n)$, is presented in the non-dimensional form $F(p)(n)/2\pi L_x \rho^2 \overline{u_0^2} U_0 = \theta_p$, where U_0 is now free stream velocity.

Discussion of Rigid-Body Results

Hunt [4] has treated the general case of turbulent flow past a bluff body and, making certain assumptions, shows that the turbulence is affected by vortex stretching, caused by the mean flow field about the body, and by a simple blocking of the turbulence since there can be no flow through the surface of the body. He shows that the modification of the turbulence by the body is fundamentally different in the two extreme cases as $L_x/D \to \infty$ and $L_x/D \to 0$. When $L_x/D \to \infty$ the flow approximates to a slow, quasi-steady variation of the direction and magnitude of the mean velocity and therefore

$$\frac{\sqrt{\overline{u^2}}}{\sqrt{\overline{u_0^2}}} = \frac{U}{U_0} , \tag{1}$$

i.e., as the body is approached the fluctuating velocity attenuates like the mean flow. When $L_x/D \to 0$ the distortion of the turbulence along the stagnation streamline approximates to that caused by uniform plane strain and the theory of Batchelor and Proudman [7] can be applied directly. The theoretical curves for the variation of $\sqrt{\overline{u^2}}/\sqrt{\overline{u_0^2}}$ with x/D when $L_x/D \to 0$ and ∞ are shown in Fig. 2.

The experimental results plotted in Fig. 2 are seen to fall between the two limiting theoretical curves with results for smaller scales generally tending towards the $L_x/D \to 0$ curve. When $L_x/D = 0(1)$ the turbulence is affected by both vortex stretching and by the simple blocking of the flow by the body. Therefore it can be expected that low wavenumbers will exhibit some of the features of $L_x/D \to \infty$ flows while high wavenumbers will be dominated by vortex stretching. This idea is well supported by spectra measurements of the "u" component which show an apparent large shift of energy to higher wavenumbers as the stagnation point is approached.

The assumption used by Davenport [3] and Vickery [1] that the fluctuating force can be separated into a drag term and a virtual mass term leads to the following expression for the fluctuating part of the drag force F.

$$F = \rho U_0 u_0 D^2 C_D + \rho D^3 \frac{du_0}{dt} C_m , \qquad (2)$$

where C_D and C_m are the coefficients of drag and virtual mass. By a similar argument the fluctuating pressure p at the stagnation point on a flat plate can be shown to equal

$$p = \rho\left(u_0 U_0 + \frac{D}{2}\frac{du_0}{dt}\right) . \qquad (3)$$

This expression is only strictly valid if $L_x/D \to \infty$ and the velocity fluctuations are irrotational. A fuller and a more physically realistic treatment by Hunt ([8], Eq.28) shows that while Eq. (3) predicts pressures of the correct order, there are further terms which should be considered.

Assuming that the record of the pressure fluctuations forms part of an infinite stationary random process it is possible to rewrite Eq. (3) in terms of the power spectral densities of the pressure $F(p)(n)$ and the approaching velocity $F(u_0)(n)$.

$$F(p)(n) = \rho^2 F(u_0)(n) U_0^2 \left[1 + \frac{1}{4}\left(\frac{2\pi n D}{U_0}\right)^2 \right], \tag{4}$$

$$\frac{F(p)(n)}{2\pi\rho^2 U_0 L_x u_0^2} = \frac{F(u_0)(n) U_0}{2\pi L_x u_0^2} \left[1 + \frac{1}{4}\left(\frac{2\pi n L_x}{U_0}\right)^2 \left(\frac{D}{L_x}\right)^2 \right]. \tag{5}$$

Equation (5) is rewritten as

$$\theta_p = \theta_u \left[1 + \frac{\tilde{n}^2}{4}\left(\frac{D}{L_x}\right)^2 \right]. \tag{6}$$

If $L_x/D \to \infty$, $\theta_p = \theta_u$ and the pressure and velocity spectra, non-dimensionalised as above, should be identical. The spectra of pressure and upstream velocity are compared in Figs. 3 and 4 and it can be seen that they agree closely at low wavenumbers whereas at higher wavenumbers the pressure spectra fall away below the velocity spectra rather than show an increase according to Eq. (6). At each scale size examined there was a definite break point where the pressure spectrum diverged from the velocity spectrum and with increasing values of L_x/D the break point moved to higher wavenumbers. At high wavenumbers the pressure spectra fell off about 1.75 times as fast as the velocity spectra. If these results are typical of the behaviour of the fluctuating pressure over the surface of a bluff body then they help to explain the success achieved by dropping the inertia term from the calculation of aerodynamic admittance.

Experiments on a Linear Mode, Cantilevered Bluff Body

In order to test further the concept of an aerodynamic admittance function some experiments were conducted on a bluff body free to rotate about its base in a linear mode. The amplitude could be controlled by varying the damping. The aim was to measure the response for a range of structural dampings and wind speeds and to compare the results with the theoretical response given by the application of the Davenport-Vickery approach. The experiments were conducted by Cox and fuller details will be appearing in his thesis.

Experimental Arrangement

The experiments were carried out in a wind tunnel with a 5 ft (1.5 m) by 4 ft (1.2 m) by 7 ft (2.1 m) long working section. Turbulent flow was again generated by a square-mesh grid at the beginning of the working section. The cross-sectional shape of the bluff body used was an equilateral triangle with one face normal to the wind. A triangular section was chosen because it offered an upstream face similar to that of the flat plate model and it could be made with sufficient structural stiffness such that its first natural frequency was much greater than that chosen for the linear mode. The width of a face of the model was 2.54 cm and its length was 38 cm giving a height to width ratio of 15. The central core of the model was made of balsa wood and the sides were cut from thin dural sheet. The model was mounted on a torsion bar assembly fixed just below the level of the floor of the working section and the torsion bar was so arranged as to allow only body movements in the fore and aft, drag direction (see Fig.5). The deflections of the body were sensed by strain gauges on the torsion bar. Controllable damping was provided by passing a plate, attached to the base of the model, between poles of an electromagnet.

The frequency N of the linear mode of the model was chosen to be 24 Hz. Since the Strouhal number for a flat plate is about 0.14 this would give a resonance with vortex shedding in the drag direction at a reduced windspeed U/ND, where D is body width, of about 3.5. This corresponds to a wind speed of 2.2 m/s whereas the experiments were carried out in the range $6.3\ \text{m/s} < U < 22\ \text{m/s}$ where $10 < U/ND < 35$ and $10^4 < Re < 3.5 \times 10^4$.

Experimental Results

At the position to be occupied by the model the turbulence structure was investigated with a hot-wire anemometer. For present purposes grid turbulence can be characterised by measurements of an intensity and an integral scale length. The intensity of the 'u' component was $\sqrt{\overline{u^2}}/U = 0.065$ and its longitudinal integral scale was $L_x = 7.6$ cm, giving $L_x/D = 3$. Thus this grid produced flow conditions similar to those approaching the rigid bluff body with either grid C or D in the tunnel. According to the spectrum of fluctuating pressure shown in Fig. 4 the effects of turbulence distortion should be small for low values of the frequency parameter $\tilde{n} = (2\pi n L_x)/U_0$. It can be seen that θ_p begins to drop below θ_u for $\tilde{n} > 3$ and therefore there should be no serious attenuation of the pressure fluctuations along the attachment line on the body, at frequencies equal to or less than that of the linear mode, if $U/ND \geqslant 6$.

The response of the body, measured as an angular displacement, was recorded for a range of values of structural damping and wind speed. Figure 6 shows the mean angular displacement plotted against the square of the wind speed. The mean drag coefficient acting on the body was found to be 1.7. This compares with a value of about 2 for a two-dimensional flat plate, the difference being caused by an inflow into the wake near the tip. In smooth flow C_D was found to be 1.6 and it is thought that the turbulent flow may have modified the local flow at the tip slightly.

A typical curve of the rms deflection, θrms, plotted against structural damping is shown in Fig.7 for a value of $U/ND = 30$. As is to be expected with a resonant system the response decreases with increasing damping. The power spectral density of the angular displacement was computed for each of the values of damping and wind speed. A spectrum of the response at $U/ND = 30$ is shown in Fig. 8. The spectral density $F(\theta)(n)$, measured in $(\text{rads})^2/\text{s}$, is plotted against the frequency parameter nD/U and shows a sharp peak at the linear mode frequency and a rapid attenuation at higher frequencies. A lower damping case is shown in Fig. 9 and the general form of the spectrum is very similar.

Comparison with Theory

The experimental results have been compared with a straightforward application of the theory of Davenport for the response of slender line-like structures. No attempt has been made to allow for any lack of correlation across the face of the body at any given height. However, with $L_x/D = 3$ this is likely to cause only a small correction term in the frequency range of interest. The instantaneous value of the drag force per unit length was taken as $\rho U_0 u_0 C_D D$ where C_D is the mean drag coefficient which is assumed to act uniformly over the entire body. The spectrum of the approaching turbulence was measured, in the absence of the model, and the form of the cross spectrum between two points in the flow was assumed to be the same as that measured by Vickery [1] in grid turbulence. The structural damping was measured by recording the decay of oscillations in still air but a better agreement with experiment was found when an allowance was made for aerodynamic damping. The form used for the aerodynamic damping force was that used by Campbell and Etkin [9] $\rho U_0 D C_D \dot{x}$ which results from the relative motion between the air flow and the body, where $\dot{x}$ is the velocity of the body. The percentage of critical damping quoted in Figs. 8 and 9 includes an allowance for the aerodynamic damping.

Comparisons with the experimental angular displacement spectra are shown in Figs. 8 and 9. It is clear that the theory predicts the response extremely well in the vicinity of the resonant frequency. At lower frequencies, however, the measured response is substantially greater than that predicted and this may be due to wake-induced pressure fluctuations on the rear face which are not included in the linear theory of Davenport. Nevertheless, the RMS response predicted by the Davenport method, shown in Fig. 7, is not far below the measured values. At high values of damping, where the contribution at the resonant frequency is less, the errors are comparatively larger.

In order to obtain a qualitative picture of the contribution to the response coming from the separated wake, the experiments were repeated without the turbulence grid in the tunnel. The corresponding results in smooth flow are shown plotted in Figs. 7, 8 and 9. The spectra show unexpected features at very low frequencies where the power spectral density of the response is greater in smooth flow than in turbulent flow. It had been hoped that the extra response in turbulent flow might have been accounted for by a simple addition of the smooth flow response. Clearly, a very complex interaction must take place between the stream turbulence and the near wake flow. At low frequencies turbulence suppresses the response whereas at higher frequencies the wake-turbulence interaction increases the response.

In Fig. 7, the rms deflections measured in smooth flow are shown plotted against the damping and show only a small decrease in level with increasing damping, indicating that the majority of the response is occurring away from the resonant frequency. If the smooth flow response is simply added to the theoretical turbulent response, assuming it to be uncorrelated, the resulting agreement with experiment is extremely good. Unfortunately, the spectra show the problem not to be as simple as this.

Conclusions

Measurements show how turbulence, carried by the free stream, is distorted as it approaches a stationary bluff body. When $L_x/D \gg 1$ a quasi-steady type of approach can be used and, along the mean stagnation streamline, $\sqrt{\overline{u^2}}$ attenuates like the mean flow. Whereas if $L_x/D \ll 1$ the turbulence is distorted by the mean flow field and $\sqrt{\overline{u^2}}$ amplifies due to vortex stretching. Measurements of the pressure fluctuations at the stagnation point show that at low wavenumbers the level of the pressure fluctuations can be predicted by simply assuming $p = \rho u_0 U_0$. In some measurements on a flexibly mounted bluff body of finite aspect ratio it is found that the response in the drag direction, at the resonant frequency, can be predicted by the method of Davenport. At low frequencies, however, the response is substantially greater than that prediced and it is thought that this may be due to wake-induced pressure fluctuations. The measurements highlight the need for some understanding of the complex interaction which must take place between the free-stream turbulence and the near wake.

Acknowledgement

Thanks are due to R.A.Cox for kindly allowing some of his experimental results to be shown in Figs. 6, 7, 8 and 9.

Reference

1. Vickery, B.J.: On the Flow behind a Coarse Grid and its Use as a Model of Atmospheric Turbulence in Studies Related to Wind Loads on Buildings. NPL Aero. Report 1143 (1965).

2. Bearman, P.W.: An Investigation of the Forces on Flat Plates in Turbulent flow. J. Fluid Mech. 46, 177 (1971).

3. Davenport, A.G.: The Application of Statistical Concepts to the Wind Loading of Structures. Proc. Inst. Civ. Engrs. 19, paper No. 6480, 449-472 (1961).

4. Hunt, J.C.R.: A Theory of Turbulent Flow over Bodies, to be published in J. Fluid Mech.

5. Roshko, A.: A New Hodograph for Free-Streamline Theory. NACA Tech. Note 3168 (1954).

6. Bearman, P.W.: Some Measurements of the Distortion of Turbulence Approaching a Two-Dimensional Body. AGARD Conference Proceedings No. 93, Turbulent Shear Flows. Paper No. 28 (1971).

7. Batchelor, G.K., Proudman, I.: The Effect of Rapid Distortion of a Fluid in Turbulent Motion. Quart. J. Mech. and Appl. Math. 7, 83-103 (1954).

8. Hunt, J.C.R.: A Theory for Fluctuating Pressures on Bluff Bodies in Turbulent Flow. Presented at this Symposium in Session B.

9. Campbell, A.C., Etkin, B.: The Response of a Cylindrical structure to a Turbulent Flow Field at Sub-Critical Reynolds Number. Univ. of Toronto, UTIAS Tech. Note No. 115 (1967).

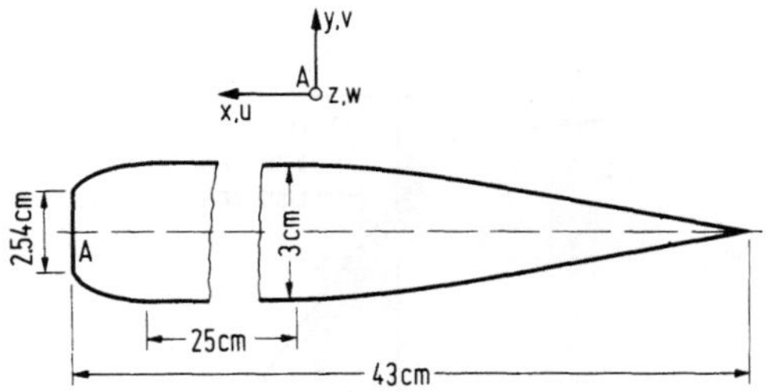

Fig.1. Flat plate free-streamline model.

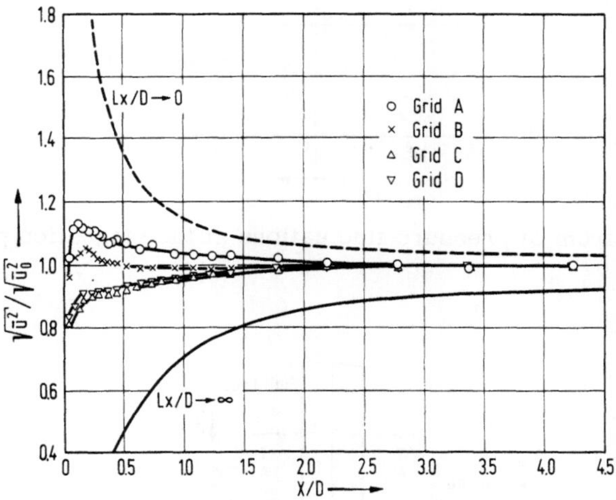

Fig.2. u-component approaching stagnation.

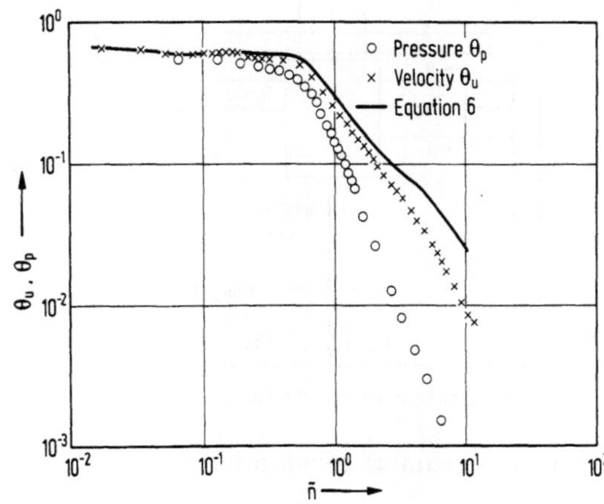

Fig.3. Spectrum of pressure fluctuations at the stagnation point, grid a.

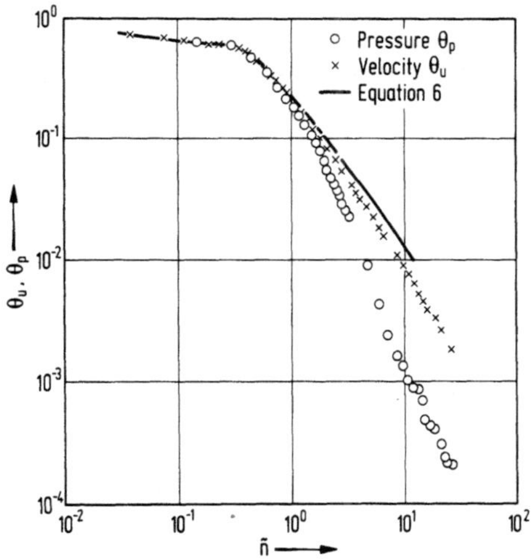

Fig.4. Spectrum of pressure fluctuations at the stagnation point, grid c.

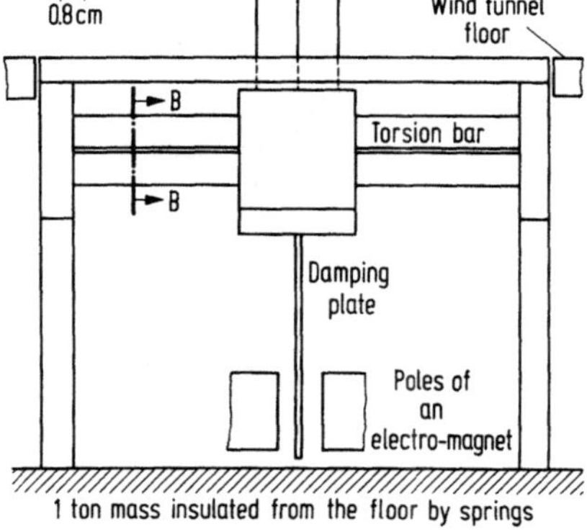

Fig.5. Diagram of torsion bar assembly.

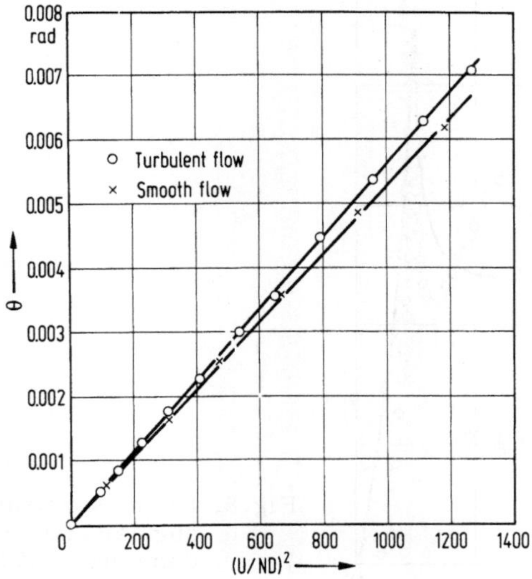

Fig.6. Mean angular displacement versus the square of the wind speed.

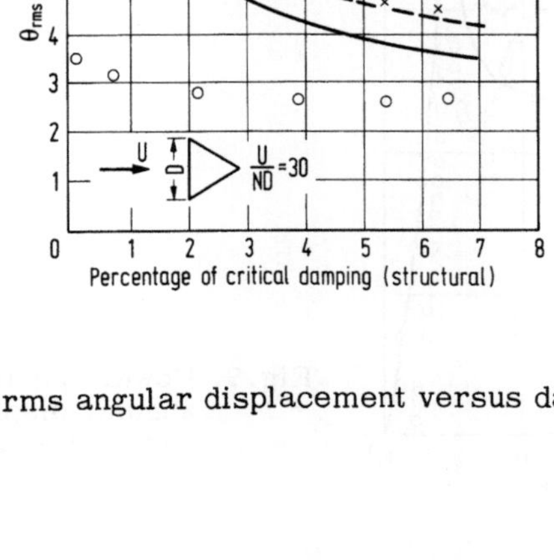

Fig.7. rms angular displacement versus damping.

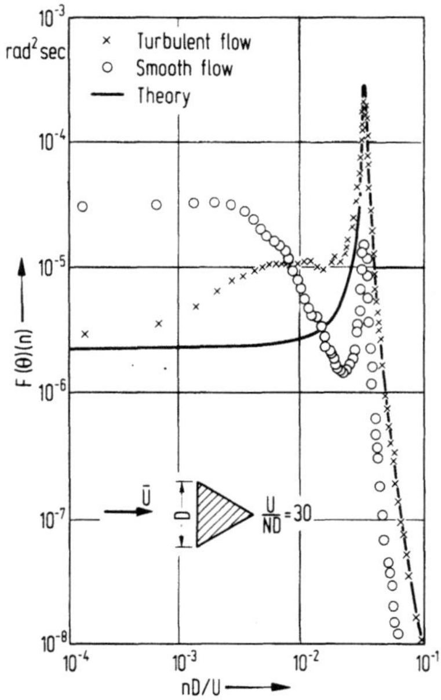

Fig. 8. Power spectral density of angular displacement; total damping = 4.7%.

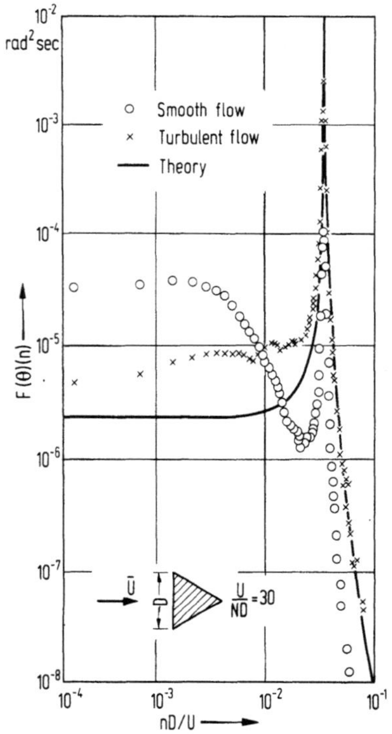

Fig. 9. Power spectral density of angular displacement; total damping = 0.946%.

Vortex Shedding from a Bluff Body in a Shear Flow

By

D.J. Maull, R.A. Young

Cambridge, U.K.

Summary

Experiments are described where the base pressure and vortex shedding on a finite-length blunt-based body have been measured in a uniform and in a shear flow. It is shown that the pressure distribution may be split into three regions; a root region, a tip region, and a central region. The vortex shedding frequency in the shear flow is shown to vary along the length of the body, and there are discontinuous jumps in frequency which may be associated with the presence of longitudinal vortices in the flow.

Introduction

Most of the investigations so far carried out on vortex shedding have concentrated on cases where the incident flow has been uniform and where the bodies have spanned a wind tunnel in an attempt to make the wake flow two-dimensional. In practice, however, a large number of flow situations occur where the incident flow on to a bluff body is far from uniform and may well contain shear and hence a vorticity vector at right angles to the flow direction.

In practical cases we also have the problem that the bluff body must have a tip and a root region where the flow will not be two-dimensional and it is of interest to discover how far along the span of the body these regions extend.

The object of this investigation was to try and understand the flow in the wake of a bluff based body of finite height in a shear flow and possibly to arrive at some means of calculating, for instance, the base pressure and the vortex shedding frequency from data on a similar body in a uniform flow.

Apparatus

The model used for this investigation is shown in Fig. 1. The nose is a semi-ellipse and is followed by a parallel sided section, the chord (c) of the model being 15.24 cm and the base height (d) 2.54 cm. The model could be tested for various projections (L) into the wind tunnel normal to the flow direction, including the case $L/d = 20$ when the model completely spanned the working section. A transition wire was fitted to the model, 3.0 cm from the leading edge.

The shear flow in the stream was generated by placing a curved gauze across the flow which produced the velocity profile shown in Fig. 2 where y is the distance from the floor of the tunnel and l is the height of the working section. The shape of the gauze required to produce this profile was calculated using a method given by Elder [2] which was empirically modified and is fully described by Maull [3]. The advantage of this method of producing a shear flow as against those using an array of rods across the stream, as for instance described by Owen and Zienkiewicz [4], is that the gauze produces lower turbulence levels. In the experiments described here, the turbulence level in the shear flow was 0.5% based on the local velocity.

All the results for the pressure distributions and the wake measurements were carried out at a Reynolds number of 2.85×10^4 based on the base height dimension d for both the shear flow and the uniform flow. In the case of the shear flow the Reynolds number was also based upon the stream velocity on the center line of the wind tunnel $(y = 10\,d)$.

Measurements of the vortex shedding were carried out with DISA hot-wire anemometers, the hot-wire signals being recorded on magnetic tape and subsequently analysed on a digital computer. No blockage corrections have been applied to the results.

Results on Pressure Distribution

Taking the case where the incident flow is uniform, Fig. 3 shows the variation of the base-pressure coefficient along the span of the model for three different cases of the projection L/d into the tunnel, the case L/d = 20 corresponding to the model spanning the tunnel. It can be seen that for the bodies not spanning the tunnel the pressure distribution may be split up into three regions, namely a tip region extending at least six base heights down from the tip, a root region extending up from the floor about six base heights and a central region of approximately constant base pressure.

It is evident from these graphs that for both the bodies not spanning the working section (L/d = 15 and 18) the base pressure in the central region does not reach the low value associated with two-dimensional flow about this cross-sectional shape. This is further shown in Fig. 4a and b where the root and tip regions are shown for a range of projections of the body into the flow. In the immediate neighbourhood of the tip (Fig. 4b) the pressure distributions are similar, as would be expected, the very low pressure being caused by a rapid acceleration of the flow along the top of the body and down the back face. There is probably some type of bubble at the top of the model extending downwards from the tip for about one and a half base heights and this can be seen in the smoke photograph of the flow shown in Fig. 10. The exact nature of this flow at the tip must be very dependent upon the geometry of the tip as has been found by other investigators.

About one and a half base heights down from the tip the pressure reaches a maximum marking the end of the bubble. From then on the pressure again decreases as it tries to attain a base pressure in the central region similar to that associated with two-dimensional flow. In the root region there is also a high pressure at the floor of the tunnel and the pressure decreases as one moves away from the root. In all the finite length cases considered it was not possible to get far enough away from the root and tip regions to achieve a central section on the model where the base pressure was as low as that associated with the two-dimensional flow.

A simple method of solving the problem of predicting the base pressure on these bodies in shear flow would be to assume that the local base pressure coefficient, that is the pressure coefficient based upon the local velocity in the free stream at the height considered, is the same as in the uniform flow. This local pressure coefficient is shown in Fig.5 for one value of L/d = 15. It is obvious from this graph that at least for this value of L/d this simple way of looking at the problem is useless in the root region and reasonably good at the tip. In Fig.6b the local base pressure coefficients are plotted for the tip region for various values of L/d and it is clear that the general form of these curves is the same as for the uniform flow case but as the height of the body is reduced then the influence of the root region pressure distribution becomes greater and the range of agreement with the uniform flow case becomes worse.

The root region pressure distribution is plotted on Fig.6a and having seen, in Fig.5, that using the local pressure coefficient is not very satisfactory, the pressure coefficient plotted here is based upon the velocity along the tunnel center line. It is immediately clear that the shape of the pressure coefficient distribution in this region, when plotted this way, is very similar to that of the uniform flow. The use of the center line velocity as a base velocity to define the pressure coefficient is of course completely arbitrary and a better quantitative agreement could have been obtained by using another velocity as a base. Since these pressure distributions are similar for the uniform and for the shear flow cases it is clear that the major influence in the root region is the boundary layer on the floor of the tunnel, which is similar for both cases, and that the dominant feature is probably a strong horseshoe vortex lying close to the root of the model and a complex three-dimensional separated region behind the model.

Results on Vortex Shedding

The next question to be answered is how does the presence of a shear flow, and its associated vorticity, effect the vortex shedding from this type of body? The velocity fluctuations outside the wake were detected with a hot-wire anemometer at a position six base heights downstream of the body and two base heights out from the wake center line. The signals were digitally analysed to produce the frequency spectra shown on Figs.7 and 8, where the spectra are given for various positions (y/d) up the span of the body for the uniform flow and shear flow cases. The Strouhal numbers, (fd/U) where f is the frequency in Hz derived from the peaks of these spectra, are plotted in Fig.9.

If we compare the two frequency spectra shown in Figs.7 and 8 certain differences between the uniform and shear flow cases are immediately obvious. In the uniform flow case, moving up from the root of the body, the main shedding peak occurs at almost the same frequency over the entire length of the body with a slight frequency shift at $y/d = 5$ and again at $y/d = 9$. These frequency shifts are not thought to be due to experimental error, such as variations in wind tunnel speed between obtaining consecutive spectra, since at $y/d = 8$ both frequencies occur.

In the shear flow case, Fig.8, a different type of flow must be occurring. From the root to $y/d = 5$ the spectra show only one shedding frequency at about 130 Hz for this tunnel speed. At $y/d = 6$, however, there is evidence of another frequency peak at about 165 Hz and as y/d increases to 8 this second peak increases in size relative to the background level and also moves to a frequency of about 170 Hz. In this range $6 < y/d < 8$ the original peak at 130 Hz decreases in size as we move away from its influence. At $y/d = 9$ another peak becomes apparent at about 185 Hz and these two peaks at 170 and 185 Hz remain in the spectra until $y/d = 13.8$, the relative magnitude of the peaks increasing and decreasing with distance from the tip.

Discussion of Results

If we now compare the frequency measurements with the pressure measurements we can see some common features. In the uniform flow case, for $L/d = 15$ the decrease in pressure coefficient in the root region tails off for y/d between 5 and 6 which is where the frequency shifts. Again between $y/d = 8$ and 9, the frequency shifts and it is in this region that the pressure coefficient starts to rise due to the influence of the tip.

Again in the shear flow case the actual base pressure (i.e., the pressure coefficient based on the center-line velocity) reaches a minimum around $y/d = 6$ which is the region where the vortex shedding frequency undergoes a major change of about 35 Hz.

To try and understand some of the features of the flow, it was visualised by bleeding smoke from the base of the body and photographing the smoke pattern. The wind tunnel speeds for this smoke visualisation were about one third of those used for the pressure and vortex shedding measurements, but this change of Reynolds number is not thought to be significant since for this body the separation points are fixed at the base. Photographs of the flow for the uniform and shear flows are shown in Fig. 10. The differences between the uniform and shear flow are now very evident. In the uniform flow case the vortices are slanting into the flow away from the base, whereas in the shear flow the vortices are slanting in the opposite direction. It is also obvious in the uniform flow that the angle of inclination of the vortices is not constant with time but in forty photographs none were found with vortices sloping the other way. In the uniform flow the vortices are reasonably straight over most of the span of the model but in the shear flow the vortices have a point of inflexion at about $y/d = 7$ near the body and this point convects downstream with the vortices and also downwards towards the root of the model. It is thought that this inflexion point is associated with a longitudinal vortex which is produced by the rolling up of the original vorticity in the stream as it is bent round the body into the flow direction.

This longitudinal vortex line acts as a boundary between two regions of vortex shedding from the body and allows different frequencies in the wake. Above this vortex the smoke becomes very confused as the tip is approached and it is difficult to find any reason for the appearance of the two frequency peaks in the spectra near the tip.

Oil-flow visualisation on the base of the model, and also some of the smoke photographs, showed that in both the uniform and shear flow cases a pair of longitudinal vortices sprung from the base of the section at the tip and were convected downstream and downwards towards the wind tunnel floor.

The possibility that the introduction of a longitudinal trailing vortex into the flow at the base could produce different vortex shedding frequencies on either side of it was further investigated in uniform flow. A slender delta wing, at incidence, was fixed to the body spanning the tunnel as in Fig. 11 and the base pressure and shedding frequency measured on either side. As is very clear in Fig. 12 there is a sudden jump in Strouhal number across the delta with a corresponding change in pressure coefficient which is, however, not as sudden.

The actual mechanism of the jump in vortex shedding frequency across a streamwise longitudinal vortex is not known but it can be qualitatively understood as follows. The longitudinal vortices, as drawn in Fig. 11 induce a velocity at the base which is towards the base center line above the delta wing and away from the base center line below the delta wing. An inwards velocity draws the separated shear layers closer together causing them to have a high radius of curvature and hence producing a low base pressure. It is well known that a low base pressure (high drag) is associated with a low Strouhal number, and thus we have a high Strouhal number below the vortex generator and a low Strouhal number above.

This last statement, however, only strictly applies for a two-dimensional blunt based body where it was shown by Bearman [1] that there is a direct relationship between base pressure and Strouhal number such that if the base pressure is raised (drag decreasing) by say fitting a splitter plate then the Strouhal number will rise. In the case of three-dimensional flow, however, this is not the case and there seems no unique relationship between the local base pressure and the local vortex shedding frequency. For instance in uniform flow for the body of length 15 d the base pressure coefficient, excluding the tip region, can change from -0.45 to -0.55 with virtually no change in Strouhal number. For the case with shear the impression is that the body can withstand large changes in base pressure without any change in vortex shedding frequency but that if the base pressure along the model changes more than a certain amount then the vortex shedding frequency must then jump to another value.

Conclusions

It is evident that the vortex shedding in the presence of a shear flow is quite different from that in a uniform flow. In the shear flow the vortex shedding frequency can vary drastically along the span of the body with the result that tests on a body free to oscillate in a shear flow may be quite different from tests on the same oscillating body in a uniform flow.

There is some evidence that the jumps in vortex shedding frequency found in the shear flow may be associated with the presence of longitudinal vortices in the flow.

References

1. Bearman, P.W.: J. Fluid Mech. 28, 625 (1967).

2. Elder, J.W.: J. Fluid Mech. 5, 355 (1959).

3. Maull, D.J.: Paper 16, AGARD Conference Proceedings No. 48 (1969).

4. Owen, P.R., Zienkiewicz, H.K.: J. Fluid Mech. 2, 521 (1957).

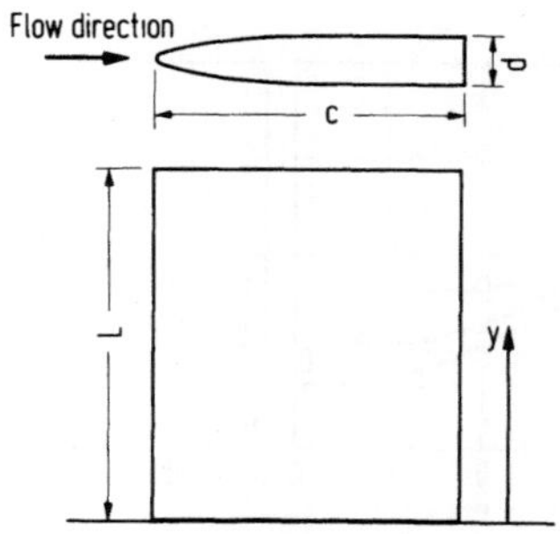

Fig.1. Sketch of model.

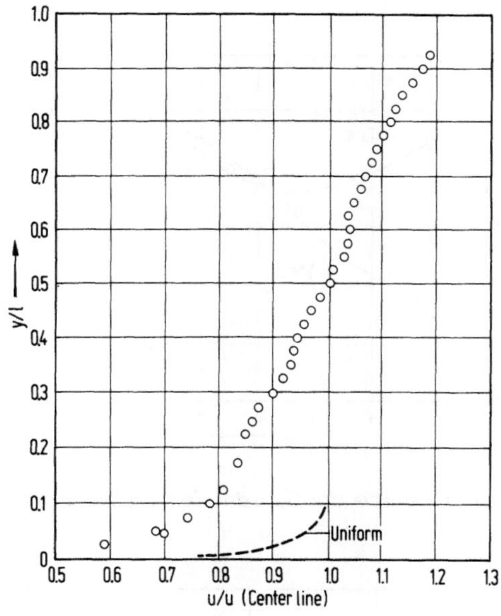

Fig.2. Velocity distribution.

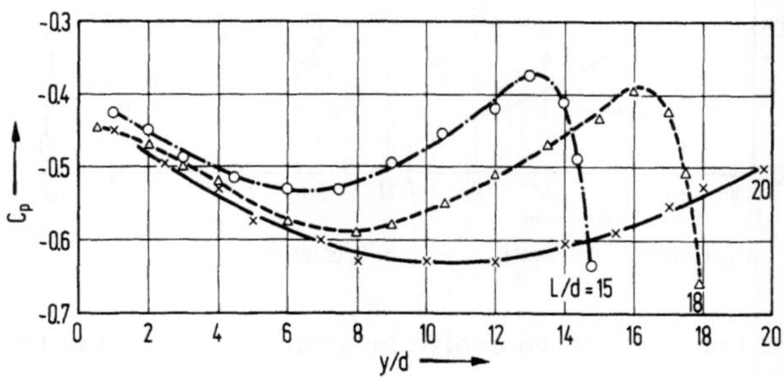

Fig.3. Variation of base-pressure coefficient along the span of the model.

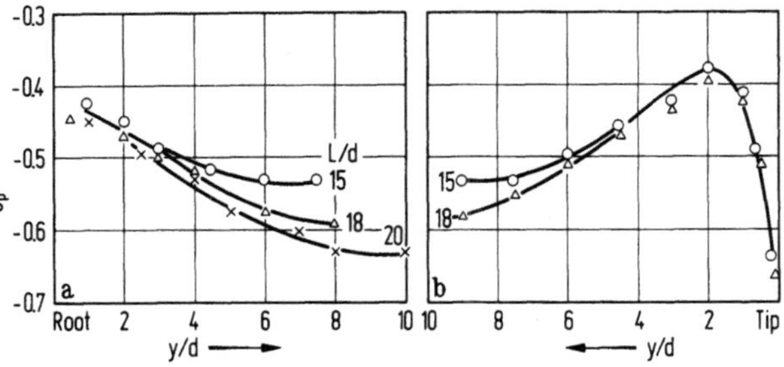

Fig.4. Pressure distributions near the root and tip regions.

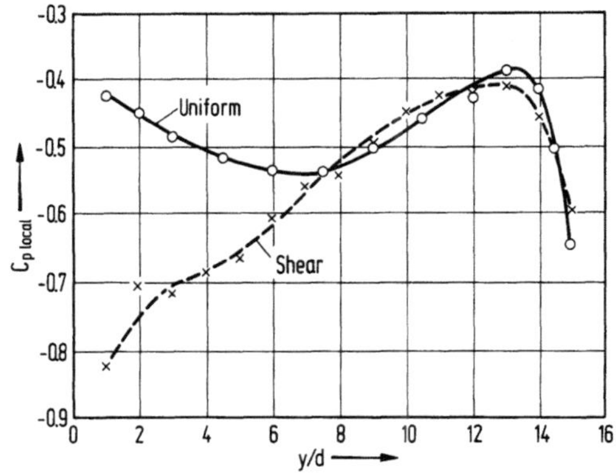

Fig.5. Local base-pressure coefficient for uniform and shear flow.

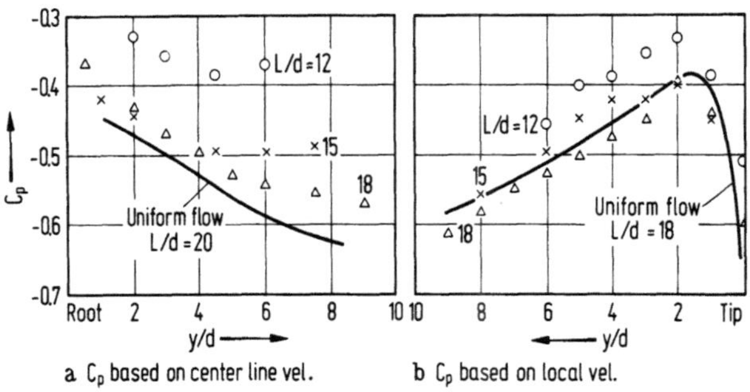

Fig.6. Local base-pressure coefficient near the root and tip regions.

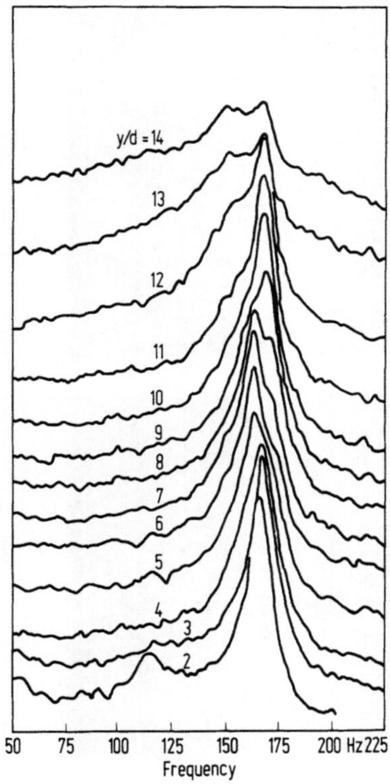

Fig.7. Frequency spectra for uniform flow.

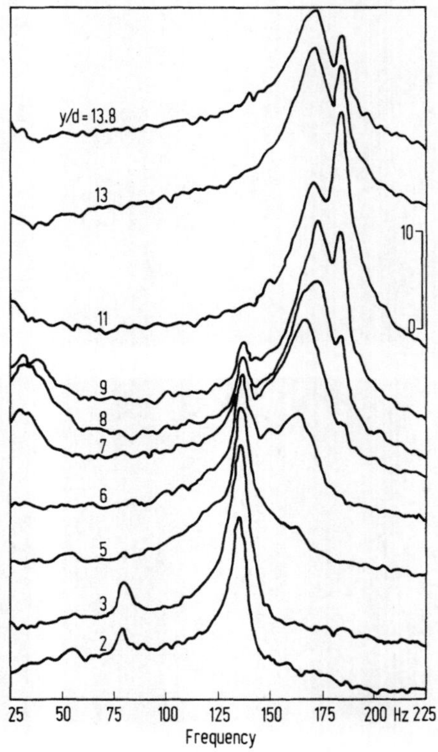

Fig.8. Frequency spectra for shear flow.

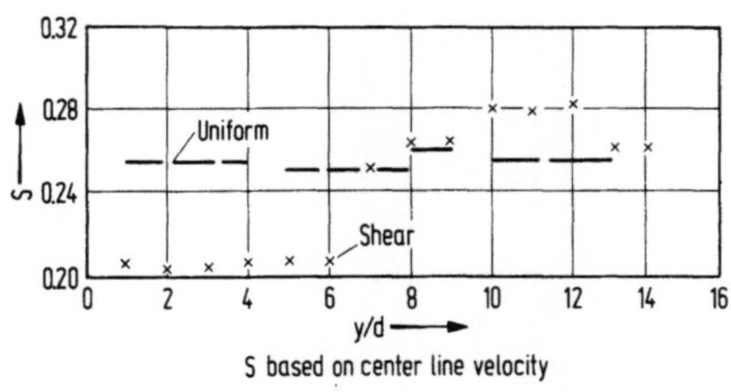

Fig.9. Strouhal numbers S based on center-line velocity derived from the spectral peaks.

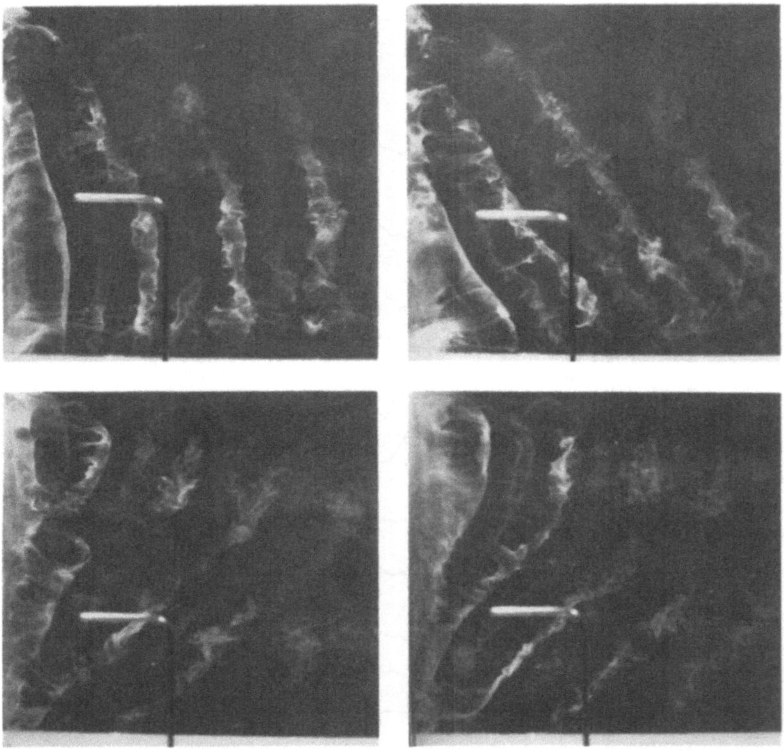

Fig.10. Smoke photographs of the flow.

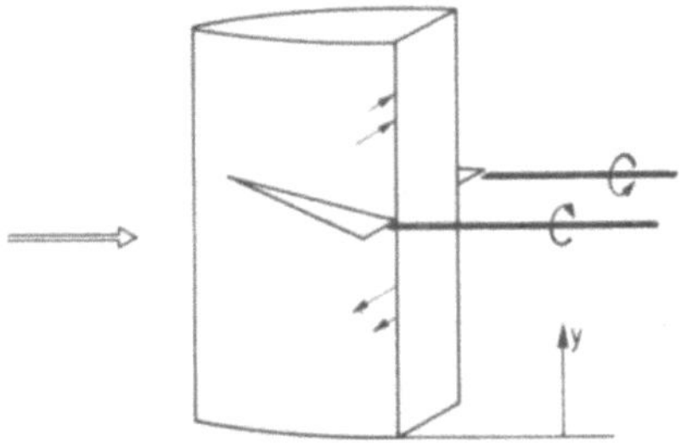

Fig.11. Sketch of slender delta wing on the model spanning the tunnel in uniform flow.

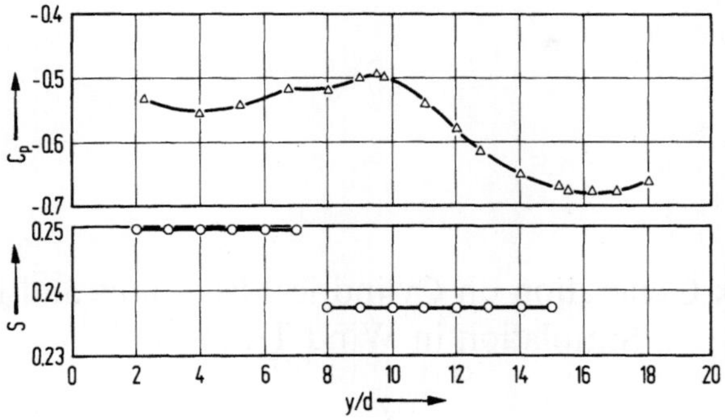

Fig. 12. Base-pressure coefficient and Strouhal number for the configuration shown in Fig. 11.

Vortex Generation on Cylindrical Buildings and its Simulation in Wind Tunnels

By

A. Naumann, H. Quadflieg

Aachen, Germany

Summary

A general criterion for the existence of alternating vortex shedding in the wake of circular cylinders is presented; in order to avoid periodic vortex formation, three-dimensional disturbances are necessary such as a non-straight-line flow separation, steplike or tapering changes of the cross section, free cylinder ends, etc. For the study of the flow behaviour past cylindrical bodies at transcritical Reynolds numbers, a simulation method is proposed which uses turbulence as well as separation wires on cylinders in the sub- and supercritical Reynolds-number range. The comparison of some flow data gained by this simulation method with those known from experiments in the true transcritical range shows an encouraging agreement.

Introduction

The formation of vortex streets in the wake of cylindrical structures is one of the most severe excitation sources for wind-induced vibrations. Failures of stacks, bridges, and cooling towers are known as consequences of vortex resonance. A relatively small structural damping coincides with a periodic flow fluctuation of sufficiently high amplitude (high wind speeds) and of sufficiently long duration. This paper merely gives a contribution to the excitation source, not to the interaction between aerodynamic wind loads and elastic forces of the structure. With regard to the trend to light-weight construction, we believe that - independent of the important problems of structural response - more work should be done to enable the prediction and possibly the reduction or avoidance of periodic wind loads. It is with this in mind that we tried to study the aerodynamic conditions for a periodically fluctuating pressure field caused by vortex shedding.

The Conditions for Alternating Vortex Shedding

It is well known that a periodic vortex shedding (von Kármán vortex street) can occur not only at subcritical Reynolds numbers but also in the transcritical range, i.e., $Re \geqslant 3.5 \times 10^6$ as reported by Roshko [1][1]. According to investigations at the Aerodynamisches Institut Aachen, an alternating vortex street can exist under suitable conditions also in the range of supercritical Reynolds numbers ($4 \times 10^5 < Re < 3.5 \times 10^6$). The empirical experience is, that a vortex street can always be formed, if the flow separates along a straight or nearly straight line of a circular cylinder; this result is independent on the Reynolds number, i.e., on the laminar or turbulent character of the separating boundary layer[2].

A non-straight-line separation leads to a contradiction to the vortex theorems for a perfect fluid and may be regarded as a three-dimensional disturbance. In a real fluid, a certain amount of disturbance (circulation difference) can be digested inwardly by the vortex filament itself; the question arises, what is the critical amount of disturbance which does not allow discrete vortex filaments to occur.

---

[1] Since the convenience of the notation "transcritical" for the upper Reynolds-number range was discussed at the Symposium, the terminology used in this paper is repeated once more: subcritical denotes $Re < 4 \times 10^5$, supercritical $4 \times 10^5 < Re < 3.5 \times 10^6$, transcritical $Re > 3.5 \times 10^6$. The notation "transcritical" for the upper range was introduced by Roshko [1] and has been commonly used since then.

[2] We thank Prof. Morkovin who pointed to a paper by Cincotta, Jones and Walker during the discussion: "Experimental Investigations of Wind-Induced Oscillation Effects on Cylinders in Two-Dimensional Flow at High Reynolds Numbers." (Meeting on Ground Wind Load Problems in Relation to Launch Vehicles, NASA Langley Research Center, June 7-8, 1966).
The oil flow photographs of non-straight separation lines in this paper are shown for supercritical Reynolds numbers; they are generally in agreement with our experience and results.

Such three-dimensional disturbances, which can lead to a non-straight-line separation, could be: the free end of a cylinder, steps in cylinder diameter, tapering cross-section, a non-uniform profile of the free stream velocity, surface perturbations or imperfections, and, finally, the history of the boundary-layer development, which is responsible for the non-straight separation line on a bare cylinder at supercritical Reynolds numbers ($Re = 5 \times 10^5$), i.e., for a turbulent wake in the classical sense. We observed - using smoke in the base region - a wavy separation line on a cylinder in incompressible flow at $Re = 5 \times 10^5$; additionally, fluctuations of the local separation points could be detected within a limited range of separation angles $\alpha_S = 135^\circ \div 140^\circ$ (measured from the stagnation point). One could object that at subcritical Reynolds numbers the separation line is fluctuating, too; it seems, however, that in this case a straight line is oscillating as a whole. Moreover, the vortex generation is much less sensitive to a disturbance at the angle $\alpha = 80^\circ$ than, e.g., at $\alpha = 130^\circ$. We will return to this point later.

Neither by flow visualization nor by hot wire anemometers a periodic vortex formation could be detected for that supercritical Reynolds number; and high frequencies as reported by Delany and Sorensen [2] at Reynolds numbers between 10^6 and 2×10^6 have not been observed at $Re = 5 \times 10^5$ on a bare cylinder. According to our concept, that only a straight-line separation is able to cause a vortex street, we forced such a straight-line separation by wires at different positions. By this way we always observed an alternating vortex shedding at subcritical as well as at supercritical Reynolds numbers. The wake is visualized in Fig.1 (by introducing smoke) for a forced separation at $\alpha = \pm 130^\circ$, $Re = 4.72 \times 10^5$, and aspect ratio (cylinder height/diameter) $t/D = 7.1$. The mechanism of the three-dimensional disturbance has been physically interpreted as a change of circulation (energy) along a vortex filament; regarding a truncated cone in uniform flow for which the local Reynolds number (according to the diameter variation) is restricted to the subcritical range, we can expect a nearly straight separation line at about $\alpha = 80^\circ$ (analogous in the transcritical Reynolds-number range at a different angle α). Inspite of the straight separation line (constant Strouhal number S along the cone axis if a vortex occurs), the different shedding frequencies again lead to a circulation difference; the mechanism of this "disturbance" is much more complicated than it is in the case of a non-straight separation line (cf. [3]). Probably one can say that it is much more difficult to destroy a vortex street the formation of which has already started than to prevent the vortex generation from the very beginning.

It must be remarked that the described alternating vortex formation must not be identical with the v. Kármán vortex street in the intrinsic theoretical sense (defined by the stability of the vortex arrangement).

Experiments Related to the Concept of Straight-Line Separation or Three-Dimensional Disturbance

First evidence of our criterion resulted from former tests about the interaction of boundary-layer separation with local shock waves on circular cylinders at high subsonic speeds [4,5,6]. Here, the strong pressure rise due to the shocks causes a separation of the laminar as well as of the turbulent boundary layer along a generating line of the cylinder.

In order to simulate a definite non-linear separation line at low velocities (incompressible flow), a test arrangement was chosen [7,8] in which stepped straight separation wires were used as shown in Fig.2; two wires, each of length "a", were fixed at the angle $\pm\alpha_2$ and a third one downstream at $\pm\alpha_1$. At the position $\alpha_1 = \alpha_2$, a well defined vortex street occurred; by increasing the angle difference $\Delta\alpha = \alpha_1 - \alpha_2$, a position was reached where the vortex street did no longer exist. Two examples are given in Fig.3. The critical angle difference $\Delta\alpha_{crit}$ determines the maximum displacement of the wires for which a vortex street will still be maintained. $\Delta\alpha_{crit}$ depends on the position α_1 as well as on the parameter 2a/t. Figure 4 shows the increasing sensitivity to such a disturbance for increasing separation angles (cf. wavyness of the supercritical separation line). From the critical angle difference, a relative critical circulation difference can be determined between α_1 and α_2. It came out as a linear function of 2a/t for all values of α_1: $(\Delta\Gamma_{crit}/\Gamma_2) \cdot 2a/t = 0.042$.

Besides other forms of non-straight separation lines not mentioned here, we investigated cylinders with steplike and tapering change of cross section and cylinders with a free end. These kinds of three-dimensional disturbances lead to non constant amounts of the circulation along the vortex filaments as well [9, 10].

The variation of the wake-flow pattern for a circular cylinder being compared on one hand to truncated cones of increasing cone angle and on the other hand with stepped cylinders of increasing diameter ratio is illustrated in Figs. 5 and 6. The vortex street becomes strongly disturbed, but the generation of a first vortex, which is not straight and sometimes branched, can be observed for the largest step as well as for the largest cone angle. From this it seems to be clear that the transition from a vortex street to a non-vortex street due to such disturbances is not a rapid one. The question, if and what fluctuating loads can act on the cylinder even if a vortex street is highly disturbed, can be solved only by force or pressure measurements (cf. the later sections).

The "disturbance" caused by a diameter step or by the free end generally extends over a certain part in direction of the cylinder axis. Thus, the aspect ratio (cylinder height/diameter) has a remarkable influence on the overall wake-flow pattern. Tests on stepped cylinders showed, indeed, that a decreasing aspect ratio has a similar effect as an increasing step. Since the transition from vortex street to non-vortex street occurs rather slowly it would be a matter of definition to determine critical data for the step or the free end.

Figure 7 shows the wake-flow pattern and the predominant vortex-shedding frequencies along the axis of a cone and of a stepped cylinder. In both cases, there does not exist *one* vortex street; generation of single vortices, however, can be observed. The wake of the cone - similar to that of the stepped cylinder - shows a kind of branched vortex filaments. Such wake flows are expected to reduce remarkably the maximum periodic loads on a body compared to cases in which *one* vortex street is present.

As an example for the interaction of two different effects, Fig.8 shows the flow around a stepped cylinder between end plates. With the smaller diameter, the Reynolds number has a subcritical value, with the larger diameter a supercritical one. There is no vortex street behind the bare stepped cylinder (upper line in the figure) because of the non-constant circulation along the cylinder axis. Forcing the separation by wires at $\alpha = \pm 124^o$ on the larger-diameter section and additionally using boundary-layer suction near the end-plate of that section leads toward the minimum of circulation difference and nearly one vortex street can be observed on the smoke picture and the hot wire oscillogram (lower line in Fig.8). The forced separation should have been a little further downstream ($\alpha > 124^o$) to increase the frequency behind the larger-diameter section.

Roshko [1] and Achenbach [11] report that the upper transition (from supercritical to transcritical Reynolds numbers) is related to the vanishing laminar separation bubble on the rear side of the cylinder ($\alpha \simeq 105^o$). The onset of vortex shedding was observed by Roshko at $Re = 3.5 \times 10^6$. Achenbach found the vanishing of the bubble by skin friction measurements at $Re = 1.5 \times 10^6$. Besides the turbulence degree and the surface roughness, the influence of model size and wind-tunnel boundaries seem to be particularly strong in supercritical cylinder flow. We believe that the frequencies reported by Delany and Sorensen could have been observed just below the transcritical range. Possibly not a vortex street, but a disturbed generation of vortex filaments near the cylinder has been detected. Nevertheless, it is significant that the frequencies of Delany and Sorensen fit very well into the general empirical relation between drag coefficient and Strouhal number: $S = 0.214/C_D$ exp. 0.6 which is shown as diagram in Fig.9 (taken from [8]). With regard to fluctuating wind loads one can say that such high Strouhal numbers S (smaller circulation of the discrete vortices) will imply smaller values of oscillating pressure coefficients than the lower values of S at sub- and transcritical Reynolds numbers (cf. Fig.12). The separation bubble located in a region of positive pressure gradient and the short distance of the turbulent boundary layer in the supercritical flow range are regarded as the source of the wavy separation line. The laminar separation bubble probably oscillates already, and there is not enough development length for a homogeneous turbulent boundary layer.

Simulation of Transcritical Reynolds Numbers in Wind Tunnels of Usual Size

With regard to drag, pressure distribution, and Strouhal number, some characteristic features of the flow behaviour at transcritical Reynolds numbers are known (e.g., Refs. [1,11]). According to the reported results, the interaction between the wake form and the vortex-shedding frequency seems to be independent of the Reynolds number. So, within a certain limitation, it must be possible to simulate the main features of the transcritical flow, if a wake form is artificially forced at sub- or supercritical Reynolds numbers in a similar way as it is naturally produced at transcritical values of Re. This is realized by the use of separation and/or turbulence wires fixed at adapted angles on the cylinder. This indirect method is proposed because of the advantage to study systematically the transcritical flow behaviour on rigid and elastic models in usual-size wind tunnels. A simulation of transcritical flow around cooling towers by surface roughness has been reported by Armitt [12]. The method, using a discrete disturbance (wire or strake, respectively) instead of a homogeneous roughness, is not only easier to handle but gives the chance to simulate the boundary-layer development by using transition as well as separation wires; they will slightly change the average values of the wind load (pressure distribution) in relation to the corresponding Strouhal number. The shedding frequency of the wake, however, is regarded to be the most significant parameter influencing the fluctuating loads.

Figure 10 serves to recall the average pressure distributions on circular cylinders for sub-, super-, and transcritical Reynolds numbers measured by different authors. The transcritical curves of Roshko and Achenbach agree quite well in the base region but not so well near the pressure minimum and near the point of flow separation.

Simulated pressure distribution curves are compared in Fig. 11 to those of Roshko and Achenbach. The curve only using a separation wire at $\alpha = 115^{\circ}$ ($Re > Re_{crit}$) fits the curve of Achenbach quite well; the corresponding Strouhal number $S = 0.262$ is very close to the average value $S = 0.267$ measured by Roshko. The curve using both transition and separation wires ($Re < Re_{crit}$) fits the Roshko curve rather well, here the Strouhal number $S = 0.247$ is a little low compared with the Roshko value. The pressure rise seen just in front of the separation wires can probably be leveled off by diminishing the wire diameter; an additional small downstream shift of the separation wires would raise the lower Strouhal number to an adequate value.

Fluctuating Pressures on Circular Cylinders

With increasing separation angles, the circulation of discrete vortices in a cylinder wake decreases, and the induced pressure fluctuations on the cylinder will diminish as well. Figure 12 shows, as functions of the separation angle α_S, the Strouhal number S and the rms-value of the pressure fluctuation at $\alpha = 90^{\circ}$ (position of the pressure transducer) for a circular cylinder between end plates. The rms-value at $\alpha = 90^{\circ}$ for a separation angle $\alpha_S = 115^{\circ}$ - which corresponds to the transcritical separation point - drops to less than half of the value at subcritical flow ($\alpha_S = 80^{\circ}$). Figure 13 illustrates that the pressure fluctuations at simulated transcritical cylinder flow are much higher for the two-dimensional than for the free-end cylinder (3 diameters away from the top). For comparison, the NPL curve by Gould, Raymer and Ponsford [13] at $Re = 5.4 \times 10^6$ is shown. The upstream turbulence degree apparently was lower for our measurements (0.2%).

The NPL investigation of pressure fluctuations near the top of a cylindrical stack with $H/D = 6$ at $Re = 5.4 \times 10^6$ is compared in Fig. 14 with our measurements using the described simulation method. We could not simulate the NPL efflux rate for this case. Nevertheless, the curves fit rather well. We found, that an increasing efflux rate generally has the effect of elongating the cylinder; the result of some measurements showed a high drop of rms pressure values near the top of a cylinder with increasing efflux velocity.

Concluding Remarks

A periodic vortex generation behind cylinders is coupled with a straight separation line and generally limited by a small vortex circulation difference; the vortex frequency depends only on the separation angle and not on the Reynolds number. The physical concept of our results, e.g., allows to understand the disturbing effect of helical strakes on vortex shedding from circular stacks, first proposed by Scruton and Walshe [14]; and it can be explained that the pitch of such strakes must be neither too large nor too small in order to be efficient.

The proposed simulation method appears to be a useful tool for systematic investigations of wind loads on circular-shaped buildings in wind tunnels which cannot provide transcritical Reynolds numbers. Because of the sharp edges at the free ends of buildings, we believe that the top effects are not significantly dependent on Reynolds number; nevertheless, more data gained from true transcritical Reynolds-number tests would be desirable. It should be emphasized that the coefficient of pressure fluctuation is remarkably lower in the transcritical than in the subcritical flow range. Since the dynamic pressure, however, increases with Re^2 (for a constant diameter) very high absolute values of fluctuating loads can occur. Results gained on elastically mounted models in subcritical flow (cf. [15]) can probably be corrected with respect to amplitudes and frequencies for the transcritical Reynolds-number range using some of the reported principles and data.

References

1. Roshko, A.: Experiments on the Flow past a Circular Cylinder at Very High Reynolds Number. J. Fluid Mech. 10, 354-356 (1961).

2. Delany, N.K., Sorensen, N.E.: Low-Speed Drag of Cylinders of Various Shapes. Nat. Adv. Comm. Aero., Wash., Tech. Note 3038 (1953).

3. Gaster, M.: Vortex Shedding from Slender Cones at Low Reynolds Numbers. J. Fluid Mech. 38, Part 3, 565-576 (1969).

4. Naumann, A., Pfeiffer, H.: Versuche an Wirbelstraßen hinter Zylindern bei hohen Geschwindigkeiten. Forsch. Ber. Nordrh.-Westf. 493 (1958).

5. Pfeiffer, H.: Strömungsuntersuchungen an Kreiszylindern bei hohen Geschwindigkeiten. Diss. TH Aachen (1961), Forsch. Ber. Nordrh.-Westf. 1062 (1961).

6. Naumann, A., Pfeiffer, H.: Über die Grenzschichtablösung am Zylinder bei hohen Geschwindigkeiten. Adv. Aeron. Sci., Bd. III, p. 185, Bd. IV, p. 1183 (1960).

7. Naumann, A., Morsbach, M., Kramer, C.: The Conditions of Separation and Vortex Formation past Cylinders. AGARD CP No. 4, Part 2, pp. 539-574 (May 1966).

8. Morsbach, M.: Über die Bedingungen für eine Wirbelstraßenbildung hinter Kreiszylindern. Diss. TH Aachen 1967.

9. Naumann, A., Quadflieg, H.: Aerodynamic Aspects of Wind Effects on Cylindrical Buildings. Proc. Symp. Wind Effects on Buildings and Structures, April 1968, Vol. 1, Loughborough/England.

10. Naumann, A., Quadflieg, H.: Über die Wirkung des Windes auf zylindrische Bauwerke. Abh. aus dem Aerodyn. Inst. TH Aachen H. 19 (1969).

11. Achenbach, E.: Distribution of Local Pressure and Skin Friction around a Circular Cylinder in Cross-Flow up to $Re = 5 \times 10^6$. J. Fluid Mech. 34, Part 4, 625-639 (1968).

12. Armitt, J.: The Effect of Surface Roughness and Free Stream Turbulence on the Flow around a Model Cooling Tower at Critical Reynolds Numbers. Proc. Symp. Wind Effects on Buildings and Structures, April 1968, Vol. 1, Loughborough/England.

13. Gould, R.W.F., Raymer, W.G., Ponsford, P.J.: Wind Tunnel Tests on Chimneys of Circular Section at High Reynolds Numbers. Proc. Symp. Wind Effects on Buildings and Structures, April 1968, Vol. 1, Loughborough/England.

14. Scruton, C., Walshe, D.E.J.: A Means for Avoiding Wind-Excited Oscillations of Structures with Circular or Nearly Circular Cross-Section. NPL/Aero/335 (October 1957).

15. Försching, H.: Zur theoretischen Behandlung aeroelastisch erregter Schwingungen kreiszylindrischer Konstruktionen bei periodischer Wirbelanregung. Z. Flugwiss. 18, H. 9/10, 347 (1970).

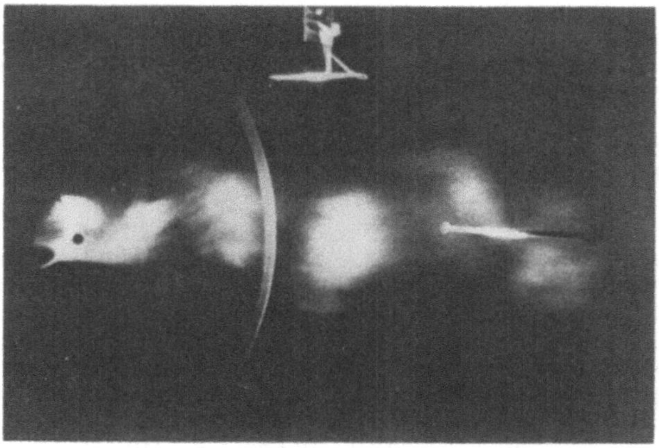

Fig. 1. Separation forced by straight wires at $\alpha = \pm 130^{\circ}$; $Re = 4.72 \times 10^{5}$.

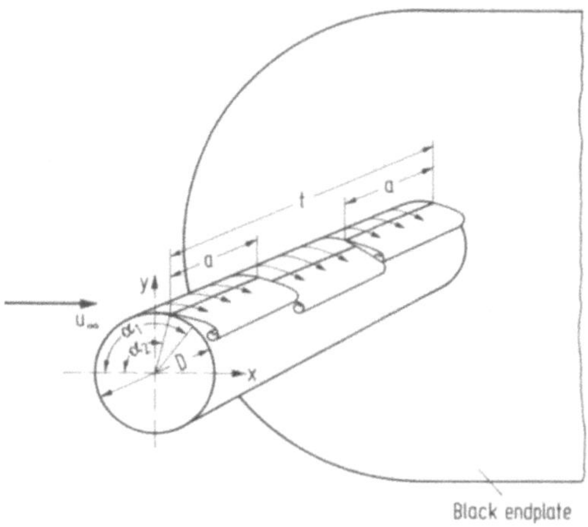

Fig. 2. Arrangement to measure the critical circulation difference.

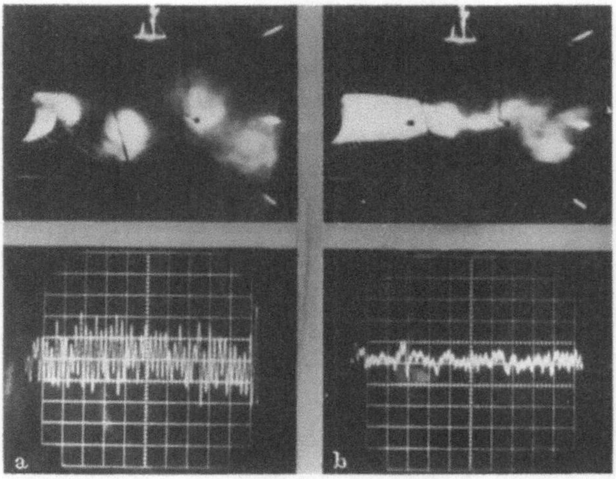

a b

Fig.3. Cylinder with forced stepped separation, Re = 5×10^5; $\alpha_1 = 115^\circ$; $2a/t = 0.5$.
a) $\Delta\alpha = 2^\circ$; b) $\Delta\alpha = 4^\circ$.

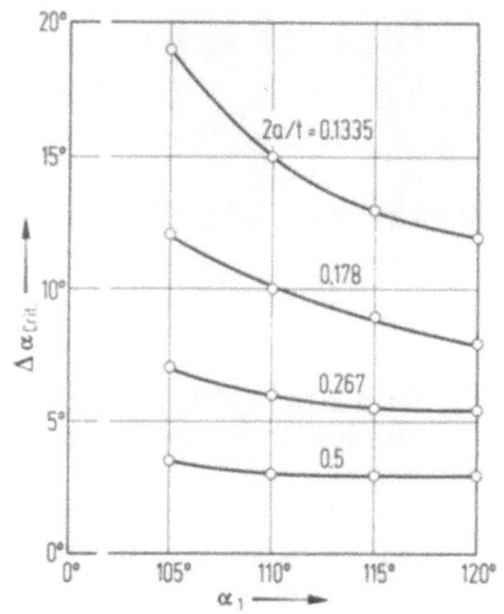

Fig.4. The critical angle difference.

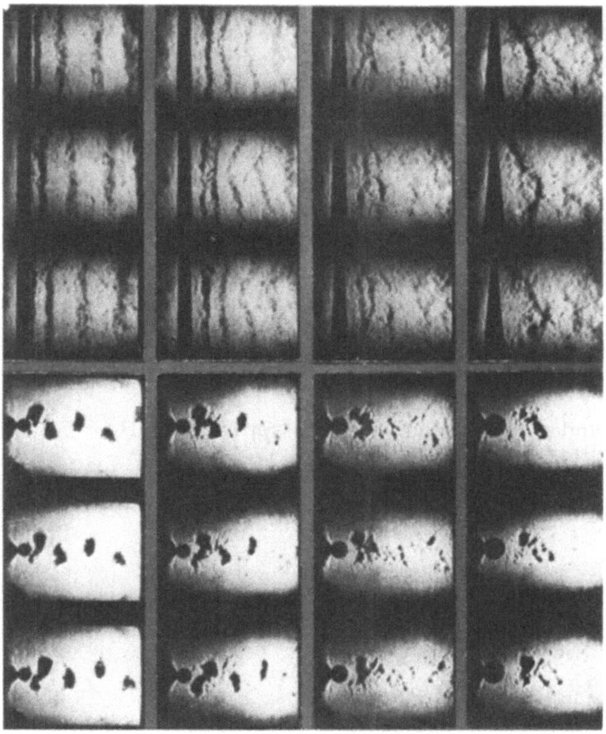

Fig.5. Schlieren fotos from a film (5750 frames/sec) of the wake: transition from cylinder to cones with increasing taper angle θ.

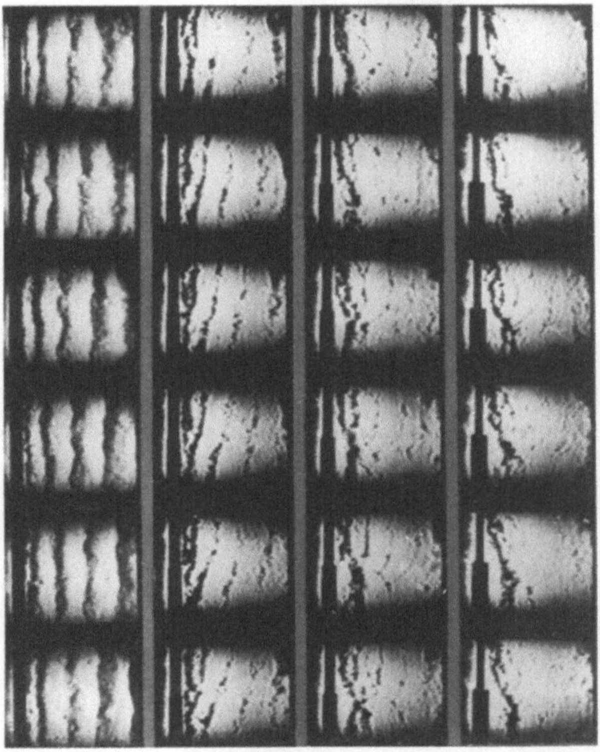

Fig.6. Schlieren fotos from a film (5750 frames/sec) of the wake: transition from cylinder to stepped cylinders with increasing diameter ratio.

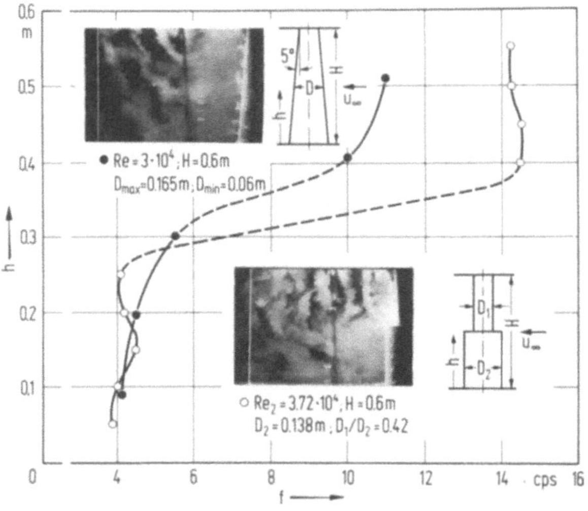

Fig.7. Shedding frequencies along the axis of a cone and a stepped cylinder.

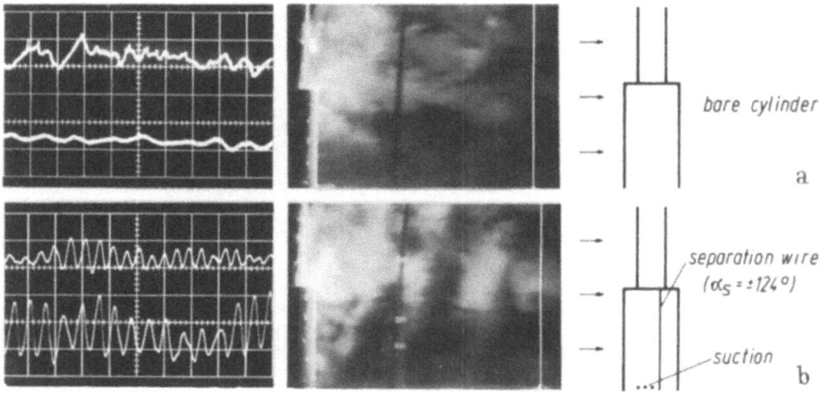

Fig.8. Wakes behind stepped cylinder.
a) Bare cylinder; b) Forced separation on larger end of cylinder (α_s = 124°) and boundary-layer suction near the end plate of that part.

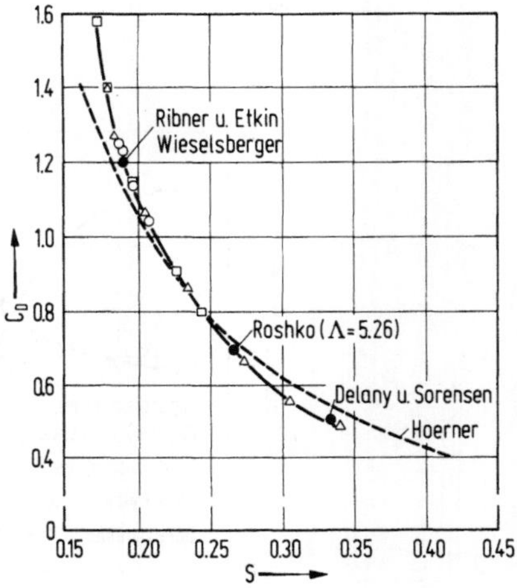

Fig. 9. Drag coefficient C_D as a function of the Strouhal-number S.

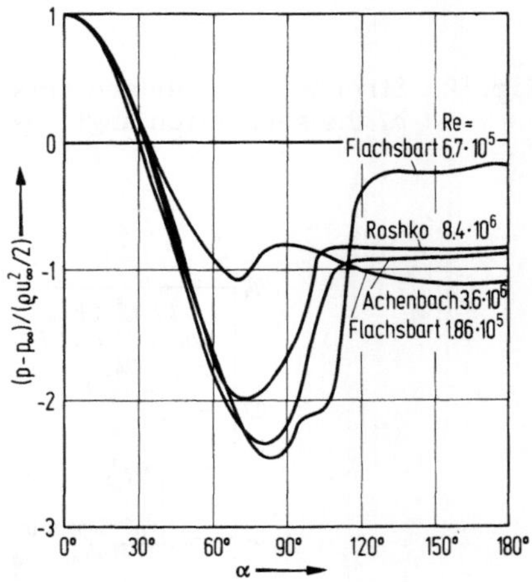

Fig. 10. Average pressure distributions on circular cylinders.

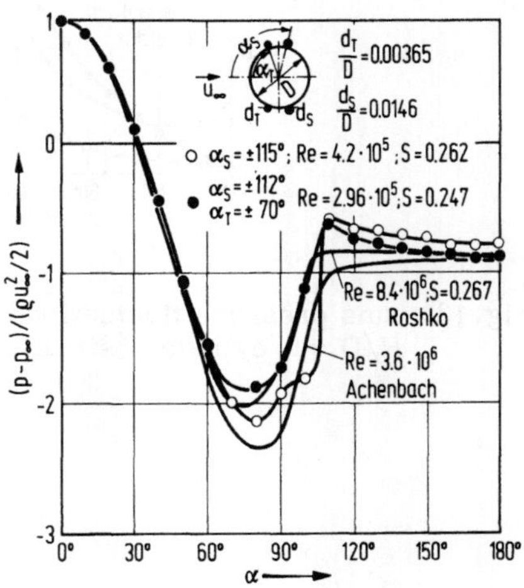

Fig. 11. Simulated pressure distributions for transcritical Reynolds numbers.

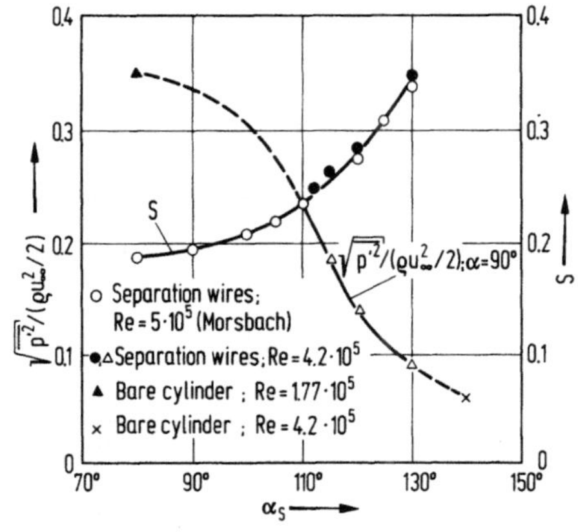

Fig.12. Strouhal-number and rms pressure fluctuations as a function of the separation angle α_S (two-dimensional flow).

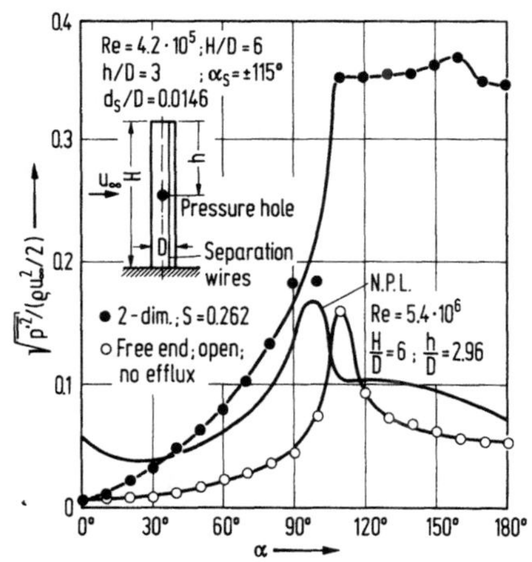

Fig.13. rms pressure fluctuations for a two-dimensional and a H/D = 6 cylinder (simulation for transcritical Re).

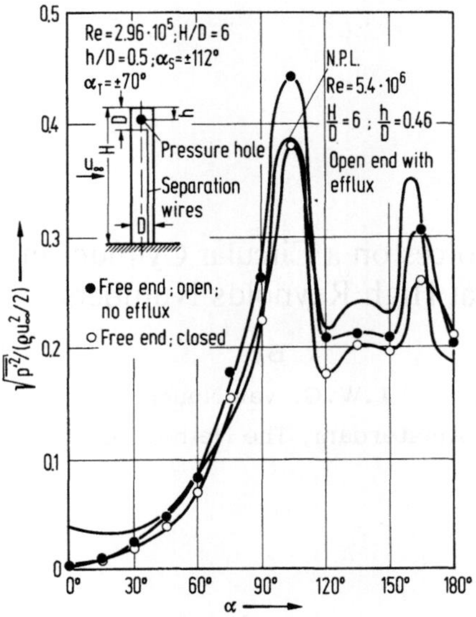

Fig. 14. rms pressure fluctuations near the top of a H/D = 6 cylinder (simulation for transcritical Re).

Pressures and Forces on a Circular Cylinder in a Cross Flow at High Reynolds Numbers

By

J.W.G. van Nunen

Amsterdam, The Netherlands

It is well known, that the steady and unsteady aerodynamic loads on a circular cylinder in a cross flow are highly sensitive to the value of the Reynolds number. Numerous investigations have been carried out to look into this problem, but in only some of them high Reynolds numbers, $Re > 5 \times 10^6$, were realised. The European Space Vehicle Launcher Development Organisation (ELDO) therefore had experiments performed at the National Aerospace Laboratory (NLR) to get more information about the aerodynamic loads on a circular cylinder at high Reynolds numbers.

To get a good insight into the behaviour of the flow around the cylinder, both steady and unsteady pressure distributions were determined. Besides, also the overall aerodynamic loads were measured by means of strain gauge balances. The test object was a circular cylinder, with a diameter of 0.3 m, spanning the wind tunnel test section. The wind tunnel used was NLR's variable density wind tunnel HST. By changing the stagnation pressure from 0.25 ata to 4 ata and by increasing the wind speed up to Mach = 0.3, a range of Reynolds numbers from Re = $= 0.48 \times 10^6$ to 7.7×10^6 could be covered.

Quite evidently, the results from these experiments contain information already gathered by other investigators, but especially because of the detailed pressure measurements also new information was gained. In the following some of the results will be briefly described. For detailed information, Ref. [1] should be consulted.

Firstly, the steady pressure distribution over a wide range of Reynolds numbers should be considered (Fig.1). At the lower value, $Re = 0.48 \times 10^6$, the presence of a laminar separation bubble can be observed together with a fairly late final turbulent separation. With increasing Reynolds number, the final separation point moves forward; at the same time the maximum underpressure on the cylinder decreases. As far as the steady drag on the cylinder is concerned, the described flow phenomenon implicates a slight increase of the drag at higher Reynolds numbers.

An idea about the distribution of the unsteady pressure at various Reynolds numbers is presented in Fig. 2. At the lower as well as at the higher Reynolds number, sharp peaks can be observed. A more close look reveals that these peaks practically coincide with the final separation point. The absence of a peak at the intermediate Reynolds number may probably be attributed to the spacing of the pressure holes.

In earlier publications (e.g., Ref.[2]) it was pointed out that at Reynolds numbers above 4×10^6, a discrete vortex shedding would reoccur in the wake behind the cylinder. In the present tests, this phenomenon was more or less confirmed by analysing signals of the unsteady lift acting on the cylinder. Moreover, it has been possible to get some idea about the origin of this periodic lift by making a spectral analysis of the unsteady pressures around the cylinder (Fig.3). Except at the point where the highest unsteady pressure is measured, the PSD's show a great energy content in the lower frequency band, together with a peak arising at a Strouhal-number of about 0.27. At lower Reynolds numbers it was found that this peak only appeared in the unsteady pressures ahead of the separation point. In these cases no predominant frequency was observed in the energy content of the unsteady lift. From this it may be deduced that the re-occurrence of the regular vortex shedding starts at the wind-ward face of the cylinder.

Interesting as to the behaviour of a cylinder, finally, is the unsteady lift and drag acting upon it. Figure 4 contains the results as gathered in the present tests and as obtained by other investigators. Over the complete range of Reynolds numbers, the unsteady loads show a great scatter. This should not be too astonishing since these loads are completely random in nature. Only at the higher Reynolds number the scatter diminishes somewhat due to the re-occurrence of the regular vortex shedding.

As to the unsteady lift coefficient, some Reynolds number influence may be observed on the amplitude and in the form of a renewal of the periodic shedding. On the other hand, the drag coefficient remains more or less constant in the considered Reynolds number range and exhibits also no tendency towards a renewed vortex shedding.

In connection with the above described results, the following observation can be added. In Ref. [5], a NASA-publication, the unsteady lift force acting on a cylinder was categorized into three regimes:

wide-band random	$1.1 \times 10^6 < Re < 3.5 \times 10^6$,
narrow-band random	$3.5 \times 10^6 < Re < 6 \times 10^6$,
quasi-periodic	$6 \times 10^6 < Re < 18.7 \times 10^6$.

In the present tests these boundaries were confirmed, although the corresponding Reynolds number may have shifted somewhat. Actually, the following results were obtained:

<u>$0.5 \times 10^6 < Re < 2.5 \times 10^6$</u>: No periodicity was observed in either the unsteady lift or the unsteady pressures. So this range resembles the wide-band random regime of NASA.

<u>$2.5 \times 10^6 < Re < 6.5 \times 10^6$</u>: In the unsteady lift, a small peak can be observed at $S = 0.20$. In the unsteady pressures, this peak only turns up at the front part of the cylinder; no peak occurs behind the separation point. So this range covers the narrow-band random regime of NASA.

<u>$6.5 \times 10^6 < Re < \ldots$</u> : A definite peak occurs both in the unsteady lift and in the unsteady pressures at $S \simeq 0.30$. This periodicity occurs both at the front part and at the rear part of the cylinder.

So this range most probably resembles the quasi-periodic regime of NASA.

References

1. van Nunen, J.W.G., Persoon, A.J., Tijdeman, H.: Steady and Unsteady Loads on a Circular Cylinder at Reynolds Numbers up to 7.7×10^6. (To be published.)

2. Roshko, A.: Experiments on the Flow past a Circular Cylinder at Very High Reynolds Numbers. J. Fluid Mech. 10, Part 3 (1961).

3. Spitzer, R.E.: Measurement of Unsteady Pressures and Wake Fluctuations for Flow over a Cylinder at Supercritical Reynolds Number. Ph.D. Thesis Caltech, 1965.

4. Schmidt, L.V.: Measurement of Fluctuating Air Loads on a Circular Cylinder. Ph.D. Thesis Caltech, 1963.

5. Jones, G.W., Cincotta, J.J., Walker, R.W.: Aerodynamic Forces on a Stationary and Oscillating Circular Cylinder at High Reynolds Numbers. NASA TR R-300 (1969).

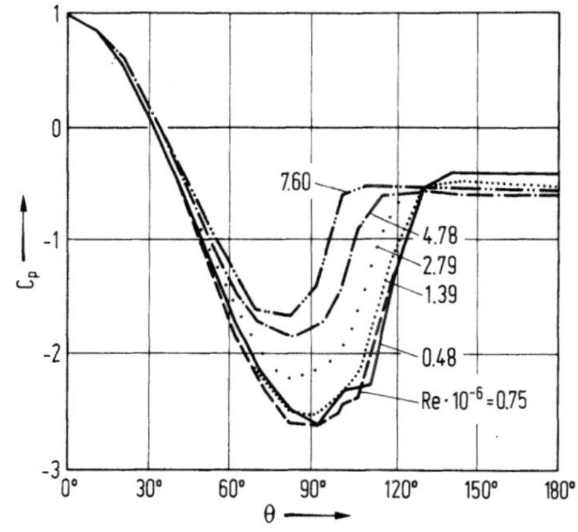

Fig.1. Pressure distribution around the cylinder for various Reynolds numbers.

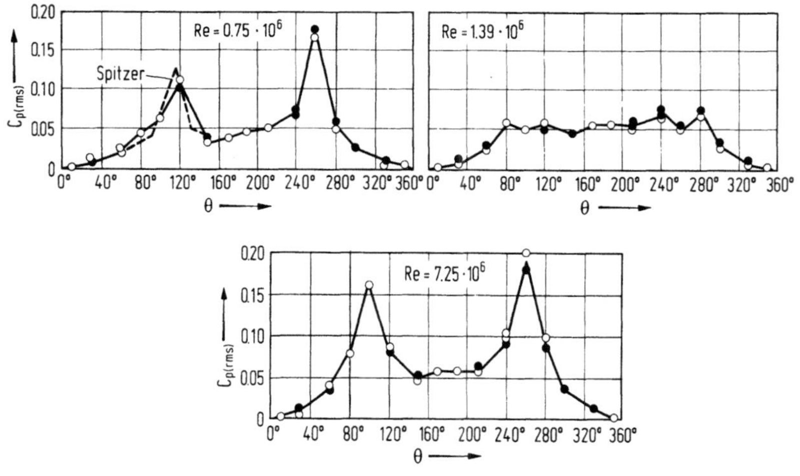

Fig.2. rms-pressure distribution around the cylinder for three Reynolds numbers.

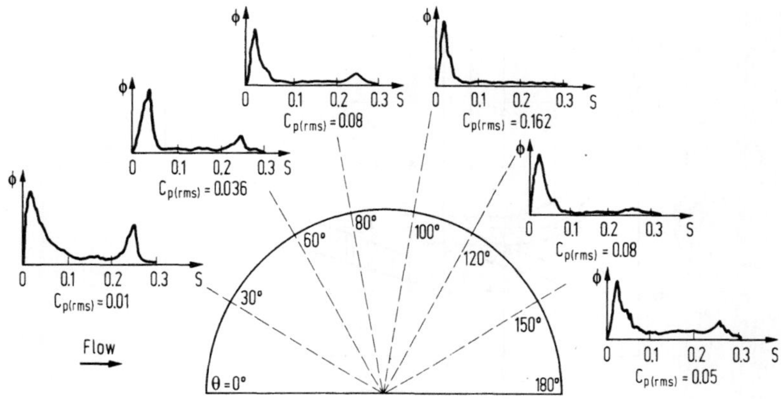

Fig.3. Power spectral densities of the unsteady pressures at Re = = 7.25 × 10^6.

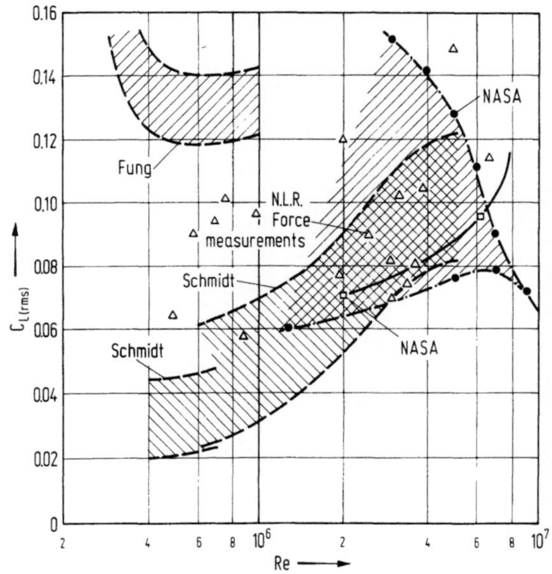

Fig.4. Unsteady lift coefficient versus Reynolds number.

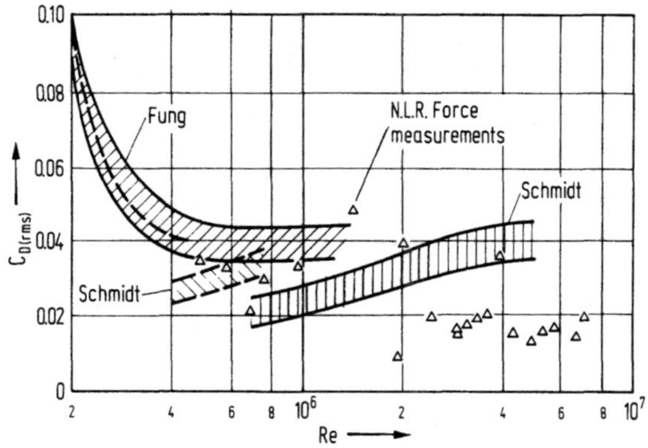

Fig.5. Unsteady drag coefficient versus Reynolds number.

Dynamic Lift on a Cylinder in High Reynolds Number Flow

By

H. Loiseau, E. Szechenyi

Châtillon, France

Summary

This paper briefly describes the experimental methods developed for a study of the unsteady forces acting on a vibrating cylinder subjected to a high Reynolds number incompressible flow perpendicular to its axis. A summary is given of the results obtained when the cylinder was clamped.

Introduction

The present study was instigated and is partly financed by the Electricité de France[1] who is concerned with vibrations and instabilities due to wind-induced unsteady forces on very large cylindrical structures such as power station chimneys. The Reynolds numbers of winds of moderate speeds around such structures (based on their diameter) can be of the order of 30 million so that a high Reynolds-number study is essential. In the present work, the range of Reynolds numbers covered is $2.6 \times 10^5 \leqslant Re \leqslant 6,5 \times 10^6$ which spans the whole transcritical region, reaching from the subcritical into the beginning of the supercritical[2].

---

[1] L. Michel, Chef de la Division Aérodynamique et Physique des Ecoulements, coordinator of the present research program.

[2] In this paper the terms "transcritical" and "supercritical" are used inversely to the norm because it appears more logical that the former should describe a transitional region while the latter should describe a region of higher values.

The initial study was envisaged in three phases: firstly, tests on a clamped rigid cylinder in order to develop a reliable measuring technique and also to gain some understanding of the flow mechanism; secondly, unsteady lift measurements on a cylinder vibrating perpendicular to the flow to gain some insight into the effects of amplitude and frequency; and finally, the formulation of a general mathematical model, based on experimental results, to permit predictions of the dynamic forces acting on a cylinder and of its resulting motion. Subsequently, the wake of a cylinder and its effect on a second cylinder in line will be considered.

To date of writing, test results of only the first phase are publishable. These can be found in much greater detail in Ref. [1].

Experimental Methods

Wind Tunnel and Cylinder

The high Reynolds number requirement seriously limits the wind tunnels that can be used. The ONERA S_2M_A tunnel with a test section of 1.75m by 1.76m was found to be the most suitable mainly for its pressure-variation capabilities (between 0.2 and 2.5 bar). A cylinder of 0.4m diameter was chosen, and the considerable test section obstruction was allowed for by a 12.5% tunnel wall permeability. As it was essential that all the tests be carried out in incompressible flow, the flow speed was restricted to Mach $\leqslant 0.3$ limiting the Reynolds number to 6.5×10^6.

Measurements

The dynamic lift measurements were made by means of twenty pressure transducers spread around the circumference of the cylinder at some discrete point along its axis (Fig.1). There were three such sections, each fitted with transducers, whose relative positions were easily changed in order to make spacial correlations between the lift forces.

The discrete pressure measurements were translated into lift by a real time summation of the lift components of the pressure (Fig.1). This method was checked against the more conventional force-transducer technique and gave identical results. Its principal advantage is that it ignores the large inertia forces of a vibrating cylinder which are necessarily picked up by dynamometers. Other advantages include the possibility of local pressure measurements and cross-correlations which are useful if one is to try to understand the flow phenomena.

Cylinder Excitation

Due to the unavoidable weight and dimensions of the cylinder, the forces required to shake it with a scotch-yoke or crank-type mechanism were too great for the wind tunnel structure on which it had to be mounted. The resonant excitation of the larger amplitude translational mode of a four degree of freedom system as used here avoided this problem. The mechanism is sketched in Fig.2. A large permanent magnet was mounted horizontally across the test section of the wind tunnel. Between the poles of the magnet was suspended a heavy, long, flat electrical coil constrained to move perpendicular to the flow. Coil springs were mounted on this coil with their axes in the plane of motion. These were evenly spaced along the width of the test section. Finally, the rigid cylindrical shell was mounted on the free ends of the coil springs.

The most obvious disadvantage of this mechanism is its fixed single useful frequency of excitation. However, since the frequency variation required is relative to the shedding frequencies that induce lift forces, flow-speed variations will give the same result. The wind tunnel used was particularly suitable for controlling the frequency ratio (excitation/shedding) by means of flow-speed changes because a constant Reynolds number could always be maintained by varying the tunnel pressure.

Flow Visualizations

In order to assess the effect of cylinder surface roughness, the flow was visualized by means of a titanium oxide based paste spread over the surface which is previously clad with a known roughness. The paste sets while subjected to air flow so that it can be subsequently studied and photographed.

Results on Fixed Cylinders

Dynamic Lift Forces

The dynamic lift coefficient measured at different Reynolds numbers can be compared to results found in existing literature (Fig.3a). Actually, the results would only be exactly comparable, if there was perfect axial correlation of lift forces, because the other experimenters measured the forces acting on the whole or a large part of a cylinder by means of force transducers. All the same, the agreement is reasonably good.

The dispersion obtained is due to measurements made at various axial positions. At the lower Reynolds numbers, the lift force is very dependent on the axial position. It was found by means of flow visualization that this is due to variations in the circumferential position of the boundary-layer separation which can be fixed by increasing the surface roughness.

Axial space cross-correlations between lift forces show an exponential decay with respect to spacing, but there is a good deal of dispersion. The correlations seem independent of Reynolds number. These cross-correlation results permit an approximate calculation of the net lift forces acting on the whole length of a cylinder.

Lift Spectra

The frequency spectra at the various Reynolds numbers show clearly the particular characteristics of the three Reynolds-number regions: the subcritical single sharp peak at a Strouhal number of about 0.2, the transcritical wide band spectrum, and the supercritical return to a sharp peak at a higher Strouhal number. In addition to this one also notes that at Reynolds numbers between 4×10^5 and 2×10^6 a peak appears at about twice the normal shedding frequency. Analyses of local pressure spectra show that this is not a shift in the normal shedding frequency but a different phenomenon altogether with a maximum pressure in the vicinity of the trailling edge of the cylinder. That double-frequency pressures should exist is normal, but that they should result in a net lift force and also occur upstream of the boundary layer separation in unexpected. Moreover the spectral density at this higher frequency varies considerably with the axial measuring position. The probable explanation for this phenomenon is that the double-frequency pressures near the trailing edge exist as always because of the alternate shedding. However, due to the randomness and irregularity of the shedding at transcritical Reynolds numbers, the pressures engendered at these frequencies on either side of the cylinder are no longer well correlated and therefore do not cancel. Consequently, they give a net lift.

The power spectra of lift and of local pressures can be used to determine the shedding frequency, known in reduced form as the critical Strouhal number (S_c). In Fig. 3b, the present results are compared with those found in existing literature. The agreement is generally good. However, it must be noted that for $4 \times 10^5 < Re < 2 \times 10^6$, the points that some authors plot as the shedding frequency are most probably the double frequency discussed earlier (S_2). Though this peak is sometimes the most prominent in the spectrum, it must not be confused with the shedding frequency.

Spectral coherences and phases as a function of axial spacing have been calculated and these generally show no axial propagation of lift forces. These functions are very dependent on the Reynolds number so that it is not possible to reduce them to a single general formulation which would permit a simple calculation of the spectra of the net lift forces on a cylinder. However for any specific case considered, the coherence and phase functions will permit an estimation of the spectral shape of these forces.

Other Results

Other results of measurements on fixed cylinders are discussed in detail in Ref. [1]. These include the circumferential pressure distribution and the correlation between the forces acting on the two half symmetrical cylinders with respect to the flow. An important point that has emerged from these tests is the influence of the cylinder surface condition. For example, two different cylinders were tested, and though they differed only slightly in surface roughness, the flow at $Re = 5.5 \times 10^6$ was found to be supercritical in the rougher case and transcritical in the other. Quantitative tests of the effect of various degrees of surface roughness on both fixed and oscillating cylinders are being undertaken.

Concluding Remarks

The technique of dynamic lift measurements using a large number of pressure transducers is subject to a very big investment in equipment. However it does seem that at the present time its numerous possibilities make it the most appropriate method for furthering the understanding of these very complicated phenomena and for making measurements in view of predictions of lift forces and vibrations induced by them.

References

1. Loiseau, H., Szechenyi, E.: Recherche Aérospatiale, No. 5 (1972).
2. Relf, E.F., Simmons, L.F.G.: RSM 917, British ARC (1964).
3. Delany, N.K., Sorensen, N.E.: NACA TM 3038 (1953).
4. Fung, Y.C.: J. Aero. Sci. 27, No. 11, 801-814 (1960).
5. Roshko, A.: J. Fluid Mech. 10, 345 (1961).
6. Schmidt, C.V.: NASA TM X - 57779 pp. 15.1 - 19.17 (1966).
7. Bearman, P.W.: NPL Aero Rep. 1527 (1969).
8. Jones, W.: ASME Publication 68 FE 36 (1968).

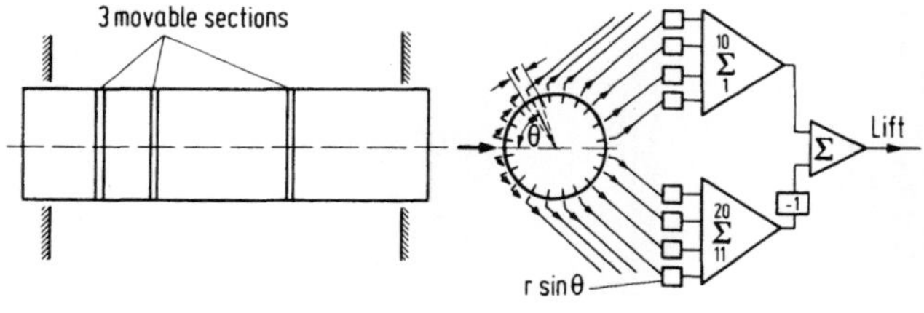

Fig.1. Lift measurement by means of pressures transducers.

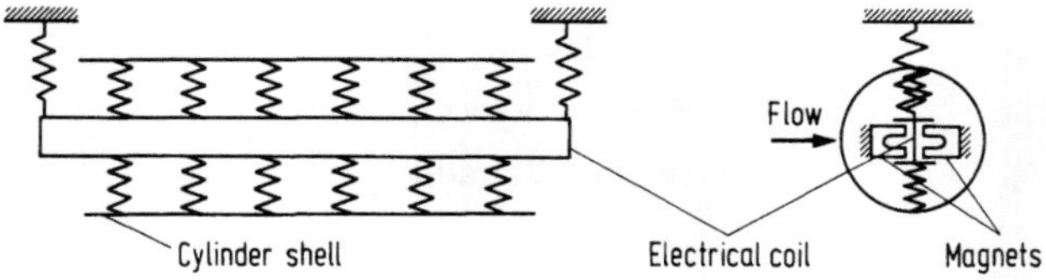

Fig.2. Sketch of cylinder excitation mechanism.

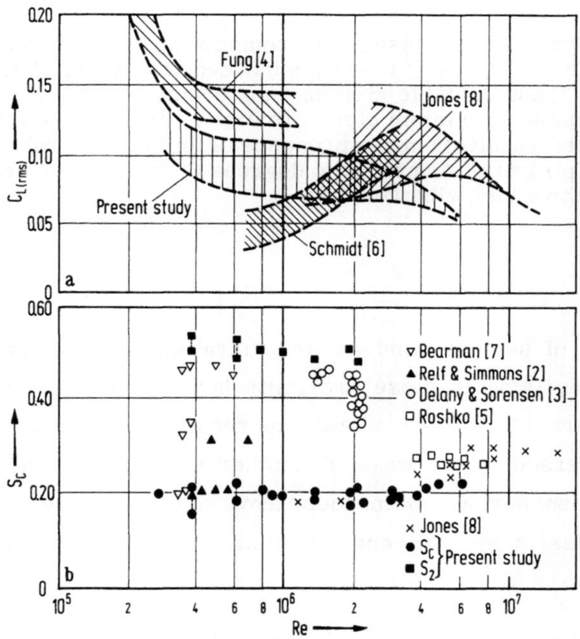

Fig.3. a) Coefficient of lift;
b) Critical Strouhal number.

Wake Galloping, an Aeroelastic Instability

By

K.R. Cooper

Ottawa, Canada

Summary

A series of dynamic response measurements were made to determine the characteristics of the wake-induced vibrations of slender bluff bodies. These unstable motions may occur for a variety of engineering structures lying in close proximity. The destabilizing aerodynamic forces result from the presence of an incident wake flow and large amplitude motions in two degrees of freedom can occur above some threshold wind velocity.

Introduction

Occurrences of serious wind-excited vibrations of civil-engineering structures lying in close proximity have been recorded in the past few years [1]. These vibrations result when a bluff structure is immersed in the wake of another bluff body. The "wake galloping" phenomenon commences above a critical onset velocity, and amplitudes of motion generally increase with increasing wind speed.

At present, the phenomenon has manifested itself most frequently in the oscillation of multiple conductor high voltage transmission lines. This conductor vibration problem has caused considerable damage to date and has yet to be controlled effectively.

It was suspected that the same phenomenon caused serious vibrations of two slender towers (diameter = 3 m, height = 73 m) in a heavy-water plant [2]. In addition, vibrations of tall steel stacks at the Oleum Steam Plant, originally attributed to vortex excited sub-harmonic resonance [3] may have been due to wake galloping. Other such incidents have been recorded in the literature, but too few data were provided to determine the exact causes of vibration; however the occurrence of wake galloping is a distinct possibility.

Section Model Dynamic Measurements

A fundamental study of the dynamic response of a circular body in a wake flow was made using a two-dimensional, spring mounted, section model spanning a 1 m by 1 m wind tunnel. This model (diameter = 41mm, density = 1.5 kg/m, lateral frequency = 1.15 Hz, longitudinal frequency = 1.10 Hz) was mounted in the wake of an identical rigidly mounted model. A family of constant-wind-speed stability boundaries was established as a function of wind-on equilibrium position in the wake. These results are presented in Fig. 1, where the inset diagram shows the test situation, defines the coordinate system, and depicts a typical elliptical path of motion.

At any wake position inside a given boundary, the model was unstable at the indicated wind speed. It can be seen that no motion occurs below a minimum velocity of 10.7 m/sec, and the regions of instability increase, particularly in the downstream direction, with increasing wind speed.

The effect of wake position (X_0/D, Y_0/D) on motion amplitude, (X/D, Y/D) and the rate of amplitude build-up (logarithmic decrement $\delta_x =$ $= \ln(X_{n+1}/X_n)$ is shown in Fig. 2, for a test speed of 12.2 m/sec. The along-wind motion was greater than the cross-wind motion, and the path of motion observed was a highly eccentric ellipse, with the semi-major axis inclined at a small angle to the wind direction. At increased wind speeds both the amplitudes and build-up rates were observed to increase, except for the closer spacings ($X_0/D<10$), where the model becomes stable at a sufficiently high speed. A greater increase occurs in along-wind amplitude than in the cross-wind component.

The instability was observed to be dependent on the elastic properties of the model. For motion to occur, the following conditions were required:

(1) The lateral and longitudinal natural frequencies must be close but not equal ($f_y/f_x \neq 1.0$).

(2) Stiffness cross-coupling, which causes a change in the X or Y elastic force with a Y or X deflection, must exist between the lateral and longitudinal degrees of freedom.

(3) For $(f_y/f_x) > 1.0$, the cross-coupling should be negative (the case shown) while for $(f_y/f_x) < 1.0$, the cross-coupling should be positive.

Static Force Measurements

A wake survey of the static aerodynamic forces on the downstream member of a set of smooth, circular cylinders was made for a range of Reynolds numbers. In addition to a drag force, a lift force was observed acting toward the wake centerline. The lift and drag coefficient distributions (based on free stream velocity), over the half wake are shown for a Reynolds number of 5.0×10^4 in Fig. 3 and Fig. 4, respectively. The lift distribution is anti-symmetric and the drag distribution is symmetric.

A striking relationship was observed between the lift measurement and the stability boundaries. If the wake boundary and the lateral location of $(C_L)_{max}$, the maximum lift coefficient, as a function of streamwise position from Fig. 3 are plotted on the stability boundaries, Fig. 1 (the crosses in Fig. 1), it can be seen that the area of the lift curve contained by these points very closely bounds the unstable wake region.

Response of Stacks

Further study of the wake galloping problem with respect to tall stacks and towers was undertaken using flexure mounted cantilevered models (diameter = 36 mm, h = 360 mm, density = 112 kg/m^3, natural frequency = 10 Hz) of both square and circular cross sections, positioned in the wakes of identical bodies. The full-scale equivalent of these models would be 4.3 m diameter, 43 m high welded steel towers whose first mode natural frequency was 0.5 Hz. The velocity scaling ratio was $V_m/V_p = 1/6$. This work was carried out in smooth flow, and some typical amplitude-response measurements are shown in Fig. 5 for a variety of separations and wind angles. The paths are approximately elliptical with the semi-major axis oriented nearly perpendicular to the wake centerline.

It should be noted that the wake galloping response of the downstream square structure was considerably more severe than that for the downstream circular body. This would suggest that the den Hartog galloping might be reinforcing the wake-induced instability.

Response of Tall Buildings

A 1:1200 scale dynamic model representing a tall building of square cross section (43 by 43 by 430 m) immersed in the wake of an identical structure was used to investigate the wake excitation of tall buildings in an urban environment. The full scale structure would have a first mode natural frequency of 0.1 Hz and a density of 160 kg/m^3. The model (diameter = 36 mm, density = 160 kg/m^3, natural frequency = 40 Hz) was tested in a turbulent shear layer representing the natural surface wind [4]. Scale wind speeds ($V_m/V_p = 1/3$) of up to 64/m/sec were investigated. No wake galloping was observed for body separations between 2 and 6 structure diameters, even though reduced velocities (V/fD) greater than those leading to motion in uniform flow were tested.

Conclusions

(1) An aeroelastic instability in two degrees of freedom may occur above a critical onset velocity for a bluff elastic body immersed in a wake flow. The paths of motion observed were approximately elliptical.

(2) Static lift and drag forces are generated on a bluff body in a wake flow. The lift force acts toward the wake centreline.

(3) The region of the wake enclosed by the wake edge and the locus of the points of maximum lift coefficient bound the area of the wake in which "wake galloping" was observed.

(4) Closely spaced, tall flexible towers and stacks, as well as other very elastic bluff bodies such as multi-bundled high voltage transmission lines and arrays of bridge stay cables are susceptible to this phenomenon.

(5) Tall buildings in urban regions would appear to avoid the instability since their stiffness results in onset velocities greater than practical design wind speeds.

(6) The combination of wind shear and turbulence leads to an increase in the non-dimensional value of critical onset velocity over that observed for a similar model in smooth, uniform flow.

(7) A necessary requirement for instability is that the participating degrees of freedom are stiffness cross-coupled and that the ratio of their natural frequencies be near but not equal to unity.

References

1. Cooper, K.R., Wardlaw, R.L.: Aeroelastic Instabilities in Wakes. International Conference, Wind Effects on Buildings and Structures. Tokyo, Japan, 1971.

2. Cooper, K.R., Wardlaw, R.L.: A Wind Tunnel Investigation of Large Amplitude Vibrations of Slender Towers at the Port Hawkesbury Heavy Water Plant. NAE, Canada, LTR-LA-35 (1969).

3. Dockstader, E.A., Swiger, W.F., Emory, I.: Resonant Vibrations of Steel Stacks. Trans. ASCE 121 (1956).

4. Standen, N.M.: A Note on Spire Generated Shear Layers. NAE, Canada, LTR-LA-94 (1972).

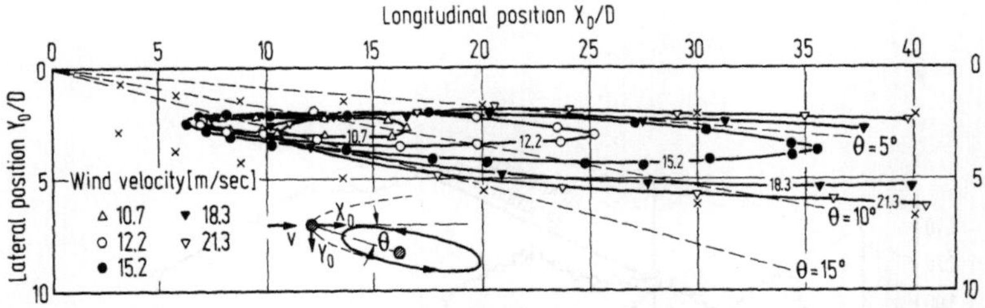

Fig.1. Wake galloping stability boundaries for a smooth circular cylinder.

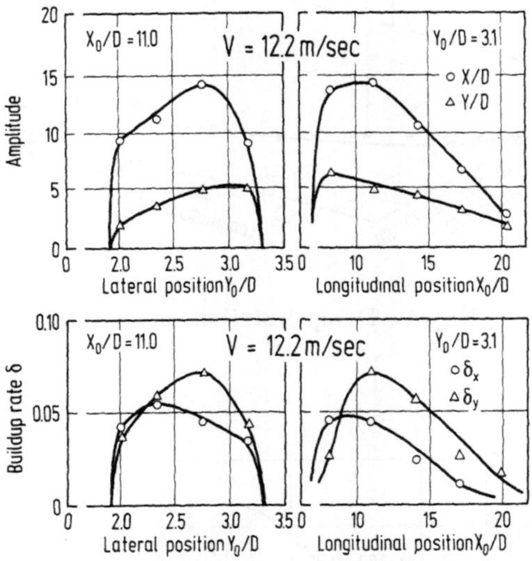

Fig.2. Vibration amplitude and build up rate as a function of relative position for two smooth circular cylinders.

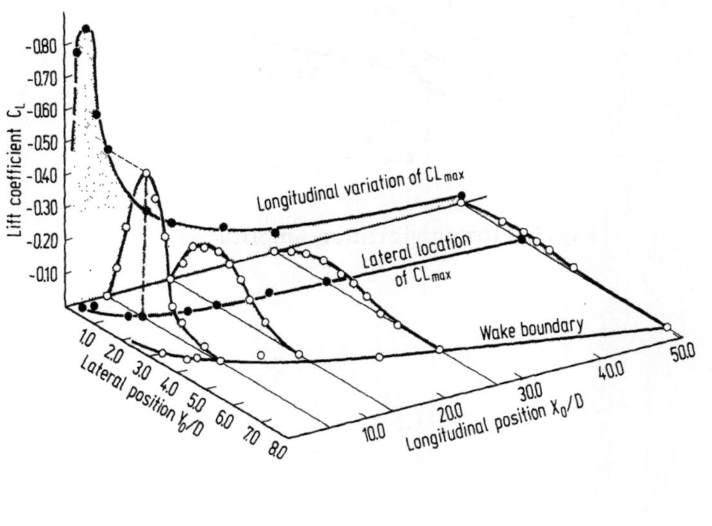

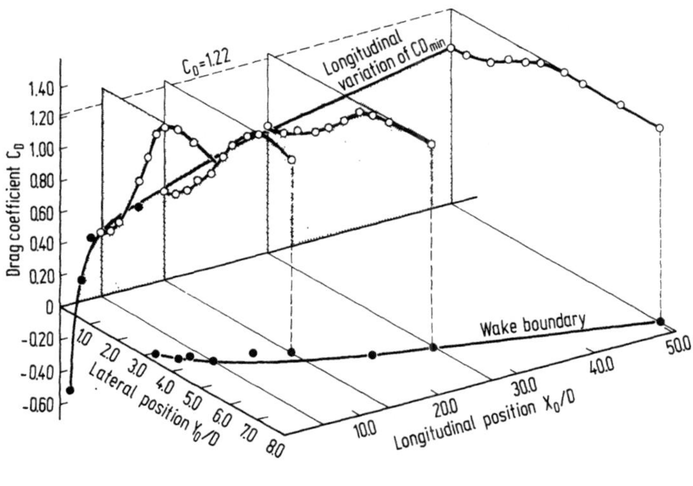

Fig.4. Drag distribution.

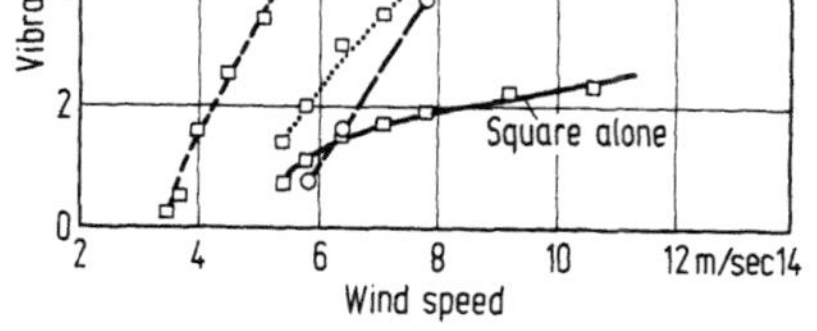

Fig.5. Tower vibration amplitudes.

Galloping Oscillations of Prisms in Smooth and Turbulent Flows

By
M. Novak
London, Canada

Galloping oscillations of sharp-edged prismatic bodies can exist in both smooth and turbulent flows. Turbulence can strongly affect quantitative as well as qualitative features of this motion to such an extent that galloping existing in smooth flow can vanish in turbulent flow or, in some cases vice versa. Some of these aspects are discussed in this paper.

The quasi-steady approach predicts the galloping oscillations very well in terms of nonlinear vibrations. Parkinson [6] was the first to prove the power of this approach for one degree of freedom in smooth flow. In Refs. [2] to [5] this approach was extended to include many degrees of freedom, unconventional instabilities defying Den Hartog's criterion, shear flow and the effect of turbulence. The latter effect can be incorporated into the quasi-steady theory with some limitations discussed in more detail in Ref. [3]. The basic limitation for the applicaition of the quasi-steady theory is that it must be applied well above the vortex-induced resonance.

The theoretical analysis is based on values of lateral force coefficients measured on a stationary model at various angles of attack. The experimental lateral force coefficient curve is then fitted by a polynomial. This can be done by means of a least squares method with a special fit for the starting part of the curve from which the most important coefficient of the first power is established. The coefficients of the polynomial are then introduced in the general equation of the response curves. The whole analysis can be programmed and the response curves readily obtained. These may then be presented in a universal form in coordinates Un/β and an/β. (U = reduced wind velcity, n = mass parameter, β damping ratio and a = = dimensionless amplitude).

Examples of such universal response curves are given in Figs. 1 and 2 in which theoretical curves are shown in full or dashed lines. These curves apply for any body with rectangular cross section and vibration modes, as indicated in the Figures (α = wind profile exponent).

Static and dynamic experiments involving turbulence were conducted in turbulent boundary layer and in a nearly isotropic turbulent flow. The boundary layers were achieved by means of surface roughness and low intake grids (Fig.3) while the isotropic turbulent flow was established behind an intake grid (Fig.4). The experiments were carried out in the Boundary Layer Wind Tunnel Laboratory. The results of experiments with vibrating prisms are plotted in Figs. 1 and 2.

Prisms with a side ratio of 2:1 gallop strongly in the smooth flow (Fig.1). However, their tendency to galloping weakens with increasing turbulence intensity and can vanish at turbulence intensities of 8.5 % or more, quite common in the natural wind. This important notion was confirmed by experiments with two different models in the boundary layer. The experimental results, plotted in Fig. 1, indicate buffeting in turbulent flow but no galloping.

With square prisms, turbulence also reduces the intensity of galloping oscillations but to lesser degree than with prisms 2:1; violent galloping oscillations are maintained even in strongly turbulent flows [3].

Behaviour of prisms with side ratio 1:2 is particularly interesting. In smooth flow, the equilibrium position is stable in accordance with Den Hartog's criterion and the prism does not gallop from zero at any wind velocity. However, steady galloping oscillations can arise above a certain minimum wind velocity if triggered by an initial disturbance larger than the unstable amplitudes shown in Fig. 2 in dashed lines (for turbulence intensity of 0 %). This behavior was theoretically predicted and was confirmed by experiments with very slender prisms.

The effect of turbulence on galloping response of prisms 1:2 is opposite to that with prisms 2:1. The tendency to instability increases with increasing intensity of turbulence as indicated in Fig. 2. A cross section 1:2, typically stable in smooth flow, can become unstable in turbulent flow. This remarkable and rather unexpected change can be attributed to the changes in the pressure distribution on the afterbody. Galloping oscillations from equilibrium positions readily built up with a very slender model at a turbulence intensity of 11 %. The model vibrating in turbulent flow can be seen in Fig. 4. The oscillations were measured with various dampings and their amplitudes (rms displacements times $\sqrt{2}$) are plotted in Fig. 2. The steady state amplitudes collapse onto a single curve in coordinates an/β and Un/β as the theory predicts. The onset from zero and drop to zero are indicated by arrows. The agreement between the theory and the experiments is quite good, particularly with larger amplitudes, considering that the models for the static and dynamic experiments were not the same. (It should be noted that the vibrations described cannot be attributed to vortex shedding or to the lateral effect of turbulence.) Further details concerning these results can be found in Ref. [5] in which the theory is also given.

The remarkable variations of galloping response with turbulence intensity were first reported in Ref. [3], were studied in detail in Ref. [5] and were confirmed by entirely independent experiments done by Laneville and Parkinson [1]. Further investigations with other kinds of prisms are underway at the Boundary Layer Wind Tunnel Laboratory. These studies clearly demonstrate that the behavior of structures in turbulent wind cannot be reliably predicted on the basis of experiments performed in smooth flow.

References

1. Laneville, A., Parkinson, G.F.: Effects of Turbulence on Galloping Bluff Cylinders. Presented at 3rd International Conf. on Wind Effects on Buildings and Structures, Tokyo, Japan, September 6-11, 1971.

2. Novak, M.: Aeroelastic Galloping of Prismatic Bodies. J. Engng. Mech., ASCE 96, No. EM1, Proc. Paper 6394, 115-142 (February 1969).

3. Novak, M., Davenport, A.G.: Aeroelastic Instability of Prisms in Turbulent Flow. J. Engng. Mechanics Division, ASCE 96, No. EM1, Proc. Paper 7076, February 1970, pp. 17-39.

4. Novak, M.: Galloping and Vortex Induced Oscillations of Structures. Proc. Third International Conf. on Wind Effects on Buildings and Structures, Tokyo, Japan, September 6-11, 1971, Paper IV-16, pp. 11.

5. Novak, M.: Galloping Oscillations of Prismatic Structures. J. Engng. Mech. Div., ASCE 98, No. EM1, Proc. Paper 8692, 27-46 (February, 1972).

6. Parkinson, G.V., Smith, J.D.: The Square Prism as an Aeroelastic Non-linear Oscillator. Quart. J. Mech. Appl. Mathematics, London XVII, Pt. 2, 225-239 (1964).

7. Parkinson, G.V., Santosham, T.V.: Cylinders of Rectangular Section as Aeroelastic Nonlinear Oscillators. Presented at the March 19-31, 1967, Vibrations Conference of the American Society of Mechanical Engineers, Held at Boston, Mass.

8. Nakamura, Y., Mizota, T.: Galloping and Vortex Excitation of Rectangular Block. Reports of Research Institute for Applied Mechanics, Kyushu University, Japan XIX, No. 64 (1972).

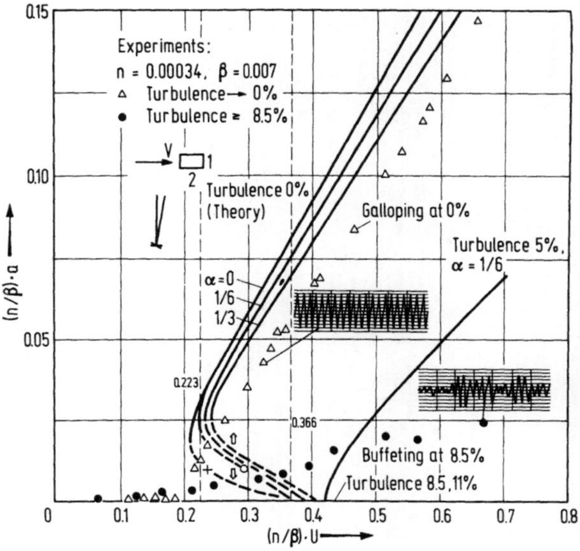

Fig.1. Universal galloping response of cantilevered prisms with side ratio 2:1.

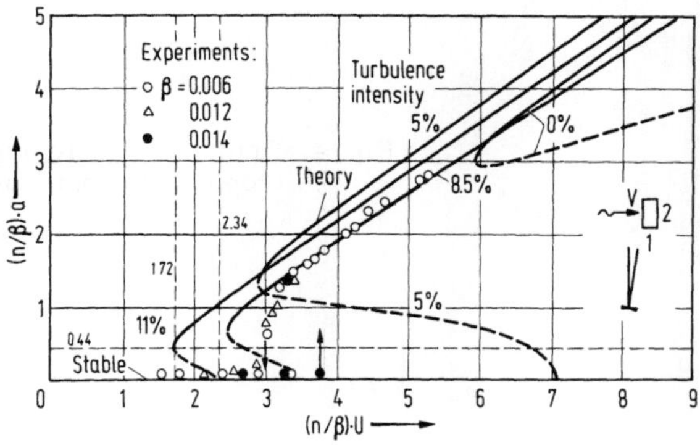

Fig.2. Universal galloping response of cantilevered prisms with side ratio 1:2.

Fig. 3. Model in turbulent boundary layer.

Fig. 4. Prism 1:2 galloping in isotropic turbulent flow.

Offsetdruck: fotokop wilhelm weihert kg, Darmstadt. Einband: Konrad Triltsch, Würzburg

MIX
Papier aus verantwortungsvollen Quellen
Paper from responsible sources
FSC® C105338

If you have any concerns about our products,
you can contact us on
ProductSafety@springernature.com

In case Publisher is established outside the EU,
the EU authorized representative is:
Springer Nature Customer Service Center GmbH
Europaplatz 3, 69115 Heidelberg, Germany

Printed by Libri Plureos GmbH
in Hamburg, Germany